Meat Inspection and Control in the Slaughterhouse

Meat Inspection and Control in the Slaughterhouse

Edited by

Thimjos Ninios
Finnish Food Safety Authority Evira, Finland

Janne Lundén
University of Helsinki, Finland

Hannu Korkeala
University of Helsinki, Finland

Maria Fredriksson-Ahomaa
University of Helsinki, Finland

WILEY Blackwell

Contents

6 *Post-Mortem* Inspection and Related Anatomy

*Paolo Berardinelli, Rosanna Ianniciello, Valentina Russo
and Thimjos Ninios*

7 Risk-Based Meat Inspection

Maria Fredriksson-Ahomaa

List of Contributors

Francesco Andreucci, Veterinary Officer, AULSS 18, Rovigo, Italy

Paolo Berardinelli, Professor of Veterinary Anatomy, Faculty of Veterinary Medicine, University of Teramo, Teramo, Italy

Aivars Bērziņš, Director, Institute of Food Safety, Animal Health and Environment (BIOR), Riga, Latvia; Associate Professor, Faculty of Veterinary Medicine, Latvia University of Agriculture, Jelgava, Latvia

Michael Bucher, Veterinary Officer, Veterinary Association JadeWeser, Wittmund Regional Office, Wittmund, Germany

Sava Buncic, Professor of Meat Hygiene and Safety, Department of Veterinary Medicine, Faculty of Agriculture, University of Novi Sad, Novi Sad, Serbia

Giuseppe Diegoli, Veterinary Officer, Veterinary Public Health Office of Region Emilia Romagna, Bologna, Italy

Terje Elias, Associate Professor, Food Hygiene Department, Estonian University of Life Sciences, Tartu, Estonia

Per Ertbjerg, Senior Lecturer, Department of Food and Environmental Sciences, Faculty of Agriculture and Forestry, University of Helsinki, Helsinki, Finland

Giorgio Fedrizzi, Specialist in Food Chemistry and Technology, IZSLER (Public Veterinary Laboratory for Control of Animal Diseases and for Food and Feed Control), Bologna, Italy

Karsten Fehlhaber, Director (retired), Institute of Food Hygiene, Faculty of Veterinary Medicine, University of Leipzig, Leipzig, Germany

Maria Fredriksson-Ahomaa, Professor of Foodborne Bacterial Zoonoses, Department of Food Hygiene and Environmental Health, Faculty of Veterinary Medicine, University of Helsinki, Helsinki, Finland

J.M. Frissen, Intelligence and Investigation Department, Food and Consumer Products Safety Authority, Ministry of Economic Affairs, Agriculture and Innovation, Utrecht, The Netherlands

María Luisa García-López, Professor, Department of Food Hygiene and Technology, Faculty of Veterinary Science, University of León, León, Spain

Geert H. Geesink, Lecturer, School of Rural Sciences and Agriculture, University of New England, Armidale, NSW, Australia

Monica Gramenzi, Adjunct Professor, Faculty of Veterinary Medicine, University of Teramo, Teramo, Italy

Joni Haapanen, Senior Officer, Border Control Section, Import, Export and Organic Control Unit, Finnish Food Safety Authority Evira, Helsinki, Finland

Marja-Liisa Hänninen, Professor, Department of Food Hygiene and Environmental Health, Faculty of Veterinary Medicine, University of Helsinki, Helsinki, Finland

Annamari Heikinheimo, Senior Lecturer, Department of Food Hygiene and Environmental Health, Faculty of Veterinary Medicine, University of Helsinki, Helsinki, Finland

Sanna Hellström, Senior Advisor, Finnish Food Safety Authority Evira, Finland

Peter Hofbauer, Senior Staff Member, The Institute of Meat Hygiene, Meat Technology and Food Science, Department of Farm Animals and Veterinary Public Health, University of Veterinary Medicine, Vienna, Austria

Rosanna Ianniciello, Veterinary Manager, ASL4 (Veterinary Hygiene of Food of Animal Origin), Teramo, Italy

Sirje Jalakas, Head of Animal Welfare and Zootehnics Bureau, Food Safety Department, Ministry of Agriculture, Tallinn, Estonia; Food Hygiene Department, Estonian University of Life Sciences, Tartu, Estonia

Suvi Joutsen, Researcher, Department of Food Hygiene and Environmental Health, Faculty of Veterinary Medicine, University of Helsinki, Helsinki, Finland

Juha Junttila, Deputy Head of Unit F6 (animal health and welfare), Food And Veterinary Office, European Commission, Dunsany, Co. Meath, Ireland

Pekka Juntunen, Researcher, Department of Food Hygiene and Environmental Health, Faculty of Veterinary Medicine, University of Helsinki, Helsinki, Finland

Tuija Kantala, Researcher, Department of Food Hygiene and Environmental Health, Faculty of Veterinary Medicine, University of Helsinki, Helsinki, Finland

Karoliina Kettunen, Researcher, Department of Food Hygiene and Environmental Health, Faculty of Veterinary Medicine, University of Helsinki, Helsinki, Finland

Hentriikka Kontio, Veterinary Counsellor, Ministry of Agriculture and Forestry, Helsinki, Finland

Hannu Korkeala, Professor of Food Hygiene/Head of the Department, Department of Food Hygiene and Environmental Health, Faculty of Veterinary Medicine, University of Helsinki, Helsinki, Finland

Päivi Lahti, Researcher, Department of Food Hygiene and Environmental Health, Faculty of Veterinary Medicine, University of Helsinki, Helsinki, Finland

Tiina Läikkö-Roto, Researcher, Department of Food Hygiene and Environmental Health, Faculty of Veterinary Medicine, University of Helsinki, Helsinki, Finland

Riikka Laukkanen-Ninios, Senior Lecturer, Department of Food Hygiene and Environmental Health, Faculty of Veterinary Medicine, University of Helsinki, Helsinki, Finland

Outi Lepistö, Environmental Health Manager, Environmental Health Care Unit Pirteva, Pirkkala, Finland

Jere Lindén, Research Coordinator, FCLAP, Pathology and Parasitology, Faculty of Veterinary Medicine, University of Helsinki, Helsinki, Finland

Miia Lindström, Professor, Department of Food Hygiene and Environmental Health, Faculty of Veterinary Medicine, University of Helsinki, Helsinki, Finland

Janne Lundén, Senior Lecturer (food control), Department of Food Hygiene and Environmental Health, Faculty of Veterinary Medicine, University of Helsinki, Helsinki, Finland

Riitta Maijala, Adjunct Professor, Department of Food Hygiene and Environmental Health, Faculty of Veterinary Medicine, University of Helsinki, Helsinki, Finland

Alessandra Martelli, Academic Researcher of Veterinary Anatomy, Faculty of Veterinary Medicine, University of Teramo, Teramo, Italy

Leena Maunula, Senior Lecturer, Department of Food Hygiene and Environmental Health, Faculty of Veterinary Medicine, University of Helsinki, Helsinki, Finland

Ute Messelhäusser, Senior Researcher, Bavarian Health and Food Safety Authority, Oberschleissheim, Germany

Anu Näreaho, Senior Lecturer, Department of Veterinary Biosciences, Faculty of Veterinary Medicine, University of Helsinki, Helsinki, Finland

Mari Nevas, Senior Lecturer, Department of Food Hygiene and Environmental Health, Faculty of Veterinary Medicine, University of Helsinki, Helsinki, Finland

Veli-Mikko Niemi, Director of Food Safety, DMV, Ministry of Agriculture and Forestry, Helsinki, Finland

Thimjos Ninios, Senior Officer/Head of Section, Border Control Section, Import, Export and Organic Control Unit, Finnish Food Safety Authority Evira, Helsinki, Finland

Niels Obbink, Intelligence and Investigation Department, Food and Consumer Products Safety Authority, Ministry of Economic Affairs, Agriculture and Innovation, Utrecht, The Netherlands

Sinikka Pelkonen, Professor, Finnish Food Safety Authority Evira, Helsinki, Finland

Miguel Prieto, Professor, Department of Food Hygiene and Technology, Faculty of Veterinary Science, University of León, León, Spain

Leena Pohjola, Researcher, Department of Production Animal Medicine, Faculty of Veterinary Medicine, University of Helsinki, Helsinki, Finland

S.B. Post, Intelligence and Investigation Department, Food and Consumer Products Safety Authority, Ministry of Economic Affairs, Agriculture and Innovation, Utrecht, The Netherlands

Eero Puolanne, Professor Emeritus, Department of Food and Environmental Sciences, Faculty of Agriculture and Forestry, University of Helsinki, Helsinki, Finland

Marjatta Rahkio, Managing Director, Finnish Milk Hygiene Association, Helsinki, Finland

Mati Roasto, Professor, Food Hygiene Department, Estonian University of Life Sciences, Tartu, Estonia

Mirko Rossi, Associate Professor, Department of Food Hygiene and Environmental Health, Faculty of Veterinary Medicine, University of Helsinki, Helsinki, Finland

Laila Rossow, Senior Researcher, Production Animal and Wildlife Health Research Unit, Research and Laboratory Department, Finnish Food Safety Authority Evira, Helsinki, Finland

Valentina Russo, Academic Researcher of Veterinary Anatomy, Faculty of Veterinary Medicine, University of Teramo, Teramo, Italy

Risto M. Ruuska, Provincial Veterinary Officer, Regional State Administrative Agency for Lapland, Finland

Satu Salo, Senior Scientist, VTT, Espoo, Finland

Robert Savage, President, HACCP Consulting Group, Fairfax, VA, USA

Peter Scheibl, Veterinary Officer, Bavarian Health and Food Safety Authority, Oberschleissheim, Germany

Liisa Sihvonen, Professor, Department of Veterinary Biosciences, Faculty of Veterinary Medicine, University of Helsinki, Helsinki, Finland

Kyösti Siponen, Head of Unit, Import, Export and Organic Control Unit, Control Department, Finnish Food Safety Authority Evira, Helsinki, Finland

Frans Smulders, Professor and Chair of the Institute of Meat Hygiene, Meat Technology and Food Science, Department of Farm Animals and Veterinary Public Health, University of Veterinary Medicine, Vienna, Austria

Jani Soini, Provincial Veterinary Officer, Regional State Administrative Agency for Southwestern Finland, Turku, Finland

Roger Stephan, Professor and Director, Institute for Food Safety and Hygiene, Vetsuisse Faculty, University of Zurich, Zurich, Switzerland

Andreas Stolle, Head (retired), Institute of Food and Meat Hygiene and Technology, Veterinary Medicine Faculty, Ludwig-Maximilians-University, Munich, Germany

Daniele Tognetti, Assistant, Ministry of Health, Rome, Italy

Marcello Trevisani, Associate Professor – Food Inspection, Department of Veterinary Medical Science, University of Bologna, Bologna, Italy

Ivar Vågsholm, Professor of Microbial Food Safety, Department of Biomedical Sciences and Veterinary Public Health, Swedish University of Agricultural Sciences, Uppsala, Sweden

Eeva-Riitta Wirta, Head of Meat Inspection Unit, Control Department, Finnish Food Safety Authority Evira, Helsinki, Finland

Gun Wirtanen, Senior Expert, VTT Expert Services Ltd, Espoo, Finland

Claudio Zweifel, Senior Lecturer, Institute for Food Safety and Hygiene, Vetsuisse Faculty, University of Zurich, Zurich, Switzerland

1

Introduction

Hannu Korkeala

Department of Food Hygiene and Environmental Health,
Faculty of Veterinary Medicine, University of Helsinki, Helsinki, Finland

The strong development of science in the nineteenth century prompted the need to develop practical solutions to prevent diseases caused by parasites and bacteria in humans based on the findings of Louis Pasteur, Robert Koch, Rudolph Virchow and Friedrich A. Zenker. It was shown in the 1860s that *Trichinella* was transmitted to humans through contaminated pork meat. Later, many bacterial and other parasitic diseases were shown to transmit to humans through meat. Due to these findings, veterinary medicine grew in importance and it was increasingly demanded that veterinarians be in charge of public health and that the importance of veterinary education be increased in the society. The role of veterinarians became significant in meat inspection. Later, veterinary public health and food hygiene became an essential part of veterinary education. Robert von Ostertag, a German veterinarian, wrote an extensive handbook on meat inspection in 1892 and created the scientific basis for meat inspection. The book was translated into English in 1904.

The proportion of meat originating from diseased animals was high in the nineteenth century and the quality of meat was poor. A huge improvement was achieved through the organization of meat inspection and the development of the scientific basis of meat inspection. Meat inspection has been a success story and it has been the most important programme in improving food safety. Huge steps were taken to prevent the transmission of pathogenic organisms and contagious diseases from animals to humans. Meat inspection

Meat Inspection and Control in the Slaughterhouse, First Edition.
Edited by Thimjos Ninios, Janne Lundén, Hannu Korkeala and Maria Fredriksson-Ahomaa.
© 2014 John Wiley & Sons, Ltd. Published 2014 by John Wiley & Sons, Ltd.

has had a major role in prevention of zoonoses. In addition, meat inspection has been a cornerstone for the development of modern food control.

The scope of meat inspection has been enlarged over the last few decades. The substantial core of meat inspection is public health and consumer protection. Public health issues, such as prevention of the transmission of pathogenic microorganisms and contagious diseases, have been important from the beginning of meat inspection. More recently, detecting residues of chemical substances and veterinary drugs and preventing their occurrence became part of meat inspection. In addition, animal welfare issues have always been key issues and the skilful treatment of animals during transport and slaughter has been an important part of meat inspection. Faults in the transport and handling of animals influence also the quality of meat. Problems related to meat quality are also major issues. The removal of poor quality and adulterated meat from the meat chain and the prevention of contamination of meat with spoilage organisms are part of high-quality meat inspection.

The scope of the work of meat inspection veterinarians ranges from animal welfare issues associated with the transport of animals to the control of meat processing. Training of the official auxiliaries and workers, and administrative and managerial duties, should also be part of the job description of meat inspection veterinarians. Extensive meat inspection training is needed in veterinary education to ensure good theoretical and practical skills to carry out high-quality meat inspection. The presence of full-time meat inspection veterinarians in slaughterhouses is needed for efficient inspection and control.

A lot of different risks can be associated with the intensified modern meat production process. Consumers are more and more interested in the origin of the food they eat. The increase in the knowledge and interest of consumers has had a special focus on meat production. Therefore, meat production and meat inspection should be transparent and should have high ethical operational principles to gain the trust of consumers. In this regard, meat inspector veterinarians have a crucial responsibility and importance.

Strong internationalization and globalization have been a typical trend in the food industry and food trade, including the meat sector. The complexity of the modern meat chain has led also to possibilities for food frauds and unfair competition. During the last two decades many food fraud cases have been reported. These intentional activities are an alarming phenomenon and have caused concern among consumers. Companies that ignore food safety regulations gain, wrongly, a competitive edge over companies that fulfil the regulations.

Modern technical solutions to ensure good meat safety objectives during slaughtering and meat inspection procedures are a cost to companies. On the other hand, food safety problems that attract negative publicity in food production may cause enormous economic losses. A good balance between food safety regulations and control and economic resources for food safety should be achieved. It is also important that meat inspection and control is similar throughout the whole of Europe, ensuring fair competition between companies.

Meat inspection is faced with new challenges. Transmission of pathogenic organisms from animals to humans can be cut off during slaughtering. However, the recognition of different organisms is a very demanding task and needs new modern techniques. New solutions and practices should be continuously developed to improve and ensure meat safety. The changes in meat production should be taken into account in order to make necessary changes in meat inspection procedures. At the same time, the information collected during meat inspection should be recognized and used to improve food safety and the prevention of animal diseases. However, it is obvious that meat inspection is needed in the future and is an essential part of food control.

Meat inspection practices and the content of the meat inspection process have been changed and enlarged due to research and the development of control measures. Due to these changes and the importance of meat inspection, an up-to-date reference is needed for veterinary students and other interested groups to provide an extensive description of meat inspection. Meat inspection procedures are regulated by the European Union and, therefore, the writers of this guide are experts representing many different European countries. The book covers meat inspection from animal transport to the official control at the slaughterhouse. The control of meat processing, which is an important task of meat inspection veterinarians at plants connected to the slaughterhouse, has not been dealt with. The editors hope that the book will attract interest in meat inspection and control, as well as encourage scientific research to improve future meat inspection procedures and food safety.

2

From Farm to Slaughterhouse

Sirje Jalakas[1,2], Terje Elias[2] and Mati Roasto[2]

[1]*Food Safety Department, Animal Welfare and Zootehnics Bureau, Ministry of Agriculture, Tallinn, Estonia*

[2]*Food Hygiene Department, Estonian University of Life Sciences, Tartu, Estonia*

2.1 Scope

The chapter aims to explain why meat quality and safety is dependent on how farm animals are raised, transported and handled prior to slaughter. Animal health and welfare aspects as well as food chain information (FCI) are discussed. The focus is on transport and lairage, as they have an important role on animal health and welfare. It is also important to note that legislation alone is not enough to improve animal health and welfare. In addition, public awareness, compliance to good practices, official control and continued research are needed.

2.2 Animal health and welfare

Recent studies have shown increasing public concern about how farm animals are raised, transported and slaughtered, as well as about food safety. Infectious animal diseases cause major losses to livestock production and many of them (zoonoses) can pose a risk to consumers via the food chain or through other

Meat Inspection and Control in the Slaughterhouse, First Edition.
Edited by Thimjos Ninios, Janne Lundén, Hannu Korkeala and Maria Fredriksson-Ahomaa.
© 2014 John Wiley & Sons, Ltd. Published 2014 by John Wiley & Sons, Ltd.

pathways, including direct transmission. Therefore, more attention is being paid to prevention of diseases in food animals; this is also emphasized in the European Union's Common Animal Health Strategy, the motto of which is 'Healthy food from healthy animals'.

According to the Farm Animal Welfare Committee (FAWC), the welfare of an animal includes its physical and mental state, and the FAWC considers that good animal welfare implies both fitness and a sense of well-being. Any animal kept by man must be protected from unnecessary suffering.

The FAWC believes that an animal's welfare should be considered in terms of Five Freedoms. These freedoms define ideal states rather than standards for acceptable welfare:

- freedom from Hunger and Thirst – by ready access to fresh water and a diet to maintain full health and vigour;
- freedom from Discomfort – by providing an appropriate environment, including shelter and a comfortable resting area;
- freedom from Pain, Injury or Disease – by prevention or rapid diagnosis and treatment;
- freedom to Express Normal Behaviour – by providing sufficient space, proper facilities and company of the animal's own kind;
- freedom from Fear and Distress – by ensuring conditions and treatment that avoid mental suffering.

European Union legislation lays down minimum standards for the protection of all farmed animals and sets welfare standards for the protection of animals during transport and also at the time of slaughter and killing. Specific European Union directives cover the protection of individual animal species and categories, such as calves, pigs, broilers and laying hens. Standards of animal welfare and animal management practices can influence the spread of food-borne diseases. The rules alone are not able to control animal health and welfare, and the paradigm 'healthy food comes from healthy animals' is still relevant.

2.2.1   Different farming systems

There are many links between pre-harvest production and the safety and the quality of the food. These interactions are not yet completely understood but it is clear that farmers are contributing to food safety when producing healthy, unstressed and clean livestock for slaughter. This is dependent on farmers' personal attitude and knowledge and also on farm management systems. Farming systems differ significantly between species and within each species and can be divided in several ways, for example integrated and non-integrated, intensive and extensive, indoor and outdoor, free-ranged and fenced, organic and non-organic and so on. Different farming systems are facing different animal health and welfare problems. These problems have been mostly associated with intensive and industrialized farming but are present in varying degrees in all livestock production systems. With organic farming conditions animal welfare is generally better than in other systems but specific health and welfare problems could rise at these systems too and can be very serious. For example, predators can pose big problems for sheep and goat farming in all systems where animals are pastured. In the case of organic production of laying hens and broilers there is risk from wild birds spreading contagious diseases but different predators also pose risk to animal welfare.

Intensive livestock farming is the major source of the world's meat supply, and the main challenge of intensive farming is to produce food profitably whilst complying with the requirements on environment, animal welfare and food safety. During recent years amongst the European public, there has been increased demand for high levels of farm animal welfare. To some extent it is in the farmer's private interest to maintain high levels of farm animal welfare because it leads to healthy livestock and a high quality product. Certain approaches to livestock rearing, for example, that provide ample space to express more animal's natural behaviour, provide welfare and environmental benefits, and may reduce the occurrence of some diseases.

It is important to understand that in all farming systems it is necessary to guarantee the acceptable level of farm animal welfare and health and to follow good practices which contribute to consumer safety and public health.

2.2.2 Good practices

Good Farming Practices (GFP), Good Veterinary Practices (GVP) and Good Hygiene Practices (GHP) need to be implemented to fulfil animal welfare demands and to minimize the spread of animal-specific pathogens and food safety related hazards. Health- and welfare-oriented production practices mean a high degree of caring, responsible management and farming to ensure good animal welfare. Managers and stock-keepers must be thoroughly trained, skilled and competent in animal husbandry and welfare, and have a good working knowledge of their system and the animals under their care. The practices should be oriented towards breaking the infection chains within the herd or flock and along the animal production chain from breeding to slaughter, which means following the biosecurity measures.

The prudent use of veterinary drugs and appropriate veterinary treatments are the major components of GVP. Proper use of antimicrobials will help to prevent the development of resistance amongst microorganisms. Antimicrobials should not be used as growth promoters or for prophylactic aims. All veterinary actions should be based on making a correct diagnosis, which requires high quality clinical training and skills as well as access to diagnostic laboratory services. GVP also involve the adequate control and prevention of animal diseases. Therefore, systems for the monitoring and surveillance of disease occurrence have been established.

Good hygienic practices (GHP) at the level of primary production should involve the health and hygiene of animals, biosecurity, records of treatments, feed and feed ingredients and relevant environmental factors. GHP should also include elements of own-check or internal audits. In primary production, the areas where zoonotic agents and other hazards may accumulate must be effectively cleaned and, if needed, disinfected. Chemicals must be stored in a manner such that they do not contaminate the environment or poison the animals directly or via feed and feed ingredients and water, and thereby pose a risk to food safety or human health. Additionally, primary producers should record relevant information on the health status of animals as it relates to the production of safe meat.

2.2.3   Biosecurity measures

Primary production should be managed in a way that reduces the likelihood of introducing hazards and contributes appropriately to meat being safe and suitable for human consumption.

Beside classical food safety risks, there are many food safety risks that cannot be controlled by meat inspection because of missing clinical symptoms or pathological-anatomical lesions in the infected animal. Visual, palpatory and incision techniques are not always suitable for detecting food-borne diseases, such as campylobacteriosis, salmonellosis, yersiniosis and verotoxigenic *Escherichia coli* (VTEC/STEC) infections. Moreover, palpation and incision can contribute to cross-contamination of the carcasses with the mentioned food pathogens. Many food pathogens can be detected and controlled in the flock or herd level (pre-harvest) instead of detection at the slaughter line (harvest). Biosecurity measures are the main preventive measures for reducing food safety risks at pre-harvest. A risk-based approach to meat hygiene will include consideration of those risk management options that may have a significant impact on risk reduction when applied at the level of primary production as well as at slaughter or processing. It is most important that farmers are familiar with basic principles of biosecurity, which includes the knowledge of the herd/flock health status, compliance with veterinary drug withdrawal times, quarantine measures for incoming animals and sanitary measures such as changing boots and overalls, proper cleaning and disinfection between production cycles, rodent and pest control, and the use of batch production or all-in all-out where possible (Table 2.1). It must be emphasized that farmers and field veterinarians form the 'first line of defence' against the threat of outbreaks of contagious diseases.

2.2.4   Prior to transport

Animal health status should be controlled and animals showing clinical signs of disease and not fit for transport should not be transported to the slaughterhouse. In some situations, such as emergency slaughter or animals from farms under animal health restrictions, transport may proceed if the animals have been specifically identified and are slaughtered under special supervision.

Table 2.1 Possible routes of transfer of infectious agents on farms.

Route	Explanatory comment
Air	Aerosols that are produced when infected animals breathe Infected faeces, dust, feathers Infected bats, insects, birds etc.
Animals of the same species	Infected animals being transferred into uninfected herds or flocks Infected animals fencing with uninfected animals Breeding schemes that aggregate animals etc.
Dead or sick animals	Other species being alternative infected or carrier hosts Rodents or other species acting as carriers of contamination
Animals of other species	Inadequate handling and disposal of sick and/or dead animals providing access to flocks and herds and to predators such as foxes, cats etc.
Feed	Feed ingredients are infected intrinsically with the disease agent, e.g. Salmonella Feed becomes contaminated by infected animals when birds, feral animals, rodents or domestic animals have access to the feed
Water	The water becomes infected from contact with infected animals such as birds, feral animals, rodents and domestic animals
Waste	Contact of uninfected animals with infected wastes from an adjacent property
Contaminated personnel, equipment and vehicles	Movement of people, clothing, footwear, equipment and vehicles between infected and uninfected premises

Faecal contamination of skin and fleeces is a serious food safety and animal welfare problem, respectively conveying food-borne pathogens and causing skin lesions and pain. Therefore, animals that have skin or fleece conditions that indicate a serious risk of contaminating the meat during slaughter should be sheared and/or cleaned before transport. Appropriate actions should be carried out immediately at farm level to prevent unacceptable faecal/soil contamination of animals. Animals should be kept at farm in clean condition and also kept clean until arrival to slaughterhouse.

2.3 Transport

Farm animals are transported to the slaughterhouse most often by road but also by rail and on ship; they are seldom transported by air. The transport of livestock includes also a pre-transport period with transport-related operations. Before the animals are loaded onto a vehicle they should be prepared for transport. This stage consists of mustering, assembling, handling and preparation of livestock, including the feeding and watering of animals. The next stage is loading, transport and unloading of animals. Before animals are loaded onto the transport means, their fitness for the intended journey should be assessed. Additional inspections of livestock fitness for journey should be made during travelling and, especially, at staging points where animals are unloaded from the transport means.

As research has shown, transport of livestock is very stressful and livestock are prone to injuries. Stress factors such as a new environment, contact with unfamiliar animals and humans, unusual feed and different watering equipment are some examples that cause stress during transport. The influence of certain stress factors caused by transport can be reduced if animals are fit for transport and have been prepared for transport. When calculating the duration of the journey, it should be born in mind that the influence of the journey time on the animal is dependent on the animal species, animal condition and, especially, on animal age. Poorly executed transport can lead to poor animal welfare but can also have a negative influence on meat quality.

2.3.1   Fitness for transport

All animals must be transported in conditions that avoid injury and/or suffering. As mentioned earlier, prior to journey each animal should be inspected by an animal handler to assess fitness to travel. Animals found unfit to travel should not be loaded onto a vehicle for transport to the slaughterhouse. If travel fitness is doubtful, the animal should be examined by a veterinarian. Animals that are injured or have physiological weaknesses or pathological processes should not be considered fit for transport (Table 2.2).

2.3.2   Means of transport

The means of transport should be designed and maintained such that the basic needs of animals for safety, thermal comfort and adequate movement can be met. Vehicles should not cause injury and suffering and should ensure the safety of the animals (Table 2.3).

Various types of vehicles are used for animal transport. Whatever the means of transport being used, the vessel, truck and/or transport tanks must be disinfected between each transport in order to avoid the transfer

Table 2.2   Criteria for unfitness for transport.

Animals are unable to move independently without pain or to walk unassisted	New born mammals in which the naval has not completely healed
Animals have a severe open wound or prolapse	Cervine animals in velvet
Pregnant females for whom 90% or more of the expected gestation period has already passed, or females who have given birth in the previous week	Pigs of less than three weeks, lambs of less than one week and calves of less than ten days of age (unless transported less than 100 km)

Table 2.3 Criteria for animal transport means.

Should not have any sharp edges or projections	Should prevent animals from putting their paws/feet/nose through any ventilation hole or door mesh
Should provide loading and unloading equipment that is adequately designed and constructed for animal species	Should protect the animals from inclement weather conditions and extremes of temperature
Should be constructed of non-toxic material	Should provide adequate lighting to allow for the care and inspection of the animals during the intended journey
Should prevent animals from escaping	Should provide access to the animals for inspection purposes
Should provide a suitable non-slip floor	Should provide access to the animals for the purpose of providing feed, water and care as necessary
Should be easily cleaned and disinfected and should be cleaned and disinfected before each animal transport	Must provide adequate space so that the animals can stand up in a natural position, turn around and lie down.
Should be leak proof	Suitable bedding must be supplied as appropriate
Should ensure air quality and temperature is appropriate to the species and number of animals being transported	

of pathogens between consignments transported in the same containers. According to requirements set up in legislation, road transport vehicles used for animal transport longer than eight hours should additionally be equipped with a GPS system, forced ventilation system, watering system and an adequate amount of feed and bedding material for the entire journey.

2.3.3 Transport practices

Transport can have serious effects on animal welfare and can lead to significant loss of meat quality and production if it not carried out properly. For example, great stress during the transport of cattle and pigs can lead to dark firm and dry (DFD) beef meat and pale soft and exudative (PSE) pork meat. Mixing of animals from different batches, different species, different ages and also horned and non-horned animals should be avoided. If unfamiliar animals are loaded into a transport vehicle it can lead to fighting between animals, causing bruising and serious injuries.

The stress factors that the animals are exposed to during transport may lead to poor animal welfare and increase the risk of infection and disease. Changes in the environment are a source of stress for transported animals. Therefore, such stressful conditions should be minimized as much as possible. If the animals are not familiar with human contact or have unpleasant memories of it, it can be a very stressful experience. Repeated humane handling of animals during rearing and prior to transport is beneficial in order to minimize aversive reactions from human contact during transport. For example, walking calmly in the rearing pens of pigs daily decreases their stress level during transport. During loading and unloading procedures it is important to ensure animal welfare as well as the occupational health of staff handling the livestock, taking into account species specific variations of animal behaviour. Animal handlers, drivers and attendants should be trained and competent and, as well, have knowledge about the animal's movement patterns. They should also understand what flight zones, social interactions and other behavioural aspects vary significantly among species, and even within species. Facilities and handling procedures that are successful with one animal species may be ineffective or dangerous with another.

Animal species vary in their responses to loading and therefore each species requires different handling procedures. For example during the loading and unloading of animals the sheep have physiological responses rather than behavioural and these are associated with the novel situation encountered in the vehicle rather than the loading procedure. Pigs, on the other hand, are much affected by being driven up a ramp into a vehicle. The genetic strain also affects responses to loading and transport, for example all modern pig strains are adversely affected by the loading and transport and some strains, such as those carrying the halothane gene, are severely affected.

Careful driving of the animal transport vehicle is important. During transport animals should be able to cope with different movements of vehicle. Up and down movements and side to side movements can be avoided by choosing good quality roads for driving. With careful driving it is possible to reduce the forward–backward movements caused mainly by the sudden starting, acceleration and braking of the vehicle. Sows and boars should be handled separately and transported in separate compartments. As mentioned earlier, cattle, other pigs and goats should be transported without regrouping. Vehicles for cattle transport should be fitted also with partitions and the animals should be transported, loaded and unloaded in small groups.

Animal transport is defined as commencing when the first animal is loaded into a vehicle and as ending when the last animal is unloaded. The same animals should not be considered to begin a new journey until a period of 48 hours has passed since the end of the previous journey. Between journeys the animals should be able to rest and recuperate in farm conditions with adequate food and water provided.

The length of transport time and resting period should be appropriate to animal species. Hence, journeys should preferably not be longer than those for which food and water provision is recommended for each species. Further transport should be carried out only if, during the rest period, each animal can be inspected, is found to be fit for further travel and can be given space and facilities for adequate rest, feeding and watering. As studies have shown, with the increasing duration of the transport time, the welfare of animals generally gets worse because they become more fatigued, more stressed, incur a steadily increasing energy deficit if they do not get sufficient food, become more susceptible to existing infections and may be exposed for new pathogens.

The amount of space allowed for animals during transport is important for several reasons. Space allowances have two factors that need attention. The first factor is the floor area available for the animal to stand or lie down in and the second is the compartment height. When four-legged animals are standing on a surface subject to movement, such as a road vehicle, they position their feet outside the normal area under the body in order to help them to balance. They also need to take steps out of this normal area if subjected to accelerations in a particular direction. Hence, they need more space than if standing still. When adopting the standing position and making movements on a moving vehicle, cattle, sheep, pigs and horses make considerable efforts not to be in contact with other animals or the sides of the vehicle. If animals are transported, especially over long distances, with high stocking densities, it will prevent the animals from lying down, which may lead to fatigue and muscle damage. However, too low a stocking density without partitions could cause injuries to animals. At high stocking densities and with low compartment height the risk of heat stress increases. Whether animals want to lie down depends also on animal species, journey length and transport conditions, including the quality of the road surface and the availability or absence of bedding material.

Depending on animal species, there are differences of common behaviour during transport. Pigs lie down shortly after loading if the vehicle is driven well and they have sufficient space. When calves or cattle are transported they either remain standing or adopt a position of sternal recumbence. Cattle are more likely to lie down if the surface of the floor is covered with straw. Studies indicate that on motorway journeys where a substantial layer of straw bedding provided, most cattle lay down after a few hours. According to recommendations from the European Food Safety Authority (EFSA), cattle should be provided with sufficient space to stand without contact with their neighbours and to lie down if the journey is more than twelve hours. Calculation of space allowances should also take into account differences between animals belonging to same species. When calculating cattle loading density, the presence of horns must be considered, too. Cattle with horns should have a 7% greater space allowance. Depending on weather conditions, space allowances should be appropriate. For example, stocking densities of broilers in containers should be limited in conditions when external temperatures exceed the proposed acceptable range (e.g. >22 °C) and on long journeys. If, due to journey time, it is necessary to give water and feed to animals, then the stocking density of animals should be lower to provide sufficient access for all animals to food and water.

Compartment height should be appropriate to the animal species being transported and to each transported animal. Species-specific differences should be taken into consideration when assessing the proper height of the compartment. Animal compartment height for cattle transport should be at least 20 cm above the withers height of the tallest animal. Pigs need sufficient headroom to ensure adequate ventilation and to maintain normal posture when standing. For large pigs the highest point on the body is the middle of the back. However, the actual amount of headroom necessary will depend on the ventilation within the vehicle, particularly when it is stationary.

2.4   Lairage

The place where the animals spend time in the slaughterhouse before being slaughtered is called lairage. Certain facilities are required for *ante-mortem* inspection of animals arriving at the slaughterhouse. Therefore, a proper light intensity, an adequate number of well-designed pens and a competent staff are needed. It is well known that the cleanliness of animals has a major influence on the level of microbiological cross-contamination of the carcass and other edible parts during slaughter and dressing. Proper lairage is not only important for animal welfare but also for meat safety. The avoidance of stress is important for economic reasons as well as for meat quality and animal welfare. To maintain animal welfare, the lairage period should be kept to a minimum and there should be sufficient time allowed to recover from the stress of the transport. Associated personnel should be trained to recognize problems and to move animals calmly and quietly.

Excessive noise can be very stressful for slaughter animals and has to be minimized by trained handling and the proper design of the slaughterhouse. Lairage constructions should be designed carefully taking into consideration all aspects related to hygienic conditions and animal welfare. The time spent in lairage influences the animal welfare and meat quality. This can be seen in higher number of bruises in animals.

For poultry there is no slaughterhouse lairage as such because they are either slaughtered immediately after arrival or after a brief waiting period on the vehicle. At the slaughterhouse poultry must be handled with care, especially during catching and shackling because bruises and bone breakages may occur, resulting in downgrading or condemnation of the whole carcass. The quality of poultry handling should be regularly assessed and handling practices improved when necessary.

The welfare of animals in lairage can be improved by dim lighting and optimal pen size. Additionally, the area should be well ventilated for all slaughter animals. Access to drinking water should be available at all times; the animals should have feed and appropriate bedding if kept overnight. Ideally, animals should not be kept next to other species. There should be a designated pen for animals that are sick or injured, and animals that are unable to walk should be slaughtered on the spot.

2.5   Food chain information

Information shared between farms and slaughterhouses is known as food chain information (FCI). This information plays an important role in identifying animal health and welfare as well as meat quality and safety issues.

If zoonotic agents are introduced into the food chain, there is a health risk for humans. This is why animal identification systems should be in place at the primary production level, so that the origin of meat can be traced back from the slaughterhouse to the place of production of the animals, to allow regulatory investigation where necessary. At the slaughterhouse level it means that all animals intended for slaughter should be adequately marked to enable the competent authority to determine their origin. To improve the hygiene and welfare level on the farms, the farmer's information systems should be in place regarding the safety and suitability of their slaughter animals/meat for human consumption. The efficient collection and transfer of accurate information between all stages in the production chain is one part of the integrated approach to meat hygiene.

The slaughterhouse is responsible for collecting the FCI. The official veterinarian must be aware of the information and make appropriate risk management decisions. According to regulations the slaughterhouses must request, receive, check and act upon FCI. Slaughterhouses must not accept animals unless they have requested and been provided with relevant FCI contained in the farm records. In ordinary cases, the slaughterhouse must be provided with the adequate FCI no less than 24 hours before the arrival of animals at the slaughterhouse. The FCI has to cover particularly:

- the status of the farm or the regional animal health status;
- the animals health status;
- veterinary medicinal products or other treatments administered to the animals within a relevant period and with a withdrawal period greater than zero, together with their dates of administration and withdrawal periods;
- the occurrence of diseases that may affect the safety of meat;
- the results, if they are relevant to the protection of public health, of any analysis carried out on samples taken from the animals or other samples taken to diagnose diseases that may affect the safety of meat, including samples taken in the framework of the monitoring and control of zoonoses and residues;
- relevant reports about previous *ante-* and *post-mortem* inspections of animals from the same farm, including reports from the official veterinarian;
- production data, when this might indicate the presence of disease;
- the name and address of the private veterinarian normally giving veterinary service in related farm.

The information need not be provided as a verbatim extract from the records of the related farm or through electronic data exchange or in the form of a standardized declaration signed by the producer. Food business operators (FBOs) deciding to accept animals onto the slaughterhouse premises after evaluating the relevant FCI must make it available to the official veterinarian without delay and, in ordinary cases, no less than 24 hours before the arrival of the animal or lot. The FBO must notify the official veterinarian of any information that gives rise to health concerns before *ante-mortem* inspection of the animal concerned. If any animal arrives at the slaughterhouse without FCI, the operator must immediately notify the official veterinarian. Slaughter of the animal may not take place until the official veterinarian gives permission.

Risk-based approach to meat inspection at all relevant stages of the meat production chain is under consideration. According to EFSA's recommendation, the extended use of FCI has the potential to compensate for some, but not all, of the information on animal health and welfare that would be lost if visual poultry *post-mortem* inspection is reduced. Therefore, the information quality and change between food chain counterparts will be more and more important. There is a need to collect and analyse FCI at the farm and slaughterhouse levels to enable risk categorization of flocks or herds and classification of slaughterhouses according to their capacity to reduce carcass contamination. The sampling of carcasses should be based on the available FCI, including results from feed controls, and the frequency of sampling at the farm level should be adjusted accordingly.

Summary

This present chapter has given only general information on aspects that should be taken into account from farm to slaughter of production animals. The demands have changed during the decades and will change further. To guarantee wholesome, environmental and animal welfare-friendly products the most important is that all relevant people are willing to follow the 'One health' concept, which means that animal health and welfare is considered as an entity. Finally, it is important to remember that veterinary medicine is a discipline responsible for the health and welfare of domesticated animals, especially food production animals, in order to provide healthy and wholesome food for consumers. Veterinary specialists are also important in the control and prevention of zoonotic diseases because their expertise covers the whole food animal production chain, starting from primary production, continuing through slaughter and meat processing to retail.

Literature and suggested reading

Aland, A. and Madec, F. (eds). 2009. *Sustainable Animal Production: The Challenges and Potential Developments for Professional Farming.* Wageningen Academic Publishers, The Netherlands.

Arney, D. and Aland, A. 2012. Human transport and slaughter of farm animals. In: *Sustainable Agriculture* (ed. C. Jacobson). Uppsala University, Sweden, pp. 344–348.

Blaha, T. and Köfer, J. 2009. The growing role of animal hygiene for sustainable husbandry systems. In: *Sustainable Animal Production* (eds A. Aland and F. Madec). Wageningen Academic Publishers, The Netherlands, pp. 23–32.

EU Scientific Panel on Animal Health and Welfare. 2004. The welfare of animals during transport. *EFSA Journal, B,* **1–36**.

European Commission. 2012. European Union Strategy for the Protection and Welfare of Animals 2012–2015. http://ec.europa.eu/food/animal/welfare/actionplan /docs/aw_strategy_19012012_en.pdf (last accessed 11 January 2014).

Farm Animal Welfare Council (FAWC). 2013. Five Freedoms. http://www.fawc .org.uk/freedoms.htm (last accessed 11 January 2014).

Gregory, N.G. 2007. Animal welfare and the meat market. In: *Animal Welfare and Meat Production*, 2nd edn (ed. N.G. Gregory). CAB International, UK, pp. 1–21.

Gracey, J, Collins, D.S., Huey, R. (eds). 1999. *Meat Hygiene*, 10th edn. W.B. Saunders Company Ltd.

Nijdam, E., Arens, P., Lambooij, E. *et al*. 2004. Factors influencing bruises and mortality of broilers during catching, transport and lairage. *Poultry Science*, **83**, 1610–1615.

Roasto, M., Hörman, A. and Hänninen, M.-L. 2012. Food-borne pathogens and public health. In: *Ecology and Animal Health* (eds L. Norrgren and J.M. Levengood). Uppsala University, Sweden. pp. 271–282.

3

Ante-Mortem Inspection

Päivi Lahti[1] and Jani Soini[2]

[1]*Department of Food Hygiene and Environmental Health, Faculty of Veterinary Medicine, University of Helsinki, Helsinki, Finland*

[2]*Regional State Administrative Agency for Southwestern Finland, Turku, Finland*

3.1   Scope

Before slaughtering, all animals are evaluated at *ante-mortem* inspection to make sure that they are suitable for human consumption. Signs of contagious or zoonotic diseases are observed as well as indications of animal welfare being compromised. The official veterinarian at the slaughterhouse is responsible for performing the inspection.

3.2   Introduction

Ante-mortem inspection is an important part of official control on products of animal origin. It covers aspects that are very important in protecting public health, animal health and animal welfare. *Ante-mortem* inspection is made to farmed animals and poultry slaughtered for human consumption. Hunted wild game is not usually inspected *ante-mortem*. In slaughterhouses, the official veterinarian will carry out inspection tasks, and official auxiliaries may assist the official veterinarian in inspection. Usually, the *ante-mortem* inspection is made in the slaughterhouse but in some cases it can be made on the farm according to legislation. For example, pigs and poultry can be inspected on the farm

Meat Inspection and Control in the Slaughterhouse, First Edition.
Edited by Thimjos Ninios, Janne Lundén, Hannu Korkeala and Maria Fredriksson-Ahomaa.
© 2014 John Wiley & Sons, Ltd. Published 2014 by John Wiley & Sons, Ltd.

instead of the slaughterhouse. It is important that *ante-mortem* inspection is always done by a highly qualified person.

Ante-mortem inspection is carried out on all animals before slaughter. It is an examination aiming at identifying sick or abnormal animals before they are slaughtered. The inspection usually takes place within 24 hours of arrival at the slaughterhouse and less than 24 hours before slaughter, or at any other time when the official veterinarian requires it. At the inspection issues that affect public health, animal health and animal welfare, such as identification, cleanliness of animals and visible abnormalities, must be observed. A very careful clinical inspection has to be made of all animals that an official auxiliary or the food business operator may have put aside for an official veterinary inspection because of some defect in animal health, identification or welfare. The official veterinarian, the official auxiliaries and the slaughterhouse staff responsible for the animals before slaughter must be trained to detect abnormalities and non-compliances in the animals.

The slaughterhouse layout and conditions must facilitate the *ante-mortem* inspections, so that each animal can be readily checked and identified. For example, adequate lightning is required. Separate pens for sick or suspect animals are necessary. In these pens, contamination of other animals must be avoided. In large animal groups like pigs and poultry, it is sometimes difficult to observe each animal separately. Poultry, which is transported in boxes to the slaughterhouse, is especially problematic. The *ante-mortem* inspection of poultry must often be made to the transport container systems, which consists of stacked cages. The cages and the transport container systems are often stacked upon each other. In addition, the inspection hall is often dark so that the birds are calmer. If the inspection conditions are poor, the *ante-mortem* inspection cannot be made properly, hindering the detection of animal diseases and signs of animal neglect.

The official veterinarian makes a decision on the basis of the *ante-mortem* inspection whether animal is suitable for human consumption and can proceed to slaughter. If some abnormalities are observed at the inspection, the animal can be declared unsuitable for human consumption, and be killed separately or it can be slaughtered separately from other animals, for example at the end of normal slaughtering if necessary. The official veterinarian has to inform of his findings to the food business operator, the original farm, or the competent authority when necessary.

The strengths in the *ante-mortem* inspection include the inspection of animals, the evaluation of animal cleanliness and clinical health status, and the identification of animals. Clinically ill animals are detected quite well at the inspection because of their clinical symptoms. Asymptomatic carriers of the zoonotic pathogenic microorganisms are problematic, because they do not show any clinical signs that can be detected at the *ante-mortem* inspection. The food chain information containing, for example, results of the National Salmonella Control Program for animals conducted in Finland, Sweden and Norway, or results of a monitoring programme concerning Campylobacter in broiler chickens, together with *ante-mortem* inspection, is used to solve this problem. Still, there is a lack of information about some aspects of the history of the animals clinically asymptomatic presented for *ante-mortem* inspection. This weakness in the current *ante-mortem* inspection system has resulted in re-evaluation of how to improve the *ante-mortem* inspection as a part of the overall reform of meat inspection. In the future, the role of the *ante-mortem* inspection will probably be more important in meat inspection and the role of the veterinarian as a specialist will be more pronounced.

3.3   Identification of animals

Identification of slaughtered animals is important in ensuring traceability in the food chain, and in cases where there is reason to suspect an infectious animal disease. Animals transported to slaughter must be identified and the slaughterhouse must be aware of the origin of the animals. At *ante-mortem* inspection, the animals are identified and it is ensured that they are marked and registered in accordance with the current regulations. Together with the food chain information, the identification of animals is an important part of the *ante-mortem* inspection.

Animals may be identified by either an individual marking or a group marking that can be linked to a specific farm or other place of origin. Individual animals may be marked using, for example, ear tags, microchips or tattoos. If the animals are not identified as an individual animal, the animals must be linked to their farm of origin. For example, cattle, sheep, goat, horse and pig (sow and boar) can be individually identified. Pigs (fattening pigs) and poultry are identified as group animals at farm level. Each slaughtered pig has usually a farm code tattooed on the skin, but poultry seldom have any specific markings. In the latter case, the farm of origin must be reliably ensured at the *ante-mortem* inspection. Markings and registration can be ensured with animal passports or national computerized animal registers, or by other reliable means. If the *ante-mortem* inspection reveals signs of infectious animal diseases, or any other irregularity, measures can be effectively directed to the right parts in the food chain. Only when the animals are properly identified, can their movements be reliably traced.

The animal should not be accepted for slaughter if it is not correctly marked and identifiable at the *ante-mortem* inspection, or if the inspection reveals that the animal or its farm of origin is under official restrictive orders. The official regulations and restrictions are used, for example, in a case of an infectious animal disease or when there is a lack in marking or registration. In these cases, the official veterinarian must remember to notify the competent authority and the food business operator. If the slaughtered animal is not reliably identified, the traceability in the food chain is lost. These animals are killed separately and declared unfit for human consumption.

3.4   Abnormalities

The *ante-mortem* inspection is intended to detect abnormalities in the animal. A general impression of the health status of an animal or animal group can be obtained. Particular attention must be paid to check for symptoms and lesions that may be indicative of a serious and highly contagious animal disease (old A-list classification of diseases notifiable to the World Organisation for Animal Health (OIE)), other contagious animal diseases or the zoonotic animal diseases. In addition, attention must be paid to animal welfare and food hygiene factors such as cleanliness of animals or abscesses. Signs of given medication should also be noticed. Time should be taken to monitor animals in place and in movement to detect symptoms such as lameness. If possible, an abnormal animal has to be separated or isolated from the others to a separate pen, where an official veterinarian will inspect it thoroughly.

With many animal diseases, infected animals are likely to present clinical signs, which can be detected at the *ante-mortem* inspection. Transmissible animal diseases that have the potential for very serious and rapid spread, are of serious socio-economic or public health consequence and are of major importance in the international trade of animals and animal products, have to be detected in *ante-mortem* inspection. For example, the foot and mouth disease epidemic in the United Kingdom in 2001 was first identified by authorities following suspicion of lameness in sows during an *ante-mortem* inspection. Global warming is likely to bring new diseases to new areas. Many of them are spread by insects. For these reasons, for example, the *ante-mortem* inspection is an important part in detecting animal diseases, and in preventing the spread of them. Table 3.1 shows the typical symptoms of diseases classified earlier as A-list in the OIE classification. These symptoms must be recognized at the *ante-mortem* inspection and an accurate clinical inspection has to be done before slaughter. If the animal is found or suspected to be suffering from a contagious animal disease, measures must be taken to prevent the possible spread of the disease. In these cases, the official veterinarian must immediately notify the competent authority and the food business operator.

Production animals can be infected with zoonotic microorganisms causing clinical signs at *ante-mortem*, or they can also carry pathogenic microorganisms in their gastrointestinal tract or coat without any clinical signs of disease. Zoonotic diseases can be harmful to either slaughterhouse workers handling live animals or spread via carcasses and animal by-products to the final consumers. Many zoonoses can be detected at the clinical stage at *ante-mortem* inspection. Some examples of these zoonoses and their symptoms are shown in Table 3.2. In these cases, the risk of zoonotic disease can be identified and slaughter and meat inspection can be arranged so that zoonotic disease risk is minimized. Asymptomatic carriers of zoonotic pathogenic microorganisms are problematic because they do not show any clinical signs that can be detected at the *ante-mortem* inspection. Many of these microorganisms are significant causes of food-borne diseases. During slaughter and dressing procedures these pathogens, including *E. coli* O157 and

Table 3.1 Typical symptoms of diseases classified as A-list in the OIE classification.

Disease	Symptoms	Susceptible species
Foot and mouth disease	Vesicle formation in the mucosa of mouth, udder or teats and skin of feet Lameness	Multiple species
Vesicular stomatitis	Vesicular lesions and ulcers and erosions in the mouth Lameness	Multiple species
Rinderpest	Oculonasal discharge Necrotic lesions in the mouth	Multiple species
Rift Valley Fever	Oculonasal discharge and hypersalivation Bloody diarrhoea Rapid respiration	Multiple species
Bluetongue	Vesicles and ulcers in the mouth Oculonasal discharge Rapid respiration	Multiple species
Contagious bovine pleuropneumonia	Difficult breathing	Cattle
Lumpy skin disease	Swellings Oculonasal discharge and hypersalivation Eruptions on the skin	Cattle
Peste des petits ruminants	Oculonasal discharge Necrotic lesions in the mouth	Sheep and goat
Sheep and goat pox	Swollen eyelids and nasal discharge Skin lesions	Sheep and goat
Swine vesicular disease	Vesicular lesions in the mouth, lips and skin of feet	Swine
Classical swine fever	Haemorrhages and cyanosis of the skin notably of the extremities	Swine
African swine fever	Hyperaemia of the skin of the ears, abdomen and legs Respiratory distress Vomiting Bleeding from the nose or rectum and sometimes diarrhoea	Swine
African horse sickness	Dyspnoea, coughing and dilated nostrils Difficult breathing Swellings in the eye	Horse
Highly pathogenic avian influenza	Respiratory or nervous symptoms	Poultry
Newcastle disease	Respiratory or nervous symptoms Cyanotic combs	Poultry

other VTEC, *Salmonella spp.*, *Campylobacter jejuni*, *Listeria monocytogenes* or *Yersinia spp.*, can be directly or indirectly transferred to the meat or to the slaughterhouse workers. *Post-mortem* inspection based on visual inspection is able to react poorly to these problems. The risk of these microorganisms must be assessed; for example, on the basis on how dirty animals are and what kind of food chain information and former meat inspection records are

Table 3.2 Some zoonoses that show clinical symptoms and can be found at *ante-mortem* inspection.

Disease	Symptoms	Species
Botulism	Paralysis	Multiple species
Brucellosis	Often asymptomatic but testicles can swell	Multiple species
Listeriosis	Encephalitis, meningitis or eye infections	Multiple species
Tuberculosis	Tissue necrosis and caseation	Cattle/multiple species
Anthrax	Sudden death Bleeding from all natural orifices Deep black blood which does not clot	Cattle/multiple species
Salmonellosis	Usually asymptomatic Diarrhoea in young animals	Multiple species
Erysipelas	Swine: reddish skin and red skin lesions, arthritis or dermatitis Turkey: septicaemic arthritis	Swine and turkeys /multiple species
Rabies	Nervous symptoms, aggressiveness Hypersalivation	Multiple species
Cryptosporidiosis	Diarrhoea in young animals	Multiple species
Bovine spongiform encephalopathy (BSE)	Nervous symptoms	Cattle

available. At *ante-mortem* inspections those animals usually do not show any clinical symptoms.

At an *ante-mortem* inspection, animals can have visible abnormalities or clinical signs, such as abscesses, swellings, wounds, infections, fractures, lameness, abnormal posture, pain, abnormal behaviour, discharges or cough. If there is no reason to suspect a contagious animal disease, the official veterinarian must make a decision to accept the animal for slaughter or discard it. If it is accepted for slaughter, it must be considered whether it can be slaughtered normally at the slaughter line or slaughtered separate from other animals, for example at the end of normal slaughtering. This evaluation has to take into account the welfare of animals and the possibility to slaughter the animal hygienically at the slaughter line without contamination of the carcasses or the slaughter line. If animals show clinical signs of systemic disease, for example septic infections, polyarthritis, neurological problems and so on, they are not to be slaughtered for human consumption. These animals must be killed separately and declared unfit for human consumption. The killing must be done under conditions that other animals or carcasses cannot be contaminated.

Sometimes a suspicion of used medication rises when animals have signs of injections or swellings at the typical injection sites. The food chain information should be examined and the animal should be separated for more careful inspection, and the decision of sample taking and possible slaughtering should be made.

The *ante-mortem* inspection is important in identifying clinical signs as indicators for diseases that are important to human and animal health. In some conditions, the possibility of detecting some diseases and important human pathogens during the *ante-mortem* inspection is low. In these cases, the food chain information, the earlier laboratory findings and the earlier meat inspection data, in addition to *ante-mortem* inspection observations, help the official veterinarian to estimate the risk the animals cause to public health.

3.5 Cleanliness of animals

Ante-mortem inspection is of great importance for detecting dirty and dungy animals that cause risk for food safety and public health. The risk of carcass contamination by, for example, human pathogens *Salmonella spp.*, *E. coli* O157 and *C. jejuni*, is high in dungy animals, since dunginess significantly increases the microbial surface contamination of carcasses during slaughter. Ventrally located dirt, especially, causes microbial contamination of cattle and sheep carcasses during dressing, since the cutting line of the hide lies on the ventral midline of the animal (Figure 3.1). In turn, dirty swine and poultry contaminate scalding water and microbial contamination of the carcasses increases. Greater care in slaughtering dirty animals cannot compensate for the higher contamination.

Dungy animals should be evaluated by the official veterinarian or official auxiliaries. Dirty animals should not be slaughtered, because of food safety

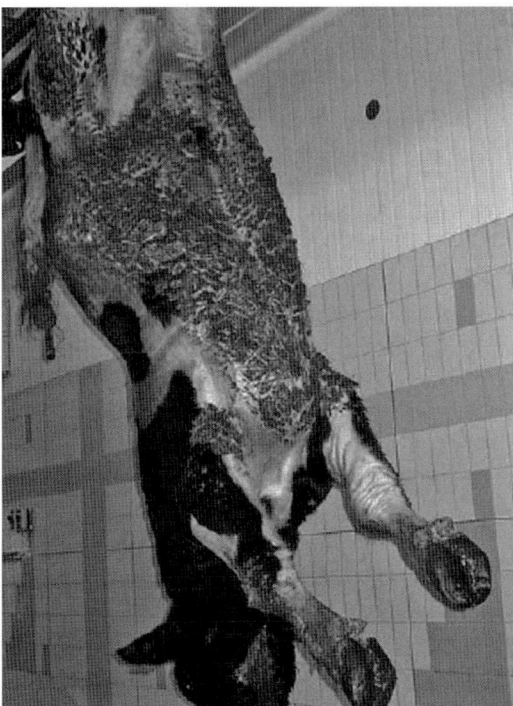

Figure 3.1 Dungy cows are a food safety and animal welfare problem. Source: Courtesy of Elisa Pitkänen.

reasons, unless the animals are cleaned before slaughter. However, if the food safety is not endangered, slightly dirty animals could be stunned and slaughtered separately with greater care in work procedures and slower speed of the slaughter line to avoid contamination of the carcass and other carcasses, equipment and the facility. Unacceptably dirty animals that cannot be cleaned are killed separately and declared unfit for human consumption.

Dirtiness indicates poor housing and management. The time of the year and housing system also have effect on the dirtiness of cattle. During the winter season dunginess is more common. In Finland, an agreement between stakeholders was made to reduce hide damage and extra labour costs as well as for food safety and animal welfare reasons. According to the agreement, farmers clean dungy animals; otherwise, the animals undergo casualty slaughter or are slaughtered separately after the clean animals with greater care, with the extra costs of the special treatment incurred by the farmers. The agreement decreased the dirtiness of cattle by 85% when adopted. Similar practices are used in some European countries and wide adoption of these practices would be beneficial.

Swine and poultry may be fasted prior to slaughter with the purpose of decreasing defecation and dirtiness during transport and slaughter. This may cause animal welfare problems if the duration of the transport is long, since fasting animals are prone to stress.

Farms may not comply with regulations of animal protection, if the animals produced on the farm are continuously dirty. In these cases, the official veterinarian must notify the competent authority, because an animal welfare inspection visit to the farm may be necessary. The possible actions are based on national animal welfare regulation.

3.6   Animal welfare

Ante-mortem inspection may be the only occasion during animals' life when they are inspected by a person other than the farmer. Therefore, *ante-mortem* is of great value in animal welfare control in general. In the slaughterhouse animal welfare is evaluated by the official veterinarian or, in some cases, by official auxiliaries. Where deficiencies in animal welfare are observed, it should be evaluated if the neglect has taken place on the farm, during transport or unloading. To stop the improper action and preventing its recurrence it is important to communicate the neglect found to the food business operator, the farm and the competent authority.

3.6.1   Animal welfare on the farm

Animal welfare on farm can be described by five freedoms:

1. Freedom from thirst, hunger and malnutrition.
2. Appropriate comfort and shelter.

3. The prevention, or rapid diagnosis and treatment of, injury, disease or infestation.
4. Freedom from fear and distress.
5. Freedom to display most normal patterns of behaviour.

Deficiencies in these freedoms may be noticed at the *ante-mortem* inspection, especially when the deficiencies are severe. The body condition of animals, size of the animal in relation to its age, dirtiness, injuries, untreated illnesses and, for example, uncared for hoofs are matters that give information of animal welfare on the farm. Abrasions caused by unsuitable bedding may occur, as well as abscesses formed due to unhygienically performed injections, untreated abrasions or injured tails. Bitten tails on swine and broken tails on cattle may indicate too high stocking density on the farm. Tail-docking of piglets is allowed in many countries, which diminishes the possibility of the official veterinarian observing the signs that may indicate deficiencies in housing conditions, for example lack of enrichment material. The inspection results are communicated to the farm, the food business operator and the competent authority if necessary.

Although animal welfare issues can usually be evaluated at *ante-mortem* inspection, some signs of deficiencies in animal welfare may be noticed based on food chain information and *post-mortem* inspection. For example, evaluation of poultry welfare at *ante-mortem* is complicated at slaughterhouses, since birds are held in cages that are stacked on each other. Therefore, poultry welfare should be evaluated based on cumulative daily mortality, which is a part of food chain information, and defined *post-mortem* findings, such as total rejections, joint lesions, respiratory problems and foot pad dermatitis. The results are communicated to the farm, food business operator and competent authority for the appropriate measures to be taken to improve the housing conditions at the farm.

3.6.2   Animal welfare in transit

Animals should be transported in appropriate vehicles, they should be handled carefully and the transport should be as comfortable as possible. Transport causes stress to animals and animals may be exhausted or injured when arriving at the slaughterhouse due to long journey duration or inappropriate transport conditions. Fresh injuries may indicate fighting during transport, rough handling when loading or inappropriate transport vehicles. Poultry, especially, is easily injured during loading and transport, since there are four to five birds in the same cage. *Ante-mortem* inspection at the farm and inspection again at the slaughterhouse would enable the evaluation of the transport injuries. Unloading of animal transport is a favorable moment to make *ante-mortem* inspection.

Animals that have died during transport should be carefully inspected. Death may have occurred because of illness or stress. It is possible that finishing pigs and poultry may die during transport due to stress, which may be worsened due to poor ventilation and overcrowded vehicles. In inspection of the animals that have died during transport, the signs of contagious diseases or animal welfare problems on the farm or during transport should be carefully evaluated.

3.6.3   Animal welfare at the slaughterhouse

Animal welfare should be secured also in the slaughterhouse. If the time in lairage is long, animals should be fed, and cows milked if necessary. Water should be available and the temperature and ventilation suitable. The handling of animals should be calm and gentle. Tired animals should have the ability to rest before slaughtering. If the animal welfare rules are not respected, the official veterinarian is to verify that the food business operator immediately takes the necessary corrective measures and prevents recurrence.

Literature and suggested reading

EFSA (European Food Safety Authority). Scientific opinions on the public health hazards to be covered by inspection of meat. *EFSA Journal* (bovine animals: 27 June 2013; poultry: 29 June 2012; swine: 3 October 2011). http://www.efsa.europa .eu/en/search.htm?text=Scientific+opinions+on+the+public+health+hazards+to +be+covered+by+inspection+of+meat (last accessed 10 February 2014).

Ridell, J. and Korkeala, H. 1993. Special treatment during slaughtering in Finland of cattle carrying an excessive load of dung: Meat hygienic aspects. *Meat Science*, **35**, 223–238.

4
The Slaughter Process

Eero Puolanne and Per Ertbjerg

Department of Food and Environmental Sciences, Faculty of Agriculture and Forestry, Helsinki, Finland

4.1 Scope

The slaughter process for pigs, beef, sheep/goats and poultry starts at the farm. It includes measures at the farm before transport, transport, lairage at the slaughterhouse and the slaughtering (moving to stunning, stunning/bleeding, dehiding/dehairing/defeathing, evisceration, grading/classification, weighing, and cooling/chilling). This chapter deals mainly with the operations after stunning and bleeding.

4.2 General

The operations of slaughtering are basically the same to all animal species: preparation for transport and slaughter, transport from farm, lairage, moving to the stunning, stunning, bleeding, dehiding/scalding+dehairing, evisceration, inspection, classification, weighing and chilling (or moving to a hot boning). Consequently, slaughtering as a process influencing animal welfare, as well as meat quality, takes about 12–24 hours or more, depending on the animal species and transport time. Roughly, one half of the duration is before slaughter and one half after. The size of the operations and the level of technology may differ and influence the slaughter process and the facilities. When the number of animals slaughtered is high, slaughtering is usually done in modern large facilities where most operations are assisted mechanically;

Meat Inspection and Control in the Slaughterhouse, First Edition.
Edited by Thimjos Ninios, Janne Lundén, Hannu Korkeala and Maria Fredriksson-Ahomaa.
© 2014 John Wiley & Sons, Ltd. Published 2014 by John Wiley & Sons, Ltd.

these operations are increasingly carried out automatically using image analyses to cope with animals of different sizes. Poultry slaughtering is nowadays mostly mechanical, due to the rather uniform size of the animals and their large numbers; thus, human labour is not much needed on the line. Pork lines are also mechanical and increasingly automated but cattle lines still mostly rely on human labour assisted by mechanical devices. Animal welfare, hygiene, occupational safety and the need to increase efficiency sets challenges and boundaries to the design of modern slaughter lines. Much effort is devoted to ensure that the hygiene requirements are fulfilled in automated lines and that the facilities function precisely without damaging the valuable parts of carcasses.

Carcasses need to be cooled after slaughter for microbial safety and keepability. The rate at which the carcass is cooled is of importance to the process efficiency as well as the meat quality. The cooling/chilling processes for different species differ widely depending on factors such as the size of the animals, amount of subcutaneous fat and the rate of muscle metabolism. Larger carcasses cool more slowly due to a smaller surface-to-mass ratio. The thermal conductivity of fat is less than that of muscle or bone and a thick fat layer will, therefore, slow down cooling. The fibre composition of muscles influences the rate of metabolism; white muscles are generally more glycolytic with a larger capacity for a rapid decline in pH and rigor development. Just before slaughter, the body temperature of pigs and ruminants is around 38–39 °C, whereas the body temperature of poultry is 40–41 °C. The body heat must be removed in the cooling process, together with the heat generated in metabolic processes. Metabolic heat is generated when ATP is produced from conversion of glycogen to lactate and protons, and heat is also produced when ATP is consumed in processes such as *post-mortem* muscle contraction and pumping of calcium out of the sarcoplasm. The temperature of pork carcasses can thereby increase to more than 42 °C during the first 30–60 minutes after death. It is generally considered beneficial to reduce the carcass temperature as quickly as possible for several reasons:

- to maximize the shelf life by inhibiting the growth of spoilage organisms and any pathogens that may be present;
- to minimize the water loss during chilling, such as the weight loss that occurs by evaporation of water from the carcass surface (a low water loss in chilling improves the process efficiency);
- to minimize the protein denaturation that occurs due to a higher temperature. As muscle pH decreases the proteins become more prone to denaturation. The combination of high temperature early *post-mortem* with a pH value approaching the low ultimate muscle pH will result in protein denaturation, leading to more pale meat with decreased water-holding capacity.

Nevertheless, carcasses can also be cooled too quickly if the emphasis is to ensure optimal meat tenderness. If the muscle is cooled below the temperature region of 7–10 °C before the onset of rigor, the muscle will contract, a phenomenon known as cold shortening. Subsequent cooking after cold shortening will result in tough meat. Ruminants have a relative slow rigor development and are, in spite of the large carcass size, more prone to cold shortening (i.e. cold induced pre-rigor shortening of sarcomeres) than pigs and poultry. For any given species, the rate of pH decline of the major muscles is, therefore, of major importance for the optimal cooling rate, which on one side should be fast enough to minimize growth of microorganisms, minimize water loss and minimize protein denaturation and, on the other side, be mild enough to ensure optimal meat tenderness. These considerations define a pH/temperature window that optimal cooling will lead the muscles through.

4.3 Pigs

An overview of the main operations in pig slaughter is presented in Table 4.1.

4.3.1 Moving the animals from the stable to stunning

From an animal welfare as well as meat quality point of view, it is of utmost importance to move the pigs peacefully to stunning. Any stress factor would launch a stress reaction that, in turn, would reduce the effectiveness of the stunning as well as trigger the glycolytic pathway in the muscles. This easily leads, in connection to physical stress, to an accumulation of lactic acid in the body of these animals with very weak aerobic capacity. A rapid pH decline in perimortem porcine muscles, in connection with high temperature *post-mortem* (>40 °C), results in a partial denaturation of structural proteins as well as precipitation of soluble proteins onto the surfaces of the former, causing a reduction in water-holding and inducing paleness of the meat (PSE meat, which is pale, soft and exudative). Due to the high glycolytic and low oxidative capacity, the accumulation of lactate will make the animal fatigued, and therefore the physical stress of the pig will usually not last long. A long-lasting stress may cause an elevated ultimate pH also in porcine muscles.

Animals should not be severely stressed before slaughter. Psychological or physical stress, or their combination, causes a reduction of muscle oxygen and creatine phosphate level, a rise in body temperature, elevated calcium level in the sarcoplasm and a high plasma adrenalin content. Psychological stress is mediated by blood circulation and, thus, influences all muscles; physical stress (exercise) has an effect on those muscles that are active. In pigs, stress triggers intensive lactate (lactic acid) production; the lactic acid formed is moved outside the fibres through monocarboxylate transporters that are able to transport lactate along the concentration gradient, either in or out. The lactate content in plasma also increases quickly, which means that the lactate content in muscle fibres starts to increase. Part of the lactate formed in glycolytic fibres is transported to oxidative fibres to be used oxidatively. The muscles of modern domesticated pigs contain very high proportions of glycolytic fibres; therefore, the oxidative use of lactate in and immediately after stress is low. This leads to a rapid perimortem accumulation of lactic acid in muscles. To prevent these consequences, all possible measures to avoid stress should be taken.

Table 4.1 Main unit operations, produce, machines and facilities in pig slaughter.

Unit operation	Produce	Machines and facilities
Preparation for transport, identification (tattooing, ear tag)		
Loading, transport	Faeces, urine, dead animals	Loading deck, truck
Unloading, receiving of animals, lairage	Faeces, urine, dead animals	Unloading deck, lairage pen
Moving to stunning		Single/double-file chute, moving wall, V restrainer conveyor, electric prod, stunning pen
Stunning		Stunning point, electrical stunners, stunning tongs, gas stunning chambers, shackle
Bleeding, blood recovery	Blood	Sticking point, vacuum blood removal knife, blood tank
Scalding	Solid and liquids from animal → waste	Scalding equipment
Dehairing (dehiding)	Hair, outermost layers of skin → waste	Dehairing machine, gambrel
Singeing	Heat to be recovered	Singeing machine
Eviscerating	Genitals, casings, stomach, oesophagus, spleen, pancreas, lungs Liver, heart, tongue, fat	By-products department, viscera department, (edible/inedible) rendering plant
Carcass splitting	Carcass sides	Splitting machine/saw
Inspection and labelling		
Removal of kidneys	Kidneys	
Classification		Muscle scoring, fat analyzer
Weighing		Scale
Cooling	Carcass, offal, by-products	Carcass cooler, chilling tunnel, carcass cold storage

4.3.2 Stunning

Pigs are moved from holding pens to stunning through a single-file or double-file race. Restraining conveyors are used to move the pigs the last few meters to electrical or carbon dioxide stunning. Group stunning is increasingly in use. The pigs are moved (pushed) with a slowly-moving plate towards a pen that will close when a proper group of pigs (usually 4–5 pigs) is present. These pigs are lowered to a carbon dioxide chamber, where they are stunned and then moved to shackling for sticking. Sticking is performed with a sticking knife or a vacuum blood removal knife if the blood is to be collected.

4.3.3 Bleeding

The purpose of bleeding is to remove quickly as much blood as possible because blood is an ideal medium for bacterial growth. The main blood vessels of the neck are severed in order to allow blood to drain from the carcass resulting in the death of the animals from cerebral anoxia causing minimal damage to the carcass. The bleeding knife must be sharp so that the cut ends of the blood vessels are not damaged. Sticking must immediately follow the stunning. In pigs, a longitudinal bleeding stick is made into the chest (in the centre of the neck just in front of the breastbone) to sever the deep vessels, i.e. event to aorta. Incomplete bleeding may cause excessive retention of blood in the tissue, which will result in early spoilage of the meat. To reduce the contamination with scalding water, the cut should be as small as possible.

The most hygienic way of bleeding is to shackle the pigs immediately after stunning and then lift them on to a moving rail. Another way of bleeding is horizontal bleeding, which is claimed to give faster bleeding rates and a greater recovery of blood, probably due to less pressure on certain organs and blood vessels that obstruct the blood vessels and restrict the flow. However, bleeding on the floor is very unhygienic. Horizontal bleeding should take place on an easily cleaned, stainless steel table. After sticking, the animal should be left to bleed until the blood flow becomes negligible.

4.3.4 Scalding

After the stunning, the carcasses are hung on a rail with a hook from one hind leg (shackling).

The traditional way of immersion to scald the pigs is to sink them fully into a scalding tank of water at 60–62 °C for 4–6 minutes. This provides an effective scalding effect, kills most of the surface bacteria and also distributes heat resistant microbes from carcass to carcass. Scalding water will thus also contains many impurities: blood, urine, faeces, slime, mud and so on. Therefore, in modern slaughtering, steam scalding is used, with the same parameters except the time needed, which is shorter due to the higher heat transfer coefficient of steam compared to water. In immersion scalding the heat transfer is uniform, but the problem in steam scalding may be the non-uniformity of the heat transfer. Problems may occur especially with heads, front legs and groin-folds. These have been more or less eliminated with most modern technologies where temperature regulation has been improved.

The surface of the skin has the following cell layers, from the surface downwards forming the epidermis: *Stratum corneum, Stratum ludicum, Stratum granulosum, Stratum spinosum* and *Stratum basale*, and a strong elastic connective tissue membrane, the *dermis*. The cells differentiate from bottom to top with an increasing amount of keratinization. Scalding weakens the cellular membranes on the skin as well as the *dermis*, which is composed of collagen, small amounts of elastin and ground substance, and where the hair bulbs are embedded. The collagen of the dermis shrinks; this pushes the hair towards the surface, which makes it easier to scrape away without breakage. Newly-formed hair is more tightly bound than older hair, meaning hair can be more easily removed in the summer than in the fall, although the difference is relevant only at scalding temperatures of 60 °C or lower. Stress may also make the dehairing more difficult, probably due to the earlier rigor mortis of the hair muscle.

Immersion scalding results in higher microbial surface contamination than steam scalding. To reduce the contamination, scalding water should be changed frequently, pigs should be as clean as possible at sticking and bleeding should be fully completed before immersion. In addition, to reduce the contamination by scalding tank water, the sticking cut should be as small as possible. The lungs of immersion-scalded pigs will also contain microbes originating from the scalding water.

4.3.5   Dehairing

Dehairing is performed with rotating scrapers that remove the outermost layer, the epidermis, from the skin surface. The scrapers also pull the hairs out of the hair follicles, due to the loosened connective tissue of dermis, without cutting them into pieces. If the scalding temperature has been too high and/or the time too long, the dermis will also be damaged which, thus, poses a hygienic risk. After dehairing, the carcass is hung by inserting a double-sided hook behind the Achilles tendon of the hind leg (gambrelling) for the further unit operations all the way to the cutting. The process then continues by singeing the remaining hairs for a few seconds with a flame. The operation is completed by intensive brushing, where the burned hair remains (black scraping), and finally by washing the carcass intensively with, eventually, another scraping (white scraping). Scalding, burning and washing reduce the microbial load on the carcass markedly, with the total counts usually at the levels of between $10^2 - 10^4$ cfu/cm^2.

4.3.6   Skin removal

If the pork skin is to be used for leather, scalding cannot be used, and the pigs must be dehided. This is carried out by first removing cloven hoofs and male genitals and then skinning the head. The skin can then be removed by pulling it upwards or downwards. Downwards pulling results in hygienically safer meat because upwards pulling causes more contamination from the skin to the carcass. There are also methods where only the ends of the carcass are steam scalded and then dehided. In this case, only the middle part of the skin is used as leather. Dehiding will allow a very fast processing of the carcass, for example if the processor wants to use accelerated processing (hot boning). In pigs, the subcutaneous fat is tightly bound to the skin. Therefore, it is difficult the remove the hide without a significant amount of fat remaining attached to the skin, which damages the fat layer as well.

4.3.7 Evisceration

Evisceration starts with loosening the rectum of its adhesions. A manual or an automatic bung cutter with a vacuum system is widely used to reduce the contamination from faeces still present in the colon. The rectum can also be sealed with a plastic bag. After sealing the rectum, the skin is opened on the ventral side from the anal opening to the throat (a brisket saw is used for the sternum). The genitals, casings, stomach and oesophagus are removed. Then the liver, lungs, spleen, pancreas, heart and tongue are removed and put on a viscera conveyor or hung on a hook. This can be facilitated by mechanical equipment. Kidneys usually stay attached to the carcass for the meat inspection, after which they will be removed. To facilitate cooling, the carcass is then split in two halves by a splitting saw or large cleaver, manually or in a splitting machine. A cleaver is not used in larger plants. Saws may be manual or mechanical, or automatically adjusted according to the size of the carcass. Finally, the inner fat from the coelom will be removed and the carcass will be finished: bloody throat muscles will be cut off, the spinal cord will be removed with a spinal cord cleaner and the rest of the abdominal fat removed with a fat removing machine. The carcass is now ready for veterinary inspection, printing of the veterinary inspection stamps on the carcass, grading and weighing, and finally cooling.

In modern slaughterhouses, most of the above unit operations are nowadays performed mechanically or automatically. Current image analysis technology and robotics allow the complex processing of pigs of varying sizes more effectively than, and hygienically as safe as, manual processing. Commercial producers of slaughter lines have published schematic diagrams and videos of their facilities and processes on the Internet.

4.3.8 Cooling/chilling

Pig carcasses are cooled to minimize both the microbial load of the raw material and the water loss during chilling. In porcine, the pH typically declines from 7.0 at slaughter to the ultimate pH of around 5.4–5.7 over a period of 6–10 hours. The pH decline can become much faster if pigs are stressed before slaughter. Significant stress just before slaughter can reduce the pH to 6.4 at the moment of stunning in some fast glycolysing muscles. The rather fast pH decline rate of porcine as compared to ruminant muscles is an important factor for the chilling process. Protein denaturation due to the combination of high temperature and low pH can be reduced by chilling quickly, which will also lower the incidence of PSE (pale, soft and exudative) pork. Towards the end of the slaughter line (35–45 minutes after death), the surface temperature (rind side) is about 25–29 °C and the centre temperature of leg, loin and shoulder is 39–43 °C.

The conventional system for chilling pigs is a batch process where the carcasses are placed in chill rooms and then cooled together overnight. The air temperature in the chill room is −5 to 0 °C with air circulation. The air is distributed in the room with a low air flow (around 1 m/s). The efficiency of the chilling is dependent on the air temperature, the air velocity, the weight of the carcasses and the loading pattern of the chill room. The carcasses should hang separately to allow air movement between them. A large quantity of heat is released from the carcasses in the initial hours of chilling and the air temperature in the chilling room may rise. This is described as heat peak and can lead to increased weight loss because of the extended cooling time. The chilling rate can be increased by spraying cold water mist onto the carcass.

More rapid cooling systems for pig carcasses reduce the weight loss and have become popular alternatives to the batch chilling system. These fast chilling systems have been referred to as 'rapid chilling', 'very fast chilling' or 'blast chilling'. Blast chilling is a continuous process in two stages: a tunnel operated at freezing temperatures followed by conventional chilling, which is described as equalizing chill. An example of blast chilling uses cold air at −20 °C with an air velocity of 3 m/s for 80 minutes. Upon exit from the chilling tunnel the centre temperature is around 35 °C in leg and shoulder and around 20 °C in the loin. In blast chilling, some areas of the carcass surface can freeze due to cold air combined with high air speeds. However, any major freezing of the lean should be avoided, as it will lead to a large increase in drip loss of the retail cut. The combined cooling process, including chilling and equalizing, normally takes 18–20 hours. It is recommended to chill pork to temperatures of less than 10 °C in 12 hours and to 2–4 °C in less than 24 hours. The temperatures of the carcasses are, in most operations, quite uniform prior to cutting.

4.4 Cattle, horses, sheep and goats

An overview of the main operations in cattle slaughter is presented in Table 4.2.

4.4.1 Moving the animals from the stable to stunning

A long-term state of stress influences the quality of beef because it results in a high ultimate pH, causing dark, firm, dry meat (DFD or dark-cutting meat, the ultimate being pH 6.0 or higher) in some muscles. Muscle glycogen is mobilized by adrenalin (mental stress) and/or free calcium (physical stress causes free calcium levels to remain high and trigger the contraction). The combination of mental stress and physical exercise results in the fastest reduction of muscle glycogen. In long-term stress situations glycogen is used while the animal is still alive, but in non-stress situations only prolonged starving lowers the muscle glycogen level. The resting level of glycogen in bovine muscles, given as

Table 4.2 Main unit operations, produce, machines and facilities in cattle and sheep slaughter.

Unit operation	Produce	Machines and facilities
Preparation for transport		
Loading, transport	Faeces, urine, dead animals → waste or destruction	Loading deck, truck
Unloading, entrance control, lairage	Faeces, urine, dead animals → waste or destruction	Unloading deck, lairage
Moving to stunning		Chute, electric prod, stunning pen
Stunning		Stunning point, bolt pistol, pneumatic stunner, shackle
Bleeding, sticking, blood collection	Blood	Sticking point, vacuum blood removal knife, blood tank
Electrical stimulation		Stimulator
Removal of horns, cloven hoofs, deheading, dehiding	Head, hide; horn, cloven hoofs → waste	Hoof remover, head splitter, dehiding machine, gambrel
Eviscerating	Genitals, casings, stomachs, oesophagus, spleen, pancreas, lungs. Liver, heart, tongue, fat	By-products department, viscera department, (edible/inedible) rendering plant
Carcass splitting	Carcass sides	Splitting machine/saw
Inspection and stamping		
Removal of kidneys	Kidneys	
Classification		Muscle scoring
Weighing		Scale
Cooling	Carcass, offal, by-products	Carcass cooler, chilling tunnel, carcass cold storage

glucose, is about 100 mmol/kg muscle (1.8%). The long-term rate of glycogen depletion in stress is usually not higher than about 10 mmol per kg and h. Studies have shown that the ultimate pH starts to stay at a higher level than about 5.5 when the glycogen level at stun is lower than about 50–60 mmol/kg. DFD properties can be identified when the glycogen content is about 40 mmol/kg (0.7%). By then the low glycogen content does not provide enough glucose for *post-mortem* lactate (and protons, i.e. lactic acid) formation and the ultimate pH stays high, 6.0 or higher. Glycogen is not resynthesized as long as stress-induced adrenalin is attached to its receptors.

After stress, full glycogen recovery in exhausted beef animals lasts days and in pigs several hours. Therefore, the main metabolic difference in beef animals in stress compared to the non-stress situation is that glycogen is degraded instead of other energy substrates, aerobically in all cases. Similarly, due to the slower and more aerobic metabolism of beef animals compared to pigs, a short-term stress just before slaughter does not have any significant effect on beef animals, that is beef animals do not easily develop PSE for that reason, while pigs will easily do so. After prolonged mental and physical stress just prior slaughter, however, they both develop DFD meat. The net effect is that in a long-term stress state beef animals also use glycogen aerobically, while domestic pigs turn quickly to anaerobic metabolism that accumulates lactic acid, and therefore they get much more easily fatigued and reduce/stop their physical activity. Therefore, a long-term stressful transport to slaughter and/or slaughterhouse lairage may cause DFD problem in beef after slaughter but the short moving period to slaughter does not. Irrespective of quality aspects, however, animal welfare is always relevant as long as the animal is alive.

Long-lasting stress during lairage at the slaughterhouse can be avoided if beef animals unknown to each other are not mixed. Animals of different sexes or of essentially different sizes should not be mixed, either.

Female animals form their hierarchy in groups in 10–30 minutes, which does not cause a marked reduction in glycogen level. Bulls in group pens, however, may need hours, especially if they are roughly of about the same age/size. This will cause a decrease of glycogen in affected muscles to values lower than 50 mmol/kg, which then results in DFD properties in these muscles. In some countries, bulls are kept in individual pens, thus inhibiting fighting, keeping the air fresher and the drinking devices cleaner. In countries where bulls are castrated, the male animals fight less and, thus, the incidence of DFD carcasses is lower.

When animals are moved, it is of utmost importance to see the situation from the animal point of view, which is strongly underlined by the animal etiologist Temple Grandin, who has designed many slaughterhouse systems and their auditing, based on animal behaviour. Beef animals have a large view angle, about 330°, and they see well below and above themselves. Therefore, unknown objects and reflections will easily make them baulk; for example, stripes across their walkway will stop them moving. Beef animals are sensitive to high frequency noises (7000–8000 Hz) and move most easily in a solid-wall curvilinear single-file race from which they cannot see out of. The movement of the animals can be controlled by operators who must be at positions in relation to the balance point of the animals' flight zone that favour the movement, that is does not stop it. Readers interested in animal welfare at slaughter are advised to study the books of Temple Grandin and Neville Gregory for more detailed information.

4.4.2 Stunning

Beef animals are most usually moved to a stunning box for stunning. The stunning can also be carried out in an open area, but this would risk the occupational safety of the operator. In the stunning box, there should be head lifter to support the head and to prohibit sudden movements of the head during the stunning. The stunning is usually with a bolt device, either a penetrating bolt or concussion bolt; electrical stunning is very rare in modern slaughterhouses. In some countries, for religious reasons, no stunning is used, the killing of the animal is done just by bleeding.

4.4.3   Bleeding

In cattle, the skin at the neck between brisket and jaw is opened through a 30-cm longitudinal cut. For hygienic reasons, a clean knife must be used and inserted at a 45° angle in order to sever the jugular and carotid vessels. In sheep, both carotid arteries will be severed. Kosher and Halal slaughter sever the carotic and jugular vessels.

4.4.4   Electrical stimulation

Once the animals have been stunned they are shackled to the rail and bled hanging from one hind leg on rail. The carcass may be electrically stimulated, either at low voltage (<80 V) or high voltage (around 300 V), to increase the rate of ATP consumption and pH fall to avoid cold shortening (Section 4.4.6). The alternating current ($15–50$ Hz) passes through the carcass, mimicking nervous stimulation, causing the pH to fall by $0.2–0.6$ pH units and the fibre membranes to be damaged to the extent that the pH fall will continue at double the normal rate. The electrode can be put individually into the nose, or in larger operations it is fixed, where carcasses slide on them when passing the stimulation zone. The rail is the ground electrode. Electrical stimulation may cause PSE-like quality if the cooling is slow.

4.4.5   Dehiding and opening the carcass

Due to the large size of beef animals, elevator platforms are used to facilitate the operations. After stimulation, the operators cut the horns with a pressure-driven horn remover, remove the hoofs and cut the rear end of rectum hide around the anus. The head is dehided and removed, and then the operator cuts the skin from anus hole to the trout with a dehiding knife. The genitals and udders are removed, and then the skin is party excised to help the mechanical dehiding. The oesophagus is closed by inserting a rubber ring, to avoid the rumen contents running out. The carcass is then fixed to the hide pulling equipment and the hide of the forelegs is fixed with hide clamps to the puller. The hide pulling equipment pulls the skin off. The equipment operates down–up, or up–down, the latter being better from a hygienic point of view, as there is a smaller risk of contamination from the skin, especially in the very end of the unit operation when the hide comes off the carcass fiercely. After dehiding, the brisket is opened with a brisket saw; the anus is detached, covered with a plastic bag and then pushed into the coelom. Then, the frontal side is opened with a knife or opening device, with the sharp edge outwards to avoid cutting the intestines. Casings, stomachs and oesophagus are removed, followed by the lungs, heart, liver, spleen, pancreas and tongue. The kidneys, again, stay in the carcass until the veterinary inspection. Spinal cord and brains are treated separately because they are materials at risk of BSE (bovine spongiform encephalopathy caused by prions). The carcass is then inspected, stamped, graded, weighed and, finally, cooled.

Sheep and goats are slaughtered in basically the same way. In larger slaughterhouses, sheep are transported to stunning using conveyors, which are usually double rail restrainer systems, where the animal is supported by the rail and the feet do not touch the ground. A bolt pistol or electrical stunning (head only for stunning or head-to-back for stunning and cardiac arrest) are most usually used for sheep and goats.

4.4.6 Cooling/chilling

Beef, sheep and goat carcasses are cooled to inhibit the growth of microorganisms and to minimize the water loss during chilling. The normal system for chilling ruminants is a batch process where the carcasses are placed in chill rooms and then cooled together. The air temperature in the chill room is 0–5 °C with air circulation in the range 0.7–1.5 m/s. It usually takes around 18 hours to reach the ultimate muscle pH in sheep and 24–36 hours in beef. A general rule is to avoid a muscle temperature below 10 °C for at least 10 hours after slaughter and the pH must reach 5.7 before the temperature falls below 7 °C. Shrinkage during cooling is an important economic factor and, for beef carcasses, water loss through evaporation is typical between 1 and 2% during the initial 24 hours of the chilling process. Shrinkage can be reduced by maintaining a low temperature, using a low velocity of circulating air and maintaining a high relative humidity. Spray chilling of beef and lamb carcasses with cold water mist in cycles during the first 3–12 hours has been shown to reduce carcass shrinkage. The cold water sprayed on the carcasses will absorb heat from the carcass as it evaporates and also replaces some of the muscle water that would otherwise be lost by evaporation. Lamb is susceptible to cold shortening due to:

- a high surface-to-volume ratio leading to faster cooling;
- a relatively high amount of red fibres – more red muscles are generally more prone to cold shortening;
- a relatively slow glycolysis rate.

Cold shortening will occur if the temperature inside the muscle falls below around 7–10 °C before the onset of rigor mortis. Under these conditions, the ATP level in the muscle will be high enough to drive a cold-induced contraction, resulting in shortened sarcomeres after rigor and more tough cooked meat. Cold shortening is also a phenomenon that must be considered when cooling beef carcasses. Small and lean animals, such as calves, are more prone to cold shortening. The large body size of mature beef animals reduces the cooling rate, but still the glycolysis is not fast enough to eliminate the risk of cold shortening. Many slaughterhouses therefore use electrical stimulation of beef and sheep carcasses before the cooling process. Electrical stimulation (as previously described) reduces the risk of cold shortening by using energy in the form of ATP for muscle contraction and for pumps that move calcium out of the sarcoplasm. Glycogen is depleted faster and the onset of rigor is accelerated. Electrical stimulation thereby allows more rapid cooling without compromising the meat tenderness. Another method to avoid cold shortening is by delaying the chilling. A slow cooling of deep beef muscles nearly always leads to a PSE-like conditions, especially if electrically stimulated.

4.5 Poultry

Poultry production from egg to table is a very integrated process. The birds are fed for 30–40 days and they often reach a weight of 2.5 kg or more. Birds usually number from thousands to tens of thousands in the same hall, which means that the animals are not treated individually. The slaughtering schedule is carefully planned. The time from farm to actual slaughter is about four hours.

4.5.1 Transport to slaughter

Before transport the birds are collected into cages of about 10 birds per cage. The collection can be made manually or using collecting machines. Dim blue light is usually used, as the birds do not see well and they stay immobile in such light. The cages are then put on a truck and after a short transport are unloaded; the birds wait in the cages for slaughter, again in dim blue light.

4.5.2 Stunning

In modern slaughterhouses, the following unit operations are performed mechanically in-line. Slaughter starts by lifting the cages onto a conveyor from where they are emptied. After the birds are electrically stunned, they are lifted manually to hang upside down from a W-shaped hook. The rail moves the birds forward so that the head of the bird sinks into a salt solution where they receive an electric shock of either 50–100 V or 120–150 V for a few seconds. Care must be taken that the number of birds in the solution is constant in order to have optimal voltage and current for each bird. It is, anyway, necessary to have an additional operator to check that each bird is stunned before bleeding.

In carbon dioxide stunning, the cages are emptied onto a conveyor. Care must be taken that the birds are not damaged when they slide down from the cages. The conveyor moves the birds to the carbon dioxide chamber, where they are stunned within about one minute. After stunning, the birds are hung on a W-shaped hook as above and moved to the bleeding. Bleeding is carried out by passing the birds on a rail with a rotating round blade with a sharp edge.

4.5.3 Scalding

After bleeding the birds are scalded. Scalding lasts about 1–3 minutes; at higher temperatures shorter times will be needed than at lower temperatures. Water baths and, more often, steam are used. There are three severities of scalding:

- Hard (58–60 °C), which gives a good scalding result but totally removes the epidermis and commonly causes damage to the skin. The surface layers of muscles may also be slightly 'cooked'.

- Medium (52–54°C) also produces a good scalding result, but the reduction of microbes is smaller, thus increasing the risk of *Campylobacter* that have penetrated into the skin.
- Mild (about 50 °C) does not cause damage to the skin, but results in more difficult defeathering and an increase in microbial hazards. The deeper parts of the epidermis will remain, providing a good quality skin which may be important when broilers or their parts are sold with the skin-on.

4.5.4   Plucking

Feathers and outer layers of epimysium are removed with a plucking machine. This consists of a rotating plate and rubber 'fingers' that scrape the feathers and parts or all of the epidermis off. Some slaughterhouses use additional singeing after mechanical plucking to burn the lowermost, soft feathers. Finally, the birds are washed.

4.5.5   Evisceration

This process begins by removing the head and legs (*metatarsus*); the carcasses are then hung back on the rail. The carcasses are moved to the evisceration area, where the neck is excised. Then, the anus is opened with a drill and gut pulled out using vacuum. Special mechanical devices remove the offal: crop, gizzard, heart, liver, spleen, gall bladders, lungs, ovary, and kidneys. The carcass is then thoroughly washed and cooled.

4.5.6   Other poultry

The slaughtering of turkeys, geese and ducks follow a similar pattern but, because the birds are larger, the lines more usually include more manual unit operations than the virtually totally mechanical broiler slaughtering lines. The scalding time and temperature are the same as for broilers, but cooling times/temperatures differ.

4.5.7   Cooling/chilling

Chilling of poultry carcasses starts a short time after evisceration. For poultry, the main objective of the chilling process is to reduce the microbial load of the raw material and to maximize the shelf life and safety of the products. A temperature of less than 4°C is normally reached within two hours *post-mortem*. Poultry is cooled by blasting the carcasses with cold air or by immersion chilling, where the carcasses are submerged in cold water. Some systems use a combination of cold air and water misting, so-called air spray chilling. In spray chilling, the water will absorb heat from the carcass as it evaporates. Air chilling systems are commonly used in Europe and are, more recently, becoming increasingly used in North America. In air chilling, the birds are individually shackled and moved through refrigerated chilling rooms for 1–3 hours. Immersion chilling can take place in a long chiller with a countercurrent flow of cold water, that is the carcasses and the water flow in opposite directions. The cooling time in the water is 30–60 minutes. In addition to the chilling, the immersion in water results some washing of the carcass. Upon exit from an immersion chilling system some water may have been taken up by the product and excess water is allowed to drip for a few minutes. The maximum amount of water in percentage of carcass weight that is allowed to be adsorbed in the chilling process is regulated in many countries. The water loss upon thawing is a good indicator of the absorbed water. In the European Union, the allowed percentages are 1.5% for air chilling, 3.3% for air spray chilling and 5.1% for immersion chilling. In air chilling systems, the risk of cross-contamination and spreading of bacteria between carcasses is less compared to immersion chilling. The development of rigor in poultry occurs rapidly and the time between slaughter and chilling is normally long enough to avoid cold shortening in chicken as well as turkey muscles. Poultry carcasses are often cut and deboned within 4–6 hours after slaughter and, thereafter, further chilled in coolers. A high relative humidity in the cooler is recommended to reduce shrinkage due to evaporation from the product surface.

4.6   Treatment of slaughter by-products

The annual world production of slaughter by-products is about 100 million metric tons (carcass meat with bones about 300 million metric tons). By-products can be divided into three categories: offal, which are edible and used as food; by-products, which are inedible and used for technical purposes; and waste, which are used as, for example, fertilizers, landfill or are destroyed in incinerator. Slaughter by-products make up a relevant environmental and hygiene risk but are, simultaneously, a significant source of food for humans and animals. To reduce the environmental and hygiene risks the by-products should be heated or/and acidified, which are expensive operations. Therefore, there is an increasing need for more effective treatment technologies and ue of slaughter by-products.

The by-products are moved from slaughtering hall by viscera conveyor or gut conveyor. When possible, the by-products are moved to the basement. The edible by-products of pig slaughtering (offal) are blood, liver, heart and tongue, but genitals, udders, lungs, spleen, pancreas and intestines are used as animal feed (although, depending on the cultural habits, some of them will be used as food as well). Some parts of the stomach are used for food but most parts will be used for feed or will be waste. The contents of the intestines are used as fertilizers or they are waste. Fatty tissues are rendered and fat is used as food, cosmetics or, increasingly, as fuel. Rendering must be done separately for edible and inedible products. Intestines are used to some extent for sausage casings but artificial casing materials have mostly replaced them.

The most used natural casing is the small intestine. The small intestines are emptied in a gut cleaner. The washed small intestines are turned inside out and the innermost layers of the casings removed. After several washings the casings are heavily salted.

After scalding, skin is used as food; alternatively, unscalded skin can be used as leather. Cloven hoofs remain attached to the carcass. The hoof wall, hair (bristle), tail, eyes and ears (sometimes used as dog feed) are waste materials. Bones are usually mechanically deboned using high-pressure or low-pressure deboning. In high-pressure deboning, in which the soft tissues (meat and marrow) of minced bones are pressed through a perforated membrane, the material easily warms up and, therefore, becomes susceptible to microbiological growth. Whole bones or the bone fraction of the mechanical deboning process can be used as the raw material of bone meal.

Edible materials are carefully washed and cooled rapidly. Inedible parts also need to be cooled and transported to rendering facilities. Recovered blood can be used as whole blood or the plasma alone can be used (after centrifugal separation).

Edible beef slaughter by-products are mostly the same as with pork and they are similarly treated, but while the head and skin stay with the carcass in pigs, beef animals are dehided and the head is removed. The hide is the most valuable by-product of beef animals. The fat and muscles that are still attached to the hide are removed and the hides are heavily salted with coarse dry salt for several weeks, during which time the salt penetrates into the corium and water and proteins exudate, making the hides storable and ready for tanning. Head meat is cut for food and the rest of the head is discharged. The cloven hoofs with tarsal/metatarsal bones are also removed, they are waste material or are used as a raw material in bone meal. The stomachs and intestines are pressed empty and washed. The stomachs and intestines are treated similarly to the respective porcine organs, as described above, but the stomachs, especially, are used more than pig stomachs. The small intestine of sheep is still commonly used as a natural casing in frankfurter-type sausages.

Poultry liver, heart and gizzard are used as food; the crop, spleen, gall bladders, lungs, ovary, kidneys are used mostly for feed. The most valuable parts of the birds, such as breast muscles, legs and wings, are removed from the carcass; the rest of the meat is mechanically separated and then used in meat products.

Until the 1980s, slaughter by-products were greatly used as a source of fine chemicals but since then the need for that has decreased due to the rapid development of biotechnology. However, pharmaceutical compounds are still extracted from by-products. Collagen is also an important animal-based product, but its use has decreased because of the BSE problem.

Literature and further reading

Danish Crown. Danish Crown in Horsens (Slaughterhouse video). http://www.danishcrown.dk/custom/horsens_uk/3755.asp (last accessed 14 January 2014).

Gregory, N.G. 2007. *Animal Welfare and Meat Production*, 2nd edn. GABI Publishing, Wallingford, UK.

Romans, J.R., Costello, W.J., Carlson, W.C., Creaser, M.L. and Jones, K.W. 2000. *The Meat We Eat*. 14th edn. Pearson College Division, New Jersey, USA.

Savell, J.W., Mueller, S.L. and Baird, B.E. 2005. The chilling of carcasses. *Meat Science*, **70**, 449–459.

Thompson, J.M., Perry, D., Daly, B. *et al*. 2006. Genetic and environmental effects on the muscle structure response *post-mortem*. *Meat Science*, **74**, 59–65.

5

Animal Welfare – Stunning and Bleeding

Michael Bucher[1] and Peter Scheibl[2]

[1]*Veterinary Association JadeWeser, Wittmund Regional Office, Wittmund, Germany*

[2]*Bavarian Health and Food Safety Authority, Oberschleissheim, Germany*

5.1 Scope

This chapter presents available methods commonly used to stun pigs, ruminants and poultry. The minimal conditions by which the method is likely to be effective from the animal welfare point of view and the criteria that could ensure that the stunning method is properly enforced are also discussed. Additionally, different sticking/bleeding techniques are explained.

5.2 Introduction

Basically slaughter is the killing of animals by exsanguination for human consumption. For this, a lot of stages must be crossed, from transport to the bleeding. Once animals enter the slaughterhouse, the transporter and the slaughterhouse operator share the responsibility for their welfare. By the time unloading is completed, the slaughterhouse operators have the sole responsibility for the welfare of the animals. The transport of the animals from the farm to the slaughterhouse, the unloading and driving at the slaughter plant, as well as the restraint during stunning are unavoidably stressful situations, as the animals are not familiar with these occurrences and

Meat Inspection and Control in the Slaughterhouse, First Edition.
Edited by Thimjos Ninios, Janne Lundén, Hannu Korkeala and Maria Fredriksson-Ahomaa.
© 2014 John Wiley & Sons, Ltd. Published 2014 by John Wiley & Sons, Ltd.

the environment. Hence, it is important to keep the excitement low and to avoid any additional stress in advance. The loading and unloading of animals, especially, must be accomplished as calmly as possible. Transport vehicles and loading ramps must be designed to support and not to impede autonomous animal movement. The same applies to the races at the slaughter plant. Avoiding optical and mechanical distractions, inclining and declining races and noise will improve animal movement. Noise reduction does not only include the calm behaviour of slaughter plant employees, but also a low-noise design of doors, hydraulic valves and similar facilities. Regarding animal welfare, animals should be handled as little as possible and the handlers should be competent.

Slaughtering, including stunning and bleeding, must ensure animal welfare. Any kind of excitement may affect the sufficiency of stunning and, thus, at least meat quality. To consider these aspects there are different permitted stunning methods available for pigs, ruminants and poultry (Table 5.1). The equipment used for stunning must be suitable for the purpose and meet the current minimum requirements. After stunning the animals, sticking/bleeding must take place without delay to ensure the animal dies through loss of blood before it regains consciousness. The bleeding procedure itself must be correctly carried out to ensure rapid exsanguination. Bleeding is performed either by throat cutting to sever major blood vessels or by chest sticking to open the blood vessel near the heart. Due to the painfulness of these incisions, the animals must be stunned prior to bleeding. Stunning is most effective in calm animals. It is most important that the plant personnel are experienced with animal behaviour, apply the right driving methods and use as few driving aids as possible.

Table 5.1 Commonly used stunning methods for pigs, ruminants and poultry.

Stunning method	Used in[a]			
	Pigs	Cattle	Sheep/Goats	Poultry
Mechanical				
Penetrative captive bolt	x	**x**[b]	**x**[b]	x
Non-penetrative captive bolt [c]	x	x	x	x
Manual percussive blow				x
Electrical				
Dry electrodes				
Head only	**x**	x	x	**x**
Head-to-body	**x**	**x**	**x**	x
Water bath				**x**
Controlled atmosphere				
Carbon dioxide (CO_2)	**x**			**x**
CO_2 with inert gases				x
Inert gases				x

[a] mostly used methods in bold letters
[b] unsuitable for specimen collection from transmissible spongiform encephalopathy (TSE) suspects
[c] should only be used when alternative methods are not available

5.3 Pigs

The stunning of slaughter pigs poses a particular challenge for slaughter plants regarding animal welfare. On the one hand, slaughter capacities are very high in comparison to the capacities for ruminants. On the other hand, the signs of effective and ineffective stunning are not always very clear, which requires extensive experience concerning evaluation. Regarding both animal welfare and meat quality, it is important to consider susceptibility to stress in some pigs. Therefore, the effectiveness of stunning is greatly dependent on the condition and the excitation status of the animals prior to slaughter. Unusual physical exertion or psychological stress, such as ranking fights, regrouping, enhanced driving or transport during hot and humid weather, affect the cardiovascular system and, thus, the efficiency of stunning. Furthermore, stress may have a negative impact on the meat maturating process.

There are several kinds of stunning methods available for the slaughtering of pigs. While electrical stunning is frequently applied by small and medium-sized slaughter facilities, big industrial slaughter plants commonly use carbon dioxide stunning. Sometimes captive-bolt stunning may also be applied in pigs but, due to the difficult handling and the strong responsive reaction in pigs, it is rarely used during regular slaughter. All methods aim to induce changes in the animal's brain physiology, ensuring loss of consciousness and loss of sensibility until exsanguination causes death.

5.3.1 Electrical stunning

Electrical stunning leads to stimulation of the entire brain, causing a long-lasting depolarized neuronal state followed by spread of action potentials. When this occurs in a larger group of neurons in the brain, a generalized epileptic fit (provoked grand mal seizure) is induced. Grand mal epilepsy is a pathological condition in the sense of neuronal synchrony, considered incompatible with intact neuronal function and, thus, consciousness. Accuracy of electrode placement and the use of sufficient amperage will lead to electrical synchronization in the brain within less than one second.

Neurotransmitters regulate interaction between neurons in the brain. Excitatory (glutamate/aspartate) and inhibitory (gamma-aminobutyric acid, GABA) neurotransmitter levels are in physiological balance. The current flow through the brain causes a dramatic increase in extracellular glutamate and aspartate levels. Thus, the physiological balance between the neurotransmitters is disrupted and cell structures are in a state of increased stimulation and uncoordinated activity. As a result, coherent information processing in the brain will be disrupted before any stimuli, caused by the current flow, may be perceived as painful. After a certain current flow time, excitatory neurotransmitter reserves are exhausted and neurons will fail to respond appropriately to any further stimuli. Ultimately, the interruption of signal processing will lead to instantaneous and prolonged loss of consciousness and sensibility. The slow release of GABA finally terminates the seizure.

Extracellular GABA levels remain high considerably longer than glutamate or aspartate levels and will lead to analgesia over a period of several minutes. As stress may also cause an increase of GABA levels in the brain, which interferes with the effect of glutamate and aspartate, and this may negatively affect the possibility of provoking a grand mal seizure. This indicates the importance of gentle treatment of the animals prior to slaughter.

Electrical stunning of pigs may be performed by head-only or head-to-chest stunning. Depending on the amperage used, head-to-chest electrical stunning may induce ventricular fibrillation beside the epileptic seizure. The separation of the two circuits for the head and the chest stun is another variation of electrical stunning. The current flow is passed through the head and subsequently through the heart. When the current flow passes the heart, an immediate and persistent drop in blood pressure is initiated, whereby the return to consciousness and sensibility can be prevented. In this way, ventricular fibrillation will provide a lasting stunning effect. Applying the heart circuit will not affect exsanguination. Another benefit of provoking ventricular fibrillation or cardiac arrest is the reduction of blood spots in the meat. Since ventricular fibrillation or cardiac arrest implies the risk of causing pain to the animals, applying the current across the brain prior to or simultaneous with the induction of ventricular fibrillation or cardiac arrest is essential. In addition to the effects on the heart, passing electricity through the body will cause exhaustion of the peripheral nervous system and, thus, immobilization of the animal. Consequently, this will lead to less physical activity and to immediate, safe exsanguination. The recommended minimum amperage required to achieve a proper stunning effect is 1.25 A in fattening pigs and 1.8 A in adult pigs. During the head stun, the animal's body stiffens due to tonic muscle contractions caused by the stimulation of the brain and the electrical impulses passing through the spinal cord. The hind legs will be tucked up and the animal will go down. The generalized contractions during the tonic phase usually continue for about 10 seconds after the removal of the current flow and are followed by the clonic phase. These clonic cramps are caused by dysfunction of certain brain structures and appear as intermittent contractions and relaxation of muscles, resulting in alternating flexing and stretching of extremities.

A broad range of electrical stunning equipment and stunning systems is available, differing in electrical parameters (amperage, frequency and current quality) and type of application. While animals are usually stunned manually by the use of stunning tongs in small-scale slaughtering, big, industrial plants are equipped with automated stunning systems with a capacity up to 600 pigs/hour. Compared to carbon dioxide stunning, electrical stunning is related to more manual handling and, thus, to a greater number of potential failures. There is a bigger possibility for an inaccurate contact point when manual stunning tongs or semi-automatic stunning tongs, which can be operated with one hand, are applied. Fully automated stunning restrainers are used in large slaughter plants for stunning pigs of the same size and weight. Regardless of the level of technology, stunning must be performed carefully and effectively at any time. Especially in fully automated stunning systems, the correct placement of the electrodes and effectiveness of stunning have to be verified regularly, since engagement of personnel with the animals is not intended. Repeated cases of improper stunning demand a full assessment of the stunning process, regardless of the stunning system, in order to eliminate the errors.

Effectiveness of stunning is impaired from the point where the animals are driven into the slaughterhouse. Depending on numerous conditions, such as animal species and group, structural conditions of the plant and intended slaughter capacity, every driving system is individually arranged. Animals should be moving on by themselves, if possible, without being over-excited. According to the size of the driving area and the intended slaughter capacity, it might require several persons to make the animals move towards the stunning point. Principally, a humane driving pace will prevent the need for using painful aids such as electric prods.

Pigs are basically social animals. For this reason, it is easier to drive pigs in groups rather than individually. Calm, steady driving, the use of voice and occasionally tapping the animal with the flat of the hand or a board, if necessary, will make animals move on. Additionally, using a board to push the animals from behind will make them move forward. Furthermore, pigs have a high motivation to explore. If not driven, they will explore the entire surrounding area on their own initiative. Noise prevention also contributes to stress-free movement of animals. Therefore, it is recommended that the driving system is arranged away from the lairage and slaughter area. The noise level is an appropriate quality indicator for driving systems.

It is mainly the separation of animals just before stunning that has great impact on the behaviour during the stunning process. At this point, the separation from other animals creates confusion. Thus, a well-designed race system is of major importance and facilitates moving animals from the holding pens towards the stunning point. Visual or acoustic distractions that impede movement of animals are caused, for example, by poorly delimited chutes, poor visibility in the race, uneven floors or walls, glaring light and reflections, drains in the floor, sharp curves, steep incline, loud noise, moving objects or people in the animals' field of view or blowing air towards approaching animals. The fewer distractions occurring at the same time, the better the animals will be able to deal with transitions. For example, a change in the structure of the wall should not appear in a curve. The entrance to the stunning box should not be located in a transitional area of the slaughter plant.

A well designed handling system will facilitate the forward moving of the animals by:

- uninterrupted and steady movement towards the stunning point;
- careful driving with little/no use of driving aids;
- optimally arranged separation of animals;
- elimination of visual and acoustic distractions.

Pigs can be restrained for stunning rather loosely. The head or the animal itself, thereby, has a certain range of movement and the person stunning must take time and wait for the right moment to place the stunner. In plants with high slaughter capacities, there is little time for handling, thus the restraining of the animals requires tighter control. In this case, in comparison to the stunning of cattle, immobilization of the animal is achieved by restraining. Also, when using captive bolts in agile pigs, head restraint must be applied. Being restrained is a burden for animals. Therefore, the duration of restraint should be as short as possible. Ideally, restraining should allow effective stunning without stressing the animals.

In order to ensure adequate stunning, sufficient current must be passed through the animal's brain. With the commonly used stunning tongs, correct placement of the electrodes is on both sides of the head just below the ears. Placement of one electrode near the eye and one behind the opposite ear is another variation; this requires perfect restraining. In practice, this method is difficult to implement and not recommended, as incorrectly placed electrodes may cause severe pain to the animal. In case that head stunning is followed by a heart-only stunning, it is recommended that the electrodes are positioned in such a way that the heart lies between the electrodes, either by side-to-side or back-to-chest. When simultaneously applying heart stunning, one electrode must be placed on the animal's forehead or below the ear and the other one must be placed on the sternum or the left side of the thorax.

Moistening the skin before placing electrodes reduces skin resistance. However, the animals should not be dripping wet, as this might cause creeping current. Current leakage will cause a decrease of the total amperage passing through the brain and may reduce effectiveness of stunning. Frontal hosing down of the animals should also be avoided, as this will cause traumatic impressions, making the animals back up in the race when a front placement of the stunning tongs is intended. Wetting the animals about 15 minutes prior to stunning, for instance by using a sprinkler, is the best option. In this way the skin is sufficiently moistened to ensure effective current flow.

Stunning devices must always be fully functional and operative. This includes well maintained stunning tongs with a sufficient opening angle that can be adapted to the size of the animal. The devices must be clean and non-corrosive to ensure proper electrical contact. Frequencies used in modern electrical stunning systems range from 50 to 1500 Hz. A reduced duration of the stunning effect is one of the disadvantages of high-frequency stunning. Hence, special attention must be paid to assessing the persistence of unconsciousness and insensibility during bleeding. The quantity of electricity passed through the brain is not correlated with the duration of the stunning. For this reason, applying a long-lasting current flow or a great amount of electricity has a counterproductive effect. This might lead to bone fractures and petechial haemorrhages in the muscles. When electrodes are not held firmly enough against the animal and slide during the head stunning, the stunning tongs must be re-placed (partial or double stunning). This may also have a negative effect on the sufficiency of stunning and may cause increased petechial haemorrhages due to repeated muscle contractions.

The basics of effective electrical stunning are:

- appropriate stunning tongs with adequately shaped electrodes;
- accurate placement of electrodes;
- fastest possible bypass of skin resistance;
- firm electrode contact;
- calm animals;
- sufficient amperage and current flow time.

Poor electrode contact, slow rise of amperage or insufficient maximum amperage will result in a delayed stunning effect and might cause severe pain if the animal is not rendered unconscious, especially during head-to-chest stunning.
Signs of effective stunning are:

- the animal collapsing (not in the restrainer);
- onset of tonic phase, followed by clonic phase (kicking or uncoordinated paddling movement of hind legs);
- apnoea (breathing arrest) during the entire tonic and clonic phase;
- rotated eyeballs;
- no attempts to rise or directed movement.

Ventricular fibrillation, as a result of effective stunning, cannot be assessed visually. The induction of ventricular fibrillation depends on the following factors:

- the pathway of current flow through the body;
- the region of the heart the current passes through;
- the phase of cardiac activity at the moment the current flow arrives the heart;
- flow time, frequency and waveform of the current.

High frequencies are not as efficient as low frequencies concerning induction of ventricular fibrillation. The recommended maximum frequency is 90 Hz. It should be kept in mind that applying heart stunning after head stunning takes time and extends the stunning–sticking interval. If the current partially passes through the spinal cord during heart stunning, this may mask seizure activity as well as signs of insufficient stunning.

For animal welfare reasons, the current flow through the brain must induce instant unconsciousness and insensibility. Electrical immobilization applied by passing electricity only through the spinal cord, without involving the brain, will cause initial stiffening of the muscles and, subsequently, exhaustion of the neurons, but will not induce unconsciousness and is, therefore, not acceptable. If done so, animals will not show the typical signs of an epileptic seizure (tonic phase followed by clonic phase) but will express sensation of pain. It needs to be taken into consideration, that vocalization during correctly applied current flow through the brain will not be possible.

Indications of insufficient stunning that may occur during the entire stunning process are:

- flight behaviour;
- missing seizure activity (caution: masking the effect of heart stunning);
- return of rhythmic breathing;
- vocalization during or after current flow through the brain;
- attempts to rear up;
- eye tracking to moving objects.

The so-called 'silent cry' is also a sign of ineffective stunning and must not be mistaken for gasping. The return to rhythmic breathing, which is often accompanied by the end of limb movement (clonic phase), indicates the end of the hypersynchronization of neurons in the brain and the gradual recovery of normal brain function. The return to sensibility can be assessed by testing the corneal reflex or by pinching the nose using clamps for instance. These tests must not be performed within 45 seconds after stunning, as the animals are hyperirritable during epileptic seizure and false-positive results may emerge. Therefore, during this period, the animals are to be exclusively observed.

It is important to immediately re-stun an animal showing any sign of stunning failure. In practice, the following deficiencies commonly cause inefficient stunning:

- improper equipment/electrodes;
- insufficient stunner maintenance (dirty or blunt electrodes, cable breaks);
- inaccurate placement of electrodes;
- inappropriate construction or handling of restraint device;
- stunning tongs with an undersized opening angle.

In order to correctly interpret animal welfare during electrical stunning of pigs, all persons involved must be well aware of the principles of stunning, such as the applied electrical parameters, the performance of stunning and the condition of the animal before and after stunning. It is certainly not acceptable only to evaluate the protocols of the stunning devices and to rely on the recordings of instruments. Animal welfare related details may be traced back to their point of origin even after slaughter or scalding. The placement of the stunning tongs causes slight burning marks on the animal's skin. Thus, the particular placement of the tongs can be examined even by inspecting the carcasses in cold storage.

Electrical stunning of heavy pigs frequently causes problems. Practical experience shows that insufficiency of stunning commonly occurs particularly in these animals. Usually, the same equipment (transformer and tongs) is used for regular slaughter pigs. The amperage of 1.3 A during the first second is apparently not sufficient for heavy animals. In addition, head-to-chest stunning is not practicable with some stunning tongs, due to an insufficient wide opening angle. Thus, electrical stunning of heavy pigs requires the application of different electrical parameters and appropriate equipment (for instance a minimum amperage of 1.8 A and stunning tongs with an extra wide opening angle).

5.3.2 Carbon dioxide stunning

Carbon dioxide (CO_2) stunning of pigs is used in many countries. As opposed to electrical stunning, which can cause irreversible ventricular fibrillation, CO_2 stunning is a gas anaesthesia and always has a reversible effect. There are still continuing discussions over this stunning method concerning animal welfare, as the gas may cause irritations of the respiratory tract. Generally there are three stages of CO_2 stunning:

1. analgesia stage: loss of pain;
2. excitation stage: loss of consciousness (uncontrolled movements);
3. anaesthesia stage: relaxation (intact respiration and circulation).

During the first stage, which lasts about 30 seconds, the pigs contact the accumulating gas and inhale CO_2 while being fully conscious. The anaesthetizing effect of CO_2 is undisputed. However, the onset of insensibility is not instantaneous and animals are exposed to stress, especially to respiratory distress, over a period of 10–20 seconds. Most animals have little or no reaction to the change in air conditions. Nevertheless, it is reported that some breeds will violently attempt to escape after contacting the gas. There seem to be genetic differences regarding reactions of pigs to CO_2, and thus the effectiveness of CO_2 stunning. Several researches have investigated the use of other gases, such as argon or nitrogen, also in mixture with CO_2, in order to avoid the disadvantages of CO_2, especially during the induction of anaesthesia. Although, in experiments, different gas mixtures appear to be better tolerated, these variants currently do not represent an alternative due to their negative effects on the meat quality (for instance bloodspots in the back).

The slightly raised level of CO_2 in the air will cause both an increase of the respiratory rate and the depth of breathing in animals. Due to the high CO_2 tension in the air, respiratory acidosis will develop and result in a lowering of the pH value in the cerebrospinal fluid. The lowering of the pH value will initiate hyperpolarization of neurons, which will consequently reduce the spread of action potentials and thus raise the stimulation threshold, while inducing loss of sensibility. Loss of consciousness is, therefore, not caused by hypoxia but by hyperpolarization of neurons. The deep breath, which is hardly distinguishable from respiratory distress, is a responsive reaction to the stimulating effect of CO_2 and is not a sign of hypoxia.

The second stage is also known as the stage of excitation. During this phase of CO_2 exposure, strong motoric reactions are frequently observed. By this time the pH value in the cerebrospinal fluid has dropped substantially. The animals must be unconscious when entering this phase.

The third stage is the so called tolerance (or anaesthesia) stage. During this phase, the motionless and completely relaxed animals are unloaded from the gondola and conveyed to the bleeding point. The animals will not respond to manipulation. The longer the animals are exposed to CO_2, the stronger the stunning effect will be. This condition is comparable to deep anaesthesia and will induce apnoea, which will consequently cause circulatory failure and cardiac arrest, resulting in a smooth transition between tolerance stage and asphyxia. However, commonly used CO_2 concentrations and CO_2 exposure times are not assumed to induce circulatory failure and cardiac arrest.

The long period of time between the initial contact with the gas and the unloading of the animals from the gondola, as well as the delayed onset of unconsciousness and insensibility, are important factors responsible for CO_2 stunning differing significantly from other stunning methods (electrical stunning and captive-bolt stunning). In practice, it is often the intention to achieve a rapid induction of anaesthesia and a short excitation phase by preferably using a higher CO_2 level in the gas mixture. Carbon dioxide stunning is not applicable to all breeds of pigs. On the one hand, the gas apparently will irritate certain breeds more than others during immersion and will cause strong motoric reactions in the animals. On the other hand, excitations during the second stage may vary considerably.

The possibilities to influence CO_2 stunning are limited. From the drive into the gondola until the unloading of the animals, only a limited impact on the process is possible. All the more important is the knowledge of the required parameters and control mechanisms, to ensure the safe and rapid loss of consciousness and sensibility in the animals. With CO_2 stunning, slaughter capacities are dependent on the number and size of gondolas, duration of exposure to CO_2, the concentration gradient in the CO_2 chamber and the interval between CO_2 exposure and sticking. The more time that passes until the sticking of the last animal in the gondola is performed, the longer the CO_2 exposure time that must be chosen.

The main advantage of CO_2 stunning certainly lies in the possibility of moving the animals into the CO_2 chamber in groups (≥ 2). Depending on the size of the pens and the gondola, groups of five or more animals can be stunned using the back-loader system. As a consequence, lining up and moving pigs in single file races will be avoided and excitement of the animals will be significantly reduced. Furthermore, animals are also driven automatically by mobile partitions in modern plants. Hence, the human influential factor can be minimized. The use of electric prods is actually not necessary anymore in these plants.

The so called dip-lift system is an elevator designed for CO_2 stunning by moving groups of animals into a gondola that descends into the CO_2 gas without stopping. Within the so-called paternoster system, the gondolas move up and down in a loop. The transportation system must be stopped for the loading and unloading of the gondolas. With commonly used paternoster and dip-lift systems, the animals are usually moved to the elevator through single-file races. Consequently, there is no difference between CO_2 and electrical stunning concerning susceptibility to stress during the driving of the animals. In case the animals are separated before moving into the gondola, the same precautions must be taken as with electrical stunning. The use of electrical prods increases excitement and has a negative impact on the stunning effect no matter what kind of stunning method is applied.

The gondola must offer enough room for at least two animals to stay without restricted ribcage expansion. Unlike electrical stunning, CO_2 stunning does not require restraint of animals. The elevator must be firmly fixed especially as the the animals enter. The noise level caused by the opening and closing of doors must be kept low to avoid excitement of the animals. Any kind of distraction, such as gaps, steps or puddles of water, can confuse or scare the animals and make them stop moving. In particular, the metal parts inside the gondola must be slip-proof to reduce the risk of skidding, otherwise the animals will hesitate or refuse to enter the gondola. Appropriate illumination of the gondola will facilitate movement of animals.

The induction of CO_2 stunning is less stressful in calm animals. The animals must remain in the required CO_2 concentration ($>80\%$) long enough. Since operating and monitoring personnel do not have access to the animals during this phase, technical parameters (CO_2 concentration, exposure time to the gas atmosphere etc.) must be precisely verifiable. Standing time and transport time of the gondola must be programmed in order to comply with the minimum exposure time to $>80\%$ CO_2. The minimum exposure time required to safely pass all stages of stunning is 100 seconds. Consequently, it is understandable that the exposure time (100 s) does not include the time required for the loading and unloading of the gondola.

In order to monitor animal behaviour during the stages of induction and excitation, the stunning system must have a window that allows observation of the animals inside the gondola. The induction phase, that is the period of time until the animals collapse, is of importance for assessing animal welfare during stunning. Strong motoric reactions may be an indication of poor stunning. Basically, stunning devices must have a function display in order to clearly signal technical malfunction. In addition, noticeable acoustic and optical warning signals must be emitted as soon as CO_2 concentration falls below the recommended minimum concentration.

When using CO_2 stunning, assessment of effective stunning is possible only as soon as the animals are unloaded from the gondola. The following signs indicate deep anaesthesia (asphyxia):

- animals are limp and floppy when lying on their side or being hoisted;
- open eyes and dilated pupils;
- no signs of rhythmic breathing.

The animals are unloaded, hoisted and bled as quickly as possible after stunning. There is no difference in the performance of these process steps between electrical and CO_2 stunning. Particularly within the back-loader system, the fastest possible exsanguination of all animals must be ensured. It may require several persons at this point to guarantee the immediate hoisting and sticking of the animals. The personnel must very carefully examine whether the animals are sufficiently stunned. For this purpose, appropriate education and training regarding investigation of effective stunning and assessment of insensibility are necessary. In the case that insufficiently stunned animals are detected, they must immediately be re-stunned.

After exit from the gondola, the animals will usually lie completely relaxed and will not react to the hoisting or sticking procedure. There will be no signs of rhythmic breathing or response to touch of the eye.

When any of the following reactions occur during exit from the CO_2 chamber or during the hoisting procedure, the animal must be closely monitored:

- animals are not relaxed;
- front legs are tucked up;
- single breathing movements;
- no dilation of pupils;
- eye-blink response to the first touch.

Animals, whose condition is classified as 'questionable' must be monitored for further reactions. When a clear stunning effect cannot be confirmed during the re-evaluation phase, the animal must be re-stunned in order to ensure animal welfare. As soon as an animal shows one or more of the following signs it must be immediately re-stunned:

- animal raises its head;
- directed movement or arched back;
- breathing along with response to touch of the eye;
- eye tracking to moving objects or spontaneous blinking;
- vocalization.

The plant must ensure enough time and personnel for close monitoring of questionable animals and the re-stunning of insufficiently stunned animals. When the person performing the hoisting/bleeding is working to capacity, he/she will not be able to accomplish these additional tasks. Moreover, appropriate equipment must be available for re-stunning. Re-stunning is frequently carried out by using electrical stunning tongs. However, the effectiveness of electrical stunning in partially stunned animals is restricted. At this point, the re-stunning can be applied by using captive bolts. Furthermore, it must be pre-determined who is capable of taking the necessary steps in case of failure. Such an example would be a lower CO_2 concentration or a reduced exposure time to the CO_2. In such a case, it must be ensured that the further stunning process proceeds in compliance with animal welfare.

5.3.3   Captive-bolt stunning

Penetrating captive-bolt stunning causes both physical destruction of the brain and concussion. The steel bolt in penetrating captive-bolt stunners is powered by either a cartridge or by compressed air. The tip of the bolt is not pointed but slightly concave and should have a sharp edge without nicks. The captive-bolt stunner must be placed in correct position and the bolt of appropriate length and diameter must be sufficiently accelerated. As a consequence, energy will be transmitted to the animal's head causing concussion. However, brain structures are also damaged by the penetrating bolt. The rapidly spreading shock waves as well as the sudden acceleration and deceleration of the relatively light brain inside the skull (shearing force and contrecoup effect) will induce instantaneous loss of sensibility and loss of consciousness.

The animal's head must be easy to reach for the stunner to allow a precise shot. Only when the captive bolt is positioned correctly, can essential brainstem areas be reached and disabled by the bolt. On a cuneiform head, the captive bolt must be placed one centimetre above an imaginary line between the eyes. The stunner must be tilted downwards to an angle of 25° from the perpendicular angle to the skull so that there is a fingerbreadth between the top edge of the unit's base and the animal's head. In this way, the bolt is not placed perpendicular on the skull surface. as is done when using captive bolts with cattle. On a steep forehead, the captive bolt is to be placed perpendicular to the skull surface 2–3 cm above a line drawn between the eyes. The captive bolt is basically to be placed on the median of the head. Large sows and boars commonly have a bony ridge along the median of the forehead. In this case, the captive bolt must be placed next to the ridge, aiming towards the mid-portion of the head.

In particular, small slaughterhouses without a stunning box or separation of animals must have handling practices to hold the animals in position by using sorting panels or physical restraint. Correctly shooting pigs running free with a captive-bolt gun is not feasible. When the captive bolt is placed correctly, the hole will be circular with even margins. The more the captive-bolt gun is tilted when placed on the head, the more oval-shaped the entrance hole will be. Furthermore, shooting on a slight angle might cause fissures and fractures in the skull surface, originating from the hole.

Instantaneous collapse of the animal and the immediate absence of rhythmic breathing are indications of effective stunning. The muscles of the back and the legs will start to cramp. The forelegs and hind legs will be flexed initially and after a few seconds the forelegs will become extended. Depending on the affected brain area, pigs may develop a strong clonic phase. These cramps are not a sign of insufficient stunning and must not be confused with attempts to rise. To ensure immediate and effective bleeding that results in rapid brain death, the pig should be rendered insensible and unconscious by a single shot.

Signs of effective captive-bolt stunning are:

- instantaneous collapse of the animal during the shot;
- tonic phase;
- no attempt to rise;
- absence of breathing;
- absence of corneal reflex.

Indications of poor stunning are:

- soft muscles;
- return of rhythmic breathing;
- rotated eyeballs.

Insufficient or delayed bleeding will result in the return of rhythmic breathing. Poor maintenance of the stunning device is a common cause of insufficient stunning. Captive-bolt stunning of pigs is frequently applied as a re-stunning method, which means that the captive-bolt gun is not in use routinely. This makes regular maintenance and cleaning of the captive-bolt stunner all the more important, as it may affect bolt velocity. When the rubber rings or springs required for full bolt retraction from the animal's head are worn, they must be replaced immediately. Ignoring this rule might result in damage to the bolt or the captive-bolt gun (for example, when the bolt is stuck in the animal's head and, consequently, damages the stunner). Low-energy shots due to a deformed or rusted bolt will not reach the required velocity. Consequently, the impact energy on the skull surface will be lower and will cause a slighter concussion. Increased frictional losses due to a dirty stunner will also reduce bolt velocity, and thus impact power.

In pigs, especially in large animals, very strong tonic cramps are observed frequently. This must be taken into consideration for occupational safety reasons. Captive-bolt stunning of pigs is, therefore, rarely used during regular slaughter.

5.3.4   Bleeding

The bleeding procedure is the most important process step, apart from effective stunning, regarding animal welfare during slaughter. It is a determining factor for recovery of consciousness. Bleeding must be performed as quickly as possible after the onset of seizure activity and must be highly efficient. The occurring blood loss will induce a hypovolemic shock and, subsequently, cause the death of the animal. At this stage, hypoxia will prevent the recovery of consciousness and sensibility. The stunning effect, in any case, must last until advanced blood loss inhibits the return of consciousness.

Exsanguination is usually performed in hoisted pigs by chest sticking. By applying this method, a knife or a hollow plunger knife is stuck into the pre-sternal hollow on the animal's chest and drawn forward in direction of the body's longitudinal axis, severing major blood vessels. When properly performed, the blood will gush immediately. It must be ensured that the incision is large enough for effective bleeding. If the incision is not long or large enough, a clearly visible swelling under the skin caused by accumulating blood will occur. This may keep the blood from flowing out rapidly enough and impede complete blood loss. Sufficient blood loss is easily verified when using a regular knife. The incision is often made too small in order to avoid carcass contamination from scalding water. The use of a hollow plunger knife requires the implementation of different mechanisms, for example monitoring the rate of blood loss to ensure effective bleeding.

Insufficient or delayed bleeding may facilitate the return of consciousness. To reliably prevent the animal from regaining consciousness, the stunning effect must be assessed before and after bleeding. Eventually, effectiveness of stunning is the result of immobilization, stunning and exsanguination.

5.4   Cattle, sheep and goats

The two stunning methods mostly used for cattle, sheep and goats are electrical stunning and captive-bolt stunning. Both methods are only stunning practices and must always be followed by a killing method.

The purpose of stunning is to render the animals insensible and unconscious. Only during this state can the extremely painful bleeding methods may be applied. Both the depth of the stunning effect, which is essential in order to reliably eliminate sensibility and consciousness, and the duration of the stunning effect, which must last until the death of the animal, must be considered essential. When the initial stunning effect appears sufficient, but decreases before death occurs, the animal may regain consciousness and will suffer severe pain caused by the stunning and bleeding procedures. Furthermore, the associated excitement may affect meat quality.

Cases of ineffective stunning are unavoidable. Hence, observing the stunning effect in every animal and the reliable detection of poor stunning is especially important. In case of an insufficient stunning effect, the particular animal must be immediately re-stunned. For this purpose, replacement equipment must be kept available at any point between the stunning box and the end of the bleeding rail to insure instantaneous and reliable re-stunning of animals. After the re-stunning procedure, it is necessary to determine the effectiveness of the stunning again and perhaps to re-stun the animal once more.

5.4.1   Captive-bolt stunning

Captive-bolt stunners are firearms that drive out a bullet-like bolt. The required acceleration force is generated by either a blank cartridge or compressed air. As opposed to ordinary bullets, the bolt does not entirely leave the gun. Furthermore, the bolt is retracted immediately after the shot by a retraction system. The complete mechanism lasts only about two thousandths of a second.

There are three modes of action in penetrating captive-bolt stunning. The first is concussion, caused by the bolt hitting the skull surface with high power. Depending on its direction and penetration depth, the bolt subsequently physically destroys brain tissue in the second mode. In the third mode, the concave-shaped tip of the bolt produces great overpressure inside the cranial cavity, which is followed by great negative pressure when the bolt is retracted. These pressure changes cause additional damage to the brain structures.

The described effects are best achieved by using a correctly maintained and cleaned captive-bolt stunner. For this purpose, the captive bolt must be completely taken apart, cleaned and lubricated after each day of use. The parts of the bolt retraction system that wear, rubber buffers and/or recoil springs must be regularly checked and replaced if necessary. Otherwise, with some models, the bolt will not be fully retracted and, consequently, will not be positioned close enough to the propellant before the shot. Due to the increased size of the expansion chamber, the explosive pressure will be reduced and the bolt will not be driven out as powerfully as from the correct starting position. Besides, a worn bolt retraction system may cause the bolt to remain stuck in the animal's head. Thus, the captive bolt can be damaged when the animal goes down.

The tip of the captive bolt must be re-sharpened as soon as sharpness decreases or nicks occur. The sharp edge facilitates punching a circular hole into the cranial bone and overcoming the resistance of the bone. A blunt end will, on the one hand, affect the stunning effect by decelerating the bolt as it hits the skull surface, and, on the other hand, frequently cause uneven fractures of the skull. When the bolt does not enter through a circular and precisely fitting hole, but through an irregularly shaped hole, overpressure and negative pressure next to the bolt will consequently disappear and, hence, reduce the stunning effect.

The entire surface of the bolt must be as smooth as glass. Gunpowder deposits, a dirty bolt, corrosion or pitting corrosion will cause a power loss in the bolt during the penetration of the skull and, thus, impair effectiveness of stunning.

The ability of the animal to move its head horizontally and vertically must be restricted to facilitate correct stunner placement. Additionally, any kind of excitement prior to stunning must be avoided, since poor stunning will become more likely as excitement increases. Sheep and goats can be easily restrained manually. Neck stanchions and chin lifters used in cattle stunning boxes are commonly designed to restrict vertical as well as horizontal movement of the cattle's head. However, none of the existing stunning box designs can be considered as ideal. Hence, a function check of the restraint device must be carried out in each single case, depending on the specific operation conditions.

For most effective stunning, the captive bolt must be placed medial and perpendicular on the skull surface. For cattle, the correct location for captive bolt placement is at the intersection of two imaginary lines, each drawn from the centre of the eye to the base of the opposite horn. In very heavy cattle, the bolt must be placed about two centimetres above the intersection point. In hornless sheep, the captive bolt is placed on the uppermost point of the head, aiming towards the throat. In goats and horned sheep, the stunner must be positioned medially behind the poll, the shot being directed towards the jaw angles.

The captive bolt must be held firmly against the surface of the head. If the pressure against the animal is too low or a shot is fired just before the bolt is placed, the energy transfer to the skull as well as the penetration depth of the bolt will be significantly reduced. Both occurrences are associated with impairment of the stunning effect.

Different cartridge sizes are available for captive-bolt stunners. The propellant must be adjusted as per the manufacturer's instructions corresponding to the size of the animal. Pneumatic captive bolts require adjustment of the pressure. Insufficient acceleration of the bolt will result in poor stunning. Excessive acceleration may cause fractures and the above mentioned consequences. Furthermore, the stunning device might wear off sooner. Commonly used bolt extensions of about 8 cm are not sufficient for stunning heavy cattle. Unless only sheep, goats and light cattle are slaughtered, penetrating captive bolts with bolt extensions of 12 cm or more should be chosen.

After a correctly performed captive bolt shot the animals show a typical behaviour with the following signs of effective stunning:

- the animals collapse immediately;
- the tonic phase occurs (legs are tucked up and the back is straight);
- onset of the clonic phase after about 10 seconds (legs start kicking);
- ears and tail hang floppy;
- the tongue hangs out limply;
- rhythmic breathing is absent immediately after the shot;
- eyes are open with a blank stare;
- eyeballs are not rotated and show no nystagmus.

After the tonic phase transitions into the clonic phase the legs are kicking and the person performing the bleeding is at risk of injury. The bleeding should, hence, be accomplished during the tonic phase.

It is absolutely necessary that the slaughter personnel observe every animal after stunning. To assess the effectiveness of stunning, the personnel must know the signs of a sufficient and an insufficient stun. Typical signs of insufficient stunning are:

- absence of tonic cramps and, hence, relaxed muscles (ears, tail and tongue, however, must be floppy);
- attempts to rise;
- arched back or neck, also when hanging on the rail;
- pricked ears, movement of ears;
- movement of tail;
- rhythmic breathing;
- eye tracking to moving objects or spontaneous blinking;
- rotated eyeballs or nystagmus;
- vocalization after the shot.

The assessment of effective stunning includes the observation of all animals and the active examination of a selected sample. In doing so, individual animals are exposed to stimuli that will not provoke a responsive reaction in sufficiently stunned animals. The touch of the cornea or the lid and a nose pinch are used as stimuli. If the animal shows no reactions in response to these stimuli, it is to assume that the animal is still insensible and unconscious. Furthermore, the location and the shape of the shot holes must be examined in the cool room.

Captive-bolt stunning is a reversible stunning method. Although individual animals are killed by captive-bolt shot, many others stay alive and regain consciousness. The animal will then experience pain from the severe head injuries as well as from the bleed cut. It is essential, therefore, that the animals are stuck immediately after stunning.

When an animal must be re-stunned, a shot through the already existing hole in the skull must be avoided, since this will only lead to soft tissue injury and will not transfer enough energy on the skull to cause concussion. Thus, pressure compensation by the already existing hole and the resulting reduction of pressure variations is acceptable. Choosing a new location for the placement of the captive bolt is the lesser of the two evils in this situation and will usually bring about a better stunning effect than shooting through the existing hole. When re-stunning is performed, focus must be placed on immediate exsanguination.

5.4.2 Electrical stunning

To reliably pass current flow through the brain, two electrodes are placed on the animal's head, either manually by using stunning tongues or automatically. As a result, neurons will release the excitatory neurotransmitters glutamate and aspartate. After about two tenths of a second, the brain will be in a state of increased stimulation and uncoordinated activity, which makes processing of signals impossible. The affected neurons are synchronized and thus inhibit regular information processing in the brain. This is comparable with a grand mal epileptic seizure and is associated with loss of sensibility and consciousness. Since this condition is reversible, electrical stunning will not kill the animals. The inhibitory neurotransmitter GABA, which is gradually released during the epileptic-like seizure, terminates the seizure. Stress will also induce the release of GABA. As extracellular GABA concentrations impair the effect of glutamate and aspartate, any kind of excitement prior to slaughter will affect sufficiency of stunning. Thus, calm and gentle handling of animals is essential from the moment the animals are transported to the slaughterhouse.

The stunning transformer must provide sufficient voltage to ensure adequate current flow through the brain. Also, the electrodes must be pointed and clean in order to bypass skin resistance. The required stunning amperage will not be reached fast enough, or not at all, if blunt or dirty electrodes are used. If the necessary amperage is not reached within a split second, the animal will feel the very painful current flow. Furthermore, the use of blunt or dirty electrodes may cause scorching of skin and hair, and thus additionally increase resistance. It is likewise important to press the electrodes firmly against the animal's head, otherwise the increase in amperage will be delayed.

The minimum amperage recommended for stunning of adult cattle is 1.5 A, better still is 2–3 A, and 1.3 A for calves up to six months of age. The current flow should last about four seconds. In fattening sheep and goats, the recommended minimum amperage is 1.0 A, in adult sheep it is 1.3 A, with a recommended minimum current flow time of two seconds.

Significantly increased current flow times will cause exhaustion of the brain and muscles and are commonly applied to improve the immobilization of animals. This will reduce the injury risk to personnel during the bleeding and hoisting procedure and, consequently, enhance working safety. However, efficiency of stunning is not improved by this measure. The drawback of this approach is that animals are disabled to show signs of poor stunning due to the immobilizing effect of the prolonged current flow and the exhaustion of muscles.

Slaughter animals are, occasionally, wetted to reduce skin resistance in the area of the electrode placement. When done correctly, this may be helpful. When animals are hosed down immediately before stunning, a considerable part of the electricity may flow from one electrode to the other via the wet skin and will, consequently, fail to go through the brain. The skin, therefore, should be moistened not less than 15 minutes prior to stunning to ensure animals are not dripping wet.

The reliable and correct placement of electrodes is essential in order to ensure sufficient current flow through the brain. As with captive-bolt stunning, this requires restraint of the animal to restrict vertical as well as horizontal movement of the head. The brain must be positioned between the electrodes so that the current flow can pass through. For this purpose, the electrodes are applied on both sides of the head between the eye and the ear. The further away from the brain a line drawn between the electrodes appears to be, the greater is the probability of an ineffective stun.

Correctly stunned animals will show the following signs:

- instantaneous collapse;
- immediate tonic cramps, rigid muscles;
- about 10 seconds later the clonic phase, paddling leg movement;
- apnoea;
- tail and ears hang floppy;
- tongue hangs out limply.

Due to its effect on the spinal cord, the current flow simultaneously causes immobilization of the animal. When the electrodes are placed too far away from the brain and the animal remains sensible and conscious, it will be disabled to respond to pain. Hence, ineffective stunning is frequently undetected and accurate placement of electrodes must be examined in every individual animal.

In pigs, incorrectly placed electrodes may be removed and immediately repositioned during electrical stunning. By contrast, in cattle and sheep, brain cells are refractory for about 15 minutes once the current flow has passed through the brain. Repeated application of current flow will, consequently, not have a stunning effect but will cause severe pain to the animal. When using electrical stunning in cattle, re-stunning must always be performed by using a captive-bolt stunner. For this reason, a loaded and ready-to-use captive-bolt stunner must be kept at the stunning point.

Beside the absence of the above mentioned typical signs of sufficient stunning, indications of insufficient stunning are:

- coordinated attempts to escape;
- vocalization;
- rhythmic breathing;
- attempts to rise after the current flow;
- eye tracking to moving objects.

During the epileptic seizure the animals are hyperirritable. They might respond to stimuli of the cornea, the lid or the nasal septum without being sensible or conscious. As opposed to the assessment of effective captive-bolt stunning, tests to verify sufficiency of electrical stunning must not be performed during the first 45 seconds following the stun to avoid incorrect results. For the first 45 seconds after removal of the electric stunner the animals must be exclusively monitored. Corneal and blink reflex as well as response to painful stimulus of the nasal septum are only meaningful after this length of time.

Cattle regain sensibility and consciousness within 20–90 seconds of electrical stunning. To avoid the sensation of pain, the chest stick must be applied within the first 10 seconds after stunning. For sheep, the recommended maximum stun-to-stick time is only eight seconds. Exsanguination by throat cutting, particularly in case of cattle, is not considered as sufficiently rapid.

The short duration of the stunning effect requires the application of current flow through the heart to trigger ventricular fibrillation simultaneously with or immediately after the electrical head stun. Ventricular fibrillation causes cerebral hypoxia and, thus, as does the stunning process itself, induces insensibility and unconsciousness. In an ideal case, the stunning effect of the current flow through the brain will coincide with the effect of hypoxia without the animals meanwhile regaining consciousness. If ventricular fibrillation continues, the return to consciousness before death is impossible. Because the high frequencies used for head stunning are not efficient enough to reliably disturb the autorhythm of the heart and cause cardiac arrest, the use of frequencies below 100 Hz is recommended for cardiac arrest stunning. During the ejection or hoisting of animals, heavy intestines such as the rumen might increase the pressure on the heart and cause the return of autorhythm. Due to the short duration of insensibility, electrical stunning of cattle, even when applying cardiac arrest stunning, always requires intense monitoring of animals from the stunning point to the end of the bleed rail.

5.4.3 Bleeding

To insure that death by bleeding occurs as soon as possible, the major blood vessels must be reliably severed. In many animal species, bleeding rapidly induces hypoxia of the brain. This is associated with irreversible insensibility and unconsciousness. However, in ruminants there are blood vessels near the spinal cord known to supply the brain with arterial blood, even after the carotid arteries and the jugular veins have been severed. They are more distinct in cattle than in sheep and appear less developed in goats. Therefore, the risk of delayed hypoxia after captive-bolt stunning is higher in cattle and sheep. Since the intermediate return of consciousness is possible, it is essential to rapidly induce a profuse blood loss in these animal species. The shorter the stun-to-stick interval, the unlikelier the return of sensibility will be. Subsequently, the effectiveness of stunning must be monitored in every individual animal throughout the bleeding operation. The bleeder will usually be too engaged in sticking the following animals to reliably accomplish this task. The slaughter plant must either give the bleeder enough time to observe the animals or provide another person to do so.

Random measurement of the blood flow immediately after sticking may remove doubts about bleeding efficiency. Blood accounts for about 7% of the body weight. Nearly half of this amount (40–60%) is lost during bleeding. The rate of blood loss is highest in the first seconds following the cut and decreases during the bleeding procedure. Thus, a bovine weighing about 600 kg will lose nearly 21 l (17–25 l) of the total amount of 42 l circulating blood. In cattle, exsanguination is considered as sufficiently effective when the rate of blood loss during the first 30 s after cutting is about 2 l per 100 kg body weight. While chest sticking results in adequate blood loss, throat cuts are frequently associated with significantly longer bleeding times and less blood loss due to the small diameter of the severed blood vessels. Additionally, the blood vessels of the neck may be narrowed by diverse mechanisms, which will result in delayed exsanguination of the head and an increased risk of return to consciousness before death occurs; therefore, cattle should be bled by chest sticking only.

5.5 Poultry

Poultry is usually stunned by electrical stunning. Large slaughter plants also use gas stunning frequently. As with other slaughter animals, pre-stun stress may affect the stunning effectiveness and, therefore, the animals must be treated as gently as possible from transport to stunning. Any kind of excitement must be avoided.

5.5.1 Electrical stunning

Head-only stunning of poultry is applied only by small holdings and rarely by large slaughter plants. Whole-body electrical stunning is a commonly used method. The animals are suspended headfirst from a metal shackle. The shackles are attached to a moving overhead conveyer belt and positioned above an electrified water bath for stunning. The animals' legs are in contact with the electrically grounded shackle line through the shackles. In the majority of cases, the shackles slide along an earth bar above the water bath for grounding. The tight contact between these components must be regularly examined. The live electrode is situated in the water bath in which the heads of the animals are immersed. The electrode must span the entire length of the water bath. Thus, the current flows from the head to the feet, coincidently passing through the brain and inducing insensibility and unconsciousness. As the current is passed through the whole body, it is possible to simultaneously induce ventricular fibrillation by applying according parameters. This will cause rapid hypoxia of the brain and, consequently, loss of consciousness. By this means, the reversible loss of sensibility and consciousness becomes irreversible. However, to avoid meat quality defects, high-frequency currents are usually applied that will not cause ventricular fibrillation. Furthermore, the stunning effect will be shorter than with lower frequencies. Hence, poultry must be intensely monitored for signs of return to consciousness on the bleed rail and, if necessary, must be re-stunned. The slaughter personnel in charge of stunning must know the parameters required for the corresponding animal species and be able to read display values off of the electrical stunning equipment at any time.

Because high-frequency current causes a reversible stunning effect but not death, animals must be bled immediately after stunning in order to be killed. Increasing the amperage and the current flow time (reduction of line speed) may also cause death of the animals. In the event of an outbreak of a disease, slaughter facilities applying electrical stunning will allow the culling of animals without opening the carcasses.

Usually, several animals are stunned simultaneously in the water bath, whereby the carcasses are parallel-connected electrically. The amount of current delivered to each bird will vary depending on each bird's electrical resistance. To reduce the resistance between the shackles and the bird's legs, shackles should be wetted at the point of contact with the leg by softly spraying water. Furthermore, the animals should be immersed to the same depth, so only homogeneous batches of birds should be slaughtered at the same time.

The strain of shackling and transporting the animals head down may affect the stunning effectiveness. Gentle handling of animals and the use of blue light in shackling will calm the animals down. Moreover, the use of a breast comforter has a calming effect on the birds and will significantly reduce wing flapping on the shackle line. The plant must be designed to keep the interval between shackling and entry to the water bath as short as possible. If the animals receive pre-stun electric shocks due to overflow of electrified water, it will cause severe pain and excitement. The sloped entry ramps to the water bath must be electrically isolated and allow overflowing water to drain off underneath in order to keep the surface of the ramps dry.

The level of the water bath must be adjustable to the height of the birds and the head of each animal must be completely immersed in the water to ensure sufficient stunning. The stunning transformer must deliver enough voltage to ensure sufficient current flow through every animal in the water bath. To estimate the required total current, the minimum current applied per bird must be multiplied by the maximum number of birds in the water bath. The head must be submerged in the water before any other body part. When other body parts, such as the wings or the chest, make contact with the water first, the animals will suffer severe pain through the current before the onset of the stunning effect.

Signs of effective stunning are:

- tonic phase immediately after immersing in water;
- no movements noticeable;
- tonic phase persists until exit from the water bath;
- open eyes;
- clonic cramps with fitful movements of legs and wings (after some time);
- absence of wing flapping during the clonic phase and during bleeding.

The animals can hardly be observed during immersion in the water bath, or not observed at all, and testing for reflexes is impossible. Due to the high line speed, the animals cannot be examined after exiting the water bath. Individual animals must be unshackled from the shackle line and closely monitored to assess stunning effectiveness.

The following examinations are possible:

- no response of the pupil to a light stimulus;
- absence of corneal and third eyelid (nictitating membrane) reflex after stimulus of the cornea;
- no response to painful stimulus of the cockscomb;
- no muscle tone in beak and neck.

Most poultry processing plants do not have the possibility of unshackling insufficiently stunned animals to re-stun them in the water bath. For this reason, personnel must be available at the exit of the water bath at the bleeding point and throughout the bleeding rail to re-stun animals by delivering a blow on head. They must be supplied with applicable striking tools and must be capable of reliably recognizing the signs of effective and insufficient stunning.

5.5.2 Gas stunning (controlled atmosphere stunning, CAS)

With gas stunning systems, animals are exposed to a gas atmosphere that renders them insensible and unconscious while they remain in transport crates. Several variations of gas mixture can be used for gas stunning. Concentrations of CO_2 or mixtures of CO_2, nitrogen and/or inert gases are commonly used. Frequently, the animals are conveyed through two gas atmospheres, of which the first one causes analgesia and the second one induces unconsciousness.

In this way, process stages associated with excitement, such as unloading the birds from the crates, shackling and transporting them overhead to the water bath, are avoided. However, the delayed onset of insensibility due to a phase of induction is a disadvantage of gas stunning. Depending on the applied gas atmosphere, the animals may feel uncomfortable or even suffer pain during exposure and have aversive reactions. When inspecting such facilities, it must be made sure that the animals will not show signs of discomfort, such as deep open beak breathing, when entering the (first) gas atmosphere. The transport crates should not be overfilled. The personnel must be familiar with the gas concentrations required for effective stunning; these must be clearly displayed in the stunning area. Deficiencies of the gas atmosphere must be immediately signalled by the measuring system. It is usually impossible to observe the animals for clinical signs during the exposure.

This method also requires bleeding as soon as possible after stunning. During the bleeding procedure none of the following indications for ineffective stunning should occur:

- attempts to rise;
- wing flapping;
- rhythmic breathing;
- vocalization.

The following applies to any kind of stunning method: when signs of insufficient stunning occur, not only must re-stunning be performed but the cause of the insufficiency must also be determined and eliminated.

5.5.3   Bleeding

The bleeding procedure is rarely carried out manually in poultry slaughter plants but by an automatic neck cutter. To ensure adequate bleeding, the interval between stunning and bleeding should not exceed 20 seconds. Since, in many slaughter plants, reliable severance of both carotids is not always possible, effective stunning may only be guaranteed if the animals are bled within 5–10 seconds after the stun. A slaughterhouse worker must stand beside the automatic neck cutter and rapidly bleed every insufficiently bled animal by performing a throat cut with a knife.

5.6   Conclusions

The slaughterhouses should always have a plan for animal welfare to ensure a good level of animal welfare at all stages of the handling of animals until they are killed. It should include specific corrective actions for specific risks that could negatively affect the welfare of animals. The conditions of animals should be assessed upon their arrival for any animal welfare and health problems. Animals should be handled in such way as to avoid harm, distress or injury at any time. The competence of the slaughterhouse workers, the appropriateness and effectiveness of the stunning methods and the maintenance of the equipment should be regularly checked even if there are no problems reported.

Literature and further reading

European Food Safety Authority (EFSA) 2004. Welfare aspects of the main systems of stunning and killing the main commercial species of animals. *EFSA Journal*, **45**, 1–29.

Grandin, T. 2013. Making slaughterhouses more humane for cattle, pigs and sheep. *Annual Review of Animal Biosciences*, **1**, 491–512.

World Organisation for Animal Health (OIE). 2012. Terrestrial Animal Health Code: Slaughter of Animals. http://www.oie.int/fileadmin/Home/eng/Health_standards /tahc/2010/en_chapitre_1.7.5.htm (last accessed 18 January 2014).

6

Post-Mortem Inspection and Related Anatomy

Paolo Berardinelli[1], Rosanna Ianniciello[2], Valentina Russo[1] and Thimjos Ninios[3]

[1]*Faculty of Veterinary Medicine, University of Teramo, Teramo, Italy*

[2]*ASL4 (Veterinary Hygiene of Food of Animal Origin), Teramo, Italy*

[3]*Border Control Section, Import, Export and Organic Control Unit, Finnish Food Safety Authority Evira, Helsinki, Finland*

6.1 Scope

The proper performance of a *post-mortem* inspection requires a good knowledge of the anatomy of the slaughtered animal species. The aim of this chapter is to offer an inspective point of view of the anatomical structures perceived during a *post-mortem* inspection and to discuss the available inspective tools.

6.2 Introduction

In particular, this chapter is focused on the inspective anatomy of:

- pigs
- large and small ruminants
- poultry.

To perform *post-mortem* inspection procedures properly, it is crucial to be familiar with the comparative anatomy of livestock carcass and organs. It is

Meat Inspection and Control in the Slaughterhouse, First Edition.
Edited by Thimjos Ninios, Janne Lundén, Hannu Korkeala and Maria Fredriksson-Ahomaa.
© 2014 John Wiley & Sons, Ltd. Published 2014 by John Wiley & Sons, Ltd.

also true that the identification of animal species, as well as the definition of the gender, is possible through a good knowledge of the differential anatomical characteristics and, in particular, of the skeletal system. To describe the anatomy of pigs, ruminants and poultry, the inspective anatomy has been divided into the following sections:

- anatomy of the head
- anatomy of viscera
- anatomy of the carcass
- anatomy of poultry.

The anatomical characteristics of pig, bovine and small ruminants are similar and are discussed together in the three first sections, while the anatomy of poultry is discussed in its own section.

6.3   Anatomy of the head

The inspection of the head can be performed after it has been removed from the carcass or when it is still attached to the carcass. The inspection includes at least an external visual analysis, aimed at describing, in particular, the muscles, and an internal evaluation, needed to capture the features of the skeleton as well as the viscera of the cranial cavity and face.

6.3.1   Skeleton structures and viscera of the cranial cavity

The view in a sagittal section allows observation of the cranial cavity containing part of the central nervous system and the remaining bones of the face that contain viscera of the respiratory and digestive systems. The cranial cavity contains the *medulla oblongata*, the *pons Varolii*, the midbrain (*mesencephalon*) and interbrain (*diencephalon*), which together constitute the brainstem (*truncus encephali*). Contained in the same cavity are also the telencephalon and cerebellum, which are not part of the brainstem. All these organs are related to the bones of the skull, such as the basal part of occipital (*os occipital*), sphenoid (*os sphenoidale*), ethmoid (*os ethmoidale*) and frontal (*os frontis*) bones. The cranial cavity is separated from the nasal cavity and placed in the face by a bone septum perpendicular to the ethmoid bone.

6.3.2   Skeleton structures and viscera of the face

The bones of the face delimit the nasal cavity dorsally and laterally and oral cavity ventrally (Figure 6.1). Both cavities converge caudally in the pharynx, a common structure of the digestive and respiratory systems. The nasal cavity placed dorsally contains the turbinates (*conchae nasales*) and is bounded laterally and dorsally by maxillary (*maxilla*), nasal (*os nasale*) and incisive

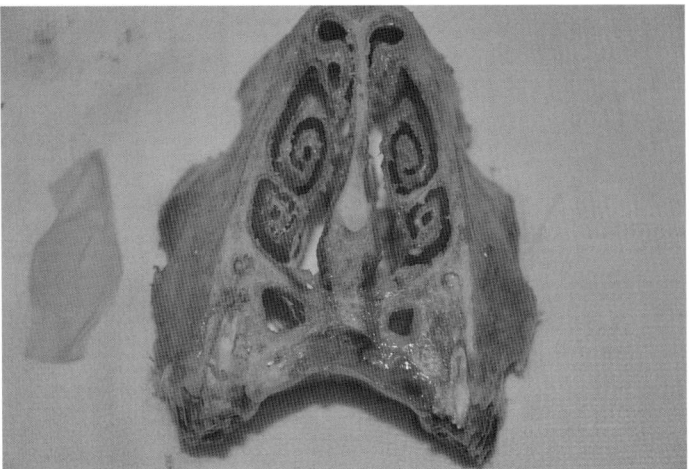

Figure 6.1 Transverse section of ovine head in the proximity of the nasal cavity.

(*os incisivum*) bones. The rostral opening is represented by the nostril and its aboral opening, which flows into the nasopharynx, is made up of the choanae (*apertura nasalis posterior*). The nasal cavity is separated from the underlying oral cavity by the hard palate, which consists of the palatine processes of the maxillary and incisive bones that continue aborally in the soft palate. The oral cavity, which occupies the ventral half of the face, is placed between the two mandibles (*mandibula*) and extends up to the isthmus of the fauces (*isthmus faucium*). It contains the tongue and teeth and is bounded dorsally from the hard palate. Caudally to the isthmus of the fauces is the oropharynx placed between the soft palate and root of the tongue. The nasopharynx is the dorsal part of the pharynx and is located dorsally to the soft palate and in contact with the nasal cavity through the choanae. On the side walls are the openings of the auditory tubes (*tuba auditiva*) with their ostia. In the most caudal part, the *pharynx–larynx* portion, can be found between the base of the *epiglottis* and the beginning of the oesophagus.

6.3.3 Lymph nodes of the head

Inspection of the head also provides information about the main lymph node formations corresponding to the parotid (*nodi lymphoidei parotidei*), submaxillary (*nodi lymphoidei submandibulares*) and retropharyngeal (*nodi lymphoidei retropharyngeales*) lymph nodes, and, if present, the palatine tonsils (*tonsilla palatina*).

6.3.4 Pigs

The slaughtered animal is mostly presented for *post-mortem* inspection as sagittally sectioned half carcasses. This allows an accurate and complete inspection of the head and its contents to be carried out, different from other species. In the median sagittal section of the head it is possible to see inside the skull bones, the central nervous system (encephalon, medulla oblongata and cerebellum) and, protected by the bones of the face, the first digestive and respiratory tracts, the lymph nodes and the muscles of the region. Depending on the slaughter technique, the tongue and tonsils can be removed from the head; they are described later (Figure 6.2).

Masseter muscle　The masseter muscle (*musculus masseter*), as in all species, is the most developed mastication muscle and provides the external shape of the head. It is divided into a large, superficial portion, formed by oblique muscle bundles in the ventrocaudal direction covered by a robust aponeurosis, which disappears in the ventral and caudal fleshy parts. The deep portion of the muscle is separated from the superficial by a well developed layer of

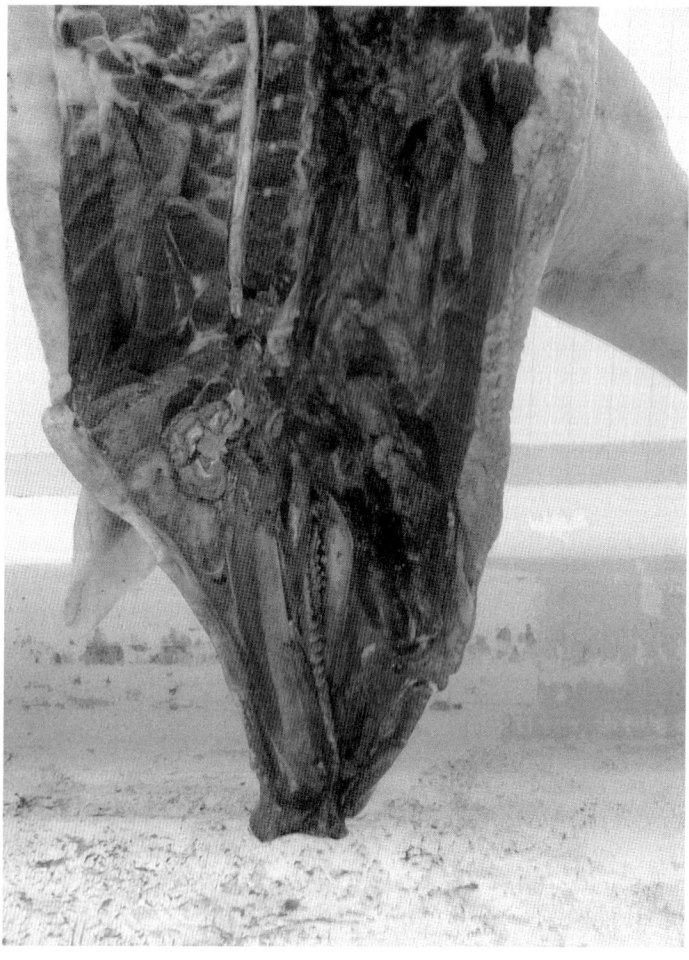

Figure 6.2　Sagittal section of swine head.

connective tissue. The morphologic evaluation of the masseter muscle is prevented by the coating skin. This muscle in the pig is much thickened, with a superficial square layer and a rostral margin almost perpendicular to the mandible. The deep layer is much thickened but it is entirely covered by the superficial layer and is not visible in the vicinity of the temporomandibular joint (*articulatio temporomandibularis*).

Lymph nodes of the head The submaxillary lymph nodes are usually present as a unique formation of large size (Figure 6.3). Sometimes there is also an accessory submaxillary lymph node. Both are located just behind the angle of the mandible (*angulus mandibule*) in front of the submaxillary gland, lying on the hyoid muscles. Caudally they are covered by the pointed portion of the parotid gland, which can go up to the intermandibular space.

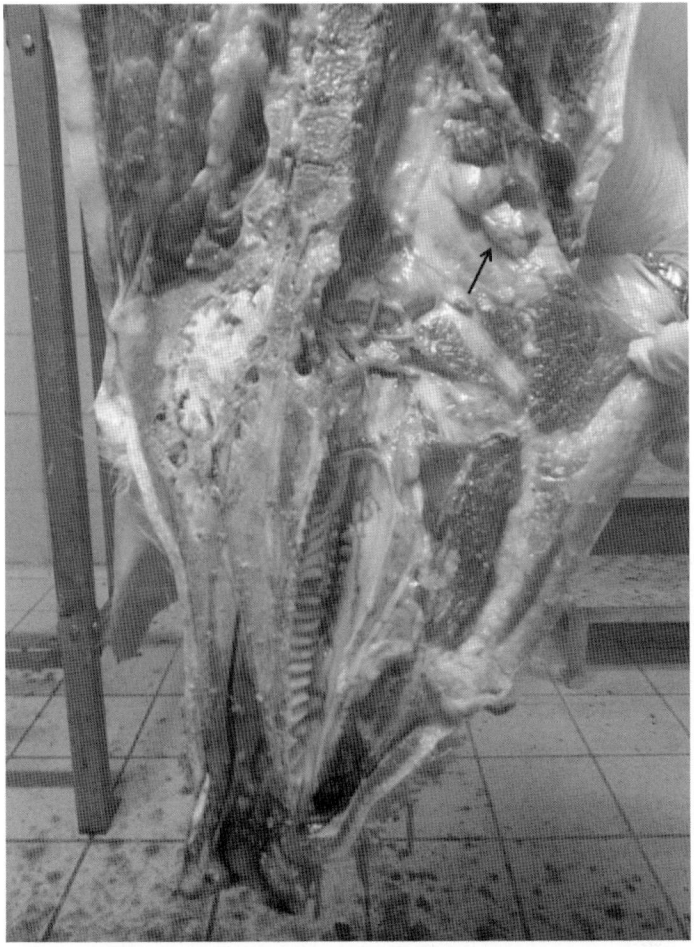

Figure 6.3 Submaxillary lymph nodes in swine.

The submaxillary lymph nodes receive lymph from the nose, lip, masseter, eyelid and orbital regions.

After the incision of the throat and the removal of the tongue, the lymph node appears in correspondence of the mandibular salivary gland. Some separation techniques of the tongue may cause removal. In this case, it is to be found in the adipose tissue, adjacent to the base of the tongue.

The retropharyngeal lymph nodes can be divided into two distinct structures:

The medial retropharyngeal lymph node is a single body in pigs, placed under the joint between atlas and occipital bone (*articulatio atlantooccipitalis*), between the spinal column and the oesophagus and caudally to the velum and the tonsils.

The lateral retropharyngeal lymph nodes are located in close contact with the cranioaboral margin of the parotid gland, from which must be carefully distinguished. The retropharyngeal lymph nodes receive lymph from the skin of the head, the lips and cheeks, the tongue, the palate, the tonsils, the submandibular salivary glands, the pharynx, the larynx and the thymus.

The parotid lymph nodes are a mass of considerable size in which three separate nodular structures can be distinguished. The first is located under the auricle (*auris externa*) at the temporomandibular joint, the second is located near the middle of the aboral margin of the mandible and the third is placed lower on the level of the angle of the mandible. They have a close connection with the parotid gland, from which they can be partially or completely hidden. The parotid lymph nodes receive lymph from the skin, the muscles of the parotid, the nuchal, temporal, auricular and frontal regions, and also from the parotid and lacrimal glands. All three lymph nodes are related to the parotid gland, from which they are more or less covered.

6.3.5 Bovine

In bovine the head is usually examined while hanging with the tongue reversed, to allow a thorough exploration of the oral cavity. The tongue can be examined *in situ* or separately (Figure 6.4).

Masseter muscle   The masseter muscle is relatively thin, with a portion almost entirely aponeurotic at its insertion to the zygomatic arch, on the facial ridge and on the facial tuberosity. The deeper portion is very thick at its base and is hardly separable from the external part (Figure 6.5).

Tongue   The tongue of bovine is consistent, thick and fleshy (Figure 6.4). In adult animals it is about 30 cm in length and weighs an average of 1.5 kg. The weight varies depending on race, sex and age (in calf it is about 500 g while in adult bovine it is 2 kg or more). The mucosa is often pigmented. The apex is thickened and ends with a blunt tip. It is devoid of the median groove

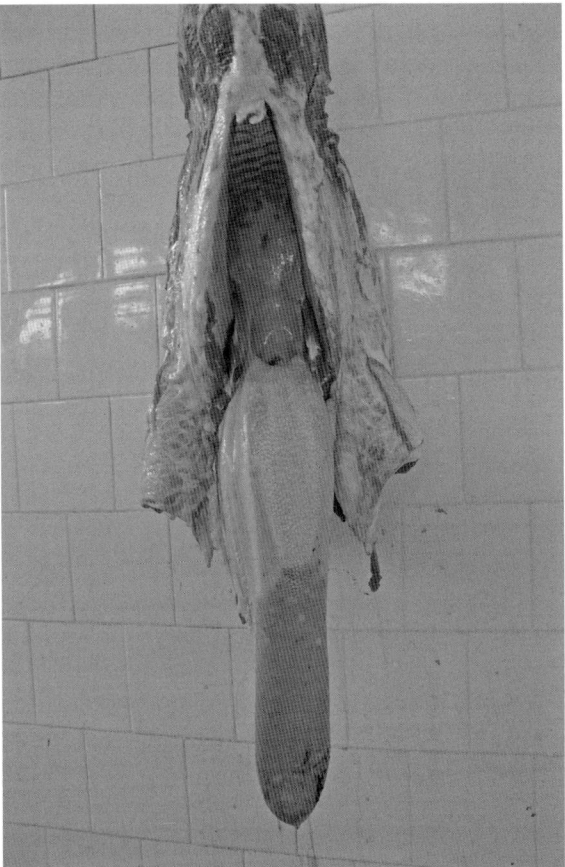

Figure 6.4 Head of bovine.

and is rich in pointed conical papillae (*papillae conicae*), keratin curved caudally, which makes them rough. The frenulum is very lax. The dorsum of the tongue is raised in its caudal half due to the presence of a voluminous convex lingual relief occupied by large conical and lenticular (*papillae lentiformes*) papillae. Cranially to the lingual relief is the lingual fossa (*fossa linguae*), a groove whose transverse median part presents a depression of varying depth. Rostrally to this depression the filiform (*papillae filiformes*) and conical papillae have the same appearance as those located in the apex. The fungiform papillae (*papillae fungiformes*) become abundant on the apex in vicinity of the margins. On the caudal side of the relief, the circumvallate papillae (*papillae vallatae*) are numerous and varying in height from 3 to 6 mm from each side, arranged in a double row that converges caudally to that of the opposite side; their number varies from 8 to 16 per side. The foliate papillae (*papillae foliatae*) are usually absent. The root of the tongue (*radix linguae*) is short and wrinkled and is provided with a twofold lingual tonsil and of numerous lingual glands.

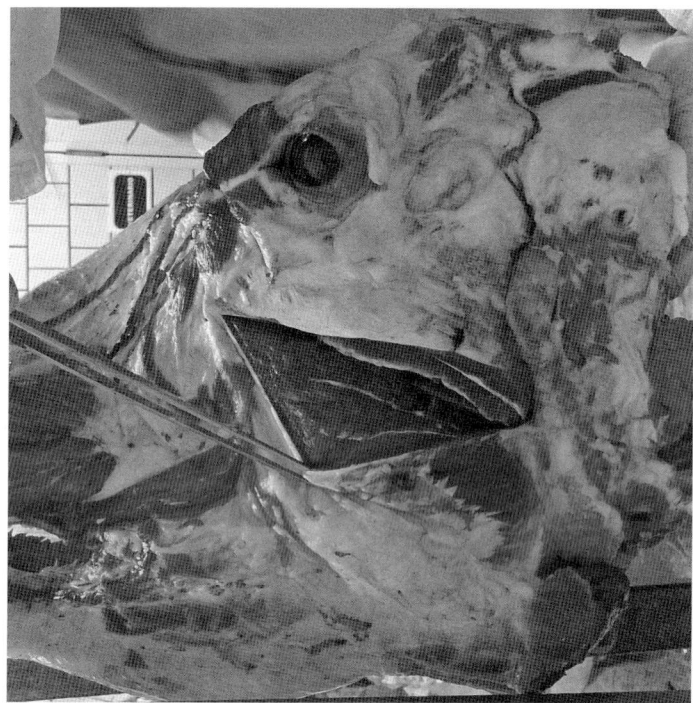

Figure 6.5   Section of bovine masseter muscle.

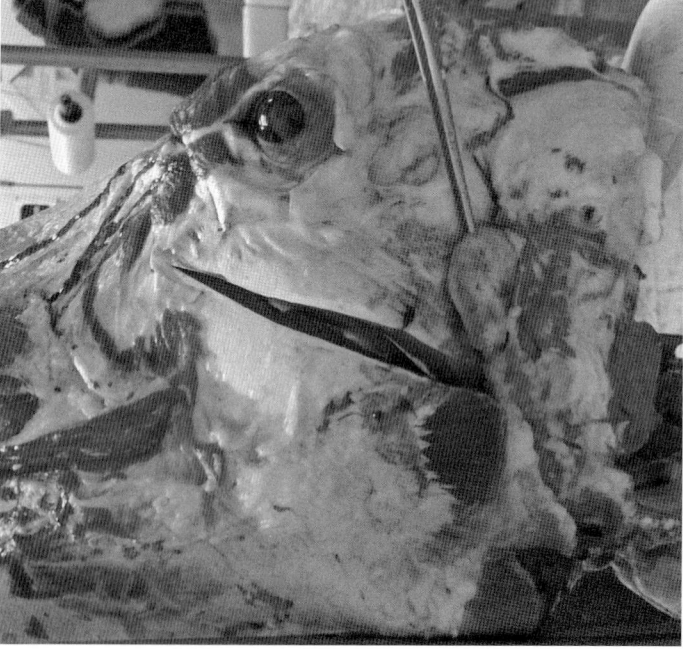

Figure 6.6   Parotid lymph node, evidenced by a section between the masseter muscle and acoustic meatus.

Lymph nodes The inspection of the head requires the evaluation of four main lymph node formations. The parotid lymph nodes (Figure 6.6) are very voluminous, about 9 cm in length, placed posteriorly and ventrally to the temporomandibular joint. The cranial part of the lymph node is covered by the subcutaneous tissue while the cuadal part is located below the parotid gland. It can be found, as seen in the Figure 6.6, in the space between the masseter muscle and the external acoustic meatus.

The submaxillary or mandibular lymph nodes (Figure 6.7) are placed laterally to the intermandibular space in the vicinity of the lower edge of the body of the mandible. They are two for each side with a size of approximately 2 cm. They are found within adipose tissue placed at the sides of the base of the tongue.

The retropharyngeal lymph nodes are divided into medial and lateral (Figure 6.8). The medial lymph nodes are located between the base of the skull and the pharynx, medially to the branch of the hyoid bone adhering to the pharyngeal muscles. When the tongue is upside down during the inspection, on the right and on left of the cut pharyngeal mucosa these formations are located right behind the palatine tonsils. The lateral retropharyngeal

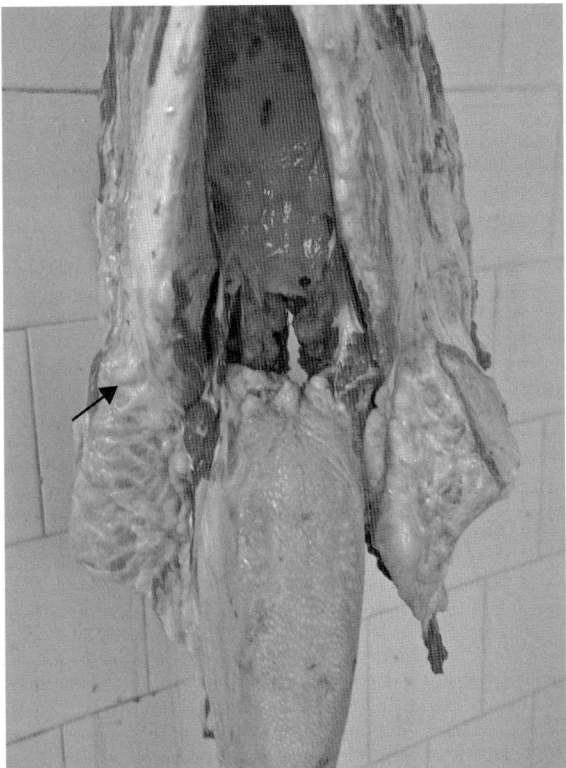

Figure 6.7 Submaxillary lymph nodes in bovine, shown by the black arrow, visible sidewise to the root of tongue.

Figure 6.8 The clamp indicates the bovine medial retropharyngeal lymph nodes close to the palatine tonsils.

lymph nodes (Figure 6.9) are voluminous, with a size of 4–5 cm, and they cannot be found when the head is detached too close to the mandible. In fact, the lymph nodes are located below the wing of the atlas in the cluster of adipose and glandular tissues located in correspondence with the ramus of the mandible.

The tonsils, once called amygdalae, are pharyngeal lymphatic formations. They are divided into three groups: one located in the fauces, one in nasopharynx and one in the laryngopharynx.

In the fauces, three clusters of lymphatic tissue are observed: palatine tonsil, lingual tonsil and soft palate tonsil (Figure 6.10). On each side of the fauces there is a palatine tonsil located in the tonsillar fossa between the glossopalatine arch and pharyngealpalatine arch; this is lacking in the pig. The lingual tonsil is located on the floor of the fauces caudally to the root of the tongue. The soft palate tonsil, as the name implies, is located in the

Figure 6.9 The clamp indicates the lateral retropharyngeal lymph nodes detected on bovine head.

Figure 6.10 Soft palate tonsil indicated by the clamp.

soft palate. In nasopharynx, there are two clusters of lymphoid tissue, the pharyngeal tonsil and tubal tonsil. The pharyngeal tonsil is placed in the vault of the pharynx while the tubal tonsils, which are equal, are placed near the ostium of the auditory tube on each side. In laryngopharynx there is a single cluster of lymphoid tissue: the paraepiglottic tonsil (*tonsilla paraepiglottica*).

Specifically in bovine tonsils are well developed. The pharyngeal tonsil forms a wide relief of about 3–4 cm. The palatine tonsils are voluminous, invaginated in the glossopalatine arch and pharyngealpalatine arch. Each one is about 3 cm thick. The lingual tonsil is developed and evident in the mucosa. The tonsils are absent and the soft palate tonsil is spread and therefore difficult to distinguish.

6.3.6   Small ruminants

For the purposes of *post-mortem* inspection the head usually remains connected to the carcass. The tongue is less thick compared to bovine tongue, with an average weight of about 250 g. It presents a less pointed apex and a wider and lower lingual relief with slender papillae. The fungiform papillae are numerous, as are the circumvallate papillae, which can be as many as 20 per side. The foliate papillae are missing, while the filiform papillae also invade the lateral sides of the tongue and the ventral face of the lingual apex. The root appears smooth.

Lymph nodes   The medial retropharyngeal lymph nodes are located dorsally to the muscles of the pharynx, while the lateral retropharyngeal lymph nodes can be visualized in the adipose tissue located cranially in the neck. The parotid lymph nodes are easily visible between the masseter muscle and the auditory canal in correspondence of the ramus of the mandible. The palatine tonsils are more voluminous than in bovine. The paraepiglottic tonsils are present, while in bovine they are not.

6.4   Anatomy of viscera

After their extraction from the carcass, the viscera need to be examined. The examination of every individual organ requires a good knowledge of the normal characteristics of this organ.

6.4.1   Viscera of the oral cavity

Tongue and tonsils in pigs   A common practice in pig slaughtering is the removal of the tongue and tonsils from the head. The tongue is elongated and without a median furrow, measuring about 20 cm and weighing an average of 500 g. The apex has an ogival edge and its ventral side is smooth and devoid of papillae. The body is provided with very high lateral faces. The back has a narrow median longitudinal relief, which can be considered as the equivalent of a rudimentary lingual relief.

The filiform papillae are soft and flexible in both the body and the apex. At the root of the tongue they are less numerous but more thickened and longer in length. The fungiform papillae are numerous on the dorsal surface and on the apex. They are also present on the lateral faces of the body of the tongue. The caudal end of the dorsal surface presents two large circumvallate papillae. The foliate papillae are developed and concentrated on each side in an area of about $1\,cm^2$ (Figure 6.11).

In pigs, tonsils are well developed with the exception of the palatine tonsil, which is replaced by a large tonsil of the soft palate, about 6 cm long and 4 wide, pink in colour and relieved on the ventral face of the soft palate. Moreover, there are also a pharyngeal and a tubal tonsil. The lingual tonsil is large, widespread and well developed, while the paraepiglottic tonsil is well developed and situated in a depression at the base of the epiglottis.

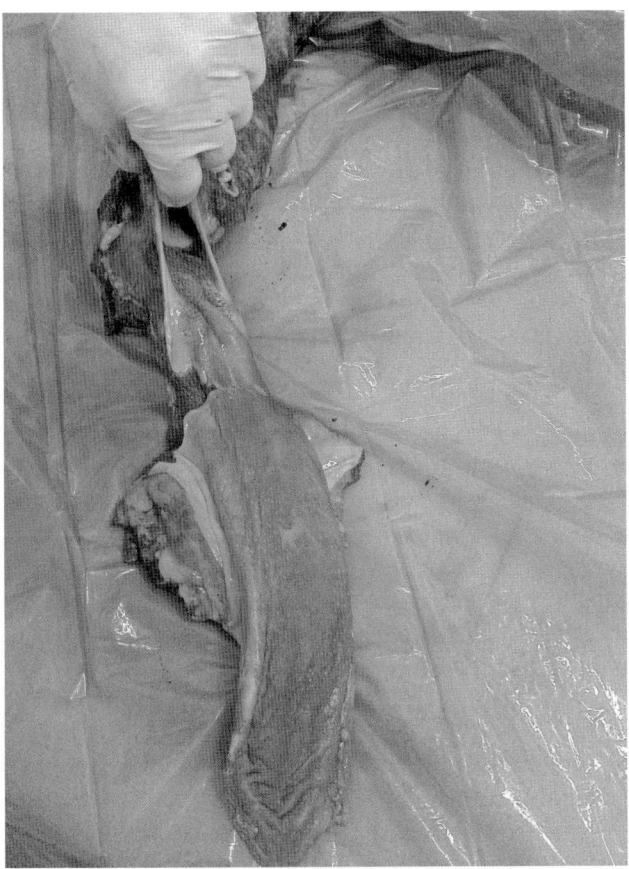

Figure 6.11 Isolated pig tongue.

6.4.2   Viscera of the thoracic cavity

Trachea and lungs

Overview

The trachea is an open tube because it is supported by a scaffold formed by cartilage rings. It follows the larynx and ends in the middle mediastinum at the tracheal bifurcation, where it splits into two main bronchi, the right and left. The trachea has a smooth inner surface and is watered by a little yellow viscous mucus that reveals the arrangement of the cartilage rings below.

The lungs are equal organs, right and left, of spongy and elastic texture. They are divided into lobes by fissures of varying depth. The lungs are surrounded by serous pleura and are suspended from the mediastinum, formed by the buttressing of the pleura on the median plane.

Topography

The trachea has one cervical and one thoracic portion, and is slightly restricted in the entry of the thorax. The lungs occupy the thoracic cavity almost entirely.

Colour

The surface of the lungs is pink and infiltrated by thin grey streaks.

Pig   The trachea has a regular cylindrical shape and the ends of the cartilages are overlapping. The tracheal muscle takes insertion attack on its inner face. It has a length of about 30 cm and is composed of 30–36 cartilages. The trachea sprouts off to the right tracheal bronchus 3–4 cm before its termination. The lungs (Figure 6.12), of pale pink colour, present a superficial lobulation less striking than in bovine; they are, however, distinctive over their entire extension. Their weight is approximately 1–2 kg, of which more than a half is due to the right lung, which is larger. The fissures are deep. The right lung is divided into four lobes: cranial, middle, caudal and accessory. The cranial lobe has, on its medial face, the penetration of the tracheal bronchus at the level of a small accessory hilum. It is separated from the middle lobe by a short fissure, which continues in the large cardiac incision. This lobe is separated from the caudal by a deep fissure that goes up to the main hilum. The caudal lobe, which is the largest, presents on the medial face of the base a well-developed accessory lobe. The left lung, the smaller one, in the cranial part of its cranial lobe presents a culmen and a lingula. The caudal lobe is almost equal to that of the right lung but lacking of the accessory lobe.

Bovine   The trachea has a length of about 60 cm and, being laterally compressed, has an outer transverse diameter of 4 cm and a dorsoventral diameter greater than 5 cm. It consists of 49–52 cartilages that lean against each other forming a dorsal ridge. The tracheal muscle is inserted in the cartilage rings. In bovine, before the tracheal bifurcation on the right side, the start of the right cranial bronchus sprouts off.

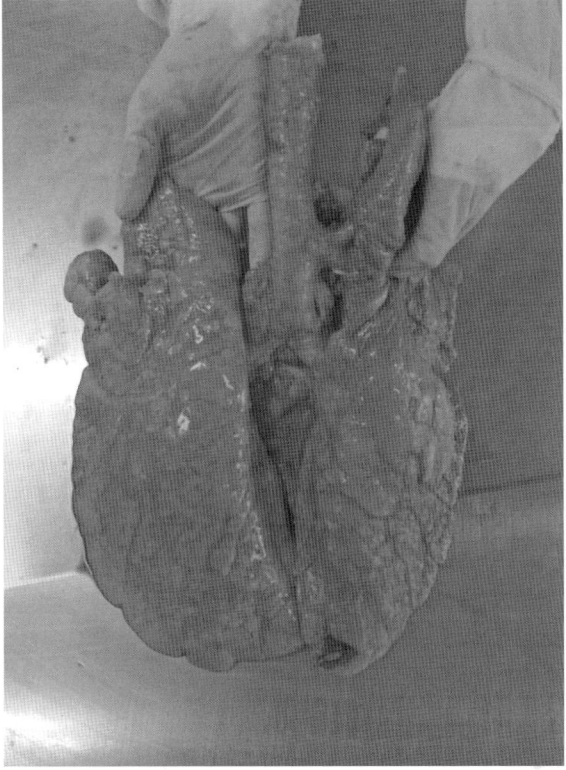

Figure 6.12 Dorsal face of pig lungs.

The lungs in bovine (Figure 6.13) present a pronounced asymmetry and the right lung is more voluminous, weighing about 2.5 kg in the adult, while the left weighs about 2 kg. The colour is pink-yellowish and the surface presents a division into lobules separated by thick connective tissue septa, evident under the pleura as polygonal features. The right lung is divided into five distinct lobes by deep fissures: the cranial, caudal, accessory, medium cranial and caudal lobes. The left lung, less voluminous, is divided into the cranial, medium and caudal lobes. The cranial and middle lobes are largely united at their base and are separated by a shallow fissure in such a way that they are described as two parts, respectively, cranial and caudal of the same cranial lobe. The caudal part is called *lingual*, while the cranial part is pyramidal, pointed and is called *culmen*.

Small ruminants The trachea is long, approximately 35 cm, and has a diameter of about 2 cm. It is nearly circular because of the succession of about 50 cartilage that overlap each other slightly and form a thin dorsal ridge. The tracheal muscle is located as described in bovine. The lungs resemble those of cattle, even if their colour is more yellowish than pink. The caudal lobe, in

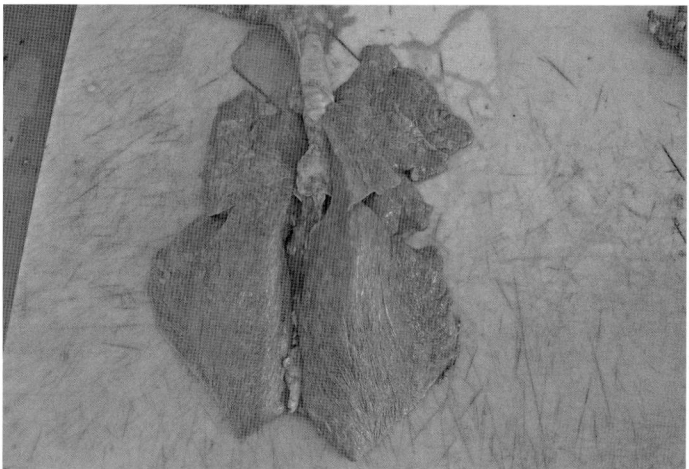

Figure 6.13   Dorsal view of bovine lungs.

both lungs, is more elongated than that of bovine. In the right lung the cranial lobe and the cranial medium lobe are divided by a thin and shallow cranial fissure. The medium caudal lobe is presented instead pedunculated. A deep fissure in the left lung divides two parts in the cranial lobe. The surface does not present clear lobules.

Thymus

Overview
The thymus is a lymphoepithelial organ that reaches its maximum development during the last periods of foetal life and in the newborn; it progressively atrophies before puberty and is replaced by adipose tissue.

Topography
The thymus is located in the visceral space of the neck, ventral to the trachea and is extended into the thoracic cavity in the cranial mediastinum. It is thus possible to distinguish a cervical and a thoracic portion.

Colour
The thymus assumes a white-pink-greyish colour and is wrapped by a capsule which deepens in the parenchyma and divides the body of the organ into lobules.

Pig   The cervical portion is particularly developed and consists of two lobes: cervical left and right. At the level of the opening of the cranial part of the thorax the two cervical lobes remain distinct but converging in the thoracic part. The cervical lobes in the neck are ventrolateral in relation to the trachea and more cranially until they arrive at the pharynx.

Bovine In cow the thoracic part is developed, particularly the cervical part, which is comprises two cervical lobes, the right and left. These two lobes lie ventrolaterally compared to the trachea, dorsolaterally to the pharynx and continue under the base of the skull as a peduncle. At the level of the opening of the cranial thorax the two cervical lobes come close and are unified in the thoracic part through the intermediate lobe.

Small ruminants The thymus of small ruminants is similar to the bovine. It weighs about 40–45 g.

Heart

Overview
The heart has the shape of an irregular cone with an apex directed ventro-caudally, more or less rounded, depending on the species. The pericardium surrounds the heart and supports it between the laminae of the mediastinum. It also includes the initial portions of the pulmonary trunk and of the aorta and the termination of the *venae cavae* and of the pulmonary vein. The peri-cardium consists in an outer *pericardium fibrosum* and a *pericardium serosum*. The *pericardium fibrosum* is a thin sac that adapts to the shape of the heart and has a whitish colour and is diaphanous. Its inner face is firmly bonded to the lamina of the parietal *pericardium serosu*, from which cannot be sepa-rated. The outer surface of the *pericardium fibrosum* is loosely bonded to the face of the mediastinal pleura. The *pericardium serosum* is a closed sac com-posed of a parietal and visceral lamina reflected around the great vessels of the base of the heart, delimiting a cavity virtual: the pericardial cavity. In the cavity there is a transudate that consists of a transparent and viscous liquid.

External conformation
Since the heart has a conical shape, it can be described as having a dorsocranial base, from which emerge the great vessels, and an apex directed ventrally, pro-jected between the sternum and diaphragm more or less directed to the left. In mammals it has flattened lateral sides so that two faces are recognizable: the left or *facies auricularis* and the right or *facies atrialis*; these are separated by two edges, one cranial (*margo ventricularis dexter*) and one caudal (*margo ventricularis sinister*). The heart surface is marked by the tips of *auriculae atri-orum* (Figure 6.14).

The atrial complex, located dorsocranially, is divided by the coronary groove (*sulcus coronarius*), where the major vessels of the organ run, from the ventricle mass that constitutes the majority. The coronary sulcus is covered by adipose tissue. The base of the heart has a semilunar form surrounding the origin of the large arteries, on the left of which the most superficial is the pulmonary trunk (*truncus pulmonalis*), covering the ascending aorta (*aorta ascendens*), which placed more deeply and which continues directly over the aortic arch (*arcus aortae*) to the left and dorsocranially. Placed cranially is the right atrium, into which the cranial vena cava opens. The caudal vena

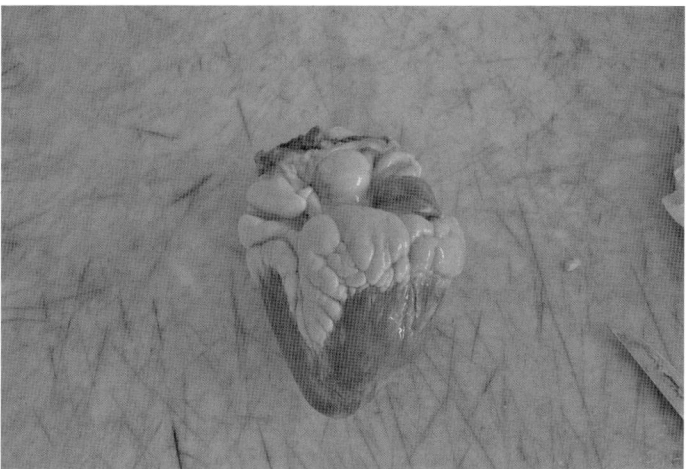

Figure 6.14 Bovine heart, *facies atrialis*.

cava ends in the caudal ventral portion of the right atrium. The atrium is extended at its end by means of the right auricular appendage that curves cranially to the left in relation to the ascending aorta and the pulmonary trunk. On the base of the heart, caudally on the left, stands the left atrium receiving the pulmonary veins caudally. This atrium extends caudally and to the left with the left auricular appendage interacting with the base of the pulmonary trunk.

The atrial face is placed on the right and presents a coronary groove that divides the atria from the ventricles. This face is divided by the deep interventricular groove (*sulcus interventricularis subsinuosus*) that extends from the coronary sulcus, ventrally and caudally to the apex of the heart, occupying the left ventricular wall. Also present in this groove, in which the cardiac medium vein and artery run, is adipose tissue. Extending cranially to the interventricular groove is the right ventricle and caudally the left one.

The auricular face is placed on the left; it has a deep coronary groove placed under the auricles, which tend to cover it. The groove is interrupted by the emergence of the pulmonary trunk. The ventricular portion is limited by the interventricular groove (*sulcus interventricularis paraconalis*) that begins caudally to the pulmonary trunk and ends on right ventricular edge, slightly away from the apex. The homonym vessels covered by adipose tissue start in the longitudinal groove. The right ventricle appears also less extended than the left one.

The right and the left margins are set at the point of union of the two faces of the heart and are divided into a right and a left ventricular (*margo ventricularis dexter* and *margo ventricularis sinister*). The right ventricular margin is placed cranially and is strongly oblique in a ventrocaudal direction. Its dorsocranial part belongs to the base of the right auricle. Ventrally to the coronary sulcus is the right ventricle, even though the apex belongs to the left ventricle. The left ventricle margin is placed more caudally and has a more vertical course, perpendicular to the caudal part of the sternum.

The apex (*apex cordis*) is constituted by the muscular mass of the left ventricle. The interventricular grooves left and right end in its vicinity.

Internal conformation
The heart is divided into two portions, right and left, by a cardiac septum and consists of two parts: the interatrial septum (*septum interatriale*) and the interventricular septum (*septum interventriculare*). They divide the two fully-delimited cavities of the right heart from the two cavities of the left heart. The interatrial septum is thin and partly muscular and is oriented obliquely. The face of the right atrium presents a depression, the *fossa ovalis*. The interventricular septum is a thick muscular wall strongly convex to the right ventricle, derived by the union of two distinct muscle layers, the one belonging to the left ventricle is about two times thicker than the one belonging to the right ventricle. The right atrial cavity has a caudal wall represented by the interatrial septum and a muscle vault with a thin wall in which the ostium of the cranial vena cava and, more caudally, the ostium of the caudal vena cava are present. In the vault of the atrial cavity, formed cranially and to the left, is a blind end: the auricula, very tortuous due to the presence of the *mm. pectinati*. The floor is entirely occupied by the right atrioventricular ostium (*ostium atrioventriculare destrum*), which is closed by the corresponding valve during the ventricular systole. The right ventricle, as a whole, is a cavity that in cross-section has a semi-lunar shape leaning to the left ventricle, placed cranially and to the right. Two walls are recognized, a marginal and a septum, a fund near the apex and a base placed dorsally. Numerous formations protrude into the cavity. These include the papillary muscles, three main thickened carnose eminences with the base inserted within the wall and the free top, which provides attachment to fibromuscular cords (*chordae tendineae*) that join the atrioventricular valves. In the medium portion of the marginal wall, the magnus papillary muscle can be found, while in the medium part of the septal wall, the subarterial papillary muscle and numerous small papillary muscles can be observed.

Other types of muscle projections are represented by free *trabeculae carneae*, which are found in the medium part. One of these has the form of an oblique columnae carneae that leads from the septum to the wall margin. It is present in all species and is called *trabecula septo marginalis*. The base of the ventricle shows the right atrioventricular ostium, on whose fibrous board takes attack an atrioventricular valve. Dorsally to the ventricular cavity of the heart against the auricular face, and separated by the right atrioventricular ostium, is the ostium of the pulmonary trunk. It has the valve of the pulmonary trunk.

The wall of left atrial cavity is thicker than that of the right atrium. At the level of the dorsocaudal part there are numerous ostia of the pulmonary veins (*ostia venarum pulmonalium*). Much of the left atrium lacks a floor, as this is entirely occupied by the left atrioventricular ostium (*ostium atrioventriculare sinistrum*). The left atrial auricula has a cavity similar to that described for the right, with a wall winding for the presence of the *mm. pectinati*.

The left ventricular cavity in cross-section has an almost circular shape and its cavity is delimited by a much more thickened wall than that of the right ventricle. The walls are smooth in the vicinity of the base, tortuous at the apex and present two papillary muscles, facing each other, at the level of the auricular and atrial face of the heart and called, therefore, the subatrial papillary muscle (*musculus papillaris subatrialis*) and subauricular papillary muscle (*musculus papillaris subauricularis*). The bottom of the left ventricle of the heart is dug in the apex of the heart and is more tortuous than the right, as it has more thickened trabeculae. The base is fully occupied by the two hosts: the ostium of the left atrioventricular closed by the homonymous valve and by the aorta ostium (*ostium aortae*) the outline of which gives attachment to its valve.

The atrioventricular or venous ostia, as well as the aortic and pulmonary or arterial ostia, are surrounded by four fibrotic rings. The two fibrotic atrioventricular rings and the aortic one form a continuum of fibrous tissue. This tissue also fills the spaces delimited by the convexity of the three rings (*anuli fibrosi*), forming two layers of vaguely triangular shape, the so-called *trigona fibrosa*. The right fibrous trigone is more developed and is placed between the three rings just mentioned. The left fibrous trigone is located at the junction between the aortic and left ventricular rings. The structure of these two *trigona* becomes denser with the increasing age of the animal and presents cartilage plaques that give rise to bones of the heart (*ossa cordis*). The fibrous rings surround the hosts, which are closed by valves: it is possible to recognize a right and a left atrioventricular valve, an aortic valve and a pulmonary trunk valve, respectively. Of the four hosts, the left atrioventricular ostium is more caudal and it is located more ventrally. It is equipped with a left atrioventricular valve defined as bicuspid (*valva bicuspidalis, mitralis*) because it is provided with only two cusps. The largest and most robust of these is the septal cusp (*cuspis septalis*), which faces the parietal cusp (*cuspis parietalis*). During the ventricular systole the cusps rise outside occluding the ostium while their *chordae tendineae*, which lead to the papillary muscles, are stretched.

The ostium of the aorta has a diameter equal to that of pulmonary trunk is located between the ostium of the pulmonary trunk and the right atrioventricular ostium. It gives insertion to the aortic valve, having three semilunar valves called the right, left and septal (*valvula semilunaris dextra, sinistra* and *septalis*). The right atrioventricular ostium has an oval shape and is positioned obliquely in a cranioventral direction. It gives insertion to the right atrioventricular valve, said to be tricuspid (*valva tricuspidalis*) due to the presence of an angular, a parietal and a septal cusp (*cuspis angularis, parietalis* and *septalis*). The *chordae tendineae* that are inserted in the margin and in the ventricular walls of the cusps attach to the underlying papillary muscles.

The ostium of the pulmonary trunk, placed against the auricular face of the heart, is located more dorsally and left of the aortic ostium. It gives insertion to the corresponding valve formed by three semilunar valves: the right, the left and the intermediate (*valvula semilunaris dextra, sinistra* and *intermedia*).

Topography
The heart is located in the mediastinum medium surrounded by the pericardium between the two lungs, which partially cover it.

Colour
The heart has a red colour similar to that of the striated muscle, although vessels running on the grooves placed on its surface are covered by adipose tissue whose colour varies from species to species.

Consistency
The consistency of the muscle is remarkable, although strongly influenced by its thickness, which varies between the right and the left portion of the organ.

Pig The pericardium (Figure 6.15) adheres to the sternum and it is reflected dorsally, covering the heart completely with the cranial vena cava and partly with the caudal vena cava. The pericardium presents also a transverse sinus

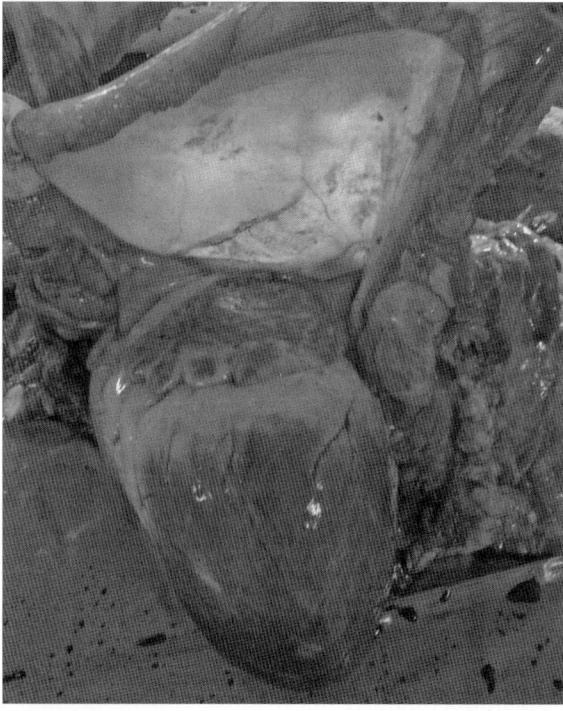

Figure 6.15 The pericardium removed from the pig heart.

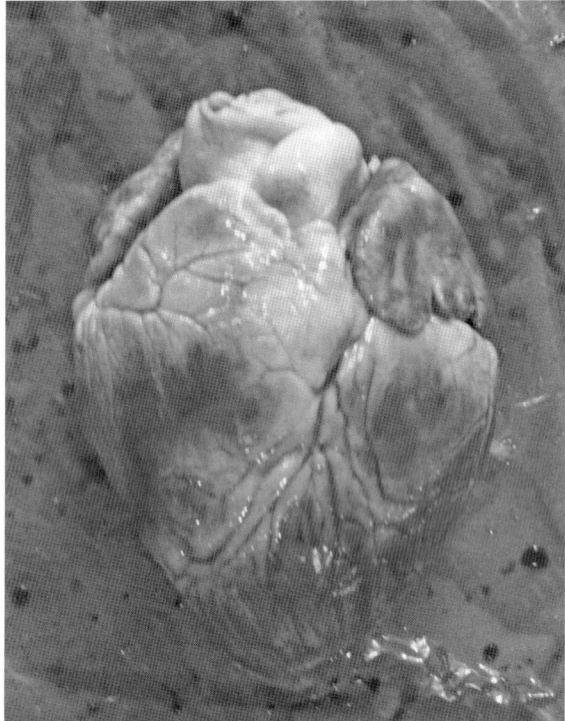

Figure 6.16   The auricular face of pig heart.

(*sinus transversus pericardii*) between the aorta and the left atrium that widens in the vicinity of the right atrium.

Weight and shape
The heart of an adult pig has an average weight of 300 g and in its major axis, which is longer, measures about 10 cm; its base has a diameter of about 8 cm. It has a ventricular capacity of 100 ml of blood. The spherical shape assumes a conical profile with rounded apex.

External description (Figure 6.16)
A white coloured fat covers the vessels running in the coronary sulcus and interventricular grooves. The ventral auricular left margin is lobed with the presence of two or three deep incisions. In pig there is a left azygos vein (*vena azygos sinistra*) that is near the caudal edge of the left atrium and reaches the great cardiac vein (*vena cardiaca magna*), beyond which it continues into the coronary sinus. The ascending aorta is continuous over the arch of the aorta from which the brachiocephalic trunk (*truncus brachiocephalicus*) and the left subclavian artery (*arteria subclavia*) come off separately.

Internal description

The oval fossa in the right atrium is very large. In the right ventricle papillary muscles are arranged normally while the septomarginal trabecula (*trabecula septo marginal*) is often very thin. In the left ventricle the two papillary muscles are inserted in the parietal wall; the accessory papillary muscles are present.

Bovine

Pericardium (Figure 6.17)

The pericardium comes with its apex on the sternum wall while dorsally the serous pericardium covers the pulmonary trunk and the ascending aorta. Its reflection reaches the emergence of the brachiocephalic trunk on the aorta.

Weight and shape

The average weight is 2.3 kg without gender differences. It measures about 25 cm in its major axis and at base level has an average diameter of about 15 cm. Each ventricle contains approximately 600 ml of blood. It assumes a conical shape, which is elongated and strongly flattened lateral sides. The apex is acute and caudally bent.

External description

Present in the interventricular and coronary grooves is adipose tissue, which is yellowish-white, hard and brittle. The ascending aorta has a calibre of about 5 cm at its origin and continues with the arch of the aorta, from which branches off the relatively long brachiocephalic trunk. The right atrial auricula is barely visible, while the left has deep incisions on its ventral margin.

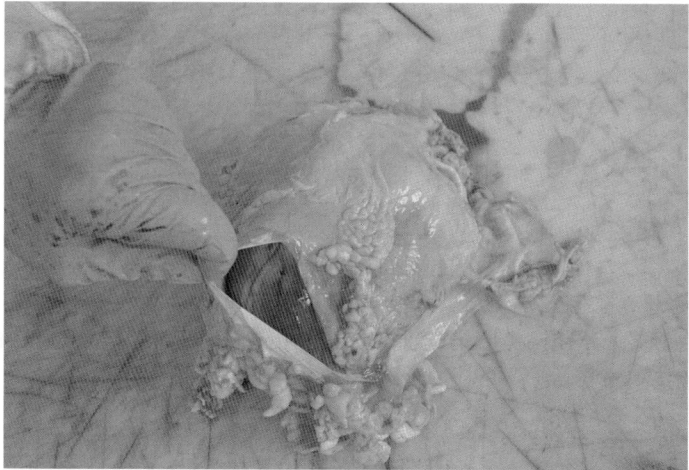

Figure 6.17 The pericardium incised above the bovine cardiac mass.

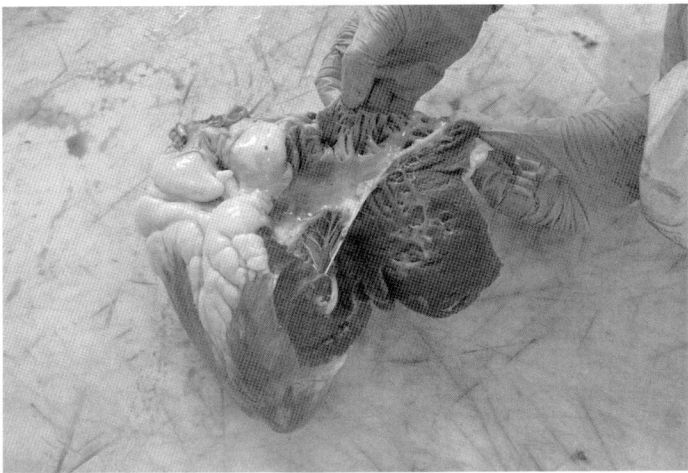

Figure 6.18   Internal view of the right ventricular cavity of bovine heart.

Internal description (Figure 6.18)

Present in the right ventricle are the great papillary muscle (*m. papillaris magnus*) and the small papillary muscle (*m. papillaris parvi*), as well as the subarteriosus papillary muscle (*m. papillaris subarteriosus*). The septomarginal trabecula is very thick. The part of the ventricle near the apex is particularly tortuous (Figure 6.19). The two left ventricular papillary muscles are very large. At the origin of the aorta are two heart bones, one right and one, that are the consequence of cartilage ossification in the heart. The right, which is more voluminous, is flat and about 6 cm long; it is formed as a triangle placed between the atrioventricular ostium and the aortic ostium.

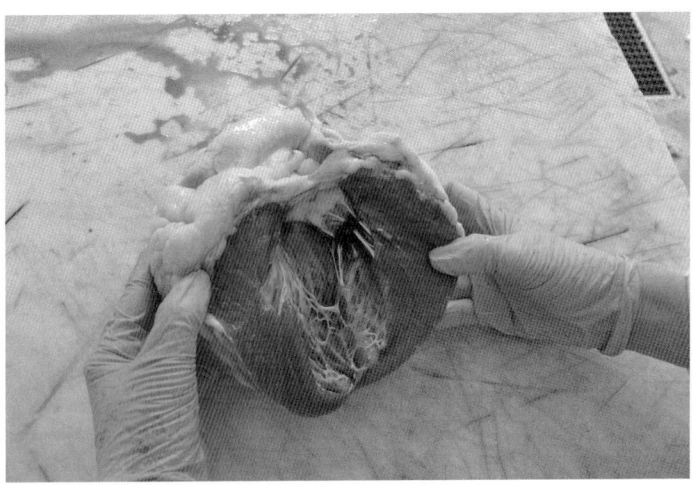

Figure 6.19   Internal view of the left ventricular cavity of bovine heart.

The left is much smaller, with an irregular triangular shape of about 3 cm long; this insinuates itself between the aortic ostium and the left atrioventricular ostium extending to the right bone although it does not reach it.

Small ruminants

Pericardium
The pericardium does not adhere profusely to the sternum to which it is joined by two sternum pericardial ligaments.

Size and shape
The heart is more elongated and cylindroid than that of the bovine with a weight of about 250 g and differs between genders. The major axis is about 10 cm long and has a capacity of 50 ml of ventricular blood.

Internal description
In the right ventricle is always present a thick septomarginal trabecula. The cartilage of the heart ossifies belatedly and incompletely.

Lymph nodes related to thoracic organs

Overview
The lymph nodes annexed to the thoracic organs (Figure 6.20) are classified into two categories:

1. those related to the bronchial lymphocentre;
2. those related to the mediastinal lymphocentre.

The bronchial lymphocentre consists of the tracheobronchial lymph nodes, respectively left, right and middle (*lynphonodi tracheobronchales sinistri, dextri* and *medii*). The mediastinal lymphocentre includes the cranial, medium and caudal mediastinal lymph nodes (*lynphonodi mediastinales craniales, medii* and *caudales*).

Pig The tracheobronchial and mediastinal lymph nodes (Figure 6.21) can be appreciated visually by separating the trachea and the right and left main bronchi (*bronchus principalis dexter* and *sinister*) from the lungs.

The left tracheobronchial lymph node is between the aortic arch and the left main bronchus. The cranial tracheobronchial or epi-arterial lymph node is located in the vicinity of the division of the accessory bronchus (*bronchus trachealis*) from the trachea, in the corner formed by the trachea and the right cranial lobe. The middle tracheobronchial lymph node is close to the bifurcation of the trachea into the two main bronchi, while the right tracheobronchial lymph node is located between the cranial lobe and the medium lobe of the right lung. The cranial mediastinal lymph nodes comprise numerous nodules that occupy the anterior mediastinum both right and left of the entrance of the thorax and are often hidden by the blood clot that occurs after the throat is cut.

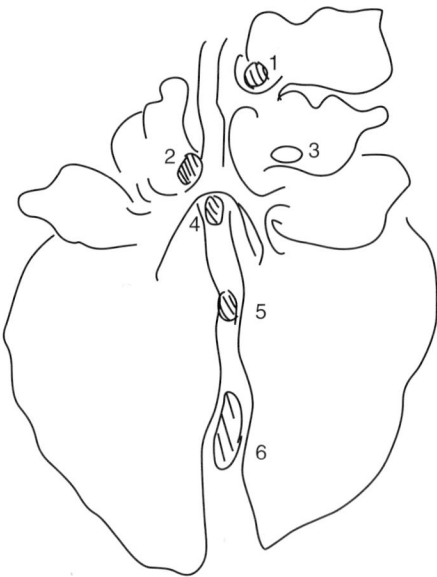

1 – cranial tracheobronchial lymph node
2 – left tracheobronchial lymph node
3 – right tracheobronchial lymph node
4 – medium tracheobronchial lymph node
5 – mediastinal lymph node
6 – mediastinal lymph node

Figure 6.20   Scheme of the thoracic lymph nodes.

Bovine   The lymph nodes are divided in two groups:

1. bronchial
2. mediastinal.

The bronchial lymph nodes are the:

- Right bronchial or inspector lymph node, a deep lymph node that it is not taken up with the cleaning of the organ. However, it may fail in 30% of subjects. It is set in a deep incisure between the right cranial lobe and the cranial medium lobe.
- Left bronchial lymph node, which is the most massive in its group. It is located deep in the corner formed by the base of the apical lobe and the trachea at the level of the bifurcation of the trachea into the two main bronchi; it is possible to visualize it by moving forcefully the aorta and oesophagus.
- Tracheobronchial or cranial right bronchial or epi-arterial lymph node, which can reach a length of 3 cm, is located in the angle between the trachea and the cranial lobe of the lung at the point where the tracheal bronchus sprouts from the lobe itself.

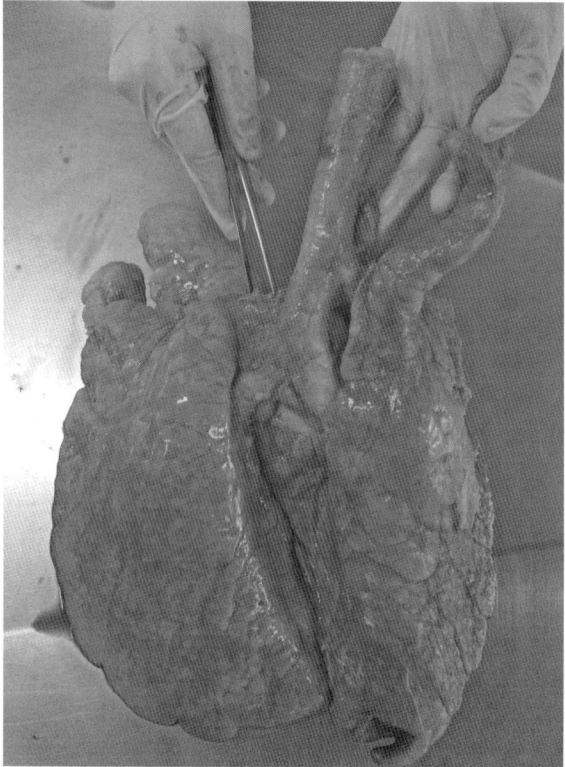

Figure 6.21 Pig tracheobronchial lymph nodes evidenced under the pulmonary tissue.

- Dorsal tracheobronchial lymph nodes, which are small formations in correspondence of the dorsal surface of the trachea at the level of its bifurcation.

Among the mediastinal lymph nodes (Figure 6.22) can be recognized the:

- Cranial mediastinal lymph nodes, which are between the mediastinal sheets in the cranial mediastinum and extend to the first two ribs. They varying in number up to ten placed around the great vessels, the oesophagus and the trachea.
- Medium mediastinal lymph nodes, which are located dorsal to the heart and right to the aortic arch ,and dorsally and laterally to theo esophagus and trachea.
- The caudal mediastinal lymph nodes, which are very large and present on the dorsal margin of the oesophagus and, in the caudal mediastinum, ventrally to the aorta extending to the diaphragm. Among these one is particularly large, up to about 15 cm, placed in proximity of the diaphragm.

Small ruminants The distribution and topography of the lymph nodes are as described for cattle.

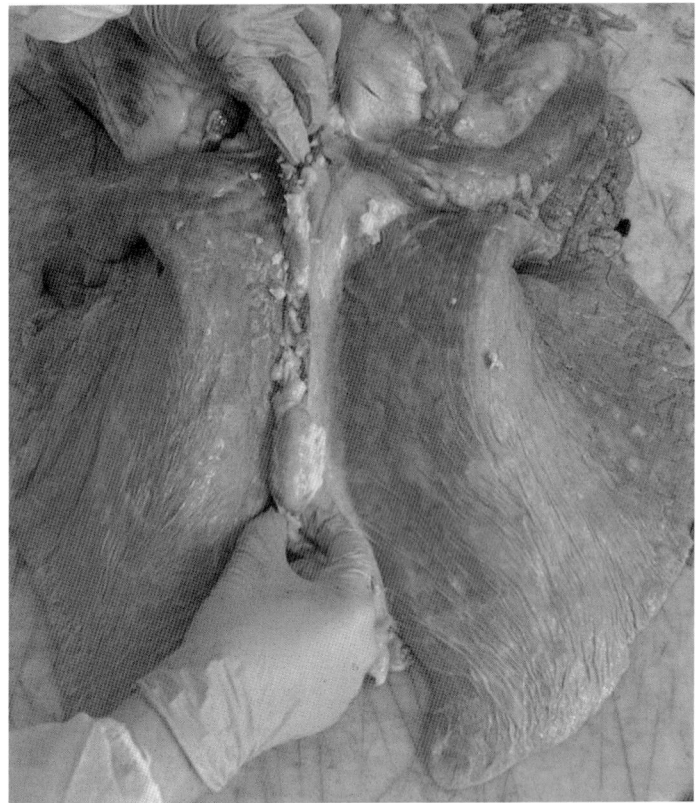

Figure 6.22 Bovine mediastinal lymph nodes located near the bronchus of the right cranial lobe.

6.4.3 Viscera of the abdominal cavity

Liver

Overview

The liver is the largest gland of the animal body and secretes bile into the duodenum. It has a diaphragmatic face and a visceral face that are separated by a dorsal and a ventral margin. It presents fissures of varying depth, depending on the species, that divide the organ into lobes. The diaphragmatic face is smooth and is covered by the peritoneum with the exception of an area positioned at the centre, the bare area. The visceral face is concave and in its middle it is possible to observe the hilum of the liver made from the peduncle of the hepatic portal vein, the hepatic artery, the hepatic nerves and the bile duct. In the margins of the hilum of the liver is the lesser omentum (*omentum minus*). The dorsal margin is thick and directed obliquely and ventrally on the left close to the diaphragm. The ventral margin, thinner and free, is oriented to the right. This margin is convex and looks indented due to lobar incisures, the depth of which vary depending on the species. The liver is attached to the diaphragm and to the lumbar region of to the other abdominal organs of the digestive system by a number of ligaments of the peritoneum. Amongst them there are:

- the coronary ligament (*lig. coronarium hepatis*), which connects the liver to the diaphragm and includes the caudal vena cava;
- the falciform ligament (*lig. falciforme hepatis*), which consists of a serous sheet that extends from the umbilicus up to orifice of the vena cava in the diaphragm where it is continuous with the coronary ligament;
- the round ligament (*lig. teres hepatisis*), a fibrous cord that represents the vestigie of the umbilical cord and extends from the liver to the umbilicus;
- the left triangular ligament (*lig. triangulare sinistrum*), an expansion of the coronary ligament that extends to the diaphragm;
- the right triangular ligament (*lig. triangulare destrum*) that goes from the dorsal margin of the liver to the diaphragm pillar;
- the hepatorenal ligament (*lig. hepatorenale*), which supports the caudate lobe of the liver reaching the cranial pole of the right kidney.

The biliary tract
The biliary tract is a system of ducts that start from biliferous ductules (*ductuli biliferi*) and end into the duodenum. They are divided into an intrahepatic tract, given by the intralobular bile ducts, bile ducts and liver ducts, and by a extrahepatic tract, comprising the common hepatic duct (*ductus epathicus communis*) and common bile duct (*ductus choledochus*).

The gall bladder
The gall bladder is a reservoir connected through the cystic duct to the two common hepatic ducts and the common bile duct.

Topography
The liver is profusely adherent to the abdominal side of the diaphragm.

Colour
The liver is a reddish brown colour and has a lobular constitution that can be also visible on the surface represented by regular polygons.

Consistency
The surface is granular, solid and inelastic with a parenchyma that can be compressed only a little but that is brittle and easily breakable.

Pig

Weight
The weight is approximately 1.8 kg (range 1.2–2.5 kg), which represents about 2% of the live weight of the pig. The major axis is 35 cm and the cross-axis 20 cm.

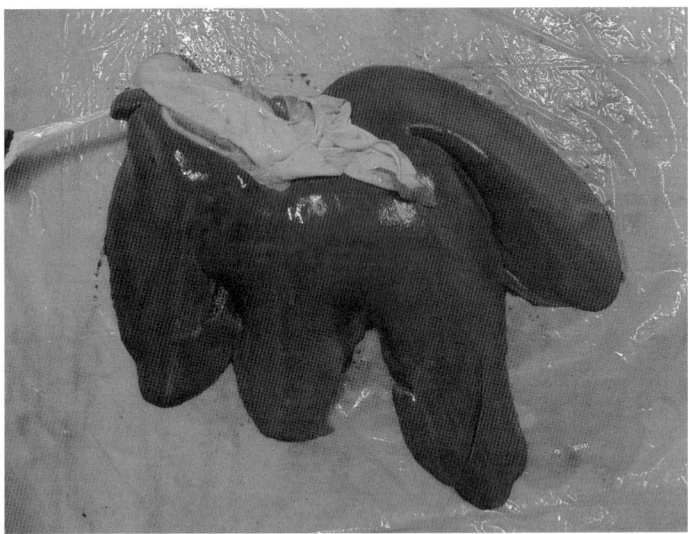

Figure 6.23 Diaphragmatic face of pig liver.

Colour

The colour is dark red; in some cases it is purple.

External shape of the liver (Figures 6.23 and 6.24)

The liver assumes a round shape. The remarkable development of the connective perilobular tissue allows the lobules below the surface capsule to be distinguished. This is so characteristic that allows species to be diagnosed even on small portions of the organ. The lobes are very pronounced and are marked by deep fissures. The incisures of the round ligament are less pronounced on the diaphragmatic face than on the visceral one and are less developed than those that separate the right and left lobe in the lateral and medial lobes. Altogether the liver has six lobes:

1. the right lateral (*lobus hepatis dexter lateralis*);
2. the right medial (*lobus hepatis dexter medialis*);
3. the quadrate (*lobus hepatis quadratus*);
4. the left medial (*lobus hepatis sinister medialis*);
5. the left lateral (*lobus hepatis sinister lateralis*);
6. caudate or Spigelian lobe (*lobus hepatis caudatus*).

On the diaphragmatic face only the lateral and medial left and right lobes are visible. On this same side the medial lobes largely cover the lateral lobes, while on the visceral face the latter cover the former. The right lateral lobe is ovoid with the major axis oblique with a ventrolateral direction. A small incision is sometimes observed on the end. The right medial lobe, equally voluminous, is joined to the quadrate lobe, which at this level is narrow and elongated and ends in a point without reaching the ventral margin. The separation

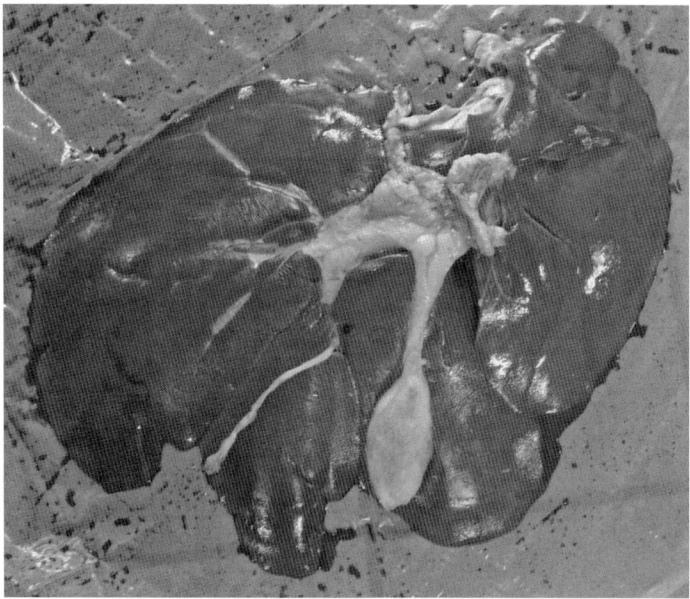

Figure 6.24   Visceral face of pig liver with gall bladder.

between the right medial lobe and the quadrate is marked only by the presence of the gall bladder. The left medial lobe is barely visible on the visceral face. The left lateral lobe is rounded and is the largest of all. The caudate lobe is small; its caudate process does not show the kidney imprint and its papillary process is shown as a tubercle that can just be distinguished located to the left of the hilum of the liver. The round ligament, thin and fragile, disappears almost completely in the adult. The falciform ligament is short and regresses until it is reduced to its hepato-diaphragmatic part. The coronary ligament is large and sturdy. The triangular ligaments are short. The diaphragmatic face is very convex. On the visceral face the gastric impression occupies the left lateral lobe.

The arteries that penetrate through the hilum of the liver are numerous and branch off from the hepatic artery. Three can generally be described: a right lateral branch that makes the artery of the caudate lobe, a right medial branch giving rise to the cystic artery and a left branch for the two left lobes. The left branch of the portal vein, bigger than the right branch, is distributed to the left lobes, to the quadrate lobe and the right medial lobe. There are three main hepatic veins: the right that drains the two right lobes, the left for the two left lobes and the intermediate for the quadrate lobe and the adjacent part of the left medial lobe. The caudate lobe is drained by numerous small accessory veins. The biliary tact resembles those of bovine.

Bile ducts
There is a common hepatic duct 2–3 cm in length. The cystic duct is about 5 cm in length and ends at an acute angle in the hepatic common duct. There are

two or three cystic–hepatic ducts. The common bile duct is 3–4 cm in length and opens into the duodenum near the pylorus in a very thick major duodeni papilla that belongs totally to it; it misses the main pancreatic duct.

Gall bladder
The gall bladder is piriformis about 4–5 cm in length. It does not reaching the ventral margin of the liver and adheres to a shallow incisure.

Bovine

Weight
The average weight is 5 kg although in some it can be as much as 10 kg. It is about 60 cm in length and 30 cm wide.

Colour
The colour of the bovine liver is reddish brown in adults and clearer in the calf.

External conformation
The liver of bovine has a rectangular shape with a vertical major axis in its left margin by the caudal vena cava. Below, to the middle of the right edge, is the shallow fissure of the round ligament. On the diaphragmatic face two lobes are observed: a small ventral lobe, which is the left lobe, and the right, quadrate lobe delimited only by the gallbladder on the visceral side. On this face the caudate lobe is massive with a dorsal caudate process and a papillary process, also big, positioned ventrally to delimit to the left the hilum of the liver that surrounds it. Since the organ is oriented vertically and moved to the right by the development of the forestomachs, the falciform ligament is reduced to a very thin mesentery. The round ligament is missing, while the coronary ligament is strong and long as the incisure of the vena cava, which delimits a long bare area on the right side. The right triangular ligament is robust and runs from the right lobe and the right part of the caudate process to the diaphragm. The hepatorenal ligament is short and wide. The left triangular ligament is thin and short.

Gall bladder
The gallbladder is voluminous about 15 cm in length and 8 cm in width. Its bottom protrudes from the right margin; it contains almost half a litre of bile.

Bile ducts
The two hepatic ducts join and form a single common hepatic duct only 3–4 cm in length. The cystic duct reaches the common hepatic duct and the common bile duct, about 5–8 cm in length, flows into the duodenum in the major duodenal papilla separately from the pancreatic duct.

Small ruminants

Weight
In small ruminants the liver weighs less than 1 kg.

Colour
The liver of small ruminants is dark red.

External shape
The liver assumes an elliptical shape very similar to that of cattle.

Gall bladder
The gall bladder lies more ventrally than in bovine.

Bile ducts
The cystic duct flows at an acute angle in the common hepatic duct and the common bile duct empties into the duodenum at a distance of 40 cm from the pylorus with the pancreatic duct in the major duodenal papilla.

Spleen

Overview
The spleen is an organ whose form varies in different species. However, it is always flattened and elongated. It is possible to recognize a parietal or diaphragmatic face in which there are the imprints of the ribs. The other face, the visceral, is divided by a major thickened surface that separates the cranial surface of the stomach from the intestinal one positioned caudally. This prominence is flanked by a depression: the hilum of the spleen, which blood vessels penetrate, and nerves on the margins of the gastrolienal ligament (*ligamentum gastrolienale*) that joins it to the stomach. It is covered by a serous, the visceral peritoneum, that adheres to the fibrous capsule that covers the entire organ with the exception of the hilum.

Topography
It is situated below the last ribs on the left side, bonded to the bottom and to the greater curvature of the stomach.

Colour
The colour varies from dark red to blue-grey and also depends on the amount of blood present in the organ.

Weight and volume
The spleen behaves as a blood reservoir and presents appreciable differences of colour, weight and volume in subjects that have died from bleeding compared to those not bled.

Consistency
Despite having some consistency on palpation the spleen is still very elastic.

Pig The spleen has a red-brown colour and is elongated and narrow (Figures 6.25 and 6.26). It is about 50 cm in length, about 5 cm in width and weighs about 200 g. The gastric and intestinal surfaces, on the visceral face,

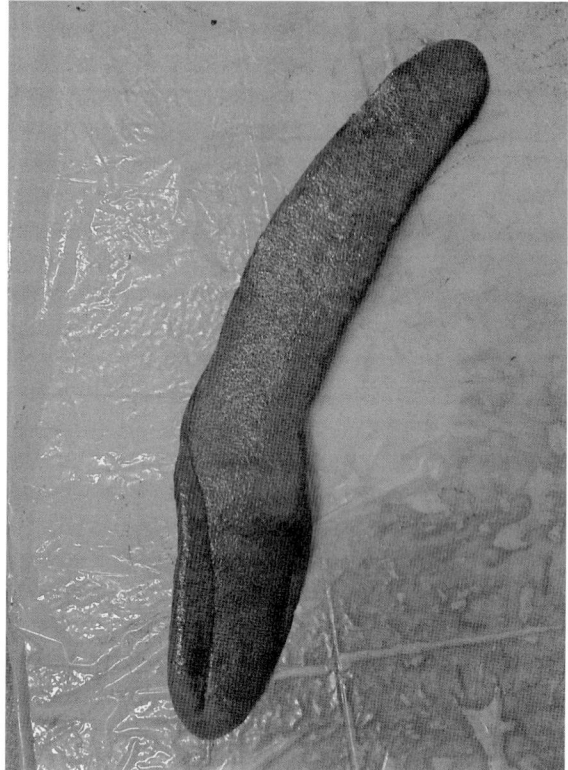

Figure 6.25 Parietal face of pig spleen.

extend along its entire length and are separated by the longitudinal ridge that is flanked from one end to the other by the hilum. Cranially to the hilum, the spleen hangs from the greater omentum and is connected to the stomach.

Bovine The spleen varies from brownish-red in colour in the calf to blue-grey-brown in the adult (Figure 6.27). In bled adults the spleen weighs about 900 g and is approximately 50 cm in length and 10 cm in width. It assumes an elongated shape with rounded ends. The parietal face is smooth as it is in contact with the diaphragm. Half of the visceral face is covered by the peritoneum, while there remaining dorsocranial visceral face is attached to the rumen. The hilum is very short and is located at the dorsal edge. The spleen has no contact with the greater omentum. It adheres to the rumen with the visceral face.

Small ruminants

Overview
In small ruminants the spleen assumes a discoid shape, is 15 cm in length and 10 cm in width.

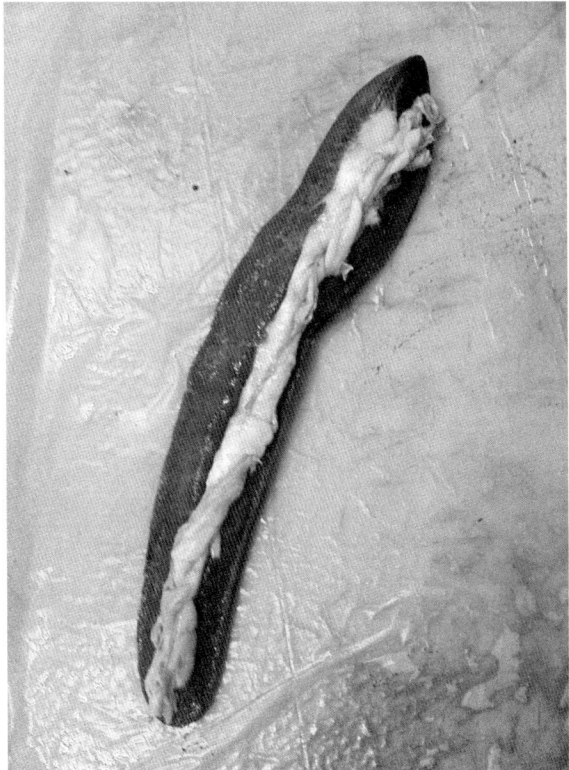

Figure 6.26 Visceral face of pig spleen.

Colour
The spleen of small ruminants is a brownish-red colour.

Weight
The average weight of a spleen is about 100 g.

6.4.4 Gastrointestinal tube, mesentery and annexed lymph nodes

Oesophagus

Topography
The oesophagus has a cervical portion and a thoracic portion that can be divided into four segments (entry thorax, cranial mediastinum, medium mediastinum and caudal mediastinum) and a very short last abdominal portion.

Description
The oesophagus is a muscular tube organ that is red in colour until a few centimetres before the cardia. The terminal part that enters the stomach is, on

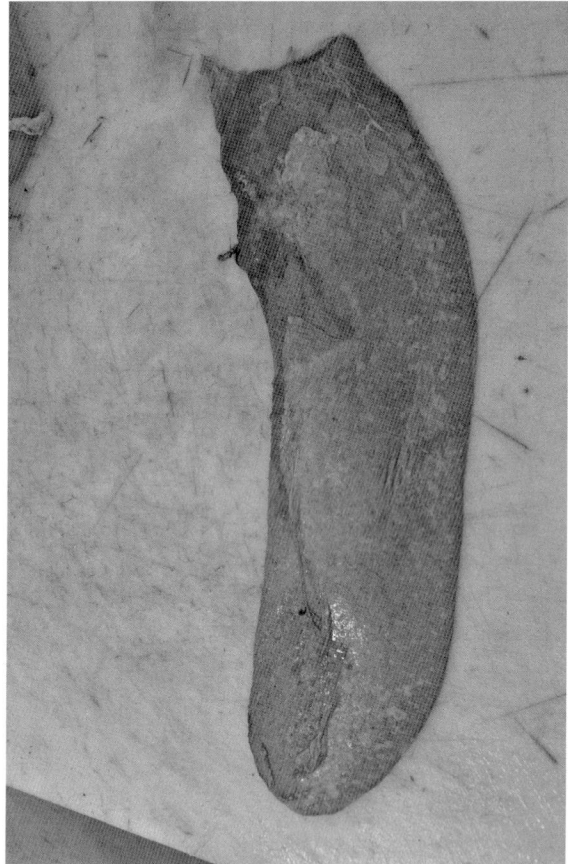

Figure 6.27   Visceral face of bovine spleen.

the contrary, white. The size is uniform and decreases significantly only at the entrance of the thorax.

Length
The organ is relatively short (<40 cm) due to the length of the pharynx and the brevity of the neck.

Consistency
It is easily compressible and soft.

Appearance of the serous
The oesophagus is surrounded by a connective tissue envelope, the adventitia, which unites it to the surrounding organs. This tunica is very loose in the neck and is thicker in the thorax, which is covered by the mediastinal pleura that forms an incomplete serous tunic. The shiny and smooth serous coating is complete in the abdominal area.

Bovine The size is not uniform, with a narrow source followed by a dilated part that corresponds to the cervical portion (about 10 cm). The thoracic part undergoes a further narrowing at the aorta level (about 8 cm). From the aorta to the diaphragm the organ resumes a wide and dilated diameter for its functional role in eructation and rumination. The oesophagus is approximately one metre long with a thoracic part that is a little longer than the cervical part.

Small ruminants The colour is bright red and the oesophagus continues directly from the pharynx without shrinking. It has a uniform size similar to that of the trachea.

Stomach
Pig

Description (Figure 6.28)
The shape of the organ is similar to an elongated sack slightly flattened antero-caudally and oriented transversely, so that the left end is the most voluminous and represents the fundus (*fundus ventriculi*) of the stomach, which in this species rises just above the cardia and has a large but short appendix curved caudally: the gastric diverticulum (*diverticulum ventriculi*). The right portion

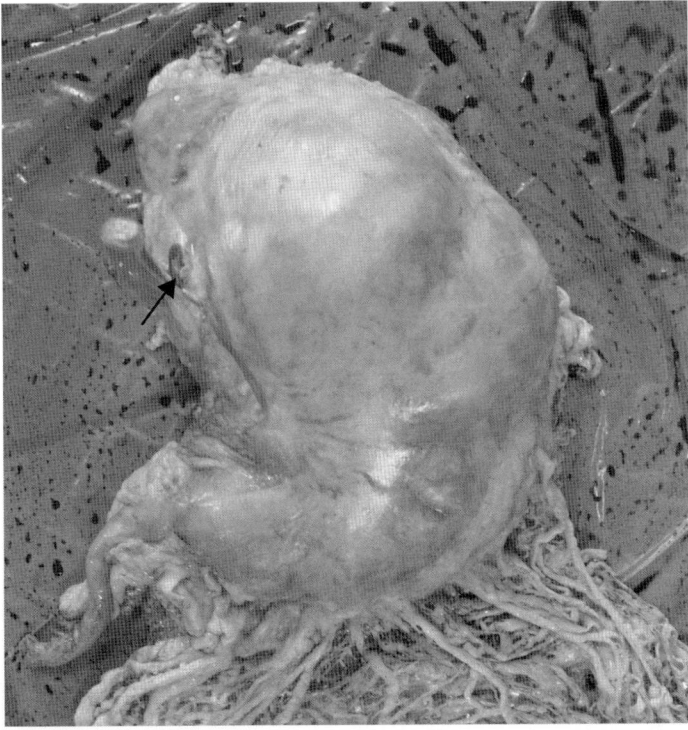

Figure 6.28 Parietal face of pig stomach. The gastric lymph nodes are located near to the termination of the oesophagus.

is the pyloric part (*pars pylorica*) and continues in the duodenum. The portion of the organ between the fundus and the pyloric part corresponds to the body (*corpus ventriculi*) of the stomach. It has two faces, a cranial parietal face (*facies parietalis*) and a caudal visceral face (*facies visceralis*), both convex and smooth coated by the visceral peritoneum. The lesser curvature (*curvatura ventriculi minor*), concave and turned right up, is short. It begins on left of the cardia and then continues up to the pylorus. The lesser curvature gives attachment to the hepatogastric ligament, part of the lesser omentum. The greater curvature (*curvatura ventriculi major*), much longer, starts at the top of the fundus and continues also to the pylorus (*p. pylorus*). It is oriented to the left and at the bottom and gives attack throughout its length to the greater omentum. The shape and size of the organ is very conditioned by the state of repletion and as a function of the consequent contraction of the tunica muscularis. In the pig iit has an average capacity of 4 litres and is quite distensible. It weighs about 700 g in adults.

Topography
An empty stomach is completely covered under the left hypochondrium and does not have any relationship with the ventral abdominal wall. With the organ replete, the greater curvature and the anterior face contract relations with the xiphoid region and the fundus comes to move caudally of two inter-costal spaces.

Texture
The fundus of the stomach is less consistent due to the reduced development of the tunica muscularis. The body has a well developed and thick musculature.

Appearance of the serous
The gastric serous consists of the visceral peritoneum and continue at the level of curvature with the mesi.

Forestomachs and stomach
Bovine

Overview
The forestomachs and stomach of adult bovine contain from 150 to 200 litres, of which 80% reside within the rumen, 5% in the reticulum, 7% in the omasum and the remaining 8% in the abomasum. The total weight of the empty forestomachs is about 8 kg, of which almost the totality is dependent on the rumen.

The rumen is elongated craniocaudally and flattened obliquely from side to side. It has a convex parietal face in relation to the wall of the hypochondrium and the side left. It has a left long longitudinal groove (*sulcus longitudinalis sinister*). The visceral face is directed dorsally and to the right and supports the omasum, abomasum and intestines; it is crossed by the right longitudinal groove (*sulcus longitudinalis dexter*), from which departs the right accessory groove (*sulcus accessorius dexter*), defining a particular area called island the rumen (*insula ruminis*). The right and left longitudinal grooves, joined together by transverse grooves at each end of the rumen, mark outside the division of the organ into two communicating compartments: the dorsal sac and the ventral sac (*saccus dorsalis* and *saccus ventralis*). The caudal end of the rumen is divided by a caudal groove (*sulcus caudalis*), from which depart two coronary grooves (*sulcus coronarius dorsalis* and *ventralis*) that combine to delimit a caudodorsal and a caudoventral blind sac (*saccus ciecus caudodorsalis* and *saccus ciecus caudoventralis*). The cranial end of the rumen presents a groove that separates the ends of the two sacs. The ventral sac ends here in the rumen recess (*recessus ruminis*), while the dorsal sac of the rumen proceeds craniodorsally in the large compartment: the rumen (*atrium ruminis*) which in turn continues with the reticulum. The boundary between the atrium and the reticulum is marked by ruminoreticular groove (*sulcus ruminoreticularis*).

The reticulum is the most cranial compartment and in bovine is the smallest of the gastric compartments. It is flattened craniocaudally and has a diaphragmatic face that is convex in shape to the diaphragmatic muscle. The visceral face, which is almost flat, is related to the rumen. The left end is dorsally in continuum with the atrium of the rumen whose boundary is the termination of the oesophagus. The right end forms a blind sac, the fundus of the reticulum (*fundus reticuli*), leaning dorsally at the base of the abomasum, with which it unites to join the omasum.

The omasum is the last compartment of the proventriculus of ruminants; it follows the reticulum and precedes the abomasum. It is almost spherical depressed on its lateral side. The parietal face is turned to the right and a little cranially. The visceral face is set against the rumen. The junction with the reticulum is by means of a peduncle called the neck of the omasum (*collum omasi*). The union with the abomasum, a little less restricted, is marked by an exterior groove between the omasum and the abomasum (*sulcus omasoabomasicus*). The abomasum is the last gastric compartment of ruminants; it is provided with a glandular mucosa as are monogastric mammals. The organ is much larger than the reticulum and the omasum and looks pear-shaped and folded back on itself. It presents two ends, of which one is the fundus of the stomach, the other is the aperture in the pylorus. The body of the abomasum (*corpus abomasi*) is the portion between the two ends.

The stomach in calves before weaning is very different from that of adults. In this evolutionary stage the abomasum is by far the more voluminous of the other forestomachs including the rumen.

Topography of the rumen
Only the dorsal sac is in contact with the neighbouring organs. The ventral sac received in the omental bursa establishes its relationship with the greater omentum that wraps it completely. The parietal face is related dorsocranially with the spleen and the left lateral and ventral walls of the abdomen. The visceral face is covered cranially from the omasum and the abomasum and

dorsally from the left kidney. The remainder is in rapport with the visceral mass of jejunum and colon. The caudal end is connected to the fluctuating portion of the jejunum and the pelvic organs.

Topography of the reticulum

The reticulum is located dorsally to the xiphoid process of the sternum between the diaphragm and the rumen and continues on the right with the omasum.

Topography of the omasum

The omasum is located dorsally, at the fundus of the reticulum, to the abomasum that is beside its base. It is located in the right diaphragmatic region from the seventh to the eleventh coast ribs.

Topography of the abomasum

The abomasum is largely located on the right in the proximity of the xiphoid region ventrally to the hypochondrium. Its long axis is oblique in the caudal direction from left to right. The parietal wall is connected through the peritoneum to the transversus and rectus abdominis muscles and the adjacent parts of the diaphragm. The visceral face is connected mainly to the ventral sac of the rumen and through the deep wall of the omental bursa with the convolutions of jejunum. It is also located cranially in relation to the reticulum, dorsally with the omasum and ventrally with the abdominal epigastric wall between the right and left costal arches.

Internal structure of the rumen

The dorsal and ventral sacs continue through the ostium intraruminale whose edge is formed by thickened reliefs or muscular pillars (*pila*) that match internally with the external grooves described previously. From the caudal pillar two branches (dorsal and ventral) come out; these are the coronary pillars (*pila coronaria*) that surround the entry of the two blind sacs. Dorsally and cranially to the cranial pillar, corresponding to the cranial groove, extends the atrium of the rumen the cavity of which continues with the dorsal sac. Its cranial limit is formed ventrally by a transverse crease given by the union of its wall to that of the reticulum: this is ruminoreticular fold (*plica ruminoreticularis*). In the vault it is possible to observe the opening of the cardiac orifice, which continues to the right in the reticular groove (*sulcus reticuli*). At the level of the pillars the mucosa looks rough but devoid of papillae. The papillae are abundant and protruding in the ventral sac and in the two caudal blind ends. They arere less numerous and developed in height in the dorsal sac.

Internal structure of the reticulum

The cavity ventrally forms, at the level of the fundus, a depression limited caudally by the ruminoreticular fold. The wall is covered by the same type of mucosa that is present in the rumen, where it is found in crests delimiting polygonal alveoli or cells of the reticulum. The interior of the cells is divided into smaller cells by secondary crests, from the sides of which even emanate tertiary crests. All the crests on the free margin are like bristles due to the presence of conical papillae with a hardened apex. These formations are replaced on the right side and on the dorsal side of the cavity by a broad and deep gastric groove that continues to the omasum; this is the reticular groove. It starts on the vault of the atrium of the rumen from the cardia and goes down and to the right, leading into the reticulo-omasal orifice within which terminates. It is surrounded by two lips of thickened mucosa bordering the fundus of the groove. It presents a slightly spiral shape. In the vicinity of the reticulo-omasal orifice are large conical papillae: the unguiculiform papillae (*papillae unguiculiformes*). In calves before weaning the lips of the reticular groove are particularly developed and can approach constituting a complete channel creating a direct connection with the reticulo-omasal orifice. This locking mechanism of the lips is accomplished by contact reflex of the milk with the mucosa of the pharynx. The reflex is extinguished at the time of weaning.

Internal structure of the omasum
The internal cavity is almost entirely occupied by flat parallel folds of mucosus membrane (*laminae omasi*) located against each other. These folds are inserted at the level of the greater curvature and on the face of the body and extend from the vicinity of the reticulo-omasal orifice up near to the entrance of the abomasum, where they are reduced in size and eventually disappear. The height of the folds varies a great deal and they are ordered according to their size in primary, secondary and tertiary folds. The cavity of the omasum is then reduced to a omasal canal represented by the space left free by the laminae which extends from reticulo-omasal orifice to the omaso-abomasal orifice (*ostium omaso abomasicum*).

Internal structure of the abomasum
The cavity is covered with a thin glandular mucosa, very vascularized. The fundus and most of the body is grey to reddish. It forms large spiral mucosal folds (*plicae spiralis abomasi*) that increase its inner surface. The pyloric portion has thin folds and its mucosa appears a light yellow-pink colour.

Small ruminants

Overview
The stomach of small ruminants has a smaller volume and differs in some details related to proportion and structure.

Rumen
In the sheep the rumen contains about 15 litres and the shape is more elongated than the bovine. The ventral sac, as well as its blind fundus, is more developed than the dorsal portion.

Reticulum
The reticulum is more voluminous than the omasum and unlike the bovine and has a total volume of about two litres.

Omasum
The omasum has a small volume (0.5 litres). It is oval and is situated deeper within the right hypochondrium compared to bovine.

Abomasum
The abomasum is a more voluminous than the bovine (3 litres).

Internal structure
The major differences compared to the forestomachs of bovine are:

- the ruminal papillae are less developed and less abundant;
- the cells of the reticulum are shallow, separated by low ridges and the division of secondary cells is less pronounced.

Small intestine   The small intestine is followed by the stomach(s) and extends from the pylorus to the ileal orifice (*ostium ileale*). It is a long tube of uniform size divided into three segments:

1. duodenum
2. jejunum (the longest part)
3. ileum.

Pig   The total length of the entire intestinal tract is approximately 35 metres, of which 20 m belong to the small intestine and 15 m to the large intestine. The duodenum is approximately 90 cm in length with a diameter of about 5 cm. It is supported in the abdominal cavity by the mesoduodenum. Inside there is the greater duodenal papilla, into which opens the termination of the common bile duct about 5 cm from the pylorus. There is also the lesser duodenal papilla situated about 20 cm in caudal direction, into which opens the accessory pancreatic duct (*ductus pancreaticus accessorius*). The jejunum–ileum is supported by a mesentery infiltrated with fat. Inside there are about 30 elongated and laminae-shaped lymph nodes, with an uneven surface excavated by small dimples.

Bovine   The length of the intestine in average is 50 m, of which two-thirds is the small intestine. The weight of the empty intestine is about 10 kg, of which 5 kg is the small intestine. The duodenum is about one metre long. Inside there is the greater duodenal papilla about 60 cm from the pylorus, into which opens the termination of the common bile duct. After 30 cm caudally there is the lesser duodenal papilla, into which opens the accessory pancreatic duct. The jejunum–ileum is around 40 m in length, even if the ileum is little more than one metre. It has a diameter of 3–4 cm. The whole intestinal mass is hung from a single and robust mesenteric fold. The inner surface presents an aggregate of lymph nodules that form plaques of lymphatic tissue slightly in relief with tiny dimples. The number of these plaques varies from 20 to 50.

Small ruminants The small intestine is proportionally longer than that of the bovine, measuring an average of 25 m and with a diameter of about 2 cm, which increases in the caudal part. Inside the duodenum is the only greater duodenal papilla, in which end both the common bile duct and the only pancreatic duct.

Large intestine The large intestine is the part of the digestive tract which follows the small intestine and ends in the anus. It can be divided into three segments:

1. caecum
2. colon
3. rectum, to which must be added the canalis analis.

Pig The caecum is about 40 cm long with a diameter of about 10 cm. It can contain approximately 2 litres. It is cylindrical and provided with sacculations interrupted by three longitudinal teniae, two dorsal (*teniae dorsalis*) and one ventral (*teniae ventralis*). The inner surface presents, in addition to the plicae semilunare and to the cavities that they delimit, a thickened cecocolic valve (*valva cecocolica*) in relief. On the top of the valve is the ileal orifice. The mucosa does not presents aggregated lymphatic nodules with the exception of a plaque which prolongs that of the ileum.

The colon, as in all species, is divided into:

- ascending
- transverse
- descending.

The ascending colon forms a spiral that takes up almost half of the abdomen; its length is 3–5 meters. Its diameter, initially equal to that of the caecum, reduces to 3 cm at its terminal part. It has a volume of about 10 litres. The transverse colon is short and poorly defined.

The rectum is expanded in its caudal half, where it forms the rectal ampulla (*ampulla recti*). The organ is wrapped in adipose tissue and lacks serous. The canalis analis is only 25–30 cm and the anus does not protrude on the outside.

Bovine The caecum is a cylindroid sac about 70 cm in length and 10 cm in width, smooth without sacculations or teniae. Inside, the ileal papilla protrudes about 1–2 cm. On the colon side of the ileal papilla there is a large plaque of aggregated lymph nodes.

The colon is long, on average about 10 m, with a diameter initially equal to that of the caecum, decreasing to 5 cm in the ascending colon, from which it does not show changes. The surface is smooth and devoid of sacculations or teniae. The descending colon has a length of about 2 m. It presents a uniform diameter for most of its course. Shortly before reaching the sacrum it increases in diameter and describes an inflection corresponding to the sigmoid colon.

The rectum is short and not very voluminous. The ampulla, poorly developed, is located in its caudal third.

Small ruminants The large intestine is longer than that of the bovine. It measures about 6 m. Inside there are about 30 plaques of aggregated lymph nodes.

Mesentery The mesentery is the largest of all the mesi, from which hang the jejunum and ileum. The mesentery fits over the entire length of the viscera from the duodenojejunal flexure up to the caecum. The mesentery sustains the jejunal branches of the cranial mesenteric artery and veins, the lymphatic vessels and the nerves that accompany them. The very short extremity that takes insertion on the lumbar region is called mesenteric root The opposite edge of the intestinal mesentery is much longer than the root and undulates to follow the length of the organ.

Lymph nodes of the abdominal organs

General description
The lymph nodes annexed to the abdominal organs are organized into four lymph centres:

1. lumbar
2. celiac
3. cranial mesenteric
4. caudal mesenteric.

The lumbar lymph centre is located in lumbar region were the aorta runs; the lymph nodes are located along the course of the aorta and the caudal vena cava but not found on the pluck and are described with the lymph nodes annexed to the carcass. The celiac lymph centre includes the celiac, hepatic, lienalis, gastric and pancreatic lymph nodes because it gathers the lymph from all the viscera vascularized by the celiac artery. The cranial mesenteric lymph centre receives the lymph from all the viscera vascularized by the cranial mesenteric artery. It is formed by the cranial mesenteric, jejunal, ileal, caecal and colic lymph nodes. The caudal mesenteric lymph centre is given by the lymph nodes distributed over the course of the branches of the caudal mesenteric artery.

Pig The gastric lymph nodes, of which there are three or four, are located between the diverticulum of the stomach and the ostium of the oesophagus (Figure 6.28). The hepatic lymph nodes, of which there are two or three, are located along the portal vein (also called periportal) and are found in the lesser curvature of the stomach in the vicinity of the pancreas if, as sometimes happens, they are not removed along with the liver. They are characterized by a reddish blue colour. They drain the deep and superficial lymphatic network of the visceral face and of the ventral and lateral parts of the diaphragmatic face of the liver. The remaining lymph vessels of the liver go to its caudal mesenteric lymph nodes, which drain at the same time the small intestine. In the pig, the lienalis lymph nodes are found along the course of its corresponding artery near the hilum of the spleen. They receive the efferent branches of the deep and superficial lymphatic network of the spleen. They are not present in ruminants. The mesenteric lymph nodes are found among the layers of the mesentery and are divided into cranial and caudal mesenteric lymph nodes. This is based on the location that the lymph nodes have respectively to the emergence of the cranial and caudal mesenteric arteries. The cranial mesenteric lymph nodes are divided into:

- the duodenal lymph nodes, which are in turn divided into two groups: the first is placed in the vicinity of the pylorus and the second where the duodenum ends into the jejunum;
- the jejunal lymph nodes, which are located around jejunal vessels near the root of the mesentery (Figure 6.29);
- the colic lymph nodes, which are found in the mesentery in the axis of Ansa spiralis coli;
- ileo-caecal lymph nodes, which are found in the mesentery near the end of the ileum into the caecum.

Amongst the caudal mesenteric lymph nodes, the anorectal lymph nodes are recognizable. The caudal mesenteric lymph nodes consist of the lymph nodes of the terminal portion of the colon and the lymph nodes of the rectal tract.

Bovine In bovine the lymph nodes of the forestomachs are present in charge of the different regions. In the rumen the right, left and cranial ruminal lymph

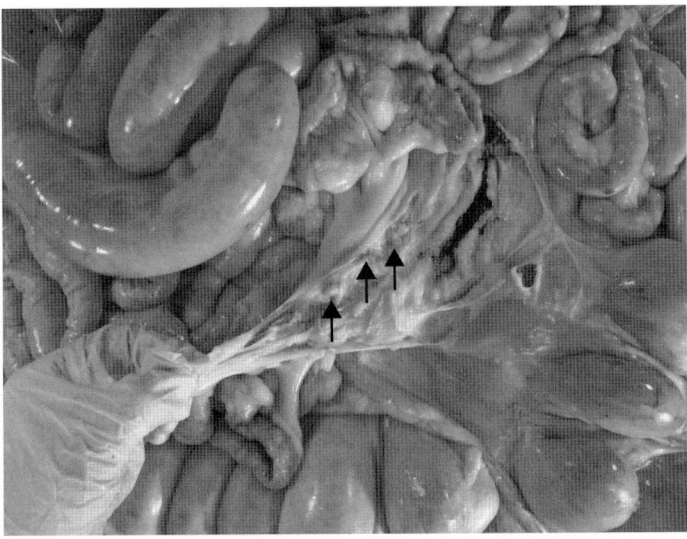

Figure 6.29 Jejunal lymph nodes along the mesentery in pig.

node chains are present in the longitudinal and in the cranial grooves. The ruminal lymph nodes on the right are more numerous than those on the left, which are sometimes not present. The ruminal lymph nodes distributed cranially placed near the cardia are sometimes called lymph nodes of the atrium.

The lymph nodes of the reticulum are located at the boundary between the reticulum and the rumen. The omasic lymph nodes, which number about ten, are located on the greater curvature of the omasum, while the reticulum abomasic and dorsal abomasic are located in the corner between the reticulum, omasum and abomasum.

There are approximately 15 hepatic lymph nodes (Figure 6.30) located in the vicinity of the portal vein. The accessory hepatic lymph nodes are distinguishable as a group located along the dorsal margin above the caudal vena cava and another group located in the adipose tissue of the hepatoduodenal ligament.

The lineal lymph nodes are missing in the bovine while the mesenteric lymph nodes (Figure 6.31) are located between the sheets of the mesentery and are divided into cranial and caudal mesenteric lymph nodes. This is based on the location that the lymph nodes have, respectively, to the emergence of the cranial and caudal mesenteric arteries. The cranial mesenteric lymph nodes are divided into:

- duodenal lymph nodes, divided between proximal duodenal lymph nodes located in correspondence to the dorsal margin of the duodenum and pancreatic duodenal lymph nodes located towards the caudal end of the duodenum;
- jejunal lymph nodes, consisting of several tens of lymph nodes (about 50) of variable size (from one to a few centimetres) placed between the sheets of mesojejunum near to the intestinal loops;

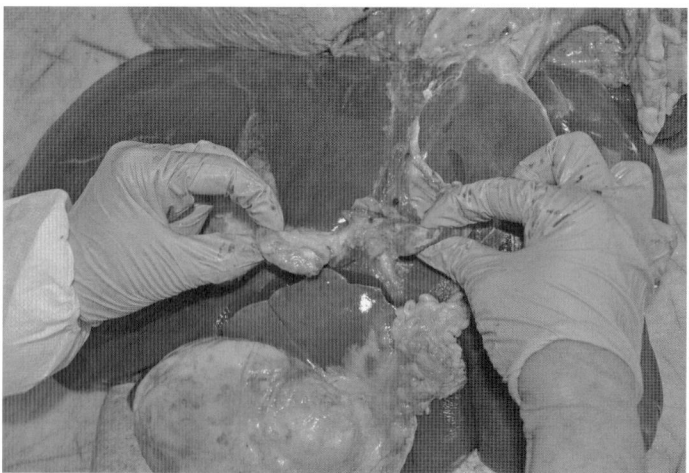

Figure 6.30 Bovine hepatic lymph nodes on the visceral surface of the liver.

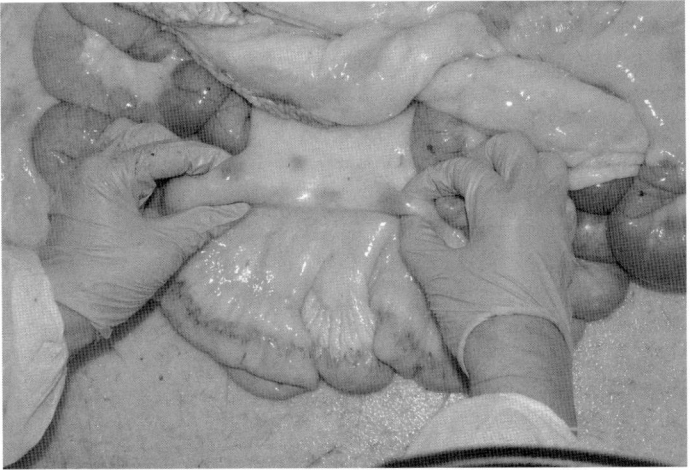

Figure 6.31 Bovine mesenteric lymph nodes in the mesentery.

- ileal lymph nodes, are located in continuation of the to the jejunal lymph nodes and they are also included in the mesentery between the ileum and the spiralis coli;
- caecal lymph nodes, located in the mesentery between the caecum and the ileum;
- colic lymph nodes: one group is located on the right face of the small colon, a second group is located to the right between the transit point between the ileum and the caecum, and a third group is located at the terminal portion of the colon;

Amongst the caudal mesenteric lymph nodes it is possible to recognize the anorectal lymph nodes, located on the lateral and dorsal surface of the descending colon and the rectum.

Small Ruminants The organization of lymph nodes is similar to that of bovine.

6.4.5 Viscera of the pelvic cavity

Located in the abdominal-pelvic cavity are organs of inspective interest such as the uterus, ovaries and bladder.

Uterus

Overview
The uterus is the part of the female genital apparatus located between the uterine tubes cranially and the vagina caudally. It always consists of two horns, a body and a cervix; organization differs from species and species.

Colour
The uterus of the non-pregnant adult is a yellow rose colour. The colour as well as the consistency may change greatly depending on the oestrous cycle. The examination of anatomical features of the uterus, together with the functional state of the breast apparatus, ovaries and age of the subject, may serve to classify the category to which they belong and to determine state reproductive state of the animal (animal prepuberal/puberal, stage of the oestrous cycle, stage of gestation, primiparous versus multiparous).

Texture
The consistency is firm and elastic.

Pig The uterus of the sow is a bicornuate and bipartite uterus entirely located in the abdominal cavity (Figure 6.32). The horns are very long and have a convoluted pattern. The length and size of the horns increases significantly after the first pregnancy. The internal mucosa is devoid of caruncles. The colour

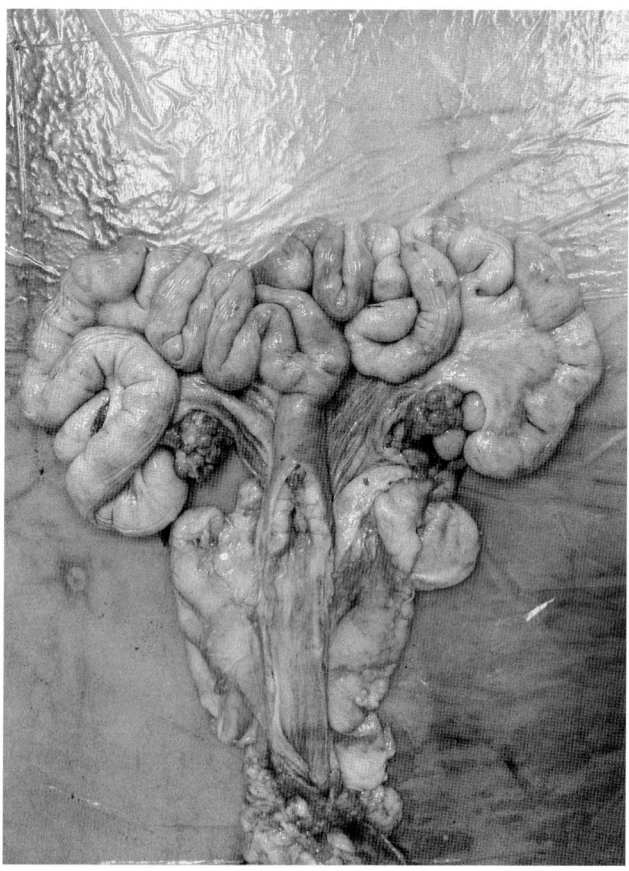

Figure 6.32   Uterus of a sow.

varies from red in the luteal phase of the cycle to bluish red in the oestrous phase. The horns caudally are joined for about 4 cm before merging into a single body length of about 5 cm. The cervix has a length of about 25 cm and externally is identifiable by a higher consistency while internally for the presence of bulky cervical tubercles stuck with each other.

Bovine and small ruminants   Bovine and small ruminants have a bicornuate and bipartite uterus with very long horns that begin with a narrowed apex that continues in the fallopian tubes. Each horn is approximately 20 cm long and presents a convexity placed dorsally and also a short body of about 4 cm. In the uterine cavity there is a highly developed median partition. It lasts to the body a cervix which presents a considerable consistency and is approximately 15 cm long. The cervix is present in the pelvic cavity while the body and horns are placed in the abdominal cavity.

Inside the uterus there are characteristic features, the caruncles, which in prepuberal animals are not very evident elevations; while in the adult cow they are about 1 cm in length and width. In pregnancy they swell and become pedunculated, increasing in number, with a hemispherical surface sprinkled with dimples. They are arranged in four rows mostly present in correspondence of the horns, rarer in the body. Their number is around 30–40 in non-pregnant females, while in pregnant cows their number becomes greater than 100. In small ruminants the uterus has characteristics similar to those of bovine but differs in that it has smaller volume and in the shape of the caruncles, which assume the shape of a cup.

The development of the foetus *in utero* can be used to infer the gestational phase the animal is going through. In cows the average duration of pregnancy varies from 275 to 290 days. After 35 days the foetal period begins. Around 45 days the length of the foetus is approximately 3 cm. After two months of gestation its length is about 6 cm and this doubles by the end of the third month. In the fifth month the length is 40 cm and in six months the foetus has very thin hair. At seven months the development of hair and pigmentation are complete. In small ruminants the duration of pregnancy is about five months. The foetal period begins after the first month. Two months later the length is about 10 cm, while at four months it reaches 35 cm, when the hairs are developed all over the body.

Ovaries   The female gonads, which are called ovaries, are a pair of organs and can be found in the sublumbar region, in the vicinity of the pubis. In the ovaries two structures with different organization can be distinguished. The first are the ovarian follicles, fluid-filled circular structures that can be recognized on the ovarian surface for their translucency and yellowish colour. Their size varies depending on the reproductive activity of the animal (prepuberal/puberal, luteal phase/oestrous cycle). The other structures that are detectable on the ovarian surface are the corpora lutea, fleshy structures whose colour and vascularity also varies depending on the stage of breeding and which are completely lacking in prepuberal animals.

Pig The ovary has an irregular surface due to the presence of numerous large follicles and corpora lutea that confer a blackberry-like appearance.

Bovine and small ruminants The ovaries are ovoid and flattened on the lateral side.

Urinary bladder The bladder is an unequal muscular membranous organ which is very distensible. In the cadaver in a state of medium distension it assumes an ovoid shape with a more voluminous pole, the apex, and is placed cranially. After it, follows the body of the organ, on the dorsal face of which emerge symmetrically the two ureters with a course that is species specific.

Pig The organ is always placed in the abdominal cavity and in a state of repletion has a spherical shape. The ostium of the ureters occurs in the vicinity of the neck.

Bovine and small ruminants In ruminants the empty bladder is situated in the pelvic cavity and overflows cranially to the pubis. In a state of repletion it is entirely in the abdominal cavity. In bovine, a full bladder is oval in shape while that of small ruminants is less globular with a pointed apex. The ostium of the ureters occurs more caudally and laterally, different to the pig.

6.5 Anatomy of carcass

6.5.1 Musculoskeletal apparatus

On carcasses presented split in half, it is easier to evaluate the number and shape of the ribs, the structure of the sternum and the particular characteristics of the vertebrae as well as the ischiopubic junction (Figure 6.33). The evaluation may provide information on the differential characteristics of the various species. In animals in which the carcass is submitted as a whole after skinning, it is also possible to see from a side view several superficial muscles.

Vertebrae The vertebrae have a dorsal arch (*arcus vertebrae*) and a ventral body (*corpus vertebrae*). The body, placed ventrally to the vertebral foramen (*foramen vertebrale*), is articulated with the body of the near vertebrae. The ventral side of the body has a pronounced median ridge. The arch is located dorsally to the vertebral foramen. Dorsally to the arch is inserted a spinous process (*procesuss spinosus*) median and unequal, while laterally two transverse processes (*processus transversus*) detach, one on each side with different shape and volume depending on the region. Also, in the vicinity of the arch are the articular processes, two pairs from each side: one cranially (*processus articularis cranialis*) and the other caudally (*processus articularis caudalis*). The articular surfaces of the vertebrae are united by fibrocartilage

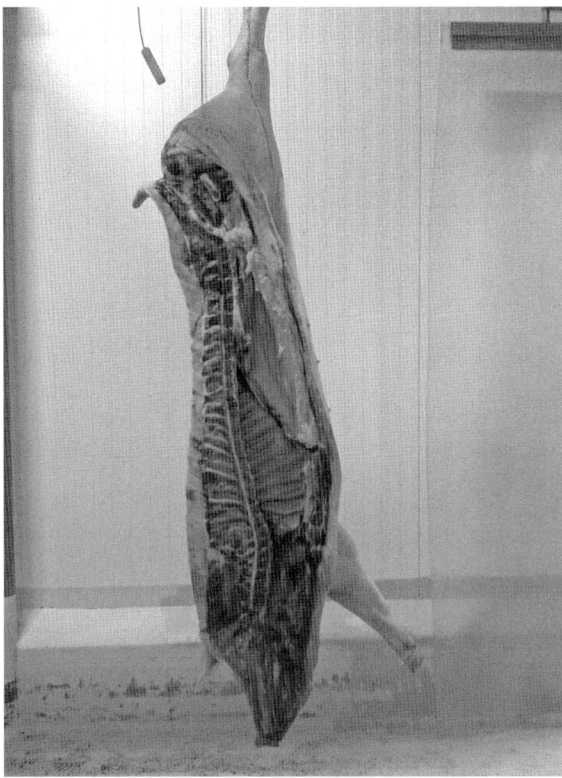

Figure 6.33 Sagittal section of a pig carcass half carcass.

interosseous ligaments called intervertebral discs (*discus intervertebralis*). They are connected to one another by longitudinal ligaments one placed ventrally and one placed dorsally (*ligamentum longitudinale ventrale et dorsale*); these contribute to join the vertebrae together throughout their length forming the vertebral column.

The rachis is the central axis of support and accommodates the spinal cord within a long canal (spinal canal). The rachis gives attachment to the skeleton of the head, of the thorax and of the thoracic and pelvic limbs. The vertebral column is reinforced also by other ligaments placed between the spaces that separate the spinous processes: interspinous ligaments (*ligamenta interspinalia*). There is also a strong and long common ligament at the top of the spinous processes, the supraspinal ligament (*ligamentum supraspinale*), extending from the first cervical vertebrae to the coccygeal region. The supraspinal ligament is stretched in the neck by a ligament consisting of two lateral portions joined in the sagittal plane; this is the cervical or nuchal ligament (*ligamentum nucae*). It is divided into two portions: a funicular portion (*funiculus nuchae*) and a laminar portion (*lamina nuchae*). The funicular portion extends from the top of the spinous processes of the first thoracic vertebrae to the external protuberance of the occipital bone. The lamina is a fibro-elastic sagittal septum that extends between the ventral edge of the funiculus and the cervical spinous processes.

There are seven cervical vertebrae in all mammals and they form the skeletal basis of the neck. The first cervical vertebra or atlas (*Atlas, vertebra cervicalis I*) is articulated cranially by two articular surfaces with the condyles of the occipital bone and caudally to the second cervical vertebra or Axis (*Axis, vertebra cervicalis II*). The atlas differs from all the other cervical vertebrae because it is devoid of body replaced by a ventral arch. It presents a considerable transverse extension conferred by the transverse processes that have the shape of two lateral wings. The vertebral foramen is very wide. The Axis serves as the axis of rotation of the head. The second vertebra, in fact, has a well structured articular process called the odontoid process (*dens axis*) on the cranial face; this allows it to rotate on the atlas. It consists of a body and an arch like all the other vertebrae. In particular, the arch is wide and convex; it is characterized by a well developed spinous process.

The other cervical vertebrae are long and strong so that they differ from all other vertebrae. They have a wide vertebral foramen and an underdeveloped spinous process. The transverse processes are not very salient laterally while are robust and bifid.

The thoracic vertebrae are characterized by their connection to the ribs that articulate between two adjacent thoracic vertebrae. Their number varies according to the species. They are less voluminous than the cervical vertebrae and have a long body, an arch that defines a tight vertebral foramen and high spinous processes with an inclination that varies according to the species. The transverse processes are very small. On each side of the body they have two articular facets: a cranial and a caudal that receive the articular head of the ribs.

The lumbar vertebrae are characterized by the great development of their bodies and their transverse processes. There are 5–7 with individual variations. The vertebral arch surrounds a large hole and the spinous processes are flattened on the lateral side and tilted cranially with an apex that ends with a tubercle. Characteristics are the very elongated and dorsoventrally flattened transverse processes. The sacrum (*os sacrum*) is an unequal bone given by the fusion of the sacral vertebrae (*vertebrae sacrales*). Cranially it is articulated with the last lumbar vertebra and caudally with the first coccygeal. It has a triangular shape with its base in the cranial portion and a narrow apex placed caudally. It has a ventral pelvic surface and a dorsal surface that shows on its median plan the spinous processes more or less welded, depending on the species, and inclined caudally and decreasing in height. Together they constitute the sacral spinae that if completely melted, as in bovine, is called the dorsal sacral crest (*crista sacralis mediana*). Laterally on each side into a large groove the dorsal sacral foramina (*foramina sacralia dorsalia*) open.

The coccygeal vertebrae constitute the basic skeleton of the tail and are variable in number, even within the same species. They are often incomplete.

Ribs and sternum   The ribs are elongated and bent, arched dorsally; they articulate with the thoracic vertebrae and ventrally join the sternum. They form pairs, and are present in a number equal to the number of thoracic vertebrae. In each rib it is possible to recognize a dorsal portion or osseus (*os costale*) oblique in the ventrocaudal direction and a cartilaginous ventral portion (*cartilago costalis*) that articulates with the osseus rib. Only the first pair of costal cartilages articulate directly with the sternum and are called sternal or true ribs (*costae verae*); the remaining ribs do not reach the sternum but are joined to the rib that precedes it, forming a continuous arch that limits the caudal part of the thorax and that takes the name of costal arch (*arcus costalis*). The dorsal end of each rib that articulates with the rachis has a head (*caput costae*) and a tuberosity (*tuberculum costae*) mutually separated by a narrowing of the neck (*collum costae*).

   The sternum is a bone cartilage formation that consists of unequal segments called sternebrae (*sternabrae*). The union of two sternebrae articulates with a pair of costal cartilages. The set of sternebrae constitutes the body of the sternum (*corpus sterni*), whose cranial extension formed by the first sternebra is the manubrium (*manubrium sterni*), while its caudal extension, consisting of a thin cartilage foil, is the xiphoid process (*processus xiphoideus*).

Pelvic girdle   The pelvic girdle is given by the union of three bone segments: ileum (*os ilium*) above, pubis (*os pubis*) and ischium (*os ischii*) ventral. The three bones fuse into a single bone, massive and irregular, the coxal bone (*os coxae*). The coxal has a middle part or acetabulum and two ends: one craniodorsal or iliac and the other ventrocaudal or ischiopubic. The middle part of the fusion of the three bones creates a joint centre (*acetabulum*) intended to receive the head of the femur, with which it articulates. Each coxal unites with the contralateral in a medionentral line in the defined ischiopubic symphysis (*symphysis pubis*). The two coxalis and the sacrum form the pelvic bone or pelvis (*pelvis*). The pelvis is divided into a roof, a floor, two lateral pelvic walls and two openings, cranial and caudal, respectively. The roof is formed by the pelvic wall of the sacrum and by the first coccygeal vertebrae. The floor of the pelvic cavity is formed by the ischium and pubis bones joined by a symphysis (*symphysis pelvina*) that becomes visible in the half carcass. Each lateral face is formed in part by the ileum and by the sacro-ischiaticum ligament (*ligamentum sacro-ischiaticum*) and by muscle formations. The cranial pelvic opening (*pelvis cranialis*) is also known as the entry of the pelvis and puts the abdominal part in communication with the pelvic cavity. It has a vertical diameter that goes from the cranial end of the pelvic symphysis to the roof of the pelvis. The transverse diameter goes from one ileus to the other near the insertion tuberculus of the psoas minor muscle. The caudal pelvic opening is occupied by the pelvic diaphragm (*diaphragm pelvis*) formed by the perineal and anal muscles and aponeuroses. Dorsally it is limited by the first coccygeal vertebrae and ventral from the ischial arch that leads from an ischial tuberosity to the other.

Distinctive characteristics of sex related to the pelvis   The pelvis of the female presents characteristics, such as size and shape, that distinguish it from that of the male. The dimensions of the pelvis as well as the transverse diameters are less in the male. As regards the conformation, the female pelvis has a more circular cranial opening and the ischial arch has a greater width. The thickness of the pubis of the male is always greater.

The gracilis muscle (*musculus gracilis*) is the anatomical basis of the medial surface of the thigh. The aponeurotic component of the muscle inserts over the entire length of the symphysis pubis and moves distally to the medial surface of the tibia together with the sartorius (*musculus sartorius*) and semi-membranosus (*musculus semimembranosus*) muscles. In the male, the gracilis muscle is covered by fat while in the female the aponeurotic component is visible.

Diaphragm muscle   The diaphragm (*diaphragma*) is a muscular aponeurotic septum that separates the thorax cavity from the abdominal cavity (Figure 6.34). It is very broad flat muscle and its muscle bundles have a radial

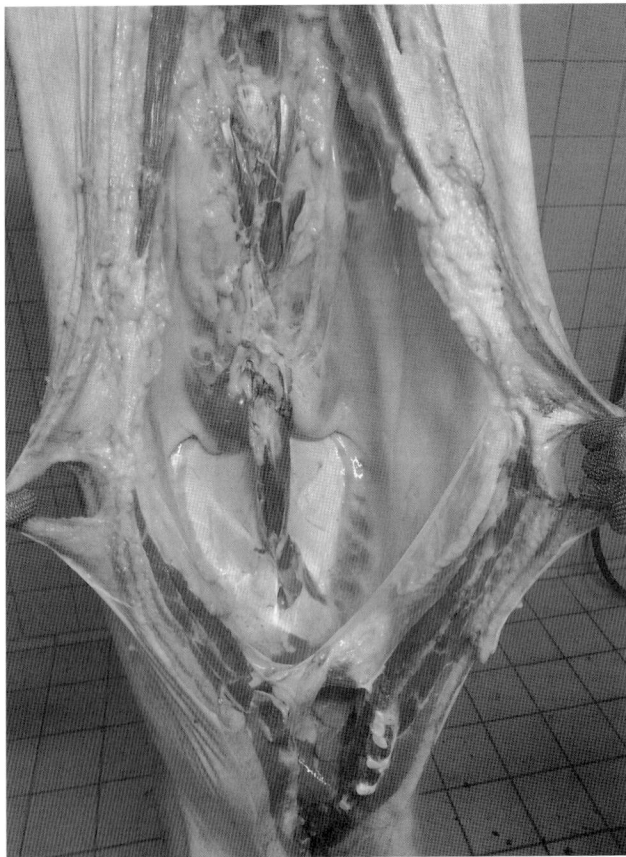

Figure 6.34   Diaphragm of swine between the thoracic and abdominal cavities.

arrangement. They originate in the sublumbar region and radiate in a fleshy part that forms the left and right pillars (*crus sinistrum et destrum*). They penetrate extending as intermediate pillars to surround in the central tendon (*centrum tendineum*) the passage of the oesophagus and, thus, help to define the oesophageal hiatus. At the periphery of the pillars starts an aponeurotic lamina, the central tendon, of pearly white colour beyond which emerge new fleshy portions that constitute the fleshy peripheral divisions.

Inguinal rings During the separation of the half carcasses the genital organs are normally removed. The analysis of the posterior hindquarters allows remaining parts of the genital organs that are still attached to the carcass or particular features resulting from their removal to be recognized. In the bull can be seen the opening of the inguinal canal represented by the superficial or external inguinal ring (*annulus inguinalis superficialis*); this is open and bounded by the aponeuroses of external and internal oblique muscles of the abdomen (*mm. obliquus externus et internus abdominis*), in which it is possible to trace the stumps of the spermatic cord and its coverings, the ileoinguinal nerve, external pudendal vein and the inguinal veins.

Thoracic and abdominal cavities The thoracic cavity has as its basic skeleton the ribs, the sternum and the thoracic vertebrae, which, with the contribution of the intercostal muscles and the diaphragm, form a closed cavity with the form of a truncated cone whose apex is placed cranially. The base closed by the diaphragm is oblique in a dorsocaudal direction. The thoracic cavity is presented laterally, delimited by smooth and concave walls covered by endothoracic fascia and by a serous membrane, the costal pleura; both cover the ribs and internal intercostal muscles whose fibres appear oriented obliquely with a ventrocranial direction. The endothoracic fascia (*fascia endothoracica*) is a thin, fibro-elastic connective tissue that covers the entire wall of the thorax and is the equivalent of the transverse abdominal fascia. It is particularly thickened in bovine. The serous that covers the thoracic cavity is represented by the costal parietal pleura, which is in contact externally with the endothoracic fascia, while its inner side is in relation with the costal visceral pleura, from which it is separated by a thin liquid film, the pleural fluid. The costal parietal pleura are reflected on the diaphragm and on the mediastinum forming blind funds or pleural recesses (*recessus pleurales*). The pleura lining the thoracic cavity (Figure 6.35) appear shiny, transparent and with a smooth appearance. Caudally to the thoracic cavity, between the costal wall and the diaphragm, the reflection is the costodiaphragmatic recess (*recessuss costodiaphragmaticus*) beyond which the parietal pleura covers the diaphragm as diaphragmatic pleura adhering particularly to the central tendon of the diaphragm.

The abdominal cavity is bounded laterally by the ventrolateral abdominal wall muscles, which are: the external oblique (*m. obliquus externus abdominis*), the internal oblique (*m. obliquus internus abdominis*) and the transverse (*m. transversus abdominis*) abdominal muscles. Ventrally the aponeurotic part of the diventrolateral abdominal wall muscles flows on the two rectus

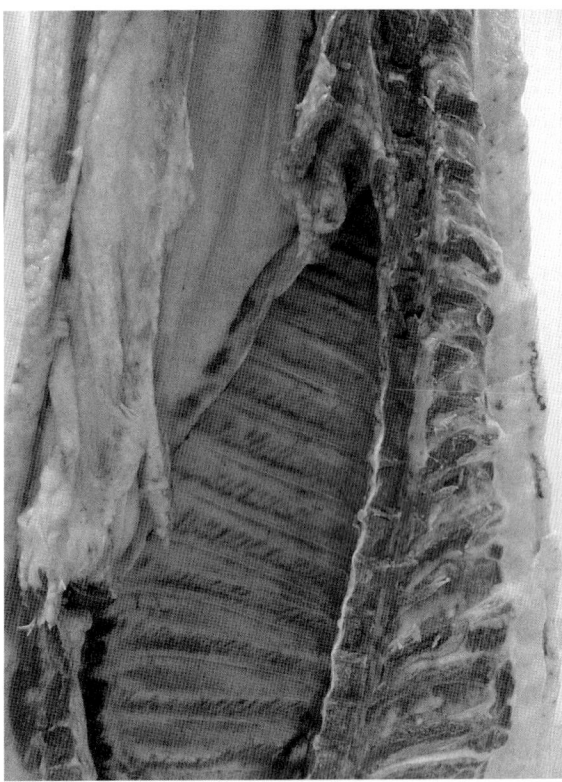

Figure 6.35 Internal view of the pig thoracic cavity in which the costal pleura is visible.

abdominis muscles (*m. rectus abdominis*) that lead from the pubic symphysis to the xiphoid process of the sternum. The two rectus abdominis muscles together with the aponeurotic parts of the ventrolateral abdominal wall muscles merge together on the median plane on each side on the *linea alba*. The vault of the abdominal wall, dorsal, is instead bounded by the thoracolumbar vertebrae and by the lumbar-iliac muscles divided into two muscular planes: superficially the psoas minor and major muscles (*mm. psoas major et minor*) and deeper, in connection with the transverse processes of lumbar vertebrae, the quadratus lumborum muscle (*m. quadratus lumborum*) and lumbar intertransverse muscles (*m. inter-transverse lumborum*). Cranially the cavity is bounded by the diaphragm whose muscle bundles originate in the sublumbar region to finish on the deep face of the hypochondriac region and on the xiphoid process of the sternum. Caudally the abdominal cavity extends into the pelvic cavity. The lateral wall of the abdominal cavity is covered entirely by a serous: the parietal lamina of the peritoneum. It departs from the linea alba and covers the walls of the abdominal cavity and some organs joined to the wall (extraperitoneal organs such as the kidneys). In correspondence of the vault of the abdominal cavity, the parietal sheets of the two sides inflect ventrally, forming the visceral layer of the peritoneum that adheres to the surface of the viscera, giving them a shiny appearance.

6.5.2 Viscera annexed to the carcass

Kidneys The kidneys are located in the cranial lumbar region on the side of the abdominal aorta and of the caudal vena cava with an asymmetric position as the right kidney is usually more cranial than the left. They are surrounded by connective tissue infiltrated with adipose tissue. They are adjacent to the psoas muscles, the quadratus lumborum muscle and the transverse abdominal muscle. The cranial portion is connected with the diaphragm. The ventral surface is covered by parietal peritoneum with the exception of a few species. On the cranial margin of the two kidneys is the corresponding adrenal gland. The medial margin has a deep groove, the renal hilum, occupied by abundant adipose tissue at the level of which the renal vessels and nerves enter/exit and the excretory duct of the kidney, the ureter, originates.

Mammary gland The mammary gland (*glandula mammaria*), present both in the male and female, assumes, however, a significant development in the female only after puberty and, in particular, in multiparous cows. It presents itself as a skin relief constituted by the presence of compound tubuloalveolar glands that in shape and topography differ in the different species. The mammary gland is topped by a papilla, the nipple, which is provided with openings called lactiferous pores.

Male genitals The external genitalia of the male are usually removed from the half carcass. The penis, the male copulatory organ, consists of erectile tissue. In it can be recognized the spongy urethra, the cavernosus bodies of the penis and the glandis (*corpus cavernosum penis et glandis*), and spongy bodies of the penis and the glandis (*corpus spongiosum penis et glandis*). In the penis a fixed and a free end can be distinguished. Between the two is the middle part or body of the penis. The body of the penis is formed for the most part from the spongy urethra and from the cavernosus bodies. The free end formed by the cavernosus bodies associated to the spongy urethra and the glandis is variable in shape depending on the species. The fixed end is called the root of the penis.

The testicle (*testis*) is the male gonad. It is an equal organ placed with the epididymis in the tunica vaginalis and scrotum. The colour is bluish white in species of large size and pink in species of small size. The texture is firm and elastic. The tension of the organ is given by the serous that surrounds it. The testicle presents in all species two sides, a lateral and a medial, a free margin and one epididymal, an extremitas capitata and an extremitas caudata, both in structural continuity with the epididymis. The epididymis is an elongated organ, cavitated and integral to the testis; on the other ends it is continuous with the vasa deferentia.

Lymph nodes of the carcass The lymph nodes are organs of the mammals intercalated along the course of lymphatic vessels to drain the lymph before it enters the bloodstream. They are flattened and have an oval or round shape with a colour that ranges from greyish-pink to grey. On the surface it is possible to observe a hollowed and limited area called hilum. The lymph nodes, isolated or in groups, can be placed deep into the cavity or on the surface in the interstices of the muscles or under the skin. Functionally, the role of the lymph centre, which is to drain the lymph from an area of the body, should be noted.

Lymph nodes of the thoracic limb The axillary lymph centre comprises the cubital lymph nodes (*lymphonodus cubitales*), which are only present in the ovine and horse, the *lymphonodi axillares proprii* and the *lymphonodi axillares primae costae*. The *lymphonodi axillares proprii* is located just medial to the common tendon of the latissimus dorsi (*musculus latissimus dorsi*) and teres major (*musculus teres major*) muscles. The *lymphonodi axillares* primae costae are a group of two or three lymph nodes on the sides of the first intercostal space.

Lymph nodes of the neck The lymph nodes of the neck are divided into two lymph centres, the superficial and the deep cervical. The superficial cervical lymph centre contains the superficial cervical lymph nodes (*lymphonodii cervicales superficiales*) located cranially to the scapula below the brachiocephalic muscle (*musculus brachiocephalicus*). The deep cervical lymph centre includes the deep cervical lymph nodes divided into the cranial, medium and caudal (*lymphonodii cervicales profundi craniales medi et caudales*). They are placed laterally to the trachea and run parallel to the vessels, the sympathetic trunk and vagus nerve.

Thoracic lymph nodes The lymph nodes of the thorax belong to four lymph centres:

1. the dorsal thoracic (*lymphocentrum thoracicum dorsale*);
2. the ventral thoracic (*lymphocentrum thoracicum ventrale*);
3. the mediastinale (*lymphocentrum mediastinale*), see lymph nodes attached to the viscera of the respiratory system;
4. the bronchial (*lymphocentrum bronchale*), see lymph nodes attached to the viscera of the respiratory system.

The thoracic dorsal lymph centre contains the intercostal and thoracic-aortic lymph nodes. The intercostal lymph nodes are placed near the vertebral end of the ribs in the intercostal space covered by the pleura and the endothoracic fascia. Because of this are also, alternatively, referred to as costocervical lymph nodes. The thoracic-aortic lymph nodes are placed between the vertebral bodies and the thoracic aorta covered by the pleura and the endothoracic fascia. The ventral thoracic lymph centre is composed of the cranial sternal (*l. sternales caudales*) and the caudal sternal (*l. sternales caudales*) lymph nodes.

Abdominal lymph nodes The lymph nodes of the abdomen belong to four lymph centres:

1. lumbar (*lymphocentrum lumbale*),
2. celiac (*lymphocentrum celiacum*),
3. cranial mesenteric (*lymphocentrum mesentericum craniale*),
4. caudal mesenteric (*lymphocentrum mesentericum caudale*).

The last three have already been described along with the abdominal viscera. The lumbar lymph centre consists of the lumbo-aortic lymph nodes located along the course of the aorta and caudal vena cava.

Lymph nodes of the pelvis and pelvic limb The lymph nodes in the pelvis and in the pelvic limb are grouped into five lymph centres:

- ileo-sacral (*lymphocentrum ilio-sacrale*),
- ileo-femoral (*lymphocentrum ilio-femorale*),
- inguinal-femoral (*lymphocentrum inguino-femorale*),
- ischiadic (*lymphocentrum ischidicum*),
- popliteal (*lymphocentrum popliteum*).

The ileo-sacral lymph centre includes:

1. medial iliac lymph nodes (*lymphonodii iliaci mediales*) placed ventrally to the origin of the external iliac artery;
2. lateral iliac lymph nodes (*lymphonodii iliaci laterales*) located near the termination of the deep iliac circumflex artery;
3. sacral lymph nodes (*lymphonodii sacrales*) placed ventrally and to the lumbosacral joint;
4. anorectales lymph nodes (*lymphonodii anorectales*) placed in the retroperitoneal connective tissue laterally to the rectum (see lymph nodes attached to the viscera of the pelvic cavity).

The ileo-femoral lymph centre, also called the deep inguinal, consists of the ilio-femoral or deep inguinal lymph nodes. The inguino-femoral lymph centre, also called the superficial inguinal, includes subiliac lymph nodes (*lymphonodii subiliaci*) and in males and females, respectively, scrotal lymph nodes (*lymphonodii scrotales*) and mammary lymph nodes (*lymphonodii mammarii*). The ischiatic lymph centre is made up of a group of ischiatic lymph nodes located between gluteal muscles and sacro-ischiatic ligament. The popliteal lymph centre comprises superficial popliteal lymph nodes (*lymphonodii poplitei superficiales*), which are missing in bovine, and deep polpliteal lymph nodes (*lymphonodii poplitei profundi*) easily explored between the fascia of the leg and the skin.

6.5.3   Specific characteristics in pig

Musculoskeletal apparatus   In pigs the cervical portion of the rachis is suspended by a reduced cervical ligament whose lamina, in some cases, is poorly
developed if not absent. The cervical vertebrae, of which there are seven, are
short and wide. The atlas is longer than it is wide. The atlas the dorsal tubercle is high. The ventral arch is much larger than the dorsal. The wings are
directed ventrally. The Axis has a large, short and conical odontoid process.
The spinous process is particular because it is very high with a rounded end
directed caudally. The remaining vertebrae have a very strong body and high
processes and inclined cranially.

There are 14 or 15 thoracic vertebrae; they have relatively long vertebral
bodies. The first thoracic vertebra has a very high spinous process which
decreases regularly in the remaining vertebrae until the twelfth, beyond
which they are of an almost equal height. The caudal inclination of the
spinous processes is reversed from the twelfth vertebra to then become
perpendicular. The transverse processes have a very wide base.

The lumbar vertebrae vary in number from five to eight. The vertebral
bodies are elongated and hollowed dorsoventrally with a length that increases
from the first to the fifth and then decreases. The spinous processes are quad-
rangular and their ends form an elongated ridge. The transverse processes
are slightly rounded at the end and are slightly curved ventrally with a length
that increases from the first to the fifth.

The sacrum has four or five sacral vertebrae that are merged to a lesser
extent than other species. It has a ventral concavity and on the dorsal face
there is a clear reduction of the sacral spinae because the spinous processes
are slightly high. Among the coccygeal vertebrae, which number from twenty
to twenty-two, the first five have a full arch.

There are fourteen or fifteen pairs of ribs, of which seven are sternal (true
ribs) and 7–8 are asternal (false ribs). The tuberosity is very pronounced in
the first eight and tends to decrease proceeding caudally. They have a wider
portion in the middle part.

The sternum consists of six sternebrae with a wider body in the middle and
narrow at the end. The manubrium is underdeveloped. The long and narrow
xiphoid process ends with an enlarged triangular cartilage.

The pelvic girdle is influenced by the shape of the coxal, which is almost horizontal with the acetabulum portion of ileum and ischium placed on the same
plane. The pelvis is narrow at the ridges of the ischium. The cranial opening is
elliptical with a vertical diameter of about 14 cm, and a transverse significantly
lower. The pelvic floor is wide and flat in the female and in the male appears
sunken. The sexual differences are evident: the cranial opening is smaller in
the boar, the pubis occurs thicker than the sow and the pelvic cavity is narrower transversely.

The diaphragm (Figure 6.35) has very strong pillars, especially the right,
whose tendon can be followed to the last lumbar vertebra. The intermediate

pillars are also highly developed reaching the right lumbo-costal vault and sometimes the left. The costal portion has only a modest contact on the last rib.

Viscera annexed to the carcass
Kidneys

Overview

The kidneys (Figures 6.36 and 6.37) are located in the perirenal fat, from which they must be released in order to be inspected. They are dorsoventrally flattened and an oval shape. In the adult they measure approximately 15 cm in length and 7 cm in width. The surface is smooth (Figure 6.36) and slightly convex, divided into two faces: a dorsal and a ventral. The kidneys are covered by connective tissue more or less infiltrated with fat; the ventral face is coated by a thin fascia and by the parietal peritoneum, while the dorsal face is set against the lumbar muscles (*m. psoas* and *m. quadratus lumborum*). The pig kidney is an extraperitoneal organ. Observed in sections are a thickened cortex that continues in the renal medulla with thin renal columns which insert in between the clearly visible renal pyramids. The renal pyramids end in about 10 conical papillae. The renal pelvis is wide and occupies most of the renal sinus. It presents short and wide major calyces placed at the cranial and caudal end of the renal pelvis. It also has minor calyces that lead to middle part of the renal pelvis and into the major calyces. The ureters are wider at their origin then along their course. The renal lymph nodes (*lymphonodi renales*) are clearly seen in the perirenal fat in the hilum of the kidney and the rachis and are located near the hilum of the kidney in relation to the renal artery and vein.

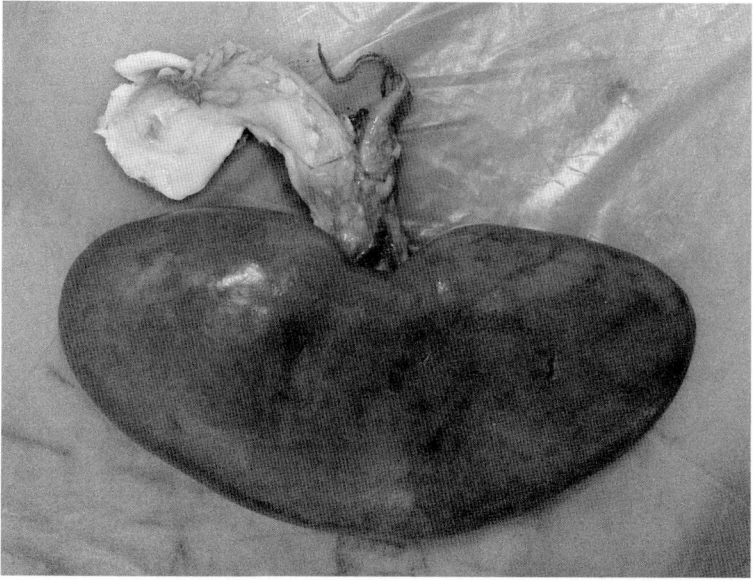

Figure 6.36 Dorsal face of pig kidney.

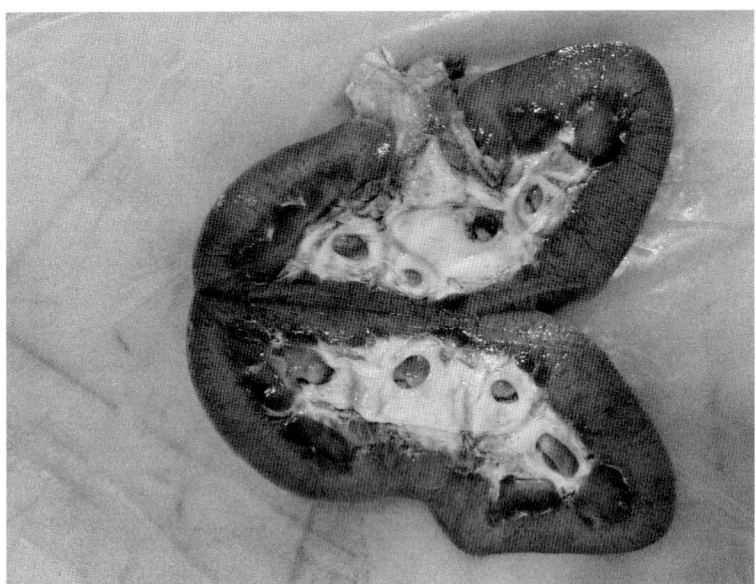

Figure 6.37 Longitudinal section of pig kidney in which is possible to see the renal cortex divided from the renal medulla.

Topography
The kidneys are located almost at the same level, even if the left one is often slightly cranial. The right is at the height of the first four lumbar transverse processes.

Weight
Each kidney weighs approximately 200 g.

Colour
The kidneys are reddish-brown in colour.

Mammary glands Attached to the carcass there are 6–7 pairs of mammary glands placed in two rows that occupy the pectoral, abdominal and inguinal regions. Often there is a numerical symmetry on the two sides. The mammary glands exhibit decreasing dimensions proceeding from the most cranial to the most caudal. The mammary glands have a hemispherical conical shape, flattened laterally with a papilla that occupies the top. The papillae are cylindroid and are about 2–3 cm long with pink and hairless skin on the end; there are 2–3 papillary ostia corresponding to the same number of lactiferous sinuses.

Male genitals The penis is long and thin, measuring about 60 cm with a diameter of 2–3 cm. The root is large and thick. The body of the penis is initially flattened dorsoventrally, becomes cylindrical in the middle part and flattened on the lateral side near the free end. The free end is half as long as the fixed portion and its extremity presents a rudiment of glandis, indistinguishable, with a helical configuration.

The testicle is very massive, with an ellipsoidal shape and an oblique major axis ventrocranially directed. The major axis is about 15–20 cm in length and the body weighs, without the epididymis, approximately 400 g.

Lymph nodes of carcass

Thoracic limb lymph nodes In the axillary lymph centre the *lymphonodi axillares proprii* are missing and the *lymphonodi axillares primae costae* are not always present.

Lymph nodes of the neck The superficial cervical lymph nodes are divided into three groups: superficial dorsal, medium and ventral. The first, the most voluminous lymph node of the neck, is located in a triangle formed dorsally by the trapezius muscle (*musculus trapezius*), ventrally by the *pectoralis cleidoscapularis* muscle and cranially by the brachiocephalic muscle. The medium placed above the jugular vein between the brachiocephalic muscle and the omohyoid muscle (*musculus omohyoideus*) can sometimes be absent. There are three ventral nodes placed above the posterior margin of the parotid gland in contact with the muscle brachiocephalic.

The deep cervical lymph nodes are small formations located along the course of the trachea. The peculiarity is that the deep caudal cervical lymph nodes placed at the entrance of the thorax may merged with the lymphonodi axillares primae costae.

Thoracic lymph nodes The sternal lymph nodes are found in a group of 2–4 voluminous lymph nodes in correspondence with the first two costal cartilages. In pigs the intercostal lymph nodes are missing.

Abdominal lymph nodes Lumbo-aortic lymph nodes are wrapped in the adipose tissue of the psoas muscles (*musculus psoas*) and are positioned symmetrically; on the left they are over the aorta and on the right above the caudal vena cava.

Lymph nodes of the pelvis and the pelvic limb The 5–6 medial iliac lymph nodes are large and placed laterally to the external iliac artery (*arteria iliaca externa*). The lateral iliac lymph nodes in pigs are often merged with the medial iliac ones and constitute a unique group. The sacral lymph nodes are located on the broad ligament of the pelvis covered by the biceps femoris (*musculus biceps femoris*) and gluteus medius (*musculus glutaeus medius*) muscles above the deep gluteus muscle (*musculus glutaeus profundus*). These lymph nodes are not, however, constantly present. The deep inguinal lymph node is about 9–10 cm located near the internal inguinal ring. Subiliac lymph nodes are located above the stifle about half of a line joining the patella with the external iliac tuberosity and they can be evidenced by the incision of the abdominal muscles cranially to the cranial margin of the tensor fasciae latae muscle

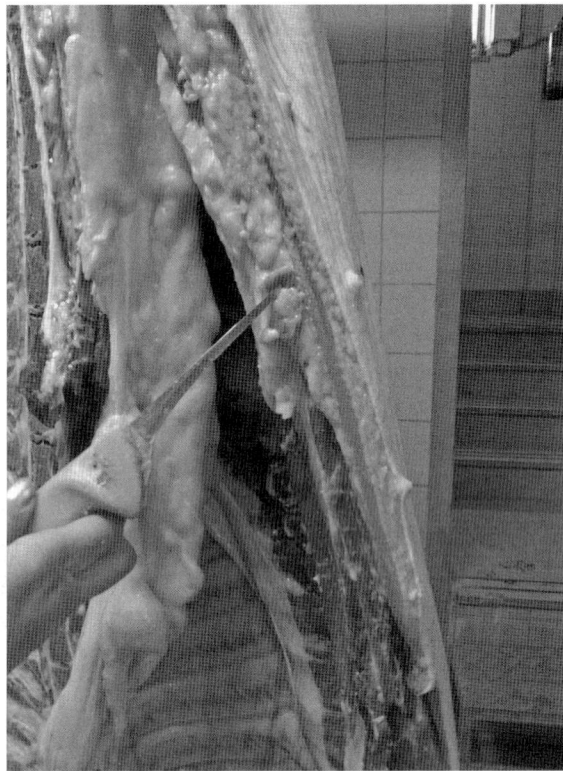

Figure 6.38   Pig mammary lymph node indicated by knife.

(*musculus tensor fasciae latae*). The mammary lymph nodes (Figure 6.38) are located inside the subcutaneous adipose tissue, located along the course of the inguinal mammary line, contained between the glandular tissue and the muscular mass of the abdominal muscles. The scrotal lymph nodes, in the male, are located dorsolaterally to the penis under the prepubic tendon. The presence of ischiatic lymph node is inconstant. The superficial popliteal lymph node can always be found in the subcutaneous fat around the middle of a line joining the top of the calcaneus with the lateral insertion of the tail. The deep popliteal lymph node is not always present.

6.5.4   Specific characteristics in bovine

Musculoskeletal apparatus   In bovine amongst the general distinguishing features of the rachis it should be noted that the shape of the vertebral column is lowered due to the reduced length of the spinous processes of the lumbar vertebrae (Figure 6.39). The intervertebral discs in bovine appear straight in section from about the middle of the thoracic region.

The cervical vertebrae are short and square-shaped. The first cervical vertebra, the atlas, lacks the body that is replaced by a ventral arch. Its transverse

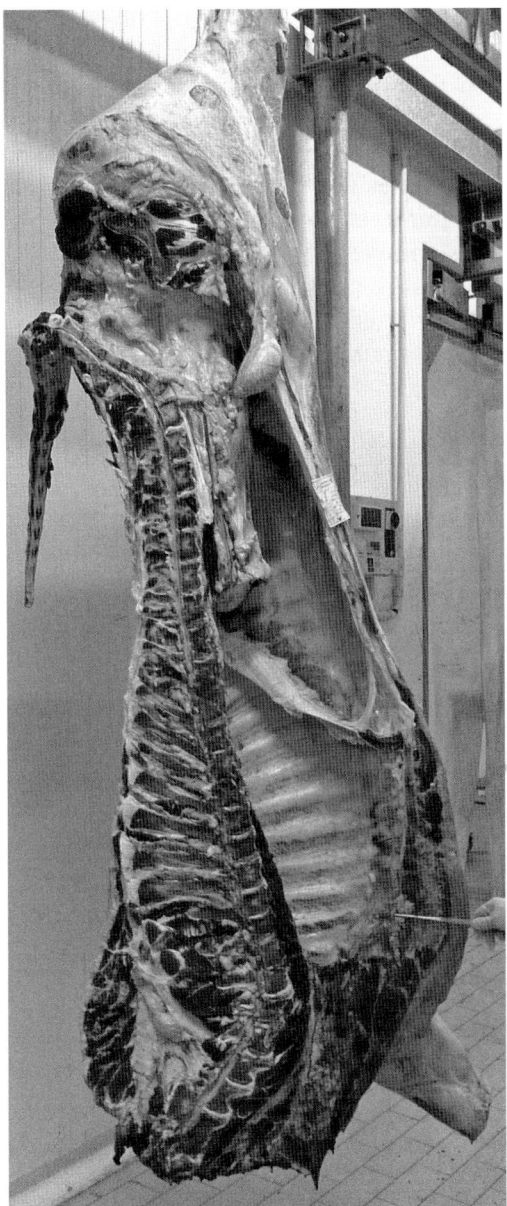

Figure 6.39 Sagittal section of a bovine carcass.

processes are very extensive and are called the wings of the atlas. The second vertebra, called the Axis, has cranially an odontoid process of semi-cylindrical shape. The spinous process has a dorsal margin straight and thin, and ends caudally with a large tubercle. The transverse processes are very small. The spinous processes, well developed from the third to the sixth vertebra, head cranially gradually increasing in height. The spinous process of the seventh cervical vertebra is very high and almost perpendicular to the body. In bovine the lamellar portion of the cervical or nucal ligament in its cranial part sends three thick cords to the spinous processes from the second to fourth the cervical vertebrae.

The thoracic vertebrae, of which there are thirteen, have a relatively long body. The spinous processes are very long and wide and increase in height from the first to the third vertebra, then lower and tilted caudally as they approach to the thirteenth vertebra whose process becomes perpendicular. The transverse processes are not very pronounced.

There are six lumbar vertebrae each with an elongated body. The spinous processes are wide and low and almost vertical, with a decreasing height from the first to the last. The very long sabre-shaped transverse processes have sharp and irregular edges of which the cranial is just concave and the caudal convex.

The sacrum is formed by the merging of five sacral vertebrae into a single bone triangular in shape with a pronounced ventral concavity. The spinous processes form a sacral crest that decreases in height proceeding caudally. On each side of the crest there are the dorsal sacral foramina.

There are 18–20 coccygeal vertebrae. The first five are well developed and those that follow see the regression of the dorsal arch. The body ventrally presents two parallel processes delimiting a groove occupied by the caudalis mediana artery (*arteria caudalis mediana*).

The sternum consists of seven bony segments, defined as sternebrae, joined by a hyaline cartilage. The first sternebra has an extension, the manubrium, while the latter is expanded in a thin xiphoid process. The sternum in the half carcass of the bovine appears reduced in longitudinal and dorsoventral extension because of the greater curvature of the ribs.

There are thirteen ribs, of which are eight sternal or true ribs. The cartilages of the five asternal, or false, ribs do not reach the sternum but are combined with the previous ones forming an arch, the costal arch. In some cases it is possible to find caudally a fourteenth floating rib. The ribs in bovine are relatively long, slightly curved and particularly flat and irregular. Usually, the ventral half of the rib is wider than the dorsal half. The head and the tuberosities are voluminous and distinct with a very curved hook shape.

The pelvic girdle is equally wide in the cranial part as well as in the caudal one. The shape of the ischiopubic or pelvic symphysis provides morphological data useful in distinguishing the sex, especially in half carcases. In cows the pubis is flat and hollow in the endo-pelvic surface. In the male the pubis is very big and has a nearly circular section. The pelvic symphysis in section is markedly concave on the dorsal surface.

The diaphragm of bovine has some distinctive features (Figure 6.40). Among these the most important is that it is in the most forward position in the thorax of any other species. For this reason, the lumbar portion is almost parallel to the rachis, so that the oesophageal hiatus is located below the projection of the eighth thoracic vertebra. It follows a nearly vertical central tendon. The pillars are very thick but relatively narrow, especially the left.

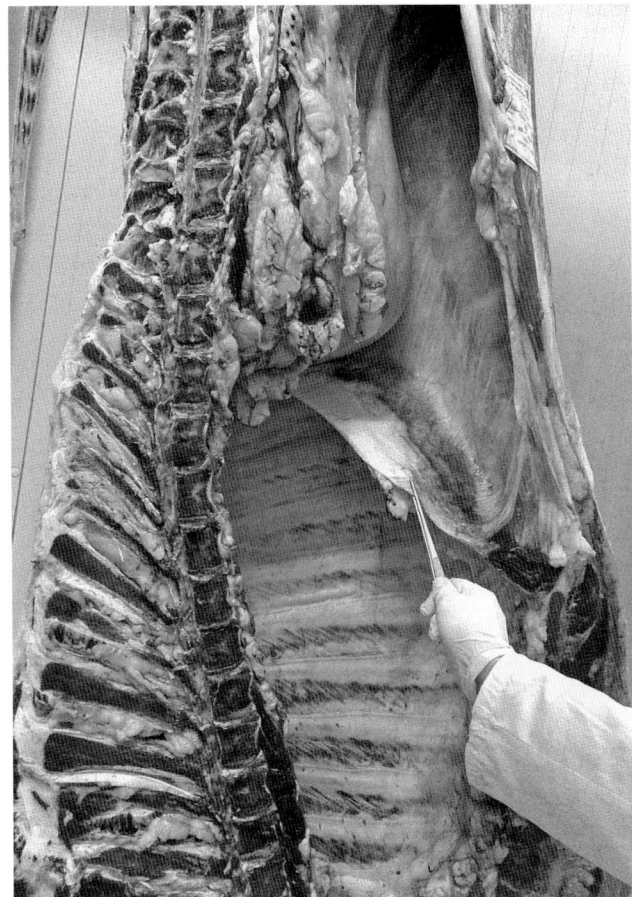

Figure 6.40 Section of bovine diaphragm.

The intermediate pillars are highly developed, thick and considerably long, constituting a sort of sphincter around the oesophageal hiatus. Both branch off at the base of the right pillar. It should be noted that the right half of the abdominal side of the diaphragm is in contact with the liver and its left half in contact with the rumen, the spleen and ventrally with the reticulum, to which it adheres with a connective adherence 20 cm large. Each costal portion is relatively wide and presents at its periphery only six or seven digits. Its insertion starts from the proximal third of the last rib and extends in a straight line until the eighth chondrocostal joint to continue on the two cartilages that precede.

Viscera annexed to the carcass Depending on the slaughter technique, some of the viscera may not be removed and remain joined to the carcass. The kidneys, the mammary gland and the male genitals represent the organs that most frequently might be presented for *post-mortem* inspection still adherent to the carcass.

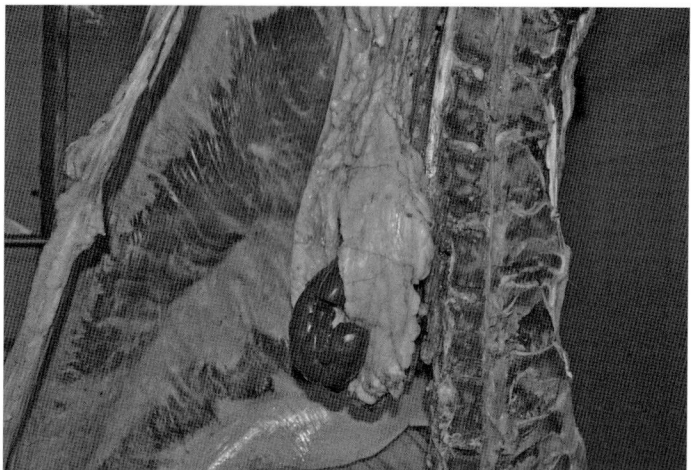

Figure 6.41   Bovine kidney hanging from the carcass.

Kidneys and renal lymph nodes
Overview

The kidneys are divided into lobes distinguished by deep grooves on the surface (Figure 6.41). The are about twenty lobes. The kidneys are asymmetrical in shape and position. The right kidney has an elliptical shape and is flattened dorsoventrally with a length of about 20 cm and a width of about 10 cm. The left kidney is usually shorter by a few centimetres and has a pyramidal shape. In section it can be appreciated that the renal pelvis surrounded by abundant adipose tissue and on its wall protrude approximately twenty conical papillae. Their number is usually greater than the lobes of the outer surface.

In bovine the renal lymph nodes constitute a group of lumbar lymph nodes placed in the adipose tissue located laterally to the right and left of the caudal vena cava and the aorta near the root of the renal vessels.

Topography

The right kidney is situated at the level of the last rib and of the first transverse processes of the lumbar vertebrae. The right kidney is covered by a thickened adipose capsule which contacts dorsally with the diaphragm and the psoas muscles and ventrally with the right lobe of the pancreas, colon and caecum. The left kidney has a particular topography as it is moved caudally and to the right, floating and suspended from a thickened mesum. It is located ventrally to the lumbar vertebrae from the second to fifth.

Weigh

The left kidney is heavier than right and the total weight is about 1 kg, of which about 500 g is the left kidney and 400 g the right one.

Colour

The colour is reddish-brown.

Mammary gland and mammary lymph nodes The cow has a pair of inguinal mammary glands, which are removed from the carcass leaving behind an extensive excavation that often leaves *in situ* residual portions of the gland and, at times, the mammary lymph nodes.

The size and weight of the mammary glands vary greatly depending on the breed and the functional state of the gland. They have an average weight in the adult during the drying-off process of about 8 kg, while during lactation they can reach 25 kg. The organ has a rounded appearance and is divided by a intermammary groove that divides it into two quarters for each side, respectively cranial and caudal. The hindquarters are bulkier than the cranial one and are placed between the thighs and, in some cases, go as far as the perineum. The forequarters are inserted near the pubis or bring themselves more cranially toward the umbilicus. Each quarter has on its top a nipple approximately 5–10 cm in length and 2–3 cm in width. The mammary gland is covered with a thin skin with hair of variable length. The coating skin, depending on the breed, may or may not be pigmented. The skin on the nipples is hairless. The excretory routes of the milk are represented by the lactiferous ducts that have an irregular conformation. In each gland, these ducts have a diameter that varies from few millimetres to few centimetres, and flow into a lactiferous sinus wide and irregular. The mammary gland is divided into a glandular and a papillary portion. The glandular part or mammary parenchyma is the most extended with a branched form drained by groups of lactiferous ducts. The parenchyma at rest appears, when cut, a yellowish-grey or pink-amber colour. The glandular clusters are separated by abundant connective tissue infiltrated with adipose tissue. The parenchyma during rest periods is consistent and gritty, while during lactation is compressible and of a light yellow or pink colour. Part of papillary mammary gland, about 5–8 cm, continues the glandular portion and is received in the nipple. The lactiferous sinus in its distal portion flows into the papillary duct on the top of the nipple. The mammary lymph nodes in the bovine are also called retromammary lymph nodes for their location: they are located under the skin near the median plane in the vicinity of the caudal end of the mammary gland at the perineal insertion. On each side there is a massive superficial lymph node about 10 cm in length and 4 cm in width. Cranially there are also more (2–3) smaller lymph nodes.

Male genitals The penis is about 1 m long with a diameter of 4–5 cm. It is located under the belly and reaches cranially the ombelicum. The root is large and thickened, the body long and thin; initially it is flattened latero-laterally between the thighs in order to become almost cylindrical cranially. The free part is about 10 cm long and cranially does not have a bulky glandis. In the carcass the removal of the penis is often incomplete and the organ is sectioned at its insertion on the ischiatic arch. On the quarter carcass or on the carcass it is recognizable at the level of the ischiatic tuberosity and arch, the sectional area of the cavernous tissue inside which the urethra is recognizable.

The testicle is about 12–15 cm in length and 8–10 cm in width; it weighs, without the epididymis, about 300 g. The long axis of the organ is vertical with the extremitas capitata located dorsally and epididymal margin placed medially. The covering serous, the tunica albuginea, is thin and below it the colour is yellow ochre.

Lymph nodes of the carcass   In bovine there is only one axillary lymph node, massive, elongated and about 3–5 cm in length located medium caudally to the scapulo-humeral joint at the level of the third rib. It is possible to see it by separating the shoulder from the trunk. A large superficial cervical lymph node elongated and approximately 8 cm in length can be found in front of the cranial margin of the scapula covered by the brachiocephalic muscle. The deep cervical lymph nodes are divided into three groups: the cranial, located cranially to the trachea near the thyroid; the medium, in the vicinity of the middle third of the cervical trachea; and caudal cervical, cranially to the first pair of ribs dorsally to the jugular vein.

Thoracic lymph nodes   A lymph node of particular interest is the cranial sternal lymph node (Figure 6.42), which is about 3–4 cm long located in the vicinity of the insertion of the first rib on the lateral side of the sternum. The remaining sternal lymph nodes are placed in correspondence of the intercostochondral spaces.

Abdominal lymph nodes   The lumbo-aortic lymph nodes are arranged along a chain ventrally to the lumbar vertebrae. There are about 25 lymph nodes of variable volume grouped close to the abdominal aorta and caudal vena cava.

Lymph nodes of the pelvis and pelvic limb   The medial iliac lymph nodes, of which there are 4–5, are placed laterally to the termination of the aorta and the caudal vena cava. A large lymph node among them, about 10 cm in length, can be easily found in the half carcass or hindquarter because it is located between the last lumbar vertebra and the first sacral. The lateral iliac lymph nodes are smaller than the medial and are placed ventrally and laterally to the medial ones. The sacral lymph nodes are located laterally to the sacro-tuberous ligament (*ligamentum sacrotuberale*).

The deep inguinal lymph node is a voluminous mass and has a length of about 10 cm; it is close to the internal inguinal ring. The subiliac lymph nodes are a single mass 12 cm in length in front of the cranial margin of the tensor fasciae latae muscle in the middle of an imaginary line between the patella and the angle of the hip (*coxa*). The mammary lymph nodes have been described in a previous section.

The scrotal lymph nodes, in the male, are located laterally to the fixed portion of the penis. They are positioned in the subcutaneous tissue lying on gracilis muscle (*m. gracilis*). The ischiatic lymph node is a large mass 3 cm in length leaning against the sacroiliac ligament below the dorsal origin biceps femoris muscle. The popliteal lymph node is a unique formation of great importance in the hind limb. This is the deep popliteal lymph node which has a length of approximately 5 cm immersed in the abundant adipose tissue placed caudally

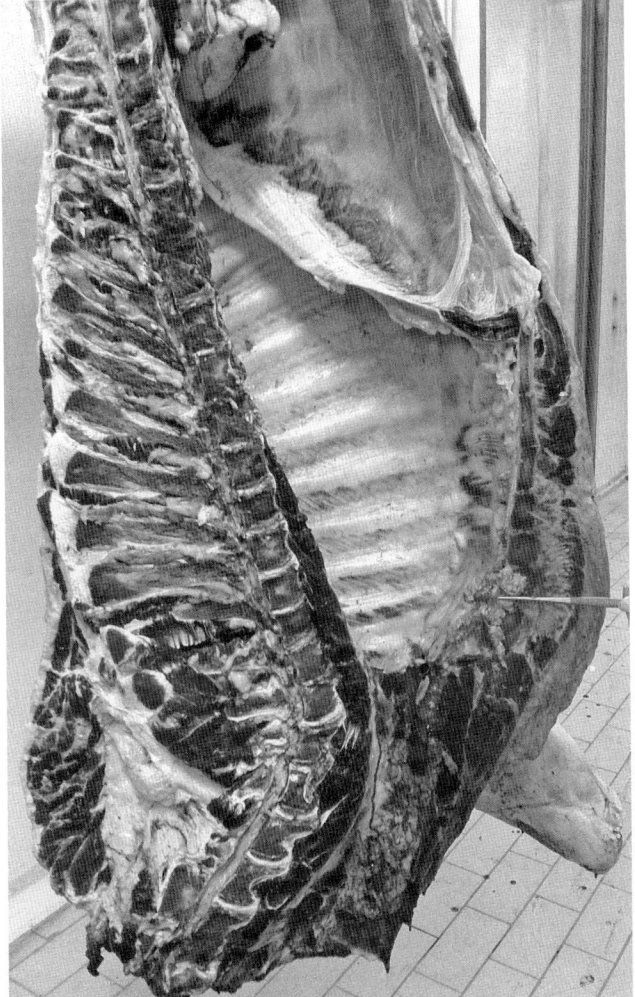

Figure 6.42 Bovine cranial sternal lymph node indicated by forceps.

to the gastrocnemius muscle. To highlight it the fleshy portions of the semi-tendinosus and biceps femoris muscles must be separated at the knee joint.

6.5.5 Specific characteristics in small ruminants

Musculoskeletal apparatus In sheep and goat the dorsal arch of the atlas is slightly convex. In the atlas of the goat ventrally there is a tubercle with a sharp edge, while in sheep the tubercle is not well developed and is rounded. The axis is similar to that of bovine but the spinous process presents a small caudal tubercle. The cervical vertebrae are much longer than in bovine. In the goat the neck is longer than the sheep. The spinous processes in the goat are longer and flatter than in sheep.

There are 13 thoracic vertebrae and, compared to bovine, they present a larger body and the spinous processes are tighter and thinner without tuberosities at their ends. The cranial margin of the transverse processes is convex. In the goat the spinous processes of the first eight thoracic vertebrae are directed caudally. In the goat the thorax is narrower and deeper than in sheep. The thoracic region of the sheep is, in fact, rounded while it is pointed in the goat.

The lumbar vertebrae are variable in number, usually 6–7, seven in sheep and six in goats. The vertebral bodies are flattened dorsoventrally with respect to the bovine. The spinous processes are broad and closely spaced to each other, while the transverse processes are all cranial in direction.

The sacrum consists of four vertebrae in the sheep and 3–5 in the goat, the last of which is always incompletely welded to the remaining part. The spinous processes remain distinct. There are about 12 coccygeal vertebrae in sheep and eight in the goat. In the goat, in fact, the tail is generally shorter than the sheep.

The 13 pairs of ribs (as in bovine), including eight sternal and five asternal. In the goat the ribs are wider and with a greater curvature at the dorsal end. However, they are less wide than those of the bovine and more convex on their outer face.

The sternum, with seven sternebrae, is flatter and wider than that of the bovine. In sheep the sternum is uniformly concave in its extension, whereas in the goat is straight with abrupt withdrawal in the front.

The pelvic girdle differs from bovine for the different form of the coxal, which is almost rectilinear given that the acetabular partions of the ilium and ischium are in the same direction. The pelvis is long and flattened dorsoventrally and only slightly inclined. The cranial opening has an elliptical shape with a vertical diameter slightly greater than a few centimetres with respect to the transverse axis. The pelvic or ischio-pubic symphysis is almost straight or slightly oblique ventrocaudally. The cranial opening of the pelvis is narrower in the goat. The pelvis itself in the goat is longer.

The diaphragm is structured as in bovine. Costal insertions are displaced less cranially, especially in the sheep the insertion on the last rib is relatively long.

Viscera annexed to the carcass
Kidneys

Overview
The kidneys are different from those of bovine. They are similar to each other, smooth with the lobes completely fused and have an elliptical shape with convex faces and rounded edges at both ends. They are about 8 cm in length and 5 cm in width. In the kidneys sinus do not appear as distinct papillae but as a single renal crest that groups the ends of approximately 10–15 pyramids.

Topography
The right kidney is placed against the lumbar wall at the level of the last rib and the first two lumbar transverse processes. The left kidney is moved to the

right of the median plane supported by a thickened mesum placed close to the dorsal sac of the rumen, in the vicinity of the third/fourth lumbar vertebra.

Weight
In small ruminants each kidney weighs about 150 g.

Colour
In small ruminants the kidneys have a reddish brown colour.

Mammary glands The sheep and goat both have a pair of inguinal mammary glands that lean on the median plane forming a single voluminous mammary complex. In sheep the average height of the gland is about 25 cm with two large nipples about 3–6 cm in length cranially directed and divergent. It has broad but shallow intermammary sulcus. In sheep at the sides of the base of the gland there is a deep depression of about 5 cm bounded by a fold of the skin: inguinalis cutaneous sinus (*sinus inguinalis*). The lactiferous sinus shows a glandular and a papillary portion: the glandular portion is large and irregular, while the papillary one is about 3 cm in length and has a regular part consisting of 4–5 large folds.

Male genitals The penis of the ram is about 40 cm in length and arranged like that of the bull. The only difference is that the glandis is more voluminous, asymmetric and curved as a hook. The testis of the ram is more spheroidal than that of the bull but arranged in the same way. Its weight is about 200 g without the epididymis.

6.6 Anatomy of poultry

6.6.1 Carcass

There is only one general cavity due to the absence of the diaphragm muscle. It is not possible to compare the general cavity of the birds to that of mammals because of the presence of air sacs connected to the respiratory apparatus and also for the special relationship that the viscera establish with the serous membranes. In the adult, the general cavity is divided into equal pleural, dorsal and ventral liver cavities. To these must be added the pericardial and the coelomic-intestinal cavities, which are unequal (Figure 6.43).

6.6.2 Viscera

Splanchnology of the digestive apparatus and annexed viscera The digestive system includes the following organs:

- oropharynx
- oesophagus and crop

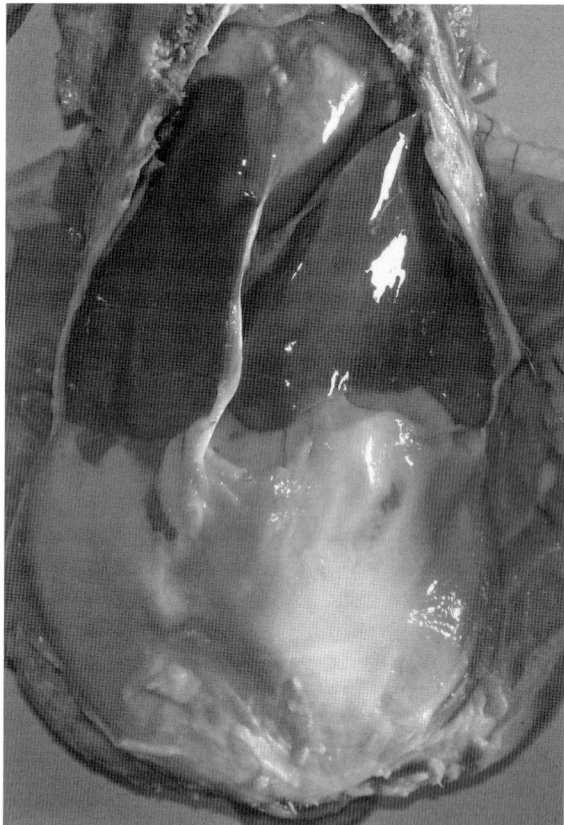

Figure 6.43   General cavity of chicken.

- stomach
- small intestine (duodenum, jejunum and ileum)
- large intestine (two caecum, colon and rectum)
- cloaca (shared with the urogenital tract).

In addition, the digestive system is connected to two major glands: the liver and pancreas.

Oropharynx   The digestive system begins with the buccal cavity, which does not have the soft palate and, thus, there is no separation between the mouth and the pharynx. Therefore, the oropharynx is a single cavity that extends from the beak to the oesophagus. The vault of the pharyngeal cavity is formed by the palate, which has a hard consistency: it is covered by a mucosa that appears raised in a longitudinal median fold and which continues aborally in two lateral folds. The pharyngeal vault continues aborally and, on the median line, there is a single elongated opening in the longitudinal direction, the choana. Caudally to the choana is the infundibular cleft, which constitutes the common orifice of the auditory tubes.

The floor of the buccal cavity is occupied by the free part of the tongue, to which it is united by a fold of the mucous called the frenulum of the tongue. In the anterior portion of the floor of the pharynx is the fixed part of the tongue or root, while observed posteriorly is the laryngeal prominence, which is characterized by a longitudinal slit, the laryngeal aditus, that misses a glottis. In birds a hyper-keratinized epidermal structure has developed; this replaces the teeth and is called beak or rhamphotheca. It consists of an upper beak with a high central portion, the culmen, and a sharp edge, the tomium. Posteriorly to the upper beak appear the orifices of the nostrils, which can be covered by opercula. The lower portion of the beak replaces the lower mandible of mammals.

Oesophagus and crop The oesophagus is a tubular organ with a muscular portion which serves as a duct for food. In birds it may also serve as a storage function due to the presence along its course of diverticula. The oesophagus, placed between the pharynx and glandular stomach, is relatively long and can be divided into a cervical and a thoracic portion. The cervical portion begins dorsally to the larynx; it starts dorsally to the trachea to then move to its right. It is very mobile and placed in a superficial subcutaneous position, thus easily palpable. In the fowl, near the entrance of the thorax on the right and leaning against the pectoral muscles, a diverticulum forms, the crop. In the adult chicken the crop and the skin that cover the right side of the thoracic inlet are strictly related. The attachment is guaranteed by two sheets of striated muscles which appear to form sling-like supports for the crop. For this reason when the crop is full in live chicken it can be easily palpated. The crop in pigeons is a ventral diverticulum of the oesophagus but differently from chickens it is divided into two large lateral sacs. The crop in both sexes of adult pigeon are enlarged structures, forming the 'milk crop'. The 'milk' is produced by the desquamation of fat-laden epithelial cells lining the organ and it is regurgitated and fed to the young during the first days after hatching. Differently, in adult duck and geese the crop is organized as a spindle-shaped swelling of the ventral portion of the cervical oesophagus. In the thoracic portion the oesophagus runs dorsally to the trachea and at the base of the heart contracts relationships with interclavicular and thoracic air sacs, then moves to the left and flows into the glandular stomach. In the duck in the terminal portion presents the oesophageal tonsil: an aggregate of lymphoid tissue.

Stomach The stomach in domestic birds has evolved as a function of dietary habits. The stomach is divided into two portions: a smaller one placed cranially, the glandular stomach, separated by a narrowing from a more massive portion, muscular stomach (gizzard). The glandular stomach, also known as proventriculus, is more developed in carnivorous birds. It continues the oesophagus without defined limits and its caudal portion is separated from the muscular stomach or gizzard by a narrowing, the isthmus. The glandular stomach is placed in the abdominal cavity ventrally and to the left of the median plane in relationship with the liver (left), the spleen (right), superiorly with the left thoracic air sac, with the corresponding gonad and caudoventrally with the sternum. The proventriculus secretes gastric juice that chemically digests food. The muscular stomach or ventriculus is particularly developed in granivorous and herbivorous species. In fowl it assumes the appearance of a squashed sack laterally placed in the medium part of the abdominal cavity in a vertical position. Itcomprises a body that expands in a craniodorsal blind sac and in a cranioventral blind sac, into which opens the beginning of the duodenum. The interior has longitudinal folds covered with a very tough membrane called koilin given by the hardening of complexed glycoproteins secreted by the glands.

Small intestine The small and large intestines occupy the caudal portion of the coelomic intestinal cavity. The small intestine includes the duodenum, jejunum and ileum. The duodenum, arranged in a U-shape, is on the right side of the abdominal wall. It has a descending and an ascending portion tightly displayed that contain the lobes of the pancreas. The duodenum in the cranial portion of the ascending portion receives the end of the pancreatic and hepatic ducts. It is joined to the gizzard and to the liver by peritoneal ligaments. Fasting in galliformes forms some wide loops, up to a metre in length supported by a long mesentery. Near the cranial mesenteric artery it is possible to view a remnant of the vitelline sac: Meckel's diverticulum. It has a thin wall that reveals the luminal content to be of a greenish colour. The ileum is short and it passes cranially near the spleen before turning caudally and continuing in the large intestine.

Large intestine The large intestine is composed of two caecum and a terminal portion indicated as colon/rectum that lead to the cloaca. It has a rectilinear course and is characterized by the presence of two diverticula, the caecum, placed at the end of the ileum. The caecum are present in all birds with an extremely varied development. From their beginning the caecum are directed cranially towards the ileum, to which they are joined by the ileo-cecal ligament. In galliformes each caecum can be externally divided into: (a) a short proximal neck portion (base), (b) a long body covered by a thin wall (body) and (c) a short distal part (apex). In the base there is a strong muscle structure that forms the caecal sphincter where large aggregates of lymphatic tissue, the caecal tonsils, are recovered. In canary and pigeon the caecum are vestigial.

The terminal section of the intestine that opens posteriorly into the cloaca is wrongly defined as the rectum. It would be more appropriate to call it the colon. The function of this terminal portion of the intestinum is the reabsorbtion of water from ureteral urine.

Cloaca This is an unequal and median organ that continues caudally the rectum and opens out into an oval orifice to a horizontal major axis. The cloaca is common to the urogenital apparatus and contains the deferent ducts or left oviduct, the ureters and the final portion of the intestine. It is divided into the coprodeum, urodeum and proctodeum. The coprodeum represents the more cranial district and is separated from the rectum by a mucosa fold and by a muscle sphincter. The urodeum in the chicken is about 0.5–1 cm in length and follows caudally to the cropodeum and in the dorsal part has the openings of the ureters and in the male, laterally to these, the opening of the deferent ducts. In the female, on its left wall, there is a wide opening pertaining to the left oviduct, the only one which is well developed in birds. The proctodeum represents the terminal part of the cloaca; located on its floor, on each side, are parts of the male copulatory organs.

Liver The liver is the most voluminous gland annexed to the digestive system and is contained in the serous liver cavities (dorsal and ventral) placed in the abdominal portion of the thoraco-abdominal cavity (Figure 6.43). It has a convex and smooth parietal surface in relationship to the portion of the abdomen leaning against the ribs, the sternum and the thoracic air sacs. In contrast, the visceral surface is concave and is related to the spleen, with the muscular and glandular stomach, with the duodenum and jejunum, as well as the right testicle or ovary. It is divided into a right and left lobe joined dorsally to the heart. The right lobe is more voluminous and in relation with the gall bladder. The right lobe is also crossed by the caudal vena cava. In ducks and pigeons the right lobe is more voluminous and in the dove lacks the gall bladder.

The left lobe is further divided by a longitudinal groove in the caudoventral and caudodorsal lobes. Since birds lack a diaphragm the liver covers ventro-cranially the heart and the pericardium. Both pancreatic and bile ducts open into the distal portion of the ascending portion of the duodenum. In adult chicken the common hepatic duct drains both hepatic lobes and it is formed by the union of the right and left hepatic ducts on the visceral surface of the right lobe. The right duct has a branch which enters the gall bladder. The gall bladder is drained by the common hepatic duct. The pigeon does not have a gall bladder, there are two common hepatic ducts. The right duct opens into the middle of the ascending portion of the duodenum and the better developed left duct opens into the proximal part of the descending portion of the duodenum.

Pancreas As in mammals, the pancreas of the birds has an endocrine and exocrine portion. It is received in the U-shaped loop of the duodenum in the thickness of the mesentery that combines the ascending with the descending portion. It is distinguished in the dorsal, ventral and splenic lobes. The latter is small and has a cranial part close to the spleen and a caudal part associated with the dorsal lobe. The pancreas is drained by three pancreatic ducts that reach the duodenum ascending in a region adjacent to the gizzard.

Splanchnology of respiratory system Respiration in birds is guaranteed by the lungs, small and inelastic, and also by numerous pairs of air sacs devoid of respiratory tissue connected to the bronchial system. The respiratory system helps to keep the homothermy, especially because these vertebrates have no sweat glands. In the respiratory tract the following organs can be recognized:

- nasal cavity
- pharynx
- larynx
- trachea
- bronchi
- lungs
- air sacs.

Nasal cavities   The nasal cavities begin with the nostrils. They are elongated openings placed at the base of the beak covered by a skin flap, the operculum. The nostrils communicate with the nasal vestibule, an uneven cavity for the presence of hollows and divided by a median nasal septum. The choana continue aborally the nasal cavities. They comprise a single median opening located on the vault of the oral cavity and a wider opening located posteriorly that opens into the vault of the pharynx.

Larynx   The larynx of birds, unlike mammals, is not involved in phonation and does not present structures similar to the vocal cords. It is as a relief on the floor of the oropharynx with an opening disposed sagittally, the glottis or laryngeal aditus. The glottis in a chicken, about 10 cm in length, leads to the laryngeal cavity bounded by the cartilages of the larynx (cricoid, arytenoid and procricoid cartilages), without the epiglottis.

Trachea   This is an unequal tubular organ formed by a series of complete ring-shaped cartilages that have a left or right half alternatively more expanded so that each ring fits with the expanded portion to the smaller part of the adjacent ring. It is divided into a cervical and a thoracic portion. In the cervical part lies ventrally to the oesophagus and then moves to the right and then again ventrally before entering the thoracic cavity. In the thoracic portion the trachea runs ventrally to the esophagus.

Syrinx   The syrinx is the vocal organ of birds. It consists of a cartilaginous skeleton made of a cranial cartilages, intermediate cartilages, caudal cartilages and pessulus. It is located where the thrachea bifurcates into the two principal bronchi and it is a modification of the bronchial and tracheal portion; the air flow causes the vibration of membranes that produces the sound.

Lungs   The lungs are equal organs that in birds are relatively small, not lobed and placed in the dorsal anterior portion of the coelomic cavity. In birds they are inextensible and connective tissue organs, rich in collagen fibres and scarce inn elastic fibres. Along the dorsal margin there are deep transverse grooves, typically five, that welcome as many ribs. Since they are inextensible respiratory acts are possible thanks to a system of air sacs that regulate air flow.

The ventral surface corresponds to the horizontal septum and does not include the heart. To understand the lung architecture it is necessary to describe the bronchi, which are divided into primary, secondary and tertiary.

From the syrinx originate two primary bronchi or mesobronchi: right and left. They have a short extrapulmonary course and a long intrapulmonary one. The primary bronchi open in the caudal extremity in the abdominal air sac through a direct connection. The primary bronchi, moreover, detach along their intrapulmonary course four series of secondary bronchi. From the secondary bronchi originate tertiary bronchi, also called parabronchi; they do not have a blind end but form between them numerous anastomoses that gives rise to a large functional unit that occupies about two-thirds of the lung. The ventromedial and dorsomedial parabronchi join to form the pulmonary parenchyma, called the paleopulmo, in relation with the cervical, cranial thoracic and clavicular air sacs. The ventrolateral and dorsolateral parabronchi form the lung parenchyma, the neopulmo, connected to the caudal thoracic and abdominal air sacs. All air sacs are connected to the lungs, except for the abdominal one, in confined areas called ostium. In each ostium a direct connection can be recognized with the secondary bronchus and more openings of smaller calibre, known as indirect connections, which are connected to parabronchi.

The air sacs perform different functions, they:

- allow the passage of air both in the inspiratory and expiratory phases, making gas exchange more efficient;
- lighten the body weight and balance the centre of gravity, adjusting it abruptly in the phases of flight and landing;
- govern the body thermoregulation.

Splanchnology of the urogenital system The urogenital system includes the urinary and genital systems that perform, in the adult poultry, different functions. However, they originate from contiguous tissues developed during the embryonic period. The organs belonging to the urogenital system are the:

- kidneys
- male reproductive system
- female genital system.

Kidneys In birds, the kidneys are equal and elongated organs positioned in two spaces extraperitoneal of the abdominal cavity bounded by the synsacrum, called renal depressions. The cranial end of the kidney reaches the corresponding lung while caudally it arrives at the caudal margin of the synsacrum. In chicken the kidney has a brown colour and is divided into three divisions, cranial, medium and caudal, of equal volume thanks to transverse grooves of varying depth. The kidney has a total length of about 6–7 cm and is 2 cm in width. Each kidney is drained by a massive ureter that leads up to the cloaca. Distinguishable in the ureter are a renal part applied to the ventral surface of the kidney and a pelvic portion, which begins at the level of the posterior end and caudally opens in the urodeum.

Male reproductive system   In birds, the male genitalia is characterized, compared to mammals, by intracavitary gonads and by less developed genital tracts. In the genital apparatus the gonads, or testes, the genital tract and the copulatory organ can be recognized. The testes have species-specific volume and shape and related to the phase of the sexual cycle of the animal; they are more voluminous in the breeding season.

In the abdomen the testis is suspended laterally to the midline of the vault of the abdominal cavity by a short mesorchium. The testicle is placed cranioventrally to the cranial division of the corresponding kidney. The male genital tract includes the epididymis, deferent ducts and the copulatory organ positioned into the cloaca.

The epididymis is a whitish tubule relatively short and twist that runs parallel to the dorsomedial margin of the gonad. It is composed of a single duct which continues caudally in the deferent duct. The deferent duct is long and convoluted with wide loops with a diameter that becomes greater proceeding caudally. It runs in the abdominal cavity to then open into the urodeum of the cloaca at the apex of a genital papilla.

Female genital system   In birds, only the left gonad and the oviduct reach complete development. The right side is present during the embryonic stage but remains vestigial. The gonad or left ovary varies in shape and volume depending on the reproductive period. The gonad is formed by ovarian follicles whose size, when mature, reach the the egg yolk. They are connected to the rest of the ovary by a peduncle. In an ovulating hen there are always 5–6 follicles of large volume. The ovary is suspended at the vault of the abdominal cavity by a short mesovarium. The left ovary is surrounded ventrally by the left abdominal air sac while dorsally it has relationships with the left kidney, part of the right one, the aorta and the vena cava. The left oviduct is a convoluted conduit with thick walls placed between the ovary and the cloaca. It undergoes changes in size according to the season. Its loops reach a length of 60–70 cm. In the oviduct the following are functionally recognizable:

- infundibulum
- magnum
- isthmus
- shell gland (improperly called the uterus)
- vagina.

The oviduct lies in the dorsal portion of the left coelomic cavity in relation to the kidney, intestine and glandular stomach. The infundibulum represents the first portion of the oviduct, is about 7 cm in length and has at its end an infundibular ostium to receive the oocyte. The infundibulum is joined to the last rib and the left abdominal air sac by a ligament. It continues in the magnum that forms large loops that in the hen reach a length of 30–40 cm. It is supported by a mesofold. Subsequently, it follows the isthmus, narrower and about 8 cm in length that enters the shell gland to finish in the vagina, the final part of the oviduct which opens with an opening in the lateral wall of the urodeum.

Splanchnology of the cardiovascular system　In birds the circulatory system, as in mammals, is double and complete. It is divided into a small (or pulmonary) and large circulation (systemic circulation). The heart is contained in the pericardial sac situated in the antero-ventral portion of the thoraco-abdominal cavity. It has a bright red colour and a cone shape. As in mammals, the right side is completely separated from the left one.

The right atrium is connected to a venous sinus, which remains distinct, within open the right cranial vena cava and the caudal vena cava. Into the right atrium flow also the left cranial vena cava and the cardiac veins. The left atrium is separated from the right by a thin interatrial septum and into it flow the pulmonary veins. The right ventricle, placed under the corresponding atrium, has thin walls and does not reach the apex of the heart. It has an atrioventricular ostium and through the conus arteriosus it continues into the pulmonary artery. The left ventricle has thicker walls and ends in the apex of the heart. It has an atrioventricular ostium, which in its cranial portion can be recognized the vestibule of the aorta, from which rises the ascending aorta. The pericardial sac is formed by a fibrous lamina internally coated by a parietal serous and the visceral serous or epicardium. The pericardial cavity containing a small amount of liquid is located between the two lamina.

6.7　*Post-mortem* inspection

The *post-mortem* inspection of the carcass and the organs of a slaughtered animal represent a fundamental step of traditional meat inspection. The use of the terms 'meat inspection' and '*post-mortem* inspection' have been frequently used incorrectly, as meat inspection is defined as the whole process of judgment of meat, which is performed in different steps, and includes the *post-mortem* inspection. The *post-mortem* inspection is precisely the routine examination of the carcass and organs of a slaughtered animal.

6.7.1　Scope of the *post-mortem* inspection

The target of the *post-mortem* inspection is to highlight any pathology or abnormalities present in the carcass and/or non-carcass components of a slaughtered animal and to provide information for the judgment of the meat. It might be crucial to:

- determine if the single finding is acute or chronic;
- determine if the findings refer to a localized or generalized condition;
- determine which are the primary and which are the secondary findings.

This information can be important for the judgment of the meat although it might not always be obtainable.

The *post-mortem* inspection should be carried out as soon as possible after slaughtering in order to improve the possibility of detecting any signs of abnormalities on the carcass and/or the organs that may prejudice the final judgment of the meat. The carcass and non-carcass components, which belong to the same animal, need to be identifiable and connectable with each other until the final judgment of the meat. This is in order to be able to reject the whole or part(s) of the carcass and/or non-carcass components according to the general picture recovered from the inspective activities. The procedures to be followed in the routine *post-mortem* inspection need to respect the requirements set by the legislation in force and may vary from country to country.

6.7.2   *Post-mortem* inspection techniques

The *post-mortem* inspection requires high professional and technical skills. Proper scientific knowledge at least of the anatomy, pathology and epidemiology related to the slaughtered species is required. The techniques used in the traditional *post-mortem* inspection are:

- **Visual inspection**, which is the observation of the carcass and organs and the evaluation of their aspect. A proper knowledge of anatomy permits certain asymmetries, abnormalities or pathologies to be detected.
- **Palpation**, which is an examination by touch of the meat. A properly performed palpation provides information concerning the consistency and integrity of the examined tissues. The use of one or two palpating hands and the grade of pressure to apply on the tissue depend on the type of tissue under examination. The use of this technique can be mandatory for various tissues according to the legislation or optional in case of suspicious findings.
- **Incision** involves the sectioning, by knife, of tissues. In some cases, such as in the presence of abscesses or faecal contamination, an incision increases the risk of contaminating or soiling the meat and it is, therefore, better to avoid this technique and choose an alternative. The use of this technique can be mandatory for various tissues according to the legislation or optional in case of suspicious findings.
- **Olfaction** is based on the sense of smell and provides information related to abnormal odours. Commonly perceived odours are the odour of urine, typically due to a breakage of the urinary bladder, and the sexual odour.

6.7.3   Visual meat inspection

Visual meat inspection is related to the visual only *post-mortem* inspection. The visual only *post-mortem* inspection is generally linked to a remarkably good epidemiological situation and a reliable primary production of animals intended for slaughtering. The lack of the inspective procedures that demand physical contact directed on the meat, such as palpation or incision, is likely also to reduce the degree of microbial contamination of the meat.

Literature and further reading

EFSA (European Food Safety Authority). EFSA opinions on meat inspection, http://www.efsa.europa.eu/en/topics/topic/meatinspection.htm; swine: http://www.efsa.europa.eu/en/efsajournal/pub/2351.htm; poultry: http://www.efsa.europa.eu/en/efsajournal/pub/2741.htm; bovine animals: http://www.efsa.europa.eu/en/efsajournal/pub/3266.htm; sheep and goats: http://www.efsa.europa.eu/en/efsajournal/pub/3265.htm (all last accessed 17 February 2014).

EUR-Lex. Details of up-to-date EU legislation on meat inspection can be found at: http://eur-lex.europa.eu/en/index.htm (last accessed 17 February 2014).

7

Risk-Based Meat Inspection

Maria Fredriksson-Ahomaa

Department of Food Hygiene and Environmental Health,
Faculty of Veterinary Medicine, University of Helsinki, Helsinki, Finland

7.1   Scope

In this chapter, risk-based meat inspection is discussed briefly. Furthermore, the connection between risk-based meat inspection with visual meat inspection, food chain information and disease monitoring is elucidated.

7.2   Introduction

Important objectives of meat inspection are to protect the consumer and to ensure good animal health and welfare. These objectives are achieved at the slaughterhouse by meat inspection including *ante-mortem* and *post-mortem* inspection and by hygienic slaughtering. Meat inspection is performed in order to recognize (i) slaughter animals that are unfit for human consumption, (ii) slaughter animals with notifiable diseases and (iii) animal welfare problems. *Post-mortem* inspection is usually based on visual examination, palpation, incision and laboratory examinations. Meat inspection detects conditions that are mostly animal health related, such as emaciation, oedema, colour changes, abscesses and bruises; less frequently it detects public health related hazards. Meat inspection has been focused on zoonotic agents that were more common decades ago than today, such as *Mycobacterium*,

Meat Inspection and Control in the Slaughterhouse, First Edition.
Edited by Thimjos Ninios, Janne Lundén, Hannu Korkeala and Maria Fredriksson-Ahomaa.
© 2014 John Wiley & Sons, Ltd. Published 2014 by John Wiley & Sons, Ltd.

Table 7.1   Preventive measures for the most important meat-borne zoonotic agents in Europe.

Agent	Preventative measure[a]					
	Herd level	Serological categorization	Meat inspection	Slaughter hygiene	Carcass Decontamination[b]	Freezing
Campylobacter	+++	++	−	+	++	++
Salmonella	+++	++	+	+	++	−
Yersinia[c]	+++	++	−	+	++	−
STEC[d]	+++	−	−	+	++	−
Toxoplasma	+++	++	−	−	−	+++
Trichinella	+++	++	+++[e]	−	−	+++

[a] great effect: +++; good effect: ++; limited effect: +
[b] by steam or/and hot water
[c] pathogenic *Yersinia enterocolitica* and *Y. pseudotuberculosis*
[d] Shiga toxin-producing *Escherichia coli*
[e] every carcasses is tested by a validated reference method, usually by digestion method

Brucella, Cysticercus and *Trichinella*, and cannot usually detect frequently occurring meat-borne hazards (such as *Campylobacter* and *Salmonella*) or chemical substances (such as residues of veterinary drugs and contaminants) because these hazards do not typically cause any clinical symptoms in live animals or pathological lesions in the carcass or/and offal. Furthermore, physical meat inspection, including palpation and incision of organs such as lymph nodes, increases the risk of cross-contamination of meat.

Dissemination of zoonotic bacteria such as *Salmonella*, thermotolerant *Campylobacter*, Shiga toxin-producing *Escherichia coli* and enteropathogenic *Yersinia*, especially, can easily occur during slaughtering through equipment and hands and during *post-mortem* inspection through palpation and incision. Thus, preventative methods should already start at the farm level by decreasing the prevalence of zoonotic microbes in the slaughter animals (Table 7.1). Risk-based meat inspection is focused on the meat chain from farm to slaughterhouse and, typically, palpation and incision are avoided to minimize the handling in order to reduce the risk of cross-contamination during *post-mortem* meat inspection.

7.3   Risk-based meat inspection

During the last decade, meat inspection practices have been developed to a more risk-based approach to protect more effectively human health against meat-borne biological and chemical hazards. In this approach, the main risks (biological and chemical risks) for public health that should be addressed by meat inspection have been identified and ranked. For biological hazards, the priority ranking is usually based on assessment of their impact on (i) incidence of disease, (ii) the severity of the disease in humans and (iii) evidence that consumption of meat of various animal species is an important risk factor for the disease. Risk ranking of chemical hazards is based on (i) the outcomes of national residue control plans, (ii) outcomes of other testing programmes and (iii) substance-specific criteria like toxicological profile.

Risk-based meat inspection is a system based on risk assessment, risk management and risk communication in the meat chain from the farm until the carcass leaves the slaughterhouse.

It is important that all persons involved in the meat production chain share the meat producers' responsibility for meat safety (zoonotic agents and residues), animal health (notifiable and production diseases) and animal welfare. The producers along the meat production chain include the feed producers, farmers and slaughterhouses. The goal is to assure production processes at the farm level resulting in healthy animals. Existing data, such as farm records and meat inspection results, are important for risk-based selection of high and low-risk operators. Food chain information (FCI), which includes information from farm to slaughterhouse in order to classify the batch/flock according to its expected safety risk, is used as an important part of *ante-mortem* inspection. Monitoring of biological and chemical hazards is regularly carried out at the slaughterhouse. Collection of this information is a basic element of risk-based meat inspection.

7.4   Visual-only *post-mortem* meat inspection

Risk-based meat inspection is typically based on a visual-only *post-mortem* inspection of individual animals. Only if necessary, palpation and/or incision of lymph nodes, offal and carcass meat are performed. This hands-off inspection reduces (i) the cross-contamination of meat with microorganisms, including zoonotic meat-borne agents, and also (ii) the exposure of meat inspectors to occupational hazards as well as (iii) the exposure of meat to microbial hazards originating from the inspectors. Typically, the slaughter animals that undergo a visual-only *post-mortem* meat inspection have been reared under controlled housing conditions. The definition usually includes only animals, typically fattening pigs and poultry, which are raised indoors and in an integrated system. An integrated system includes criteria regarding feed and bedding, detailed information about the animals from birth to slaughter and their management conditions. Furthermore, it requires FCI to be transferred backwards and forwards between the farm and the slaughterhouse.

One important prerequisite for visual-only meat inspection is that the animals at slaughter are healthy and hands-off inspection can be carried out. The time saved from omitting palpations/incisions can be used towards better exploitation of FCI and a greater focus on slaughterhouse process hygiene controls. Furthermore, the decrease of monotonous manual tasks and increase of more stimulating tasks may increase the satisfaction of meat inspectors. However, visual-only *post-mortem* meat inspection may have negative impact on the meat quality due to a reduced number of meat inspectors, which could result in visible lesions not being detected.

7.5   Food chain information (FCI)

In a risk-based meat inspection system, FCI is an essential tool for risk management and is used to plan the slaughtering. It should include information about the farm (whether it belongs to an integrated system and whether it fulfils the criteria of controlled housing conditions), the occurrence of diseases that may affect the meat safety, the health status of the animals (in the herd of origin and at the region level), the medical or other treatments, previous meat inspection reports (repeatedly high frequency of pathological lesions) and the monitoring results (zoonoses and residues) among other information. FCI is the basis to differentiate animal batches needing a more intensive meat inspection from batches needing only a hands-off inspection (visual-only). A more thorough meat inspection (typically including palpation and incision and, if necessary, also laboratory tests) is needed for slaughter animals if macroscopic lesions are expected or if the animals cause a meat safety risk. This information provides meaningful information for the official meat inspector. Risk assessment and the risk-based decision of official veterinarians and slaughterhouse operators relies on a functional system of FCI.

7.6   Monitoring of diseases by serology in the slaughterhouse

It is important for risk-based meat inspection to obtain data on the occurrence of subclinical infections in slaughter animals (and in the herd of origin) with relevance for meat safety and for animal health and welfare. Zoonotic diseases, production diseases and notifiable diseases can easily be monitored at slaughter by serology using meat juice. Valuable data for slaughterhouses, official veterinarians and farms will be obtained, especially if meat juice samples are collected on a permanent basis. These data provide a useful tool to improve food safety and animal health as a part of the risk-based meat inspection.

7.7   Conclusions

The primary goal of official meat inspection is to prevent human health hazards from entering the meat production chain. Meat inspection also provides valuable information on animal health and welfare issues. A risk-based meat inspection is food chain-orientated and typically it is a visual-only inspection (palpation and incision procedures are omitted). The benefit of visual-only inspection is related to (i) a presumed reduction in microbial cross-contamination of the carcasses and offal and (ii) a potential saving in time and resources. A risk-based meat inspection, which is based on healthy animals, aims to provide safe and high quality meat.

Literature and further reading

Blaha, T. 2012. One world-one health: threat of emerging diseases. A European perspective. *Transboundary and Emerging Diseases*, **59**, 3–8.

Hill, A.A., Horigan, V., Clarke, K.A., *et al*. 2013. A quality risk and benefit assessment for visual-only *post-mortem* meat inspection of cattle, sheep, goats and farmed/wild deer. Animal Health and Veterinary Laboratories Agency, UK. http:// www.foodbase.org.uk//admintools/reportdocuments/798-1-1416_FS245028_RA _RedMeat.pdf (last accessed 1 February 2014).

8

Meat Inspection Lesions

Jere Lindén[1], Leena Pohjola[2], Laila Rossow[3] and Daniele Tognetti[4]

[1] Pathology and Parasitology, Faculty of Veterinary Medicine, University of Helsinki, Helsinki, Finland

[2] Department of Production Animal Medicine, Faculty of Veterinary Medicine, University of Helsinki, Helsinki, Finland

[3] Production Animal and Wildlife Health Research Unit, Research and Laboratory Department, Finnish Food Safety Authority Evira, Helsinki, Finland

[4] Ministry of Health, Rome, Italy

8.1 Scope

The aim of this chapter is to summarize a few of the most prevalent lesions of some meat-producing animal species that can be detected in a practical meat inspection setting and to give a short and practical introduction to the morphology and ethiopathogenesis of the lesions.

8.2 Introduction

In this chapter morphological diagnoses are used to classify the lesions and to help referencing literature. However, the pathological discussion has been kept minimal. Most of lesions dealt with have an infectious etiology and the morphological diagnosis is thus based on inflammatory nomenclature:

Meat Inspection and Control in the Slaughterhouse, First Edition.
Edited by Thimjos Ninios, Janne Lundén, Hannu Korkeala and Maria Fredriksson-Ahomaa.
© 2014 John Wiley & Sons, Ltd. Published 2014 by John Wiley & Sons, Ltd.

nature of the inflammatory exudate and chronicity of the ailment. Technical slaughter- and processing-induced injuries are not covered.

The findings are presented in the form of a simple three-column table: The first column contains a short morphological description, common name and/or a macroscopic morphological diagnosis of a lesion or disease. The second column lists some suggested etiologies and comments (if feasible) on the pathogenesis, while the third column gives a short macroscopic description of the major findings. The figures exemplify selected typical lesions.

8.3 Bovines

Generalized conditions

Sepsis/pyaemia	***Etiology***	
	Sepsis	Sepsis
	Many gram-negative and gram-positive bacteria.	Petechial haemorrhages on serous membranes, especially on kidney surface and in kidney cortex; generalized reddening of the carcass and parenchymal organs. Splenic hyperemia (congestion), may be extreme in acutely fatal cases – Anthrax.
	• Anthrax: *Bacillus anthracis*	
	Pyaemia	Pyaemia
	• *Trueperella (Arcanobacterium) pyogenes*	Pinpoint to few millimetres purulent ill-defined foci in parenchymal organs: Kidneys usually affected (white spotted kidney), lungs, liver and spleen.
	• *Streptococcus* spp.	
	• many others	

Generalized conditions

Lymphoma	***Etiology***	EBL, BLV-associated type
	Two types: bovine leukemia virus (BLV) – associated type and sporadic type (idiopathic). BLV belongs to the family *Retroviridae* causing a disease called Enzootic bovine leucosis (EBL).	Lymphomatous tumours are located in the superficial, pelvic and abdominal lymph nodes and in the abdominal wall, retrobulbar space and vertebral canal. Organs most frequently involved are the abomasum, right auricle of the heart, spleen, intestine, liver, kidney, omasum, lung and uterus.
	The non-BLV associated lymphoma, 'sporadic type, sporadic bovine lymphoma (SBL)', is subdivided into thymic, multicentric, calf and cutaneous forms.	Thymic lymphoma (sporadic) Massive thymic enlargement. Multicentric lymphoma (sporadic)
	Age of affection and distribution of the lesions depend on the type (EBL or SBL) and form of the disease. In EBL, tumours are seen typically in animals over 3 years of age; only 0.1–10% of the infected animals develop tumors (30–70% persistent lymphocytosis). The calf form of SBL affects calves at birth and up to 6 months of age. Thymic lymphoma is most prevalent in beef cattle 6–24 months of age.	Widespread symmetric lymph node enlargement, tumours in the liver, spleen, kidneys (diffuse involvement) and, sometimes, in skeletal muscle. Calf form (sporadic) Visible lymphomatous infiltrates in all lymph nodes, liver, spleen, kidneys and bone marrow. See swine lymphoma for description of the lesion morphology.

(*continued overleaf*)

Generalized conditions

Cachexia	***Etiology***	
	Any chronic disease; e.g. systemic neoplasia, malabsorption.	Loss of subcutaneous and intra-abdominal fat deposits, serous atrophy of perirenal, mesenterial and epicardial fat.
	Often a combination of old age, chronic mastitis, skin lesions and/or subcutaneous localized purulent processes and/or leg ailments without any cardinal disease.	Muscular atrophy and shrunken lymph nodes. Finally serous atrophy of bone marrow fat (Figure 8.1).

Thoracic cavity

Pleural adhesions with variable pulmonary affection (Chronic fibrotic pleuritis or pleuropneumonia)	Microbiology usually negative if no abscess formation.	Fibrous adhesions between visceral and parietal pleura.
	Etiology	Mild to no pulmonary fibrosis or scarring in lung parenchyma.
	Generally sequel to suppurative (lobular) bronchopneumonia after the lung lesions have healed.	Macroscopic findings are generally not etiology specific.
	Suppurative bronchopneumonia usually results from bovine enzootic (viral, mycoplasmal) pneumonia, 'calf pneumonia', complicated by secondary bacteria:	Severe suppurative broncopneumonias may lead to lung abscesses or scarring and/or marked pleural adhesions, see also Pneumonic Mannheimiosis.
	• *Trueperella (Arcanobacterium) pyogenes* • *Pasteurella multocida* • *Histophilus somni*	
	Without purulent activity not a food hygiene hazard.	

Thoracic cavity

| Pneumonic Mannheimiosis (Shipping Fever) | **Etiology**

• *Mannheimia haemolytica*

Respiratory pathogen in cattle and sheep. Often a secondary invader after viral infection.
Severe toxaemia and marked respiratory signs. Stress predisposes to infection.
Contagious bovine pleuropneumonia, caused by *Mycoplasma mycoides* ssp. *mycoides,* exhibits similar lesions. | Prototypic fibrinous broncopneumonia with extensive fibrinous pleuritis. Lesions are cranioventral. The interlobular septa are distended by yellow gelatinous edema and fibrin. Classically cut surface shows a "marbling" pattern resulting from intermixing areas of coagulation necrosis, interlobular interstitial oedema and congestion.
Lung abscesses, encapsulated sequestra, chronic pleuritis and abundant pleural adhesions are typical chronic findings. |

Abdominal cavity

| Peritonitis | **Etiology**
Usually traumatic reticulitis (hardware disease): localized lesions near reticulum. Infection spread through diaphragm may lead to pericarditis. Any breach in forestomac, abomasal or intestinal wall integrity may lead to peritonitis. | Chronic or chronic purulent/ suppurative.
Fibrous adhesions between reticulum, diaphragm and possibly liver. Sometimes abscess formation in the peritoneal cavity or in liver and/or empyemic pockets in the peritoneal cavity. |

(*continued overleaf*)

Abdominal cavity

	The intensive fibrin production, typical for cattle, generally efficiently restricts infection, but leads often to substantial fibrotic adherence formation. Peritonitis and pericarditis, liver involvement and/or extensive peritoneal changes may lead to pyaemia or sepsis.	Acute Admixture of copious friable fibrin deposits and pus. Variable amount of cloudy ascites. Localized peritonitis around liver (e.g. due to abscesses) is often called perihepatitis.
Hepatic steatosis	***Etiology*** In ruminants the final cause of hepatic lipid accumulation is generally an excessive entry of fatty acids to the liver, either due to increased mobilization high energy demand (typically lactation or late pregnancy) or surplus energy intake. In severe cases defective hepatocyte function also results in decreased export of lipoprotein. In dairy cows 'bovine fatty liver disease' is mechanistically similar to ketosis and occurs in obese animals in late gestation or peak lactation. Severe cases often precipitated by illness-induced anorexia.	The liver is enlarged, yellow and friable, and the acute edges variably rounded. In extreme cases, liver size is considerably increased and texture greasy.

Abdominal cavity

Liver flukes	See small ruminants.	
Liver abscesses	Bacteria reach liver in adults through portal vein, hepatic artery (bacteraemia), biliary system and by direct extension from adjacent tissues, such as the reticulum. Often result from damage to the ruminal mucosa that allows microflora to enter portal circulation. The mucosal damage may be caused by acute lactic acidosis (toxic rumenitis) or long-term low ruminal pH. ***Etiology*** • *Trueperella (Arcanobacterium) pyogenes* • *Fusobacterium necrophorum* • other bacteria • fungi In *acute salmonellosis* the bacteria may spread via portal vein leading to multiple foci of necrosis and later histiocytic inflammation, 'paratyphoid nodules'.	Multiple liver abscesses or purulent inflammatory pinpoint foci – often not evenly distributed. *Fusobacterium necrophorum* and fungi may cause extensive necrosis. The abscesses may rupture into the hepatic vein or vena cava and induce embolic secondary abscesses to the lungs. The miliary focal necroses and paratyphoid nodules in salmonellosis are usually only 1–2 mm in size and have no capsule. In salmonellosis the liver lesions are typically accompanied by fibronecrotic ileotyphocolitis (terminal small intestine, caecum and colon).
Teleangiectasia	Dilatation of sinusoids in areas where hepatocytes have been lost.	Variable sized red blue foci in liver parenchyma that vary from pinpoint to several centimetres in size. No inflammation or fibrosis.

(continued overleaf)

Abdominal cavity

Liver fibrosis/ cirrhosis	Liver fibrosis is a common manifestation of many chronic liver diseases. It has to be very extensive to lead to signs of hepatic dysfuntion. Cirrhosis is the final irreversible result of several hepatic diseases: chronic toxicity, cholangitis or obstruction, chronic hepatic congestion, chronic hepatitis. Cirrhosis generally leads to signs of hepatic failure.	Fibrosis Firm to hard liver with increased amount of fibrous tissue. Individual areas or lobes often variably affected. Cirrhosis Diffuse process involving whole liver. Consists of both fibrosis and conversion of normal liver architecture into structurally abnormal lobules; collection of fibrosis-restricted collection of regenerative nodules.
Embolic nephritis	***Etiology*** • *Eschericia coli* • *Salmonella* spp. • *Brucella* spp. • *Trueperella (Arcanobacterium) pyogenes* Hematogenous bacteria lodge in glomerular tufts and capillaries. The initial microabscesses are replaced by chronic inflammation and finally fibrosis.	'White spotted kidney' Miliary small white nodules throughout the cortex. Later fibrotic scars and capsular adhesions.

Musculoskeletal system

Arthritis	**Etiology**	The affected joint is swollen
	• *Trueperella (Arcanobacterium) pyogenes* • *E. coli, Streptococcus* spp. (generally in neonatal animals after septicemia) In adult animals mostly introduced by direct inoculation.	and the joint capsule variably thickened. In acute infections, the macroscopic joint lesions other than swelling may be inconspicuous. Synovial fluid is usually reddened and thin. It may turbid and contain streaks of fibrin. See swine for description of subacute macroscopic joint lesions.

The integument/muscle

| Abscesses and cellulitis | **Etiology**
Cattle are prone to develop muscular and subcutaneous infections caused by pyogenic bacteria, most commonly *Trueperella (Arcanobacterium) pyogenes*.
Infection routes: direct contamination of wounds or injection sites or haematogenous spread generally due to short-term bacteraemia. | Hind legs most often affected.
Lesion extent varies from isolated encapsulated intramuscular abscesses to diffuse purulent cellulitis extending down the tissue and fascial planes (Figures 8.2 and 8.3).
Abscesses are encapsulated and filled with thick yellow-green pus. In cases of cellulitis, pus dissects along fascial planes and perimyseal sheaths inside muscle, inducing myonecrosis and later replacement by fibrous tissue. |
| Subcutaneous eosinophilia | **Etiology**
Onchocerca spp. microfilariae
Diff. diagnosis: *Parafilaria bovicola* | Subcutaneous greenish (eosinophilia) vaguely-bordered areas that have a metallic odour. Most prominent over flanks, lateral thigh areas, and over joints. |

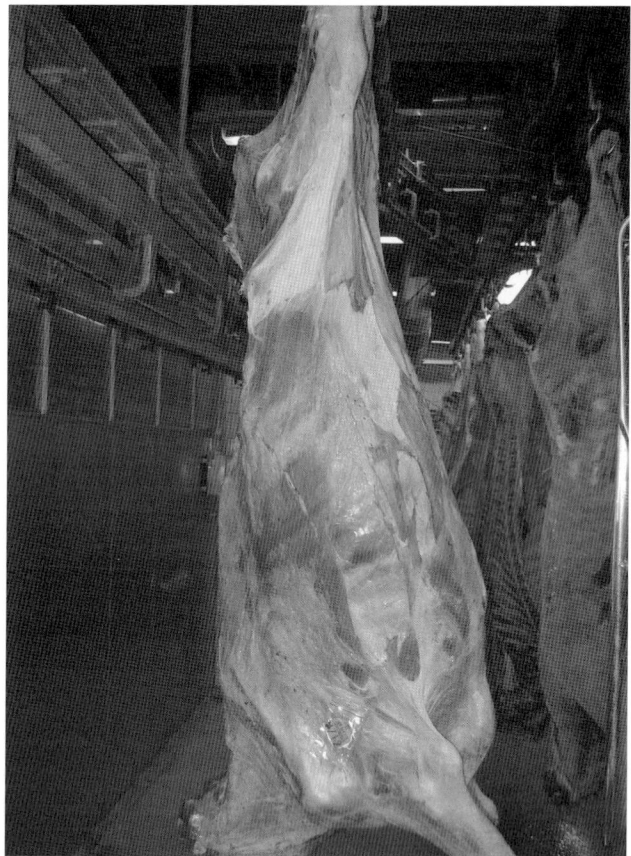

Figure 8.1 Cachexia. Carcass, bovine. Source: Courtesy of Elisa Pitkänen.

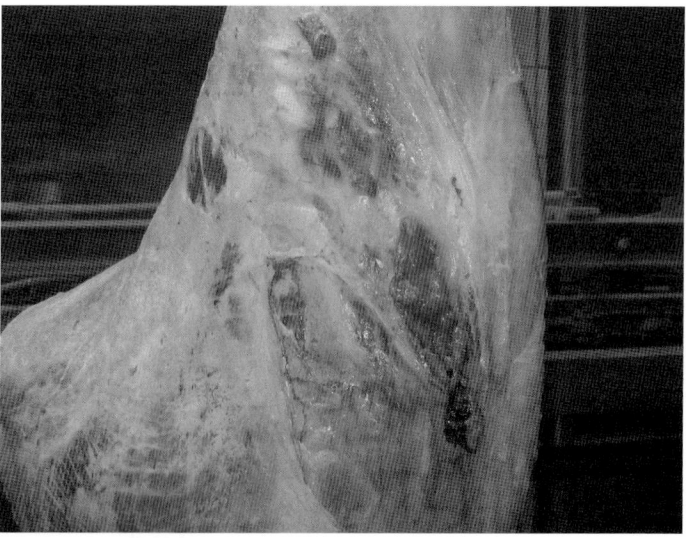

Figure 8.2 Purulent cellulitis and abscess formation. Thigh area, bovine. Source: Courtesy of Elisa Pitkänen.

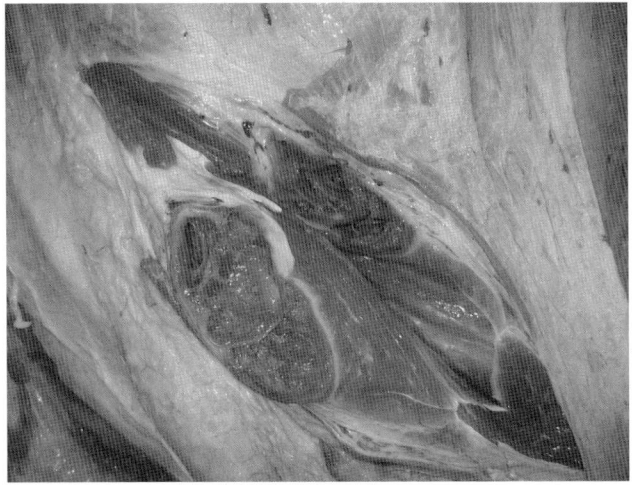

Figure 8.3 Abscess, necrosis. Typical injection site lesions. Muscle, bovine. Source: Courtesy of Elisa Pitkänen.

8.4 Domestic swine

Thoracic cavity

Pleural adhesions with minimal pulmonary affection (Chronic fibrotic pleuritis or pleuropneumonia)

Generally sequel to suppurative (lobular) bronchopneumonia after the lung lesions have healed.

Etiology

Microbiology usually negative.

See below for possible microbial etiologies.

Arguably the most prevalent lesions in swine meat inspection (up to one-third of carcasses in some lots in Finnish material). Very common in sows.

Without purulent activity not a food hygienic hazard.

Fibrous adhesions between visceral and parietal pleura (Figure 8.4).

Minimal or no pulmonary fibrosis or scarring in lung parenchyma.

Macroscopic findings are generally not etiology specific.

(*continued overleaf*)

Thoracic cavity

Widespread pleural adhesions with possibly suppuration, accompanied by pulmonary affection (Chronic purulent necrotizing pleuropneumonia)	A continuum with the above lesions. Generally sequel to fibrinous broncho-pneumonia or porcine pleuropneumonia. The latter has typically multiple pulmonary abscesses and large sequestra (see below). *Etiology* Microbiology of little value in routine meat inspection. See below for possible microbial etiologies. Lesions have to be differentiated from embolic abscessation and the *possibility of pyaemia considered.* Leads usually to extensive local condemnation.	Advanced pleural adhesions to empyemic pockets in the pleura. Scarring of the lung parenchyma, variable abscess formation and/or purulent infiltrate and/or sequester formation (necrotic, fibrosis-encapsulated lung tissue) (Figure 8.5).
Suppurative (lobular) broncopneumonia	*Etiology* • *Trueperella (Arcanobacterium) pyogenes* • *Streptococcus suis* and other Streptococci • *Bordetella bronchiseptica* • *Esherichia coli* • *Haemophilus parasuis* and other *Haemophilus*	Cranioventral consolidation and variable hyperaemia of the lungs with purulent or mucopurulent exudate in the airways. The lesions are generally confined to individual lobules leading often to an emphasized lobular pattern.

Thoracic cavity

	Often several of the above commensal bacteria in combination. These are typically secondary invaders after viral infections, *Mycoplasma hyopneumoniae* and/or environmental stress (e.g. air ammonia). Commonly becomes chronic but in favour- able conditions the infection may recede and the pathological changes disappear in one to two weeks.	In more chronic lesions the affected lung tissue is pale grey and firm, 'fish flesh' appearance. Final chronic stage is variable pulmonary scarring and/or pulmonary abscessation and/or pleural adhesions – or there may be only minimal changes.
Fibrinous (lobar) bronchop- neumonia	A continuum with suppurative bronchopneumonia. **Etiology** A severe mixed infection with the aforemen- tioned suppurative bacteria. • *Pasteurella multocida*, frequent secondary commensal invader (see above): Porcine pneu- monic pasteurellosis. More advanced lobar pulmonary injury (necrosis) and pleural involvement than in suppurative broncho- pneumonia. Restitution of the lesions is unlikely and the acute infection may lead to lethality due to sepsis or toxaemia. A possible septic or toxaemic state should be excluded.	Cranioventral consolidation and hyperemia of the lungs with fibrinous exudate and ground glass (gray-yellow-tan) appearance of the pleural surface and fluid in the pleural cavity. The inflammatory process involves several contiguous lobules affecting entire lobes and may lead to pulmonary necrosis. In chronic lesions, there is usually pulmonary scarring, necrosis may develop into sequestra and there may be abscess formation. Pronounced pleural changes with fibrotic adherences and sometimes suppuration are generally present.

(continued overleaf)

Thoracic cavity

Porcine pleuropneumonia	**Etiology** • *Actinobacillus pleuropneumoniae* Considered to be a primary pathogen with several virulence factors, also a secondary invader. Acute clinical signs are variable. Survivors often carry the infection in their tonsils. • *Salmonella choleraesuis* Usually causing enterocolitis and septicaemia in swine. May cause necrotizing fibrinous pleuropneumonia similar to porcine pleuropneumonia.	Severe fibrinous bronchopneumonia with extensive pleuritis. Consolidated areas have usually irregular, well-demarcated necrotic foci. Commonly, dorsal areas of the caudal lobes are affected. Fibrinous necrotizing pleuropneumonia involving the caudal lobe is considered almost pathognomonic for the disease. Multiple abscesses, large (2–10 cm) sequestra and pleural adhesions are typical chronic findings.
Porcine enzootic pneumonia	**Etiology** • *Mycoplasma hyopneumoniae* A highly contagious primary pathogen of worldwide distribution. Probably the most economically significant respiratory disease of pigs. The severity of the disease is much influenced by the immune status of the animals, management factors, ventilation, humidity and temperature. Mortality is generally low but complicating infections are common and lead to more severe pathology (see above).	Suppurative or catarrhal bronchopneumonia with mucoid to purulent exudate. The consolidated cranioventral lung lobules are dark red and firm, 'meaty'. Their outlines are accentuated by interlobular oedema. There may be a mild fibrinous pleuritis. In more chronic stages – without secondary infections – the affected tissue becomes paler, even pale grey, finally leading to variable fibrosis.

Thoracic cavity

Embolic pneumonia	***Etiology*** Haematogenous septic emboli that lodge in pulmonary arterioles or alveolar capillaries. • *Trueperella (Arcanobacterium) pyogenes* • *Streptococcus* spp. • *Erysipelothrix rhusiopathiae* • *Fusobacterium necrophorum* and other anaerobes etc. In swine, the source of septic emboli is most often tail biting lesion. Other skin lesions or valvular endocarditis are much less common causes. Embolic pneumonia may indicate pyaemia.	Multiple focal lesions randomly distributed in all pulmonary lobes. Early lesions are small white foci surrounded by haemorrhagic halo. Rapidly progress to abscesses.
Endocarditis	***Etiology*** • *Erysipelothrix rhusiopathiae* - swine erysipelas • *Streptococcus* spp. Large masses lead to congestive heart failure, usually chronic pulmonary congestion. The risk of bacteraemia should be assessed by culture and/or examining the kidneys for acute embolic nephritis: see Swine erysipelas.	Friable yellow-to-grey blood-streaked fibrinous masses adhering to affected, usually left atrioventricular, valves.

(*continued overleaf*)

Thoracic cavity

Glasser's disease	***Etiology*** • *Haemophilus parasuis*, part of the normal flora of the upper respiratory tract. Generally initiated by stress. Low morbidity, high mortality. Also other bacteria, e.g. *Streptococcus suis* and *Mycplasma hyorhinis*, may induce similar type of disease.	Characterized by fibrinous polyserositis; pleuritis, pericarditis, peritonitis, arthritis and leptomeningitis in various combinations.

Abdominal cavity

Capsular and portal liver fibrosis ('milk spots')	***Etiology*** • *Ascaris suum* larval migration. Differential diagnosis: liver mycobacteriosis.	Fibrous grey-white scars in the migration tracks and in adjacent portal areas (Figure 8.6).
Swine mycobacteriosis	***Etiology*** • *Mycobacterium avium* complex Pigs are infected through feed and the lesions thus concentrate on the liver and lymph nodes of the alimentary canal. Domestic swine are an unusual host for *Mycobacterium bovis*; however, wild boar may act as an important reservoir of bovine tuberculosis. Tuberculosis granulomas are most prevalent in the lymph nodes of the head, while generalized involvement is also common. The lesion size	Miliary grey-white pinpoint to few millimeters, granulomatous lesions in the liver: disseminated granulomatous hepatitis (Figure 8.7). Enlarged retropharyngeal, submandibular, cervical and/or mesenterial lymph nodes. Cut surfaces diffuse white-grey or may contain granulomatous lesions with yellowish caseotic material. Usually there is little capsule formation and no calcification. Kidney, spleen and/or lung affection with pinpoint granulomas is possible but uncommon.

Abdominal cavity

	varies from pinpoint to several centimetres and the granulomas are often caseotic and calcified. Differential diagnosis: multicentric lymphoma in the visceral lymph nodes.	
Multicentric lymphoma	***Etiology*** Swine lymphomas are idiopathic. Visceral, multicentric and mediastinal forms.	Visceral lymph node enlargement. Multiple round (1 cm) nodules in the liver. Spleen, kidneys and bone marrow may also be affected. The lymph nodes and tumour nodules have usually a bulging grey-white amorphous cut surface. Mediastinal form Mediastinal lymph node enlargement

Musculoskeletal system

Abscesses	Bacteria usually emerge form tail biting lesions. Hematogenous spread often leads to pyaemia and possibly widespread abscessation. See embolic pneumonia ***Etiology*** See embolic pneumonia for a list of potential bacterial species. Bacterial spread to vascular loops beneath growth plate cartilage resulting in embolic osteomyelitis.	Embolic osteomyelitis in the vertebrae and abscesses in adjacent soft tissues point to haematogenous spread. Abscesses in the pelvic cavity or caudal abdomen point spread via lymphatic system.

(*continued overleaf*)

Musculoskeletal system

Arthritis	**_Etiology_** Haematogenous spread (e.g. tail biting), penetrating wounds: • _Streptococcus_ spp. • _Erysipelothrix rhusiopathiae_ • _Haemophilus parasuis_ • mycoplasmas • _Trueperella (Arcanobacterium) pyogenes_ Gram-positive bacteria induce generally purulent and gram-negative fibrinous arthritis. However, they are macroscopically quite similar.	The affected joint is swollen and the joint capsule variably thickened. Suppurative arthritis causes generally cartilage erosion and ulceration, fibrinous less so. Both may lead to fibrous ankylosis. In both suppurative and fibrinous arthritis the amount of synovial fluid is generally reduced and it contains variable amounts of fibrin and purulent exudate. Chronic fibrinous arthritis (_E. rhusiopathiae_, mycoplasmas) results often in pronounced villous hypertrophy and pannus formation.
Osteochon-drosis	A failure of enchondral ossification leads to a focal retention of (necrotic) growth cartilage and defects to overlying articular cartilage. **_Etiology_** The etiology appears to be multifactorial: genetics, rapid growth rate, vascular factors and trauma.	Generally symmetrical affection. Predilection sites in pigs are distal femoral articular cartilages (medial condyle), humeral head and distal condyles. The intra-articular main lesions are cartilage erosions, ulcers, defects and detachments. The amount of articular fluid is increased; it is often bloody and may be turbid or flaky. Joint capsules may be thickened and/or oedematous. Growth cartilage defects may lead to malpositions.

Musculoskeletal system

		Osteochondrosis generally leads to degenerative joint disease with extensive cartilage damage or loss, osteophyte formation and joint capsule thickening

The integument

Swine erysipelas	**_Etiology_** • Erysipelothrix rhusiopathiae Can occur in four forms: acute cutaneous and septicaemic forms, chronic arthritis and endocarditis.	Cutaneous form: Square-to-rhomboidal firm, raised, pink-to-dark purple areas caused by vasculitis, thrombosis and infarction (Figure 8.8). Septicaemic form: Macroscopically acute embolic nephritis: multifocal regions of pinpoint cortical haemorrhages or multiple tan-to-white inflammatory infiltrates. Arthritis: Chronic fibrinous polyarthritis develops after septicaemia. Endocarditis: A subacute or chronic condition resulting from septicaemia. See endocarditis. Sometimes also pulmonary embolic abscesses.

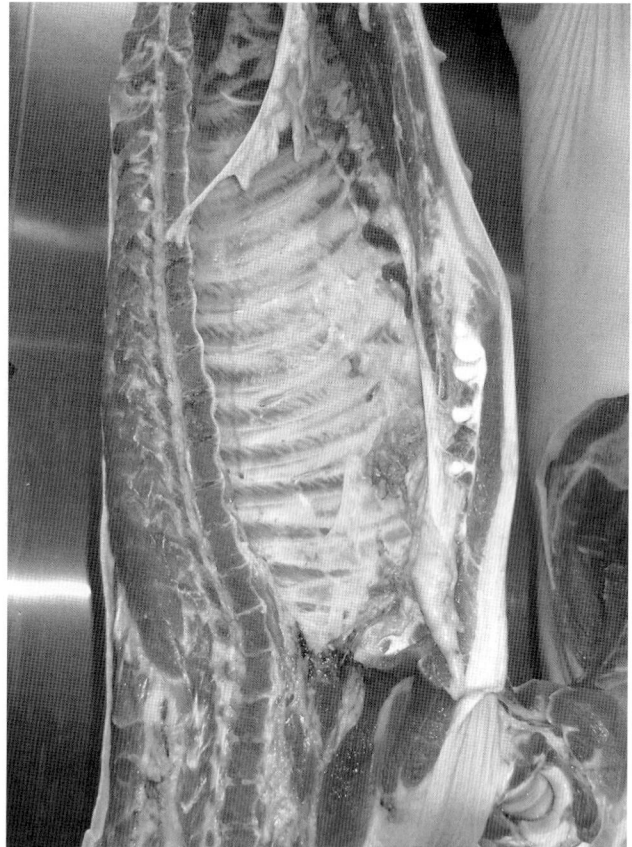

Figure 8.4   Fibrous adhesions between visceral and parietal pleura. Some lung tissue remnants. Pleura, swine.

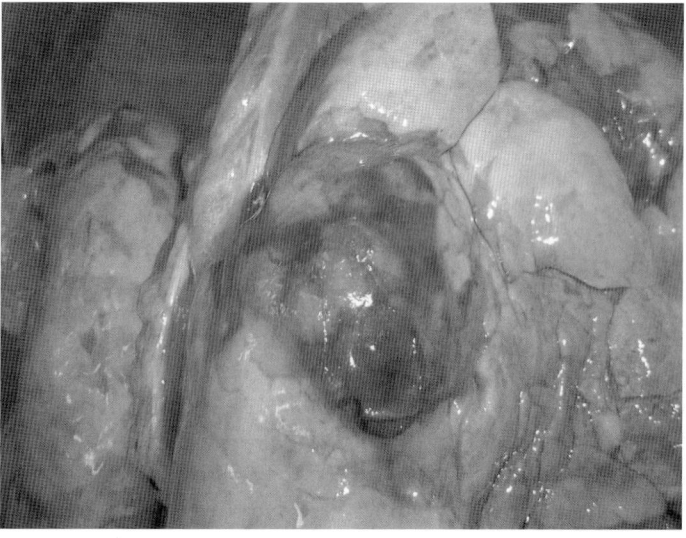

Figure 8.5   Pulmonary abscess and consolidated 'meaty' lung lobules. Lung, swine.

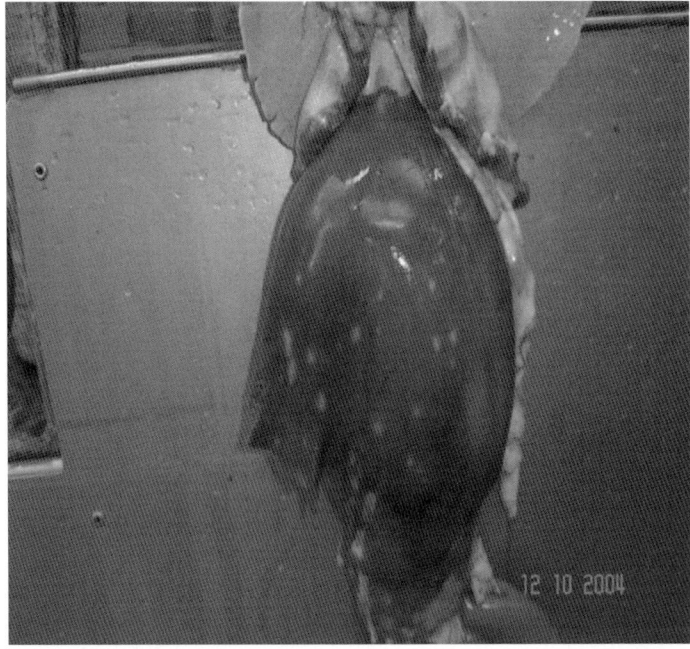

Figure 8.6 Fibrous grey-white scars caused by Ascaris suum migration. Liver, swine.

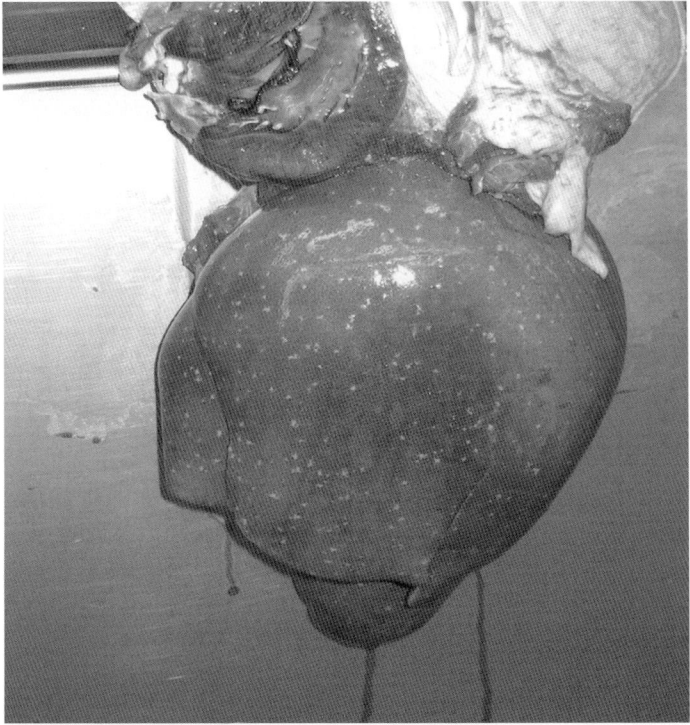

Figure 8.7 Pinpoint granulomatous foci. Liver, swine.

Figure 8.8   Square-to-rhomboidal, pink-to-dark purple areas caused by swine erysipelas infection. Integument, swine.

8.5   Small ruminants

Generalized conditions

Cachexia	See bovines

Thoracic cavity

Pleural adhesions with variable pulmonary affection	***Etiology*** Generally sequel to suppurative bronchopneumonia and fibrinous broncopneumonia.	Fibrous adhesions between visceral and parietal pleura with minimal affection of lung parenchyma.

Thoracic cavity

(Chronic fibrotic pleuritis or pleuropneumonia)	• *Mannheimia hemolytica* • *Pasteurella multocida*	Lung abscesses or scarring and/or marked pleural adhesions may be induced by suppurative or fibrinous broncopneumonias; see pneumonic Mannheimiosis. Macroscopic findings are generally not etiology specific.
Ovine pneumonic Mannheimiosis	***Etiology*** • *Mannheimia haemolytica* Pathogenesis resembles that of the bovine disease.	Lesions are characterized by fibrinous broncopneumonia with fibrinous pleuritis. Chronic changes pulmonary abscesses and fibrous pleural adhesions.
Chronic enzootic pneumonia	***Etiology*** A clinically defined disease, may be caused by several infectious agents: • viruses (adenovirus, reovirus, RSV etc.) • mycoplasmas, *Mycoplasma ovipneumoniae* Secondary invaders • *Mannheimia hemolytica* • *Pasteurella multocida* Generally mild-to-moderate clinical signs in absence of secondary infections.	Mild-to-moderate (cranioventral) broncointerstitial pneumonia. Macroscopic changes are subtle; the lungs may fail to collapse normally and have a rubbery texture and oedema. If not healed, the inflammation may progress to local atelectasis, diffuse fibrotic changes and peribronchial lymphoid hyperplasia. These changes are also macroscopically inconspicuous. Secondary bacterial invasion may lead to suppurative or fibrinous broncopneumonia and later to pleural adherences. See above.

(continued overleaf)

Thoracic cavity

Hydatid cysts Echinococcosis	***Etiology*** • *Echinococcus granulosus*	Hydatid cysts can be found in the lungs, liver and other viscera. A large number of cysts is usually present. The cysts are parenchymal and usually 5–15 cm in diameter. They are filled with clear fluid and have numerous daughter cysts, each having several protoscolices, attached to their inner wall.

Abdominal cavity

Peritonitis	See bovines. Traumatic reticulitis rare.	
Liver flukes	***Etiology*** The principal liver fluke disease of sheep and cattle is caused by *Fasciola hepatica*. Depending on the region other species may also be prevalent, for example *Dicrocoelium dentriticum* in Finland.	Mature flukes generally reside in larger bile ducts and cause chronic intrahepatic cholangitis, periductular fibrosis and sometimes cholestasis. Mineralization may occur. The contents of the bile ducts is often dark brown and viscous. Migration of immature parasites through liver parenchyma produces haemorrhagic necrotic tracts that often are repaired by fibrosis.
Liver abscesses	See bovines	
Subcutaneous abscesses	See bovines	
Fatty liver	Ketosis and fatty liver of sheep usually occurs in late gestation: pregnancy toxaemia. See bovines	

Abdominal cavity

Liver mycobac-teriosis	See swine	
Liver fibrosis	See cattle	
Cysticercosis	***Etiology*** • *Cysticercus tenuicollis* • *Taenia hydatigena*	Subperitoneal, thin-walled cysts, usually 2–3 cm in size, on liver surface, diaphragm and/or mesentery. Cysts are never in the liver parenchyma.
Urolithiasis	Composition varies by diet. Generally unnoticed alive unless obstructing ureters or urethra. Often point to water depletion or to a wrong mineral balance in the diet.	Solid casts, stones or sand in renal pelvis.
Pyelonephritis	***Etiology*** Ascending bacterial infection involving generally gram-negative bacteria from the skin or bowel, usually *E. coli*. *Corynebacterium renale* is an obligate urinary organism and potential pathogen.	Acute pyelonephritis is rare. Chronic pyelonephritis induces pus accumulation in the pelvis (in calyces in bovines), finally leading to radially distributed inflammatory infiltrates and fibrosis and scars that extend from pelvis to cortical surface.
Cortical infarcts	***Etiology*** Emboli (septic or non-septic) that occlude several small cortical arteries (e.g. interlobular arteries). Generally have little functional significance unless more than two-thirds of the kidney tissue has been lost.	Acute infarcts are haemorrhagic and swollen, become pale in 1–2 days. Chronic infarcts are pale, shrunken and fibrotic, resulting in distortion of the renal contour.

8.6 Poultry

Generalized conditions

Abnormal colour	***Etiology*** Many different etiologies can cause abnormal coloration of the carcass. In addition to many pathological conditions, abnormal coloration can be associated with the slaughter process. Coloration of the carcass can vary greatly depending on the species, breed, age, sex and nutrition of the bird.	Dark/bluish: sepsis, cachexia, ascites Yellow: hepatitis, contamination by faeces or bile Red: acute sepsis, uncut birds.
Emaciation and cachexia	***Etiology*** In emaciation the wasting is due to malnutrition. In cachexia the wasting is due to a pathological condition but specific pathogens are normally not detected in these birds.	It is impossible to separate the conditions from each other in meat inspection. Regression of body tissues. Prominent sternum, as the pectoral muscles are wasted. The coronary and body fat deposits are missing/gelatinous. The carcass is dark and dehydrated.
Septicaemia/ toxaemia	The presence of the pathogens and/or their toxins in the blood system. ***Etiology*** • *Escherichia coli* • *Pasteurella multocida* • *Staphylococcus aureus* • *Streptococcus* spp. • *Erysipelothrix rhusiopathiae* (turkeys) • *Riemerella anatipestifer* (ducks) • *Listeria monocytogenes*	The carcass is usually dark and dehydrated. Petechial haemorrhages in the muscles and serous membranes of the liver, heart, lungs and spleen. Airsacculitis, perihepatitis, pericarditis, arthritis, splenomegaly, pale kidneys, focal necrosis of liver.

Body cavity

Ascites/ oedema	**Etiology** Most commonly caused by increased portal pressure secondary to right ventricular failure. Multifactorial disease; genetics and environmental factors. Worldwide problem in fast-growing broiler chickens.	The colour of the carcass is normal or cyanotic (dark/bluish). The abdomen is distended and the vent area is enlarged. Serum-coloured fluid with or without fibrin clots in the abdomen. Hydroperi-cardium, right-sided cardiac enlargement and liver lesions.
Airsacculitis	Sum of infectious and non-infectious factors. The most important non-infectious factor is poor air quality (dust, ammonia). **Etiology** Infectious factors: • *E. coli* • *Aspergillus fumigatus* • *Mycoplasma synoviae* • *Mycoplasma meleagridis* • *P. multocida* • *Ornithobacterium rhinotracheale* • infectious bronchitis virus	In mild cases inflamed air sacs are thicker than normal and appear white or opaque. In chronic cases purulent or caseous material accumulates within the cavity of the air sac. Often associated with pericarditis, perihepatitis, cachexia.
Coccidiosis	**Etiology** Protozoan parasites of the genus *Eimeria.* Several different *Eimeria* species *E.tenella* and *E.maxima* being the most common in chicken and *E. adenoeides* and *E. meleagrimitis* being the most common in turkeys. Species specific. Most common cause of enteritis in poultry.	The mucosa of the intestine is thickened and distended with blood, mucus and tissue debris. Different *Eimeria* species cause gross lesions of varying degree in different parts of the intestine.

(continued overleaf)

Body cavity

Necrotic enteritis (NE)	***Etiology*** *Clostridium perfringens* toxins. *C. perfringens* is ubiqitous in soil, feed and litter. It is commonly found from the intestinal tract of healthy birds. Predisposing factors are required for *C.perfringens* to cause the disease: • Coccidiosis • Diet • Stress Disease outbreaks in 2–6 week old broiler chickens.	Gross lesions are found in the jejunum and ileum and also duodenum and caeca can be affected. Intestines are thin walled, distended with gas and contain thick foul smelling, brownish exudate. The mucosa is covered with a necrotic diphtheric pseudomembrane. Hepatitis with necrotic foci.
Haemorrhagic enteritis (HE)	***Etiology*** • Turkey haemorrhagic enteritis virus (adenovirus) Turkeys.	The small intestine is distended and filled with bloody contents. Splenomegaly.
Pendulous crop	The etiology is unclear but it has been associated with nutritional and hereditary factors.	The crop is distended and full of foul-smelling material (feed/litter). The lining of the crop may be ulcerated. The bird is usually emaciated.

Body cavity

Hepatitis	***Etiology*** Infectious factors: • *E.coli* • *C. perfringens* (see NE) • Adenovirus (Inclusion body hepatitis) • Avian hepatitis E virus (hepatitis-splenomegaly syndrome) • *Salmonella pullorum* • *Salmonella gallinarum*	The liver is enlarged and swollen. Often a green discoloration. White or yellow lesions in the liver.
Liver necrosis	***Etiology*** • *E.coli* • *C.perfringens* • *Staphylococcus aureus* • *H. meleagridis* (turkey)	Pale spots of necrosis on the liver surface and parenchyma. In histomonosis (*H.meleagridis*) always associated with caecal lesions.
Liver tumours	See Marek's disease	Either nodular tumours or diffuse infiltration of the liver: a coarse granular appearance.
Hepatic lipidosis	The etiology is nutritional. Turkeys.	The liver is swollen. Scattered, well-demarcated pale foci on the liver.
Multicentric histiocytosis	Multicentric histiocytosis is a neoplastic condition of young broiler chickens.	Splenomegaly and hepatomegaly. In spleen, liver and kidneys small white or yellow 1–3 mm masses.

(continued overleaf)

Body cavity

Perihepatitis	***Etiology*** Associated with systemic infections (principally *E.coli*)	Fibrinous layer on the surface of the liver. The condition is often associated with peritonitis, pericarditis and airsacculitis.
Pericarditis	***Etiology*** Associated with systemic infections (principally *E.coli*).	Inflammation of the pericardium, which is seen as a fibrinous layer on the heart. The condition is often associated with peritonitis, perihepatitis and airsacculitis.
Endocarditis	***Etiology*** Associated with systemic infections *E.rhusiopathiae* septicaemia in turkeys.	Vegetative endocarditis.
Aspergillosis	***Etiology*** • *Aspergillus fumigatus* • *Aspergillus flavus* Develops easily in poor quality bedding or contaminated feeds.	Lesions consist of greyish-white caseous nodules primarily in air sacs and lungs. Thickened air sacs membranes. Associated with cachexia and/or septicaemia.
Ornithobacterium rhinotracheale (ORT)	A gram-negative bacterium causing respiratory disease in broiler chickens and turkeys. The severity of the disease depends on other bacterial/viral infections. Mixed infection with *E.coli*.	Airsacculitis, pneumonia, pleuritis. Also associated with arthritis and encephalitis.
Mycoplasma gallisepticum	Bacterial disease of chickens and turkeys respiratory tract. Mixed infection with *E.coli*.	Airsacculitis, pericarditis, perihepatitis.

Body cavity

Viral diseases causing lesions in respiratory tract	**Etiology** • Infectious bronchitis (IB) • Infectious laryngotracheitis (ILT) • Avian metapneumovirus (aMPV) The three most common viral diseases causing slaughter downgrading. Secondary infection with E.coli.	Pericarditis, perihepatitis and airsacculitis. IB and aMPV infections are also associated with lesions in reproductive tract. These viral diseases can cause gross lesions throughout the respiratory tract and may resemble other viral diseases such as low pathogenic avian influenza and Newcastle disease.
Salpingitis, salpingoperi-tonitis	**Etiology** • E.coli Most commonly an ascending infection from the cloaca. Seen in layers and breeders.	The infected oviduct is grossly enlarged and may be impacted with a caseous, cheese-like exudate which accumulates also in the body cavity. Often associated with foul odour, septicaemia and cachexia.
Marek's disease	**Etiology** • α-herpesvirus A common transmissible lymphoproliferative disease of chickens and turkeys. Four forms: • Neural form • Visceral form • Ocular form • Skin form	Enlarged peripheral nerves. Tumours (lymphomas) in variety of organs and skin. Ocular changes (pearl eye). Differential diagnosis: Avian leucosis and reticuloendotheliosis are rare viral neoplastic diseases that can cause visceral tumours.
Visceral gout	**Etiology** • Water deprivation • Infectious bronchitis virus • Kidney disease	Abnormal deposit of uric acid on viscera. White deposits on serosal surfaces of liver, spleen and epicardium.

(continued overleaf)

Musculosceletal system

Viral arthritis/ tenosynovitis	**Etiology** • Avian reovirus Most important in meat-type chickens.	Swelling above joints. Small amount of serum-coloured or blood-tinted exudate in the joints. In severe cases associated with rupture of the gastrocnemius or digital flexor tendons and erosion of articular cartilage.
Bacterial arthritis/ tenosynovitis	**Etiology** • *Staphylococcus aureus* Other bacterial agents possible: *E.coli, Salmonella spp., P.multocida*, streptococci.	Arthritis, periarthritis and synovites most commonly in tibiotarsal and stifle joints. Purulent exudates in the joints. Associated with osteomyelitis, femoral head necrosis and green livers.
Infectious synovites	**Etiology** • *Mycoplasma synoviae*	Swelling of one or more joints with viscous to caseous exudate in synovial membranes. Enlarged liver and spleen and swollen and pale kidneys. Sometimes associated with airsacculitis and sternal bursitis.
Rupture of gas- trocnemius tendon	**Etiology** • A spontaneous rupture of the tendon Avian reovirus infection or bacterial infection can be a predisposing factor. Seen in meat-type chickens and rarely in turkeys.	Swelling and haemorrhage on the posterior surface of the leg just above the hock. In older ruptures the lesion turns green.

Musculosceletal system

Tibial dyschon-droplasia (TD)	The cartilage below the growth plate persists instead of calcifying. ***Etiology*** Etiology and pathogenesis are unclear, associated with genetic and nutritional factors. Seen in meat-type chickens and turkeys.	Abnormal mass of cartilage below the growth plate in the posterior medial portion of the proximal tibiotarsus. Associated with fractures and osteomyelitis.
Malformations	***Etiology*** Some pathological changes in the skeleton are hereditary. In non-hereditary cases the cause is yet poorly defined. Skeletal defects are an important problem in meat-type birds.	Valgus and varus of the long bones, spondylolisthesis, rotated tibia, crooked toes, spraddle legs.
Rickets	***Etiology*** Vitamin D deficiency.	Fragility and bending of long bones. Skeletal distortion.
Traumas	Bruising, fractures and joint dislocations. Can occur *ante-mortem* (in the farm or during catching and transport) or *post-mortem* (electrical stunning, plucking).	*Ante-mortem* traumas are associated with red/blue haemorrhage (Figure 11.9). In older traumas the colour is greenish. *Post-mortem* traumas are not associated with bleeding.

(continued overleaf)

Musculosceletal system

Deep pectoral myopathy (Oregon disease, green muscle disease)	An ischemic necrosis due to inadequate blood supply of the deep pectoral muscle. Associated with exercise such as wing flapping.	In early cases swollen, pale and oedematous deep pectoral muscle. In older lesions green paleness, necrosis and atrophy of the muscle. The lesion can be unilateral or bilateral.

The integument

Cellulitis	Associated with scratches caused by other birds during the grow-out period. Secondary bacterial infection enters through the damaged skin and causes infection. ***Etiology*** *E. coli* is the most commonly detected pathogen associated with this condition. Two forms: • Wet • Dry	Locally restricted inflammation of the subcutaneous tissues in the abdominal and inguinal area. The skin is swollen at the site of an inflammation and sometimes scratches and/or scabs can be detected. The wet form of cellulitis consists of subcutaneous deposits of jelly-like caseous exudate. The dry form of cellulitis consists of a subcutaneous sheet of yellow, fibrinocaseous plaques.
Scabby hip	Contact dermatitis associated with secondary bacterial infection.	Ulcers, erosions and dry scabs of the skin covering the thigh and hip area.

The integument

Gangrenous dermatitis	***Etiology*** • *Clostridium perfringens* type A • *Clostridium septicum* • *Staphylococcus aureus*	Dark necrotic patches of the skin in wings, breast, abdomen or legs. Under the affected skin there is oedema and, in some cases, emphysema. The underlying musculature is grey (cooked appearance) and crepitant.
Pododermatitis (foot pad dermatitis)	Contact dermatitis of the footpads. ***Etiology*** Associated with poor litter conditions such as wet litter and coarse bedding.	Hyperkeratosis, erosions, dark scabs filling ulcers on the plantar footpads and digits.
Dermal squamous cell carcinoma (DSCC)	Malignant tumour of the epidermis.	Crater-like ulcers originating from the feather follicles in the skin of broilers.
Skin tumours	See Marek's disease	Lymphomas of the skin and feather follicles.
Sternal bursitis	The fluid accumulates in the sternal bursa when constant pressure occurs on the sternum area. ***Etiology*** A multifactorial disease in which diet, rapid high body weight, leg problems and high temperature are known to play a role. Secondary infectious causes of sternal bursitis include *Mycoplasma synoviae*, *Staphylococcus spp.*, *E.coli* and *Pasteurella spp.* Mainly in turkey toms.	The bursitis results from the accumulation of fluid in the sternal bursa located over the keel bone. The fluid in the cyst is usually amber or reddish in colour and in most cases is sterile. The size of a cyst varies from small to 10 cm in width.

Literature and further reading

Berg, C.C. 1998. Foot-pad dermatitis in broilers and turkeys. Prevalence, risk factors and prevention. Doctoral thesis, Swedish University of Agricultural Sciences, Uppsala, Sweden.

EFSA (European Food Safety Authority). Opinions on meat inspection procedures. Available from: http://www.efsa.europa.eu/en/topics/topic/meatinspection.htm (last accessed 28 February 2014).

Grist, A. 2004. *Poultry Inspection. Anatomy, Physiology and Disease Conditions.* Nottingham University Press, UK.

Herenda, D.C. and Franco, D.A. 1996. *Poultry Diseases and Meat Hygiene: A Color Atlas.* Iowa State University Press, Ames, IA.

Jensen, H.E., Leiffsson P.S., Nielsen O.L. *et al.* 2008. *Köttkontroll – Patologanatomiska Grunder.* Biofolia, Fredriksberg, Denmark.

Quinn, P.J., Markey, B.K., Leonard, F.C. *et al.* 2011. *Veterinary Microbiology and Microbial Disease*, 2nd edn. John Wiley & Sons Ltd, UK.

Saif, Y.M. 2008. *Diseases of Poultry*, 12th edn. Blackwell Publishing Professional, Ames, IA.

Thorp, B.H. 1994. Skeletal disorders in the fowl: a review. *Avian Pathology*, **23** (2), 203–236.

Timbermont, L., Haesebrouck, F., Ducatelle, R. and Van Immerseel, F. 2011. Necrotic enteritis in broilers: an updated review on the pathogenesis. *Avian Pathology*, **40** (4), 341–347.

Zachary, J.F. and McGavin, D.M. (eds). 2012. *Pathologic Basis of Veterinary Disease*, 5th edn. Elsevier, St. Louis, MI.

9
Sampling and Laboratory Tests

Riikka Laukkanen-Ninios

Department of Food Hygiene and Environmental Health,
Faculty of Veterinary Medicine, University of Helsinki, Helsinki, Finland

9.1 Scope

The scope of this chapter is to describe sampling and laboratory methods that can be used in support of meat inspection decisions by the official veterinarian at the slaughterhouse. Other common sampling and laboratory methods used at the slaughterhouse, for example for surveillance and in-house checking systems are mentioned, but not discussed in detail. Basic laboratory procedures, such as culture or streaking methods or identification of organisms, are not in the scope of this chapter, the reader should refer to a manual in veterinary microbiology instead.

9.2 Introduction

Sampling and laboratory tests at the slaughterhouse can be used for different reasons, such as: to facilitate meat inspection in deciding whether meat is fit for human consumption; for monitoring and control of zoonoses and transmissible animal diseases; for diagnosis of transmissible spongiform encephalopathies (TSEs); for detection of residues; and to monitor microbial contamination of carcasses and the production environment. The samples and

Meat Inspection and Control in the Slaughterhouse, First Edition.
Edited by Thimjos Ninios, Janne Lundén, Hannu Korkeala and Maria Fredriksson-Ahomaa.
© 2014 John Wiley & Sons, Ltd. Published 2014 by John Wiley & Sons, Ltd.

analyses can be part of official controls, surveillance programmes or the food business operator's own checking system. The mandatory analyses should always be referred to from the current legislation of the country in question.

In many cases, rather than performing the sampling and analyses himself or herself, the official veterinarian ensures that sampling takes place and that samples are appropriately identified, handled and sent to an appropriate laboratory. It is, in any case, necessary for the veterinarian to understand the principles of sampling and analyses to ensure proper procedures at the slaughterhouse.

Sampling protocols and laboratory analyses used can be based in, for example, international agreements, international and national standards as well as legislation or methods developed in accordance with scientific protocols. The reliability of laboratory results is important for the decision making. Therefore, official laboratories should use methods validated as fit for purpose and work under internationally accepted (e.g. ISO 17025) quality management principles. The laboratory performing the tests should also be accredited, that is the competence of the laboratory to perform analyses has been recognized by an impartial party based on internationally agreed criteria. The methods should be practical, accurate, reproducible, sensitive and selective. Any alternate methods should be validated against standard methods, if such are available. Here, some general instructions and examples are given of how sampling and laboratory analyses for meat inspection can be performed. The current legislation and guidelines have to be referred to for required analyses in each country. In addition, international trade may require additional analyses or the use of specific methods.

9.3　Collecting and packaging samples

Samples should be taken aseptically and be sufficiently large or large enough for the analyses. The sampling site depends on the reason why the sample is needed; for example, the sample should be collected from the predilection site of the agent, if a pathogen is suspected, and from inside of meat or organ, if generalized infection is suspected. When surface contamination of the carcass (with a pathogen or, e.g., indicator organism) is investigated, samples can be obtained either by swabbing the carcass or by cutting aseptically pieces from the surface of the carcass. If the sampling is, for example, for the monitoring of diseases and there is no individually suspected carcass, samples need to be taken according to a separate sampling plan; this is not discussed in this chapter. Further information on sampling plans can be found elsewhere, for example the International Commission on Microbiological Specifications for Foods (ICMSF) (see the section Literature and Further Reading). Samples are placed in a clean (or when needed, sterile) container. When samples are sent outside the slaughterhouse, packaging should have the following features:

- **Primary receptacle**: Each sample should be packed in a separate leak-proof receptacle.

- **Secondary packaging**: Sample(s) should be placed in a secondary container to protect the primary receptacle(s) and to retain any leaks.
- **Absorbent material**, such as paper towels or cotton wool, should be placed between the primary receptacle and the secondary packaging. The material must be sufficient to absorb the entire contents of all primary receptacles.
- **Outer packaging**: The secondary packaging should be placed in a rigid outer package that protects it and its contents from damage while in transit.
- An **itemized list** of contents should be placed inside the outer packaging.
- Samples that are transported in a cool package need to be **precooled** near + 0 °C prior to packaging. Samples should not be frozen without permission from the laboratory, since freezing can hamper analyses.
- In **commercial transport**, proper labelling of the package is important and the package should state the correct UN number issued by United Nations Committee of Experts on the Transport of Dangerous Goods that identifies the hazardous substances in the package. Instructions for shipping the samples to laboratory should be confirmed from the receiving laboratory and transport service. Make sure also, that your package does not get stuck for the weekend or bank holidays but is delivered in a timely manner.

9.4 Boiling test

The boiling test is used to evaluate possible abnormal odours in carcasses. The test can be used in uncertain cases of abnormal odours but is unnecessary if the malodour is clear in the raw carcass. Strong, unusual odour is generally unacceptable for consumers, although it may not represent a public health risk. Abnormal odours can be caused by, for example, feeding, use of drugs, products of abnormal metabolism or sexual odour.

Sampling at the slaughterhouse:

Usually muscles with their fascia are sampled, from both forequarter (*Musculus triceps brachii*) and hindquarter (*Musculus adductor* and *Musculus gracilis*) and the sample for laboratory tests should be representative and sufficiently big (about 200 g, at least 2 cm thick). The samples should not contain bruises or other visible changes, since the generalized condition of the carcass rather than local changes is of interest.

Performance and interpretation of the test:

Uncut meat samples are placed in a pot and cold water is added to cover the meat. Samples from each carcass are cooked in separate pots. Samples are boiled with the lid on for 20–30 minutes until cooked. After cooking, the odour of cooking liquid is assessed immediately after opening the lid (without burning oneself in the hot steam) and the odour of the meat is assessed from a cut in the meat immediately after cutting. The assessors score the odour and if the odour is abnormal it needs to be described. If there is great variation in the scores from different assessors or the mean of the scores is on the borderline of unacceptable, the test needs to be redone.

The assessors need to test that the sensitivity of their sense of smell and taste are normal. For about an hour before the test, the assessors should not eat, smoke or drink other than water. Also, perfumed cosmetics or similar can disturb the test and should not be used.

The pH of the carcass should be measured together with boiling test, since pH can affect the odour of meat.

9.5   Measurement of pH

The pH value of meat provides information on the shelf life of the meat and of technical processing characteristics as well as the success of electric stimulation of carcass. Since pH is affected by the treatment of the animals before slaughter, it can also indicate welfare problems.

The pH of carcass is near neutral (7.2) immediately after exsanguination but starts to decline reaching an ultimate pH (pHu) of 5.3–5.5 within 24 hours of the slaughter. However, pHu can remain above normal and facilitate bacterial growth, thus causing the meat to spoil prematurely. In addition, high pHu changes the appearance of meat (dark, firm and dry, i.e. DFD meat) most commonly affecting beef. A pH above six can be considered to indicate reduced shelf life. Too rapid a fall of pH (down to 5.4–5.8 within 1 h) causes condition known as pale, soft, exudative (PSE) meat, particularly in pigs. The rapid fall of pH causes a reduced water holding capacity of the carcass and lowers the technological quality of the carcass. The pHu can also fall too low, causing similar technological problems as too rapid a fall.

Sampling at the slaughterhouse:

Usually muscles with their fascia are sampled, from both forequarter and hindquarter; the sample for laboratory tests should be representative and sufficiently big (at least 200 g). It is good to remember that pH is not uniform throughout the carcass but depends on the muscle type and sampling site.

Performance and interpretation of the test:

The pH can be directly measured from the carcass at the slaughterhouse or from a meat sample at the laboratory using a pH meter. The pH meter needs to be calibrated before use. The pH can be measured directly from the meat or from the meat extract.

Direct measurement from meat sample or carcass:

A hole is pierced in the sample with a knife or a sharp pin and the electrode is inserted into the hole. Alternatively, a pH meter with a spear point electrode designed for meat samples can be used. The pH meter has to have temperature compensation or the sample has to be 20 °C. The pH is read to the nearest 0.01 pH unit and the measurement is repeated at the same point of incision. The result is given as the mean of the two measurements from the same point and reported to the nearest 0.05 pH unit. If the pH measurement is made from different locations in the carcass or meat sample, the average pH of each location is reported separately.

Measurement from meat extract:

The sample is homogenized using, for example, a food chopper (rotational cutter) and 10 times as much deionized water or, more preferably, 0.15 M potassium chloride solution is homogenized with the sample. The pH is measured from the extract while stirring with a magnetic stirrer. Note that if the sample is collected before the pH has reached pHu, the pH continues to change during storage of the sample. The pH meter has to have temperature compensation or the sample has to be 20 °C. The pH is read to the nearest 0.01 pH unit and the result is reported to the nearest 0.05 pH unit.

9.6   Bacteriological examination of carcasses

A general bacteriological examination of a carcass may be considered when suspicion of systemic infection or presence of zoonotic agents in the carcass arises. If the carcass or organs show marked pathological changes or severe systemic infection, bacteriological examination is not needed and the meat inspection decision can be made based on *post-mortem* findings. However, in these cases, bacteriological examination can be used to identify the infective organism, if needed.

Sampling at the slaughterhouse:

Both muscles and organs can be used as samples. For example, muscles with their fascia from both foreleg (e.g. *Musculus triceps brachii*) and hindleg (e.g. *Musculus adductor* and *Musculus gracilis*) and spleen with capsule intact are sampled. In the healthy and physiologically normal animal, the inside of organs and muscles is virtually sterile. However, in systemic infection, pathogens invade muscles and organs. Therefore, aseptically taken deep samples from muscles and organs are used to show systemic infection. If a specific pathogen is suspected, further sampling, for example from other organs or lymph nodes, and examination specific to that agent should be carried out in addition to general bacterial examination.

Performance and interpretation of the test:

The surface of the sample is sterilized by cauterizing with an iron heated on a Bunsen burner. A deep cut is made with a sterile scalpel into the sterilized surface and a sample is taken by rubbing a sterile loop into the walls of the cut.

Care must be taken in order to avoid contamination by organisms on the surfaces. The sample is then streaked onto blood agar and incubated in aerobic and anaerobic atmospheres at +37 °C for 24 hours. If no growth is observed, the incubation is continued for another 24 hours. Pure culture on either of the plates indicates systemic infection in the carcass. Note, that the pathogen is not necessarily zoonotic. The pathogen can be identified using standard identification procedures. If mixed culture is present on either of the blood agar plates, it is most likely contamination and the examination should be repeated.

It is usually advisable to study antimicrobial residues in meat simultaneously with bacteriological examination, since any suspicion of a disease in the animal can also result in the suspected use of antimicrobial drugs. Antimicrobial drugs also inhibit the growth of bacteria in the sample, thus giving false negative results. Boiling and pH tests should also be performed together with bacteriological examination, since these parameters can also be altered.

9.7   Zoonotic agents

Animals can harbour many organisms that can be transmitted to humans via consumption of meat. There are a number of bacteria and parasites that are known food-borne agents transmitted via consumption of meat. Zoonotic viruses transmitted via consumption of meat are a new area of research due to difficulties in detection and, in general, are presently not investigated routinely as part of meat inspection or surveillance programmes at the slaughterhouse.

9.7.1   Bacteria

There are a number of different zoonotic bacteria that can infect humans via consumption of meat; these are discussed in Chapter 13. In general, bacteria are traditionally detected using culturing methods or, for example, direct microscopy, but other methods, such as polymerase chain reaction (PCR) are becoming more and more routine in the laboratory. Here, examples of detection methods for some of the most important zoonotic bacteria are shown. Note, that isolation methods depend on the sample type: methods used for meat samples are not necessarily effective for clinical, for example faecal, samples. Similarly, pre-treatment of samples for PCR analysis depends heavily on the sample matrix.

Salmonella

Sampling at the slaughterhouse:

This may be done for different reasons: to determine whether an animal is infected by or a carrier of the microbe; to estimate slaughter process hygiene (when *Salmonella* is common in food process); or to determine the suitability of meat for consumption. Note that in countries where *Salmonella* is uncommon in food animals (Finland, Norway and Sweden), the use of

Salmonella as process hygiene indicator is non-informative and other more meaningful indicators should be used to evaluate process hygiene.

For the carriage and infection status of an animal, samples from (mesenteric) lymph nodes and organs can be collected at the slaughterhouse. To investigate carcass contamination, a neck skin sample or whole carcass rinse can be used for poultry and carcass swabs or skin samples for larger animals.

Performance and interpretation of the test:

A 25 g sample is mixed with phosphate buffered saline (PBS). Complete poultry carcasses are mixed with 225 ml PBS and shaken, after which the carcass is removed. Samples are pre-enriched in PBS for 18 h at 37 °C and then enriched on selective Rappaport–Vassiliadis medium with soya (RVS) broth (0.1 ml of pre-enrichment into 10 ml of RVS) and possibly onto another selective broth such as Muller–Kauffmann tetrathionate novobiocin (MKTTn) broth (1 ml of pre-enrichment into 10 ml of MKTTn). RVS broth is incubated at 41.5 °C and the MKTTn broth at 37 °C for 24 h. *Salmonella* are isolated by plating onto a selective agar plate. Usually, xylose lysine deoxycholate agar (XLD agar) is used with another *Salmonella* selective plate, such as Brilliant green agar (BGA), or more and more commonly onto chromogenic agars. Typical colonies of *Salmonella* on XLD agar have a black centre and a lightly transparent zone of reddish colour. However, there are *Salmonella* hydrogen sulfide (H_2S) negative variants (e.g. S. Paratyphi A) that grow on XLD agar as pink colonies with a darker pink centre. Lactose-positive *Salmonella* grown on XLD agar are yellow colonies with or without blackening. On BGA *Salmonella* species are normally red colonies surrounded by bright red agar. Strains of *Salmonella* Typhi and Paratyphi may not grow on BGA. Identification is based on a reaction on triple sugar iron (TSI) (red slant, black precipitation, yellow butt), urea (negative) and serotyping.

Campylobacter jejuni and *Campylobacter coli*

Sampling at the slaughterhouse for screening:

In poultry, the caeca are usually used for the detection of *Campylobacter*. Samples from cattle, sheep and pigs are collected from the intestines by aseptically opening the gut wall or by taking rectal swabs. Campylobacters are very sensitive to environmental conditions, including dehydration, oxygen, sunlight and high temperature. Transport to the laboratory and subsequent processing should, therefore, be as rapid as possible and the samples must be protected from light. Freezing or high temperatures (>20 °C) can reduce viability and fluctuations in temperature must be avoided. If the samples cannot be analysed immediately they should be stored at 4 °C.

Performance and interpretation of the test:

Faecal samples can be streaked directly onto selective agar, for example modified charcoal cefoperazone deoxycholate agar (mCCDA) agar and incubated at 42 °C or at less selective 37 °C for 48–72 h in microaerobic atmosphere.

If it is necessary to isolate *Campylobacter* from carcasses, selective enrichment is used; 25 g of a skin sample (or 25 ml wash liquid from complete poultry carcasses, see Section 9.7.1.1) is mixed with 225 ml of Bolton broth and incubated at 42 °C for 20–24 h in a microaerobic atmosphere. In case of carcass swabs, 90 ml of Bolton broth is used. A loopful of enriched sample is streaked onto, for example, mCCDA and incubated at 42 °C or at 37°C for 48–72 h, for the isolation of *Campylobacter*. *Campylobacter* appear on mCCDA as light grey or sometimes metallic flat colonies that can spread on wet agar to cover the whole plate. Identification is based on Gram staining (slender, curved gram-negative rods), catalase (weak positive) and oxidase (weak positive) tests and hippurate hydrolysis (*C. jejuni* positive, *C. coli* and *C. lari* negative).

Shiga toxin-producing *Escherichia coli* (STEC)

Shiga toxin-producing *Escherichia coli* (STEC, synonym verocytotoxigenic or verocytotoxin-producing *E. coli*, VTEC) is very difficult to isolate, since it contains multiple serotypes that are difficult to separate from other *E. coli* as well as some other bacteria. *E. coli* O157:H7 has been the predominant zoonotic STEC and diagnostic methods have been developed for it. However, non-O157 serotypes lack efficient diagnostic methods.

Sampling at the slaughterhouse:

Ruminants are the primary reservoir for STEC, but other animals can carry it as well. Carcasses can be sampled to assess the food safety. In addition, faecal samples, for example, can be collected to estimate the prevalence in animals.

Performance and interpretation of the test:

A PCR method for the detection of Shiga toxin genes *stx1* and *stx2* from an enriched sample can be used for presumptive detection of STEC.

Isolation of STEC O157:

Selective enrichment is performed in tryptic soy broth supplemented with bile salts and novobiocin (mTSB-N) at 41.5 °C for 6 h and 18–24 h. Part of the enrichment culture is then treated with an O157-specific immunomagnetic separation (IMS) reagent based on paramagnetic beads coated with antibodies against the lipopolysaccharide antigen of *E. coli* O157 (commercially available) and plated onto sorbitol MacConkey plate with cefixime and tellurite (CT-SMAC). Most *E. coli* O157 isolates are sorbitol non-fermenters and can be distinguished from sorbitol-fermenting *E. coli*. In addition, cefixime and tellurite inhibit the growth bacteria other than *E. coli* O157. The isolates are identified using indole and positive isolates are investigated serologically by slide agglutination. Identification can also be done using serotype specific PCR.

Isolation of non-O157 STEC:

There is no universal selective agar for non-157 STEC, since they vary in metabolic and antimicrobial resistance features. For the isolation of non-O157 STEC, multiple colonies must be picked and identified from plates, such as

MacConkey, permitting the growth of *E. coli*. IMS beads are commercially available for most common serotypes (but not all) and other methods, such as colony hybridization method targeting, for example, the *stx* genes, can also be used to pick presumptive non-O157 colonies.

Enteropathogenic *Yersinia*

Sampling at the slaughterhouse for screening:

Pig tonsils are a predilection site for *Yersinia enterocolitica* and *Yersinia pseudotuberculosis* and should therefore be sampled.

Performance and interpretation of the test:

Isolation of *Y. enterocolitica* and *Y. pseudotuberculosis* is difficult and time consuming and can include direct isolation, selective enrichment, for example, in irgasan-ticarcillin potassium chlorate (ITC, 1/100 dilution of the sample at 25 °C for 48 h) and enrichment at 4°C (cold enrichment, 1/10 dilution of the sample) for 1–3 weeks in non-selective phosphate buffered saline with bile salts and mannitol or sorbitol (PMB or PSB). *Yersinia* strains are isolated on *Yersinia* selective cefsulodin-irgasan-novobiocin (CIN) agar but, for example, MacConkey or SSDC (*Salmonella Shigella* agar with desoxycholate and calcium chloride) agar for the isolation of gram-negative enteric bacteria can be used in addition to CIN, since CIN can inhibit the growth of some *Yersinia* strains. Agars are incubated at 30°C for 24–48 h – *Y. pseudotuberculosis* in particular grows slowly – so extended incubation is needed for the colonies to appear. Enteropathogenic *Yersinia* forms a small 'bull's eye' appearance on CIN where the red centre of the colony is surrounded by a white or transparent zone. On MacConkey and SSDC agars *Yersinia* appear as small white or colourless colonies. Preliminary identification of strains as *Yersinia* can be done with urea (positive), Kligler (yellow but, red or unchanged surface, no gas or H_2S) and oxidase (negative). *Y. enterocolitica* strains are biotyped and both *Y. enterocolitica* and *Y. pseudotuberculosis* serotyped. Not all *Y. enterocolitica* strains are pathogenic and, therefore, pathogenicity of the isolates needs to be confirmed using, for example, PCR designed for virulence genes.

Erysipelothrix rhusiopathiae

Erysipelothrix rhusiopathiae is a bacterium infecting mainly pigs but can also be found in other animals, especially poultry. Human infection occurs as a result of contact with contaminated animals, their products or wastes or soils, rather than via ingestion of contaminated meat and erysipeloid is an occupational hazard for people working in meat industry. The incidence of clinical *Erysipelothrix* infection in pigs has been declining due to vaccination, but uncertain clinical cases at the slaughterhouse can be confirmed using a simple isolation method.

Sampling at the slaughterhouse:

Liver, spleen, heart valves or synovial tissues are collected from suspected pigs in addition to meat sample studied in bacteriological examination.

Table 9.1   Preliminary identification of *Erysipelothrix rhusiopathiae*.

Test	Result
Colony morphology on blood agar	After 24 h: Non-haemolytic, pinpoint colonies. After 48 h: A narrow zone of greenish, incomplete (α) haemolysis. Smooth colonies 0.5–1.5 mm in diameter, clear circular with even edges. Rough colonies are slightly larger, flat and opaque with irregular edges.
Growth on MacConkey	No growth.
Gram stain	Gram-positive, small, slender, slightly curved rods with rounded ends (smooth form) or long filaments in chains (rough form).
Catalase test	Negative.
Coagulase test	Positive.
Triple sugar iron (TSI) slant	H_2S production.

E. rhusiopathiae is rarely recovered from skin lesions or chronically affected joints.

Performance and interpretation of the test:

Samples are cultured on blood agar and incubated aerobically and anaerobically at 37 °C for 24–48 h. Selective media, containing antimicrobials, sodium azide and/or crystal violet, may be used for contaminated samples. Preliminary identification can be done as described in Table 9.1.

Mycobacteria

Mycobacterium spp. contains multiple species pathogenic both for animals and humans. *Mycobacterium avium* complex and *M. intracellulare* cause avian tuberculosis in birds and can also infect an extensive range of different animal species, such as swine, cattle, deer, sheep and goats. *M. bovis*, the causative agent of bovine tuberculosis, affects cattle and other domesticated animals. *M. caprae* is a fairly recently identified species in *Mycobacterium tuberculosis* complex. It affects, particularly, goats but also other animals, particularly in Europe.

Sampling at the slaughterhouse:

At the slaughterhouse, if mycobacterium is suspected, samples of affected organs can be collected. Fast delivery of specimens to the laboratory enhances the chances to isolate *M. bovis*. If delays in delivery are anticipated, specimens should be refrigerated or frozen to inhibit the growth of other bacteria and to preserve the mycobacteria. Care must be taken when taking samples at the slaughterhouse. In the laboratory, handling of open live cultures or of material from infected animals must be performed with adequate biohazard containment.

Performance and interpretation of the test:

Detection of acid-fast rods in smears or sections from affected organs stained with Ziehl–Neelsen by microscopy can be used to demonstrate mycobacteria in a sample.

Ziehl–Neelsen stain: A smear of the tuberculosis material is prepared and air-dried. The smear is covered with carbolfuchsin stain and carefully heated by passing a flame under the slide until steam rises. The preparation is kept moist with stain and steaming for five minutes. Make sure that the stain is not boiling. The colour is poured away and the slide rinsed with water until no colour appears in the effluent. The slide is washed with decolorizing solvent (containing ethanol and HCl) and rinsed with water. The decolorizing and washing are repeated until the stained smear appears faintly pink and the rinsing water is clear. To counterstain the sample, the smear is flood with malachite green for one minute, washed with water and air-dried. The sample is examined under oil immersion under microscope. Acid-fast mycobacteria appear red and non-acid-fast background is green.

The isolation of mycobacteria and their subsequent identification by cultural and biochemical tests or DNA techniques (e.g. PCR), are used to confirm the infection and identify the mycobacterium species in question.

Bacillus anthracis

Sampling at the slaughterhouse:

Anthrax affects herbivores in particular, but other animals as well. If an animal is suspected of dying at the slaughterhouse of anthrax, blood from ear veins or other peripheral veins and exudate from orifices is collected with a syringe. Cotton wool soaked in 70% alcohol should be applied to the site after collection to minimize leakage of contaminated blood or fluid. For horses and pigs, samples can also be taken from oedematous fluid or superficial lymph nodes in the neck region.

Care must be taken when handling samples from suspected anthrax carcasses: gloves and other protective clothing should be worn when handling specimens and rubbing the face or eyes must be avoided. Clinical specimens and cultures of *B. anthracis* should be handled at World Organisation for Animal Health (OIE) biosafety level 3.

Performance and interpretation of the test:

Visualization of the encapsulated rods in a blood smear stained with polychrome methylene blue (MacFadyean reaction) is fully diagnostic, according to OIE. *Polychrome methylene blue stain*: A small drop of blood or tissue fluid is used to make a small smear. The sample is fixed by heating or dipping the smear in 95–100% alcohol for one minute and dried. The smear is stained by placing a small (about 20 µl) drop of polychrome methylene blue stain on the smear and spread over it with an inoculating loop and left for one minute. Then the stain is washed with water, blotted, and air-dried. Under the microscope, the smear is examined initially using the ×10 magnification under which the short chains appear like short hairs. These are examined under oil immersion (×1000) for blue staining rods surrounded by a pink capsule. The cells are square-ended ('box-car' appearance) and in pairs or short chains. Since the bacterium multiplies in the blood just before death, microscopy is not applicable in the diagnostics of anthrax of living animals and, since the amount of capsular material diminishes with time after the death, the capsule may be difficult to detect if the animal has been dead more than 24 h.

Antimicrobial resistant microbes

Antimicrobial resistance of microbes is a fast increasing problem and animals can be sampled at the slaughterhouse to monitor the prevalence of antimicrobial resistant zoonotic and indicator bacteria in food animals. Resistance for antimicrobials can be determined by, for example, broth dilution or agar diffusion test where the growth of the organism with the antimicrobial agent shows resistance or by showing genes coding resistance to antimicrobials by, for example PCR.

9.7.2 Parasites

Most parasites are detected visually during meat inspection and no tests are used on a regular basis other than that for *Trichinella*, which is tested for in carcasses of domestic pigs and horses, wild boar and other farmed and wild animal species susceptible to *Trichinella* infestation. *Trichinella* can be detected by direct detection of first-stage larvae in striated muscle using tissue digestion. The digestion method is recommended for the inspection of individual carcass of food animals. Indirect detection of infection by tests for specific antibodies using particularly enzyme-linked immunosorbent assay (ELISA) are also available. At present, ELISA can be used for surveillance or verification of *Trichinella*-free pig herds or regions but should not be used for the inspection of individual carcasses. ELISA may also be used for surveillance or verification of other parasites, such as *Toxoplasma* or *Cysticercus*-free herds or regions, but is not used for the inspection of individual carcasses.

Digestion test for *Trichinella*

Sampling at the slaughterhouse:

Trichinella larvae are usually localized in predilection muscle sites, particularly in low-level infections. The predilection sites are host specific and need to be sampled to maximize the test sensitivity. Samples should be stored at 4 °C and tested as soon as possible, since the number of larvae detected in samples declines after prolonged storage, putrefaction and freezing of the sample. The sensitivity of digestion method for a 1 g sample has a sensitivity of approximately three larvae/g of tissue and using a 5 g sample increases sensitivity to one larva/g of tissue. For wildlife, larger samples (≥10 g) should

Table 9.2 Suggested sampling sites and sample sizes for *Trichinella* detection in the EU.

Species	Sampling
Domestic pigs	At least 1 g is cut from a pillar of the diaphragm at the transition to the sinewy part. From breeding sows and boars, at least 2 g, is needed. In the absence of diaphragm pillars, a sample of twice the size is collected from the rib part or the breastbone part of the diaphragm, jaw muscle, tongue or abdominal muscles.
Wild boar	At least 5 g samples from foreleg, tongue or diaphragm are digested.
Horses	At least 5 g samples from the lingual or jaw muscle or, in the absence of those, a larger-sized sample from a pillar of the diaphragm at the transition to the sinewy part are digested.
Game animals such as bears, carnivorous mammals and reptiles	10 g of muscle at the predilection sites, or larger amounts if those sites are not available, are digested. Predilection sites are: (i) diaphragm, masseter muscle and tongue in bear; (ii) tongue in walrus; (iii) masseter, pterygoid and intercostal muscles in crocodiles; and (iv) muscles of the head in birds.

be used to compensate for a possible decrease in sensitivity due to variation of predilection sites (Table 9.2). Samples should be pure muscle tissue. The superficial layer of the tongue in tongue samples must be carefully removed, since it is indigestible and can prevent reading of the sediment.

Performance and interpretation of the test:

Often pools of 100 g of samples at a time are examined and the following protocol by European Commission can be used: 16 ml of 25% hydrochloric acid is added to a beaker containing 2 l of water, preheated to 46–48 °C. A stirring rod is placed in the beaker and the stirring is started on a preheated plate. 10 g of pepsin (1:10 000 NF (US National Formulary)) is added. 100 g of samples is chopped in the blender, the homogenized meat is transferred to the beaker containing the water, pepsin and hydrochloric acid and the beaker is covered with aluminium foil. The magnetic stirrer is adjusted to a constant temperature of 44–46 °C and the speed of the stirrer is adjusted so that the fluid creates a deep whirl without splashing. The digestion is continued for approximately 30 minutes until the meat particles disappear. The digestion fluid is then poured through a sieve (mesh size 180 μm) into the sedimentation funnel. The digestion process is satisfactory if not more than 5% of the sample weight remains on the sieve. The digestion fluid is allowed to stand in the funnel for 30 minutes to allow the larvae to settle to the bottom. Circa 40 ml of digestion fluid is quickly run off into the measuring cylinder and allowed to stand for 10 minutes. 30 ml of supernatant is carefully removed using a pipette, leaving a 10 ml sample of sediment, which is, in turn, poured into a larval counting basin or petri dish. The cylinder is rinsed maximum 10 ml of water, and added to the sample. The sample is examined by stereo-microscope at a 15–20 times magnification and suspected larvae are further examined under higher magnifications of 60–100 times. If a pooled sample produces a positive or uncertain result, a large sample is taken from each animal and tested in smaller batches until an individual positive animal is identified. Parasite samples can be stored in 90% ethyl alcohol for conservation and identification.

9.7.3   Transmissible spongiform encephalopathy

Transmissible spongiform encephalopathies or prion diseases are discussed in detail in Chapter 13A.

Bovine spongiform encephalopathy (BSE)

Sampling at the slaughterhouse:

The head is removed from the body by cutting between the atlas vertebra and the skull and placed on a support, ventral surface uppermost (Figure 9.1). The caudal end of the brainstem (medulla oblongata) is visible at the foramen magnum (Figures 9.1 and 9.2). The sampling spoon is inserted through the foramen magnum between the dura mater and the medulla with the bottom (convex side) of the spoon close to the bone of the scull and pushed rostrally. The sampling spoon is pushed with a side-to-side rotational action, thus cutting the cranial nerve roots but leaving the brain tissue undamaged. The spoon is pushed rostrally for approximately 7 cm and then angled sharply toward the brainstem to cut and separate the brainstem from the rest of the brain. Finally, the spoon is pulled from the skull in an angled position to remove the sample.

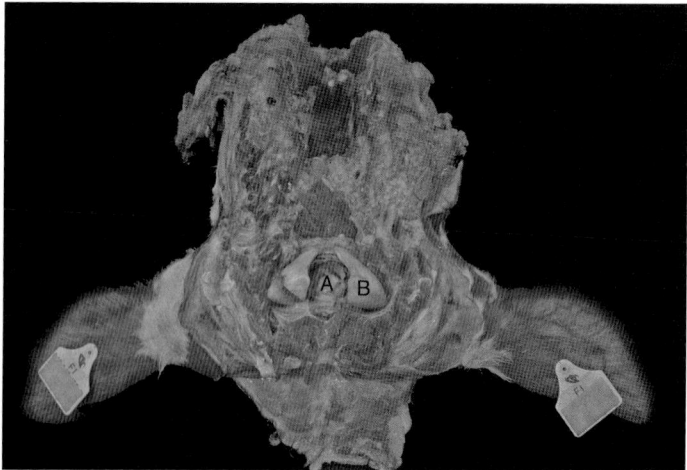

Figure 9.1   A bovine head, ventral surface uppermost. Medulla oblongata (A) is visible at the foramen magnum between occipital condyles (B). Photograph: University of Helsinki, Faculty of Veterinary Medicine, Pathology, Satu Tukiainen-Leppänen.

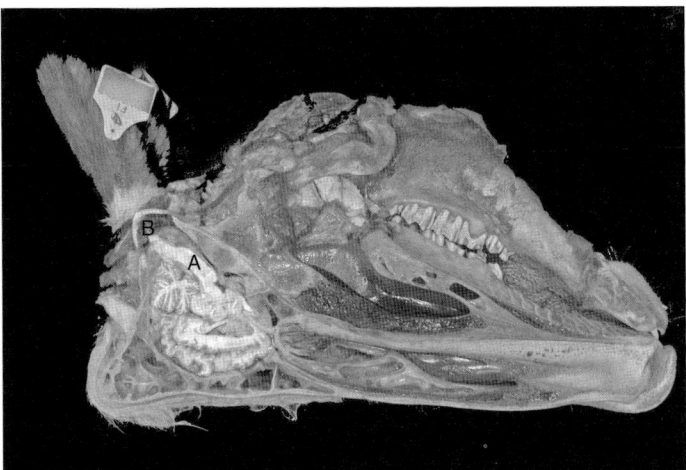

Figure 9.2 A cross-section of a bovine head. Ventral side up showing cross-section of the medulla oblongata (A) and foramen magnum (B). Photograph: University of Helsinki, Faculty of Veterinary Medicine, Pathology, Satu Tukiainen-Leppänen.

The sample is then placed into a sample jar and the ear tag of the animal is attached to the sample.

Performance and interpretation of the test:

Commercial rapid Western blot, lateral flow device and ELISA techniques have been developed for the screening of large numbers of brain samples and these techniques are based on the demonstration of prion protein. However, BSE-infected animals should be detected and confirmed by a combination of at least two test methods. Confirmation of a diagnosis of BSE ideally requires examination of fixed brain by immunohistochemistry or use of Western blot. The sampling and processing of tissue for any rapid test should be carried out precisely as specified by the manufacturer and the sampling site has to take into account the subsequent method of confirmation. At least a hemi-section of the medulla at the level of the obex should be kept intact for fixation for immunohistochemistry or histology, should a positive result in preliminary test require confirmation.

Scrapie

Sampling at the slaughterhouse:

The minimum sampling is the brainstem at the level of the obex for typical scrapie and also the cerebellum for atypical scrapie. The brainstem sample should be about 4–5 cm long and the cerebellum sample about 2 × 2 cm and contain cortex of the cerebellum. The sample can be taken through the foramen magnum, as in BSE sampling. Firstly, the sampling spoon is inserted between the dura mater and the medulla with the bottom (convex side) of the spoon close to the bone of the scull and pushed rostrally so that the bowl of the spoon is entirely inside the scull. The spoon is turned 90 degrees left and right to cut the cranial nerve roots and then angled sharply toward the brainstem to cut the brainstem. The spoon is pulled from the skull in an angled position. The spoon is pushed again into the foramen magnum and turned left and right, but this time the spoon is pushed lower to catch a part of the cerebellum in the spoon. Again, the sample is pulled from the cranium. The brainstem and cerebellum samples are placed in a sample jar and the ear tag of the animal is attached to the sample. The head should be stored until the results of the rapid test are ready.

Performance and interpretation of the test:

As with BSE, rapid tests can be used to screen a large number of samples but suspected positive samples need to be confirmed using other methods, such as immunohistochemical methods or Western blot.

9.8   Animal diseases

Although transmissible animal diseases are not a public health risk, carcasses can be deemed as unsafe for animal health reasons. In addition, sampling for monitoring and control of animal diseases are often done at the slaughterhouses. The animal diseases that are monitored and tested depend on the animal disease situation in each country and area. Mandatory samples are, again, specified in the national legislation and guideline, but World Organisation for Animal Health (OIE) publishes the Manual of Diagnostic Tests and Vaccines for Terrestrial Animals, which aims to provide internationally agreed diagnostic laboratory standards for all OIE listed diseases and for several other diseases of global importance for trade in animals and animal products. It is noteworthy that OIE is one of the three standard-setting organizations referenced in the World Trade Organization (WTO) Agreement on Sanitary and Phytosanitary Measures (SPS Agreement).

9.9   Chemical residues

Chemical residues containing sampling plans and detection of environmental contaminants, veterinary drugs and residues of feed additives are extensively discussed in Chapter 13C. Here, only methods typically used in the meat inspection from animals with the suspicion of antimicrobial residues are discussed briefly.

9.9.1   Detection of chemical residues and contaminants

Sampling:

The sampling of chemical residues and contaminants can be a part of official sampling and the sampling plan is thus provided by the central authority,

Table 9.3 Examples of methods used to analyse chemical contaminants and residues.

Method	Sample
Atomic absorption spectrometry (AAS)	Toxic heavy metals, such as mercury (cold vapour AAS (CVAAS)) and arsenic (hydride generation AAS (HGAAS)).
Inductively coupled plasma techniques with mass spectrometry (ICP-MS) and optical emission spectrometry (ICP-OES)	Toxic heavy metals such as lead, cadmium and tin.
Gas chromatography with mass spectrometry (GC-MS)	Industrial wastes and burning by-products such as dioxins, polychlorobiphenyls (PCBs), polybrominated diphenyl ethers (PBDEs), perfluorinated organic compounds, polycyclic aromatic hydrocarbons (PAHs).
Liquid chromatography with mass spectrometry (LC-MS)	Mycotoxins (ultra-high LC with tandem mass spectrometry (UPLC-MS/MS)), pesticides (UPLC-MS/MS), pollutants, veterinary drugs (LC with tandem mass spectrometric detection (LC–MS/MS)).

or it may be part of a slaughterhouse operator's own checking system's sampling plan. Samples must not become contaminated during sampling or sample preparation; all equipment in contact with the sample should be made of inert materials and should not contain substances to be determined. At the slaughterhouse, both meat and organ samples can be collected for the detection of chemical contaminants.

Packaging of the samples:

Each sample is placed in a clean, inert container with protection from contamination, from loss of analytes by adsorption to the internal wall of the container and against damage in transit. Samples should be packaged so that the composition of the sample does not change during transport or storage. Each sample needs to be recorded in a way that it can be traced back to the source after analysis.

Detection methods:

In routine screening, a more robust method may be used identify positive samples and another more sensitive method to quantify and confirm the positive sample. Examples of methods used for the quantification and confirmation of chemical residues are shown in Table 9.3.

9.9.2 Detection of antimicrobial residues in carcasses

A traditional way to screen for antimicrobial substances is to use microbiological methods based on the inhibition of a test organism by the antimicrobial substance in the sample. Microbiological methods are relatively inexpensive and easy to perform, and they detect a broad range of antimicrobial substances. Another possibility is to use methods such as enzyme linked immunoassay (ELISA) or chromatographic methods such as liquid chromatography combined with tandem mass spectroscopy (LC-MS/MS) methods. However, these methods usually detect only one group of antimicrobials at a time and rely on additional information of possible antimicrobial used or otherwise multiple tests need to be done.

Microbiological method for detection of antimicrobial residues in carcasses
In meat inspection, detection of antimicrobial residues is attempted when the use of antimicrobial substances is suspected. The method is based on the inhibition of the growth of a test organism, such as *Bacillus subtilis* BGA or *Bacillus stearothermophilus*, that is sensitive to various antimicrobial drugs. There are two main test formats, the plate or agar diffusion method and the tube test. In the agar diffusion test samples are applied on top of the agar containing test organism. Bacterial growth will turn the agar into an opaque layer, in which a clear growth-inhibited area around the sample is formed if the sample contains antimicrobial substances. The tube tests, such as the commercial Premi®Test, consists of a growth medium inoculated with (spores of) a sensitive test bacterium, supplemented with a pH or redox indicator. The bacteria start to grow and produce acid, which causes a colour change. If antimicrobial residues are present in the sample, they will prevent or delay bacterial growth, and thus inhibit or delay the colour change.

The susceptibility of a test organism varies between species and one species usually is not equally susceptible to all antimicrobial drugs. Therefore, the sensitivity of the test is different for different antimicrobials. Different bacteria can be used in tandem to benefit from the differences. However, the use of multiple bacteria makes the method more complex and time consuming to perform.

9.10   Process and slaughterhouse environment controls

Food business operators need to monitor and control process hygiene as well as the cleanliness of the processing areas and equipment. Process hygiene at the slaughterhouse is most often monitored by sampling carcass surfaces and examining the number of indicator organisms, such as aerobic colony count, *Enterobacteriaceae* or *E. coli*. In addition, common pathogenic bacteria, such as *Salmonella*, are used as well. General guidelines and recommendations on sampling are provided by the Codex Alimentarius and International Commission on Microbiological Specifications for Foods (ICMSF) (see Literature and further reading).

For environmental control multiple different methods can be used. The reference method is usually detection of live microbes (aerobic colony count, *Enterobacteriaceae*) on surfaces, for which validated commercial test kits are available. However, faster on-site methods such as ATP (adenosine triphosphate) bioluminescence or chemical methods such as protein or sugars trace tests can also be used. The ATP bioluminescence test detects organic matter on the surface by showing production of light by luciferase from ATP. The produced light is detected using a luminometer and the amount light produced is proportional to the amount of organic matter in the sample. ICMSF provides excellent further reading also on sampling of the processing environment (see Literature and suggested reading).

Literature and further reading

Codex Alimentarius Standards. Retrieved from http://www.codexalimentarius.org/ (last accessed 3 February 2014).

Commission regulation (EC) No 2075/2005 of 5 December 2005 laying down specific rules on official controls for Trichinella in meat. *Official Journal of the European Union*, L 338, 60–82.

ICMSF (International Commission on Microbiological Specifications for Foods). 2002. *Microorganisms in Foods 7, Microbiological Testing in Food Safety Management*. Kluwer Academic/Plenum Publishers. ISBN 0306472627.

ICMSF (International Commission on Microbiological Specifications for Foods). 2011. *Microorganisms in Foods 8, Use of Data for Assessing Process Control and Product Acceptance*, 1st edn. Springer. ISBN 978-1-4419-9373-1.

ISO (International Organization for Standardization). Multiple standards are relevant. http://www.iso.org/iso/home.htm (last accessed 17 February 2014).

OIE (World Organisation for Animal Health). 2012. Manual of diagnostic tests and vaccines for terrestrial animals, 7th edn, Vols 1 and 2. Retrieved from http://www.oie.int/ (last accessed 3 February 2014).

Quinn, P.J., Markey, B.K., Leonard, F.C. *et al.* 2012. *Veterinary Microbiology and Microbial Disease*, 2nd edn. John Wiley & Sons Ltd, Chichester, UK.

Wang, Q., Chang, B.J. and Riley, T.V. 2010. *Erysipelothrix rhusiopathiae*. *Veterinary Microbiology*, **140**, 405–417.

WHO (World Health Organization). 2012. Guidance on Regulations for the Transport of Infectious Substances 2013–2014. Retrieved from http://www.who.int/en/ (last accessed 24 March 2014).

10
Judgment of Meat

Thimjos Ninios

Border Control Section, Import, Export and Organic Control Unit, Control Department, Finnish Food Safety Authority Evira, Helsinki, Finland

10.1 Scope

The scope of this chapter is to review the steps which need to be followed when it comes to perform a judgment concerning the meat of a slaughtered animal. The procedures of judgment may vary according to the legislation in force in different countries or unions.

10.2 Meat inspection

The main purpose of meat inspection is to make sure that meat is safe to consume. Meat safety is ensured through a multistep procedure that includes the food chain information, the surveillance on animal welfare, *ante-* and postmortem inspection, laboratory testing and management of the various risks present in the various steps of the slaughter process. As a result of the inspection undergone, an animal can be accepted, or not, for slaughter and the meat of a slaughtered animal can be accepted, or declared unfit, for human consumption. At this point it is worthwhile to recall and review step-by-step the operations that leading to a final judgment of the meat:

Meat Inspection and Control in the Slaughterhouse, First Edition.
Edited by Thimjos Ninios, Janne Lundén, Hannu Korkeala and Maria Fredriksson-Ahomaa.
© 2014 John Wiley & Sons, Ltd. Published 2014 by John Wiley & Sons, Ltd.

- **Verification of food chain information**
  - ensure that information about animals procured for slaughter is supplied;
  - ensure that animals procured for slaughter fulfil the requirements set by the legislation in force.

- **Surveillance on animal welfare**
  - verify the compliance with animal welfare requirements regarding the protection of animals during transport, lairage and slaughtering;
  - ensure that the animals do not present any signs indicating that welfare has been compromised.

- ***Ante-mortem* inspection**
  - ensure that animals with pain and suffering are slaughtered urgently;
  - identify sick animals;
  - identify abnormal animals;
  - identify reportable animal diseases;
  - identify dirty animals;
  - the inspection can be more accurate if done both at rest and in motion.

- ***Post-mortem* inspection**
  - perform a visual inspection, palpation and incisions according to the legislation in force;
  - identify diseased carcass and/or organs;
  - identify abnormal (other than diseased) carcass and/or organs;
  - determine if the single finding is acute or chronic;
  - determine if the findings refer to a localized or generalized condition;
  - in the case of a generalized condition it might be useful to determine which are the primary and which are the secondary findings;
  - use the sense of olfaction if and where needed.

- **Laboratory testing**
  - testing to reach a definitive diagnosis;
  - testing to detect the presence of an animal disease, residues or contaminants;
  - testing to detect a non-compliance with microbiological criteria.

10.2.1   Management of risks

Risk management within the slaughter process and meat inspection is crucial to provide the consumer with safe meat. Meat inspection shall also consider and intent to control the presence of hazards not identifiable by the traditional inspective procedures, to guarantee a proper level of control on the slaughter process. New inspective tools might be necessary for the monitoring or control of such hazards. For example, the traditional meat inspection and the *ante-* or *post-mortem* inspective tools have not been aimed at providing information concerning the presence of certain pathogens like *Salmonella, E. coli, Campylobacter* or *Yersinia* in meat. However, those pathogens are likely to reduce the meat safety and create outbreaks harmful for human health.

10.3   Evaluation of the meat

10.3.1   How to evaluate

The evaluation leading to a judgment of meat needs to take in consideration at least the following parameters that are related to the nature of the findings detected within the inspective process:

- suspicion or diagnosis of a disease;
- suspicion or diagnosis of an abnormality (other than disease);
- suspicion or diagnosis of a condition that may represent a hazard for human health;
- suspicion or diagnosis of an acute condition;
- suspicion or diagnosis of a chronic condition;
- suspicion or diagnosis of a localized condition;
- suspicion or diagnosis of a generalized condition.

As meat inspection is a complex of various activities it might not always be simple to decide on the approval of meat for human consumption or the condemnation of it. The professional skills of the staff involved with meat inspection are eminently important and influence the results of the inspective activities. A functional collaboration between the members of the inspective staff, including official veterinarians, auxiliary staff and laboratory staff, may result in optimal outcomes. Efficient and fluent cooperation alongside the slaughterhouse personnel is always constructive.

From time to time the judgment of meat inspection findings is a sort of challenge. The findings do not always represent a black versus white situation where it is easy to decide to condemn the meat or not. However, there are some generic rules for condemnation only of a part of carcass and/or organs as per:

- localized chronic finding,
- localized disease,
- localized abnormality (other than disease, for example stained meat).

In addition there are some generic rules for condemnation of the whole carcass and organs as per:

- generalized disease,
- generalized abnormality (other than disease, for example stained meat),
- any other generalized condition that represents a hazard to human health.

10.3.2   Conclusion of inspecting activities

The results of the previous steps need to be evaluated together and the result of the evaluation is usually one of the following:

- the carcass and/or the organs are declared fit for human consumption;
- a part of the carcass and/or the organs are declared unfit for human consumption and need to be trimmed off;
- the carcass and/or the organs are declared unfit for human consumption and condemned.

The requirements set for the evaluation of the carcass and the organs are imposed by legislation and, therefore, a proper performance of inspective activities requires a proper knowledge of the up to date legislation.

10.3.3   Health mark

The main purpose of this mark is to distinguish meat fit for human consumption from meat unfit for human consumption. In addition, the mark identifies the slaughterhouse where the animal has been slaughtered. As a result, the health mark represents an important element in the traceability of meat.

The approval of a carcass for human consumption is expressed by stamping the health mark on the surface of the carcass. If carcasses are cut into half carcasses or quarters, or half carcasses are cut into three pieces, each piece needs to be marked in a way that every single part of the carcass leaving the slaughterhouse is identifiable. The mark needs also to be applied on the packages containing the organs approved in meat inspection.

In poultry the health mark is replaced by the identification mark, which is applied directly to the product or, alternatively, to the wrapping, packaging, label or tag. The health mark and the identification mark, whilst they are similar and contain similar information, differ in their legal significance.

10.3.4   Examples of evaluation and judgment

Example 1

Slaughtered animal: Swine.

Findings in *ante-mortem* inspection: subacute/chronic tail biting, abscess of tennis ball dimensions in the right elbow.

Findings in *post-mortem* inspection: small abscesses in the base of the tail, small abscesses in the vertebral column, abscesses in the lungs.

Evaluation of the findings: the amount of abscesses distributed in various organs indicates the possibility of a generalized infection. The abscesses represent probably the secondary lesions caused, possibly, by the tail biting representing, possibly, the primary lesion.

Judgment regarding the carcass: condemned because of a high probability of a generalized infection.

Judgment regarding the organs: condemned because of a generalized infection.

Example 2

Slaughtered animal: Sheep.

Findings in *post-mortem* inspection: *Fasciola hepatica* flukes in the liver.

Evaluation of the findings: only one finding is reported and it suggests a localized infestation.

Judgment regarding the carcass: fit for human consumption

Judgment regarding the organs: the infested organ (liver) needs to be condemned but the rest of the organs can be used for human consumption.

Example 3

Slaughtered animal: Cattle.

Findings in *post-mortem* inspection: faeces in the pelvis.

Evaluation of the findings: zero tolerance policy for visible fecal contamination is an important food safety standard. However the finding is localized and the contaminated area could be removed without any risk for the rest of the carcass.

Judgment regarding the carcass: after the contaminated area is hygienically trimmed off the remaining carcass is fit for human consumption.

Judgment regarding the organs: if the fecal contamination does not affect the organs they can be declared fit for human consumption.

Example 4

Slaughtered animal: Swine.

Findings in *post-mortem* inspection: faeces everywhere over the carcass and all the organs. This finding is probable, for example, when during the splitting of the carcass lengthways into half carcasses down the spinal column the large intestine breaks down by mistake.

Evaluation of the findings: generalized faecal contamination. Due to the extension of the finding any hygienic trimming of the contaminated areas is impossible.

Judgment regarding the carcass: the carcass needs to be condemned.

Judgment regarding the organs: the organs need to be condemned.

10.4   Record keeping in meat inspection

Comprehensive documentary evidence is needed for all the inspective activities. This is to be able to demonstrate what has been done and why. The decision concerning the judgment of meat needs to be documented and the responsibility for the decision to be assumed correctly by the inspector.

Detailed information on the up-to-date requirements concerning the documentation of the inspective activities is usually available in legislation. The competent authorities may dispose ready-to-use templates. However, there are some useful tips on how to complete documentation:

- put down the date in a clear, understandable and complete format (dd/mm/yyyy or similar);
- put down the time in a clear, understandable and complete format (hh:mm or similar);
- put down the information identifying the unit under examination (animal, carcass and organs or sample);
- describe shortly the performed inspective activity (*ante-mortem* inspection or other);
- describe in detail and clearly the results of the performed inspective activity;
- express the judgment related to the performed inspective activity and explain in detail where needed;
- sign the document and take the responsibility concerning the performed inspective activity.

Literature and further reading

EFSA (European Food Safety Authority). EFSA opinions on meat inspection, http://www.efsa.europa.eu/en/topics/topic/meatinspection.htm; swine: http://www.efsa.europa.eu/en/efsajournal/pub/2351.htm; poultry: http://www.efsa.europa.eu/en/efsajournal/pub/2741.htm; bovine animals: http://www.efsa.europa.eu/en/efsajournal/pub/3266.htm; sheep and goats: http://www.efsa.europa.eu/en/efsajournal/pub/3265.htm (all last accessed 17 February 2014).

EUR-Lex. Details of up-to-date EU legislation on meat inspection can be found at: http://eur-lex.europa.eu/en/index.htm (last accessed 17 February 2014).

Herenda, D. 2000. Manual on Meat Inspection for Developing Countries. UN Food and Agricultural Organization. Retrieved from http://www.fao.org/docrep/003/t0756e/t0756e00.HTM (last accessed 17 February 2014).

11
Classification of Carcasses

Rosanna Ianniciello[1], Paolo Berardinelli[2], Monica Gramenzi[2] and Alessandra Martelli[2]

[1]*ASL4, Teramo, Italy*

[2]*Faculty of Veterinary Medicine, University of Teramo, Teramo, Italy*

11.1 Scope

The carcasses, half carcasses of quarters of slaughtered animals comprise a significant component of the food business market. The market needs an evaluation of the amount and the quality of the meat intended for trade. Trade activities have required specific rules to standardize the classification of carcasses based on certain parameters, such as the muscularity and the fat content of the carcasses. The scope of this chapter is to review some classification systems concerning beef, pig, sheep and poultry carcasses.

11.2 Classification of beef carcasses

The carcasses, half carcasses or quarter carcasses of bovine subject to classification should be evaluated according to legal standards. The classification of beef carcasses identifies the qualitative and commercial properties of the carcasses and defines their price within trade activities.

Slaughterhouses with the capacity to slaughter a large number of cattle per week are keen to classify the carcasses. The classification procedures need to be carried out by properly trained staff. The classification is not compulsory for slaughterhouses slaughtering a small number of animals or for slaughterhouses where deboning of carcasses takes place after slaughter.

Meat Inspection and Control in the Slaughterhouse, First Edition.
Edited by Thimjos Ninios, Janne Lundén, Hannu Korkeala and Maria Fredriksson-Ahomaa.
© 2014 John Wiley & Sons, Ltd. Published 2014 by John Wiley & Sons, Ltd.

The procedures used for classification and any limit values determining where classification is compulsory need to conform to the legislation in force.

11.2.1   Classification grid of the European Union

The classification of the carcasses of adult bovine animals needs to be carried out as soon as possible and no later than one hour after the animal has been stuck. It should be based on the accurate evaluation of certain parameters:

- category
- conformation class
- class of fat cover.

11.2.2   Category

A carcass should be identified and then categorized in one of the following categories:

- Z: carcasses of animals aged greater than or equal to 8 months but less than 12 months;
- A: carcasses of uncastrated male animals aged more than 12 months but less than 24 months;
- B: carcasses of uncastrated male animals aged over 24 months;
- C: carcasses of castrated animals aged over 12 months;
- D: carcasses of female animals that have calved;
- E: carcasses of other female animals aged over 12 months;

The allocation of a carcass to a specific category is related to the identification of the sex and age of the slaughtered animal. The following definitions are used within the classification:

- a carcass is the whole body of the slaughtered bovine as presented after bleeding, evisceration and skinning;
- a half carcass is the half obtained after the separation of a carcass symmetrically through the middle of each cervical, dorsal, lumbar and sacral vertebra and through the middle of the sternum and the ischiopubic symphysis.

Identification of sex   The distinction of sex can be achieved paying attention to the anatomical features related to the different skeletal and muscular conformation of the half carcasses in male and female bovines. A relevant distinctive feature is represented by the different conformation of the pelvis bone.

The carcasses of male animals have the ischiopubic symphysis arched and the pubic tubercle developed and rounded. In the carcasses of females the floor of the pelvis is flat, concave and elongated while the pubic tubercle is less developed than in males.

The differences in the presence and development of muscles in the half carcass allow the identification with above par precision of the sex of the slaughtered animal. Particular attention should be paid to the anatomical region of the thigh and to the development of the muscle *biceps femoris* and *parameralis*, which in male cattle has a triangular form, while in female the uncovered part of the muscle has a rectangular form.

The anatomical region of the neck in the male is more developed compared to the corresponding one of the female and assumes a rectangular form due to the development of the muscles of the upper part of the neck, which protrude beyond the margin of the nuchal ligament and converge within the front limit of the region. In females the muscles of the upper part of the neck are thinner and protrude slightly from the nuchal ligament without ever exceeding in length the nuchal margin and the region of the neck. The form of the muscles in the females is triangular. Moreover, the muscles of the anatomical region of the shoulder are more developed in male than in female animals.

Another relevant differential characteristic is the development of the front shank, which donates to the internal carpal joint region a considerably larger aspect in the male than in the female. In the female the configuration appears more tapered.

Slaughter techniques may possibly influence the appearance of the carcass. However, certain distinctive characteristics of the male sex can be detected, such as the presence of components of the penis, which in castrated males appears atrophic, or the remaining portions of the ischiocavernosus muscle, which in non-castrated males is obviously present while in castrated males it is very small. Additionally, residues of the cremaster muscle with the presence of surrounding fat, which appears wrinkled, can be found at the internal inguinal ring while in castrated males the inguinal ring is completely covered with fat and the muscles of the shoulder are less developed.

Identification of the age The age of slaughtered cattle is reported in the documentation accompanying the animal to the slaughterhouse. The information reported in the documentation needs to be congruent with the appearance of the animal.

The age of the slaughter animal can be estimated by evaluation of the skeletal development. Attention should be paid particularly to the amount of cartilage and the degree of ossification of the bone tissue of the visible bones of the half carcass. Thus, looking at the bones of the pelvis of a young animal, less than two years old, the tubercle and the upper part of the ischiopubic symphysis consist almost exclusively of cartilaginous tissue, whereas in older animals the tubercle has undergone a process of ossification and the cartilage is only marginally present.

Another considerable aspect when it comes to determininng the age of an animal is the presence of cartilaginous tissue in the spinous processes of the vertebrae. Between the spinous processes of the sacral vertebrae of young animals, especially, a certain thickness of the cartilaginous tissue is still evident, while in adult animals the cartilage that remains after ossification is not relevant. Additionally, in young animals the spinous processes of the dorsal vertebrae, from the first to the fourth, present only minor initial ossification, while from the fifth to the ninth vertebrae there are no signs of ossification and, therefore, they still consist completely of cartilaginous tissue. The spinous processes of all the dorsal vertebrae of animals older than two years present clear signs of ossification.

One more method that allows an evaluation of the age of the animal is the observation of the sternebrae (Figure 11.1). Attention should be paid, in particular, to the first sternebra, which is mostly cartilaginous in its front part in young animals and completely ossified in adult animals. The thickness of the cartilage interposed between the remaining sternebrae can be relevant or less relevant in relation to the age of the animal; this applies to both males and females.

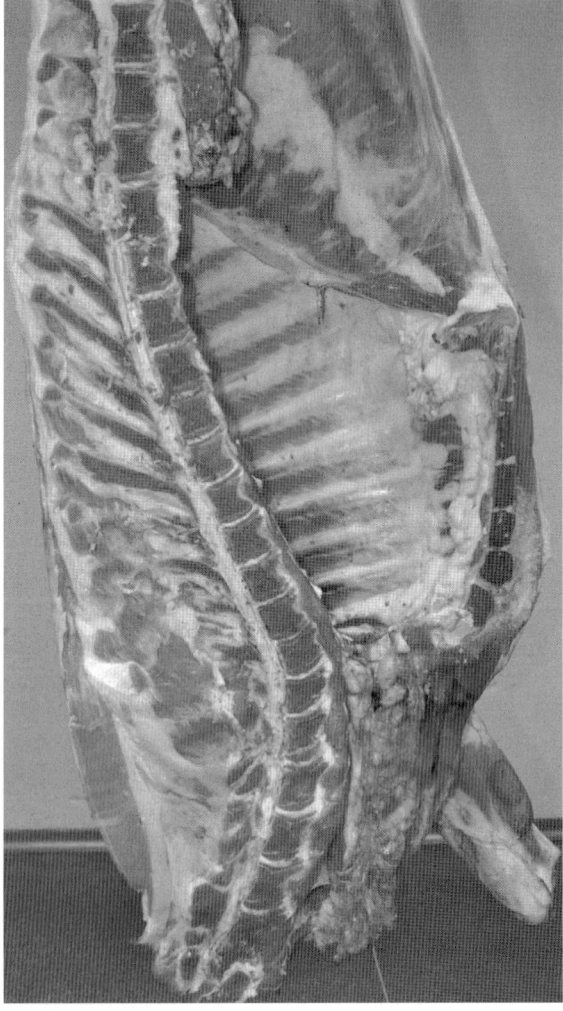

Figure 11.1 Medial view of carcasses of young bovine. Age determination based on the presence of ossification centres. The sternebrae are almost entirely cartilaginous.

11.2.3 Conformation class

The conformation class of a beef carcass can be defined by observing at the dorsal, lateral and medial profiles of the half carcasses and evaluating the development of the muscular mass, which determines the above mentioned profiles. The identification of the conformation class takes in consideration specifically the profiles of the major muscle groups putting together the anatomical regions of the thigh, the back and the shoulder. The outcome of the evaluation divides the carcasses into six different classes.

S: Superior

All the profiles are extremely convex as a result of an outstanding muscular development:

- the thigh is very rounded and the muscles protrude from the ischiopubic symphysis;
- the back is very wide and thick with very clear curves and lasts up to the region of the shoulder;
- the shoulder is very rounded.

The EUROP rating system has introduced the conformation class S for the classification of specific species of cattle which present the specific anatomical characteristic of 'double muscling' (Figures 11.2 and 11.3).

E: Excellent

All the muscular profiles are between convex and very convex and the muscular development is excellent:

- the thigh is very rounded and the muscles protrude from the ischiopubic symphysis;
- the back is wide and very thick and lasts up to the region of the shoulder;
- the shoulder is very rounded.

U: Very good

The muscular profiles remain convex and muscular development is very good:

- the thigh is still rounded and protrudes from the ischiopubic symphysis;
- the back maintains an evident thickness and the development up to the shoulder is still configured in its entire width;
- the shoulder still has a rounded configuration.

R: Good

The muscular profiles are less convex and the overall muscle development, even being well-developed, is more linear and the carcass features more rectilinear (Figures 11.4 and 11.5):

- the thigh is still well developed but less rounded than the above classes;

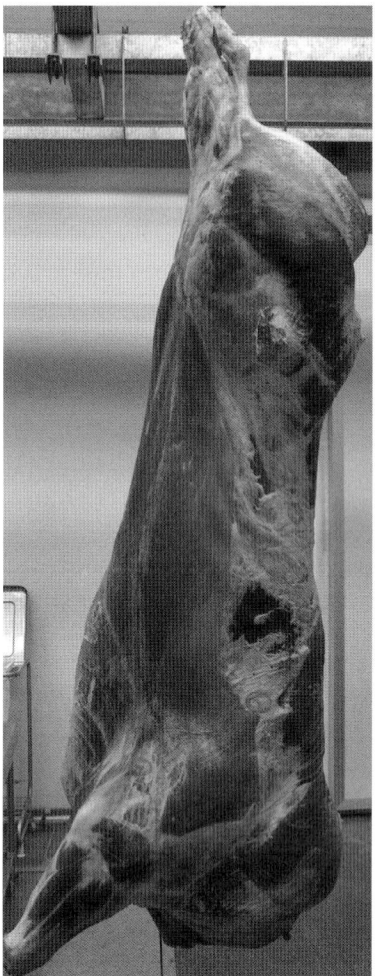

Figure 11.2   Lateral view of an S-class bovine carcass.

- the back is still thick but it is more rectilinear considering the height of the shoulder;
- the shoulder is still fairly well developed.

O: Fair

The muscular development is less evident and, therefore, the carcass profiles appear linear and occasionally might assume a slightly concave profile:

- the thigh consists of underdeveloped muscular groups and, therefore, it is almost flat;
- the thickness of the back is between medium and almost insufficient;
- the shoulder is flat and underdeveloped.

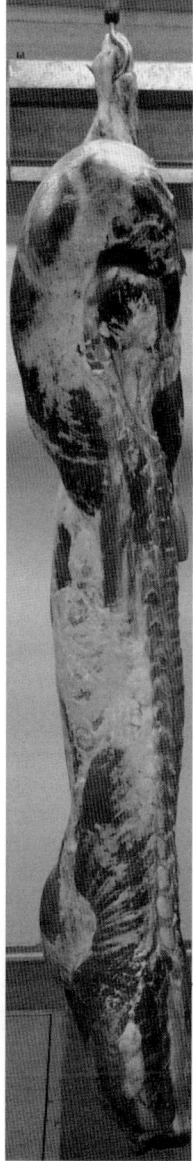

Figure 11.3 Dorsal view of an S-class bovine carcass.

P: Poor

All the muscle groups are poorly developed, the carcass profiles assume a concave appearance and the bones of the carcass become visible under the muscular profiles:

- the thigh is poorly developed and flattened while the ischial tuberosity and the hip bone become observable;

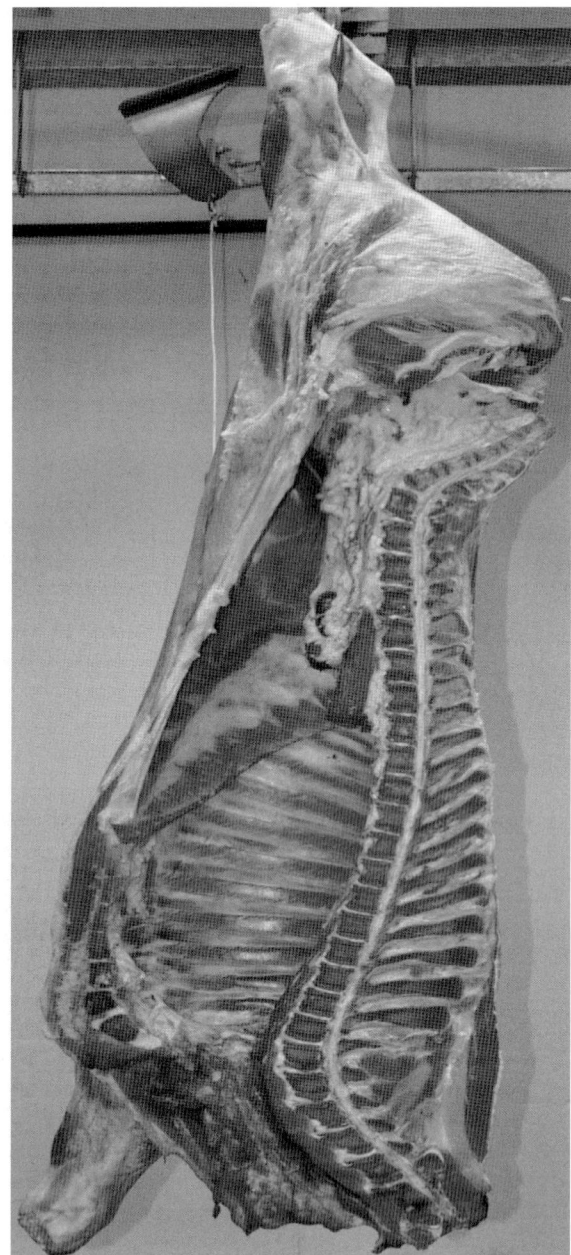

Figure 11.4   Medial view of an R-class bovine carcass.

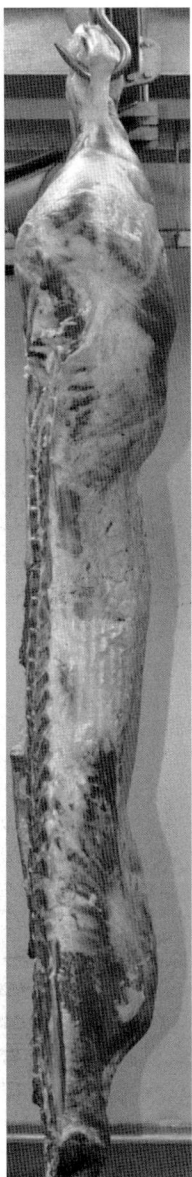

Figure 11.5   Dorsal view of an R-class bovine carcass.

- the back appears narrow and concave, the spinal column is clearly visible and the spinous processes of the single vertebrae are evident;
- the shoulder is flat due to the poor muscular development while the underlying skeletal structures and, in particular, the humerus and the acromial plug of the scapula can be clearly recognized below the muscular tissues.

11.2.4   Class of fat cover

The class of fat cover is determined by the amount of fat present on the lateral profile of the half carcass and the degree of fat infiltration in the intercostal muscles of the chest. There are five standard classes of greasiness numbered from 1 to 5.

1: Low
There is almost no presence of fat. The muscle groups of the thigh, the shoulder and part of the anterior part of the back are clearly visible and distinguishable while there is no evidence of fat in the thoracic cavity.

2: Slight
Some fat is slightly noticeable on the lateral profile of the half carcass. The muscle groups of the region of the thigh, the shoulder and the anterior part of the back conserve their distinctive characteristics and the intercostal muscles are clearly visible in the thoracic cavity because they are covered only by a thin layer of fat.

3: Average
The muscles of the carcass are almost completely covered by fat with the exception of certain muscle groups in the region of the thigh and the shoulder, where the muscles are still visible. The intercostal muscles are still visible while a modest presence of fat can be noticed on the pillars of the diaphragm and the arches of the ribs (Figure 11.6).

4: High
There is a fat cover over the entire surface of the carcass and the distinction of the muscle masses related to the regions of the thigh, the shoulder and the back is not possible anymore. The alignments of the thigh are affected by certain remarked muscle groups due to the presence of fat interposed between the muscles. The diaphragm, the ribs and intercostal muscles are infiltrated by fat.

5: Very high
The thickness of the fat that covers the carcass almost entirely does not allow the distinction of the muscular groups of the thigh, which is entirely covered by a layer of fat. The veins of the fat are no longer visible. In the thoracic cavity the fat covers the ribs and the intercostal muscles are extensively infiltrated by fat.

11.3   Classification of pig carcasses

The classification of pig carcasses is used to determine the appropriate commercial value of the carcasses. The classification takes place through an estimation of the content of lean meat in the carcass, measuring instrumentally

Figure 11.6   Average class of fat cover.

the thickness of both muscular and adipose tissue in one or more parts of the carcass.

11.3.1   Steps of the classification

The classification needs to be carried out by properly trained classifiers in slaughterhouses that slaughter a number of animals determined by legislation. The classification of a carcass is composed by the following steps:

- calibration and verification of the functionality of equipment for the measurement;
- presentation of the carcass at the weighing station of the slaughter line;
- weighing of the chilled carcass;
- use of a measuring instrument for the classification;
- identification of the carcass.

11.3.2   Calibration and verification of measurement equipment functionality

The calibration and verification of the correct operation of the instruments for the measurement need to be made prior to slaughter and should be repeated subsequently after any interruption (e.g. breaks of slaughtering, lunch break etc.).

Verification is carried out using a test block and following the technical instructions of the equipment manufacturer. The classifier responsible for the measurement is required to record on the document the classification of certain relevant information, such as the date, time, license number of the classifier, data on the performed verification activities and any eventual notes regarding the verification.

11.3.3   Presentation of the carcass at the slaughter line weighing station

For the purposes of the classification 'pig carcass' means the body of a slaughtered pig, bled, whole or separated in halves, eviscerated, without tongue, bristles, nails, genitals, kidney fat, kidneys and diaphragm. In some countries the procedures may vary according to national requirements, such as in Italy where the presence of perirenal fat in the carcass is allowed. These variations need to be considered in the calculation of the percentage of fat.

11.3.4   Weighing of the chilled carcass

According to the standards of the European Union (EU), the carcass shall be weighed chilled and within 45 minutes after the pig has been stuck. In the case that a slaughterhouse does not meet the 45 minute condition, there shall be applied a deduction of 2% of the total weight for every further 45 minutes.

11.3.5 Use of the measuring instrument for the classification

The instrumental measuring of the thickness of the fat covering the dorsal region of the back and the thickness of the *longissimus dorsi* muscle should be made at the time of weighing. The measuring instrument can be, for example, one of the following:

- Fat-O-Meater (FOM)
- Hennessy Grading Probe (HGP)

Fat-O-Meater (FOM) This device is provided with a sensor 6 mm in diameter containing a photodiode capable of an operating distance between 5 and 115 mm. Measurement occurs by exploiting the different reflectances recorded when a beam of infrared light is introduced into the various tissues. The amount of reflected light is superior when the probe passes through adipose tissue to when it crosses muscular tissue. The measured thickness values concerning muscle and fat are converted into an estimate of lean meat content.

The measured values are reported under the following acronyms:

- SL: thickness of fat between the third and fourth last lumbar vertebrae, at 8 cm from the dorsal line;
- SR: thickness of fat between the third and fourth last rib, 6–8 cm from the dorsal line;
- F: thickness of the meat detected simultaneously when calculating the SR.

Hennessy Grading Probe (HGP) The HGP uses the same principle as the photodiode used in the FOM. The device is provided with a sensor of diameter 5.95 mm containing a photodiode operating in a distance between 0 and 120 mm.

The measurement should be done:

- at the same point for the detection of both thickness of fat and thickness of muscle;
- on the left half carcass;
- at a distance of 8 cm from the mid-line;
- between the third and fourth last rib.

The combination of two measured values permits determination of the percentage of lean meat. The value is calculated automatically and allows the categorization of the carcass according to a commercial class provided by the EUROP grid.

For each carcass the following values should be recorded:

- dead weight (kg);
- lean meat (%);
- fat cover thickness (mm – SR);
- thickness of the loin (mm – F);
- class of fleshiness (EUROP).

11.3.6   Identification of carcasses

After classification the carcasses should be labelled and the label should provide information on the weight category:

- H = Heavy, when the weight is more than 110.0 kg
- L = Light, when the weight is between 70 and 110 kg

Alternatively, the capital letter H or L can be followed by the percentage of lean meat.

The classification procedures may vary slightly even between the Member States of the European Union. For example a Decision of the European Commission authorized at the time methods for grading pig carcasses in Italy.

The main changes relate to the:

- Updating of the equations for estimating the lean meat content on the instruments Fat-O-Meater I and Hennessy Grading Probe.
- Introduction of four new tools for the classification for which they have been authorized for the relevant equations:
    - (i)   Manual method ZP (for this instrument will be branched specific guidelines to regulate the use)
    - (ii)   Fat-O-Meater II
    - (iii)   CBS-Image-Meater
    - (iv)   Autofom III.

The main goal of these changes is to update estimating equations, and for Italy the carcass presentation in derogation to the Community standard, without kidneys, with lard and the remaining diaphragm.

11.3.7   Classification according to EUROP

Based on the proportion of detected fleshiness or the percentage of lean meat measured in the carcass, there are five commercial categories:

- E: lean meat content more than 55%;
- U: lean meat content between 50 and 55%;
- R: lean meat content between 45 and 50%;

- O: lean meat content between 40 and 45%;
- P: lean meat content less than 40%.

The class E includes carcasses presenting a notable muscular development remarked by a good muscular structure and a low amount of fat cover. The following classes, the U to P, are characterized by a gradually decreasing percentage of muscular mass and a gradually increasing amount of fat cover.

11.3.8 Labelling

The mark or label should contain the information concerning the classification. The mark can be applied by stamping a mark on the skin at the level of the hindquarter with indelible ink, which needs to be heat resistant and non-toxic. The same information can be provided on labels firmly fixed on the carcass. Labelling needs to comply with the legislation in force.

11.4 Classification of sheep carcasses

The classification of sheep carcasses needs to be carried out not later than one hour after the animal has been stuck. As for cattle and pigs, the classification should be performed by experienced and qualified personnel, trained to perform a properly functioning classification.

The classification of sheep carcasses needs to be carried out after the conclusion of the evisceration and trimming stages. The classification should be based on the following parameters:

- category
- conformation class
- degree of fat cover.

11.4.1 Category

Sheep carcasses can be divided in two categories based on the age of the slaughtered animals:

- L: carcasses of sheep under 12 months old (lamb);
- S: carcasses of other sheep.

11.4.2 Conformation class

According to the European classification grid there are different conformation classes for sheep carcasses; these are related to the development of the muscular masses. The conformation class is defined through an evaluation of the muscular profiles of the anatomical regions of the shoulder, the back and the hindquarters of the carcass.

S: Superior

All the profiles are extremely convex due to an exceptional muscular development characterized by a formation defined as 'double muscling':

- the shoulder has very convex profiles for the exceptional development of the underlying muscular groups;
- the back is exceptionally convex, thick and wide;
- the hindquarter has extremely convex profiles due to the impressive growth of the muscular groups (double muscling).

E: Excellent

All muscular profiles appear convex or super convex:

- the shoulder is very convex and very thick;
- the back is very convex, very thick and very wide, lasting up to the shoulder;
- the hindquarter is well developed and often presents a very convex profile.

U: Very good

The profiles are still convex due to the abundance of muscular tissue (Figure 11.7):

- the shoulder is thick and convex but less rounded than the upper classes;
- the back is well structured and lasts up to the shoulder;
- the hindquarter is slightly thinner but it maintains the convex profile, characteristic of a good muscular development;

R: Good

The muscular development is good, the texture and the thickness of the muscles are still well visible but the wholeness of muscular profiles appears more linear (Figure 11.8):

- the shoulder still appears well developed but thinner;
- the back is less wide and does not last up to the shoulder;
- the hindquarter is characterized mainly by a linear muscular profile.

O: Fair

All the muscular groups are moderately developed and appear linear or concave:

- the shoulder tends to be narrow and the thickness is insignificant;
- the back is not very wide due to poor muscular development;
- the muscular profiles of the hindquarter tend to be concave.

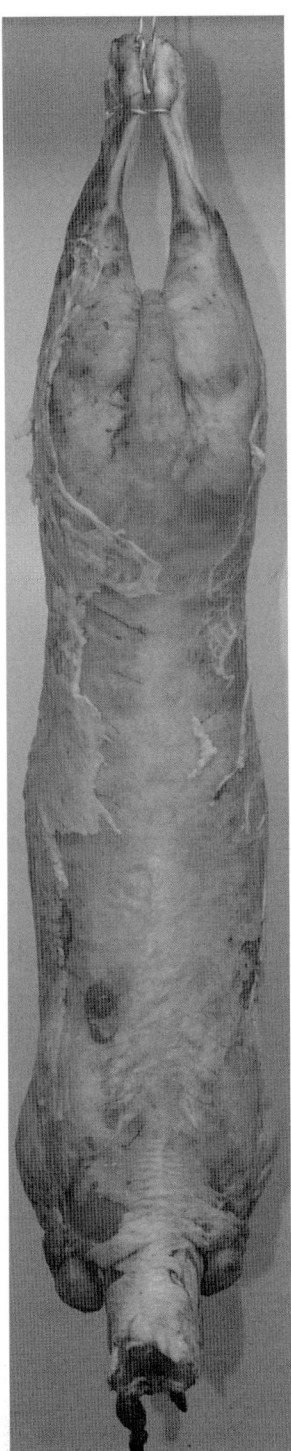

Figure 11.7 Adult sheep carcass (category U) rich in fat.

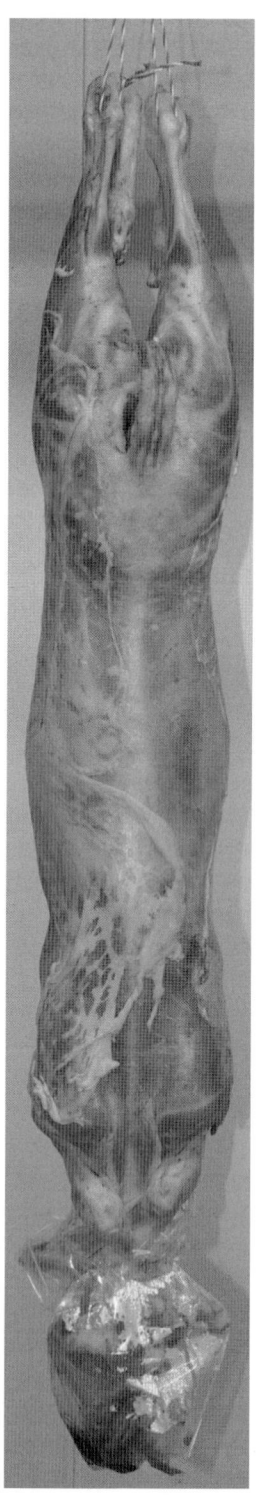

Figure 11.8   Lamb carcass belonging to category R, average class of fat cover.

P: Poor
A poor muscular development gives to the muscular frame a meagre aspect that is manifested by relatively concave profiles:

- the shoulder is narrow and flat while the profiles of the underlying bones are apparent;
- the back is narrow and concave while the skeletal structure of the region is visible;
- the hindquarter assumes concave or very concave profile.

11.4.3 Degree of fat cover

The degree of fat cover is evaluated through the thickness and the quantity of fat present on the internal and external surfaces of the carcass, with particular reference to the abdominal and thoracic region. The degree of fat cover can be divided into five classes.

Low
There is no fat cover or it is present only in traces:

- the outer surface of the carcass goes without fat or presents traces of visible fat;
- the inner surface near the kidneys (perirenal fat) does not have fat cover but sometimes there are traces of barely detectable fat;
- the thoracic region the ribs and the intercostal muscles fat can be observed while there is no intercostal fat.

Slight
The fat cover is thin and the muscles are mostly visible:

- the outer surface is covered by a thin layer of fat but it might be less evident on the limbs;
- the inner surface presents a thin layer of fat partially covering the kidneys;
- the intercostal muscles are not infiltrated by fat and they are clearly visible between the ribs.

Average
There are low deposits of fat in the thoracic cavity and the muscles are almost everywhere covered by fat, making exceptions for those of the hindquarter and the shoulder:

- the outer surface is mainly covered by a thin layer of adipose tissue and the base of the tail contains thicker fat accumulations;
- the kidneys in the inner surface are covered by a thin layer of fat that surrounds them, most of the times completely but occasionally only incompletely;
- in the thorax the intercostal muscles are still clearly visible and distinguishable between the ribs.

High

The muscles are covered by fat but are still partially visible at the level of the hindquarter and shoulder while some accumulations of fat are visible in the thoracic cavity:

- the outer surface is mostly covered by a thick layer of fat that is thicker on the shoulders and thinner on the limbs;
- in the inner surface the kidneys are completely covered by fat;
- the intercostal muscles are slightly infiltrated by fat while there are visible fat deposits on the ribs.

Very high

The carcass is profusely covered by fat and the thoracic cavity contains abundant accumulations of fat:

- the outer surface is entirely covered by a thick layer of fat and there are thicker accumulations of fat;
- in the inner surface the kidneys are completely covered by a thick layer of fat;
- the intercostal muscles are infiltrated by fat and fat accumulations on the ribs become visible.

11.4.4   Carcasses of lambs of less than 13 kg

According to EU requirements the lamb carcasses of less than 13 kg can also be classified using criteria based on the weight, the meat colour and the class of fat cover. The weight of the carcass needs to be measured while it is still warm after slaughter and there are corrective formulas for the calculation of the weight in the case that the temperature of the carcass has decreased. The colour of the meat is determined by evaluation of the muscle *rectus abdominis* and according to a standardized colour chart:

- clear pink for lamb carcasses less than 7 kg
- clear pink or pink for lamb carcasses between 7 and 13 kg

The degree of fat cover is calculated using the same criteria adopted for carcasses of adult sheep.

11.4.5   Labelling

After classification the sheep carcass should be labelled or marked; the label or mark needs to provide information indicating the category, the conformation class and the degree of fat cover of the carcass. The application of the mark can take place by stamping, using an indelible and non-toxic ink. The marking can eventually be replaced by unalterable and firmly attachable labels.

11.5 Classification of poultry carcasses

The commerce of poultry meat is widely standardized to ensure uniform requirements concerning meat quality. According to EU requirements, poultry meat can be classified in two classes, class A and class B, in consequence of the aspect and the conformation of the carcasses or the cuts derived from it.

11.5.1 Definitions

Poultry species Poultry carcasses belong to one of a number of species.

Domestic fowl (*Gallus domesticus*)

- Chicken, broiler: fowl in which the tip of the sternum is not ossified and, therefore, flexible.
- Cock, hen, boiling fowl: fowl in which the tip of the sternum is ossified and rigid.
- Capon: male fowl castrated before reaching sexual maturity and slaughtered at a minimum age of 140 days.

Turkeys (*Meleagris gallopavo domesticus*)

- Turkey (young): bird in which the tip of the sternum is not ossified and, therefore, flexible.
- Turkey: bird in which the tip of the sternum is ossified and rigid.

Ducks (*Anas platyrhynchos domesticus, Cairina muschata*)

- Duck (young), Muscovy duck (young): bird in which the tip of the sternum is not ossified and, therefore, flexible.
- Duck, Muscovy duck: bird in which the tip of the sternum is ossified and rigid.

Geese (*Anser anser domesticus*)

- Goose (young): bird in which the tip of the sternum is not ossified and, therefore, flexible. The layer of fat that covers the carcass is thin or moderate; the fat of a young goose may have a colour determined by a special diet.
- Goose: bird in which the tip of the sternum is ossified and rigid and a layer of moderate to thick fat covers the whole carcass.

Guinea fowl (*Numida meleagris domesticus*)

- Guinea fowl (young): bird in which the tip of the sternum is not ossified and, therefore, flexible.
- Guinea fowl: bird in which the tip of the sternum is ossified and rigid.

Poultry carcasses　Poultry carcass means the whole body of a bird, belonging to one of the above mentioned species, after the slaughter activities of bleeding, plucking and evisceration. The poultry carcasses may be presented for sale in one of the following forms:

- partially eviscerated, from which the lungs, heart, liver gizzard, crop and kidneys have not been removed;
- eviscerated with giblets;
- eviscerated without giblets.

Poultry cuts　The cuts of poultry meat are related to the conformation and structure of the muscular tissue. Poultry meat can be presented for sale in one of the following forms:

- Half: half of the carcass, obtained by a longitudinal cut in a plane along the sternum and the spine.
- Quarter: obtained by a transversal section of a half in order to get a leg quarter and a breast quarter.
- Wholeness of leg quarters: both leg quarters united by a portion of the back, with or without the rump.
- Breast: the sternum and the ribs, or part of them, distributed on both sides of it, together with the surrounding muscles. The breast may be presented as a whole or halves.
- Leg: the femur, tibia and fibula together with the surrounding musculature. The two cuts should be performed at the joints.
- Chicken leg with portion of the back: chicken leg with a portion of the back not exceeding the 25% of the total weight of the cut.
- Thigh: the femur together with the surrounding musculature. The two cuts need to be made at the hip joint femur.
- Drumstick: the tibia and fibula together with the surrounding muscles. The two cuts need to be carried out at the joints.
- Wing: the humerus, radius and ulna, together with the surrounding musculature.
- Unseparated wings: both wings united by a portion of the back. The portion of the back is not exceeding the 45% of the total weight of the cut.
- Breast fillet: the entire breast or half of it deboned. In the case of turkey breast, the fillet may include only the deep pectoral muscle.
- Breast fillet with wishbone: the skinless breast fillet with the clavicle and the cartilaginous point of the sternum only. The weight of clavicle and cartilage has not to exceed 3% of the total weight of the cut.

11.5.2　Classification

The classification of poultry carcasses and cuts takes into account the appearance, conformation and fat cover of the carcasses, focusing at least on the following factors:

- structure and development of the carcass;
- presence of fat;
- extension of injuries or bruises.

Appearance In order to be classified as class A, carcasses and cuts should comply with the following criteria:

- The surface needs to be undamaged, clean and free from any visible dirt, foreign matter, bruises or blood. Some insignificant damages, contusions or discolorations could eventually be tolerated if not present on breast or legs. The wing tip may be missing and a slight redness is acceptable in wing tips and follicles.
- The meat needs to be free of abnormal smell and to match the characteristic odour of the specie.
- The carcass needs to be complete without fractures or presence of protruding broken bones.

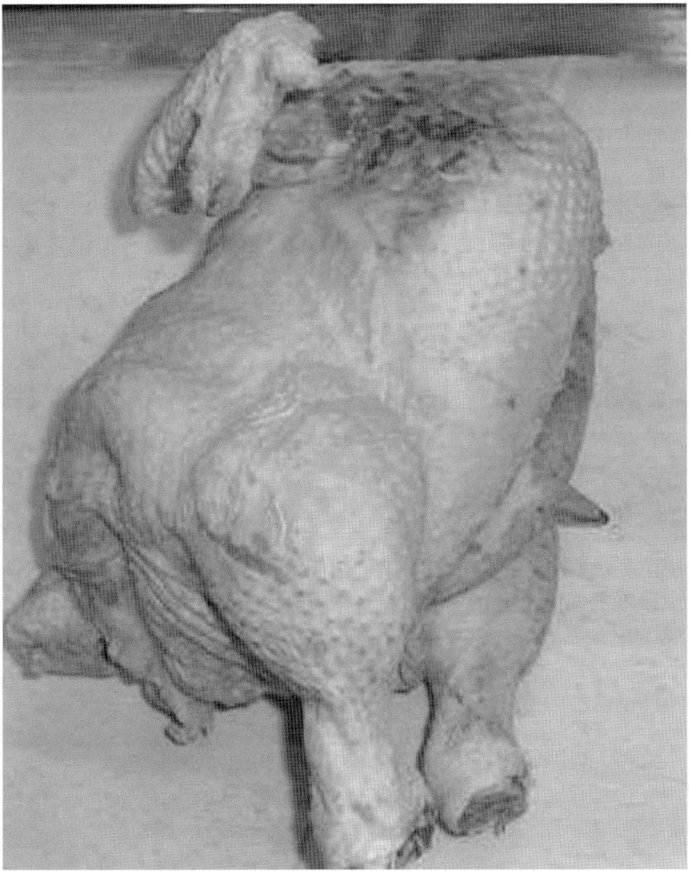

Figure 11.9 Haematoma at the breast of chicken.

- A limited number of small feathers, stubs and filoplumes may be present on the breast, legs, back, foot joints and wing tips. In the case of casserole fowl, ducks, turkeys and geese, a few of the above mentioned objects may also be present in other parts.
- The fresh poultry should not present traces of prior chilling. In the case of frozen or quick-frozen poultry there should not be traces of freezer-burn like pallor, rancidness, dryness or sponginess, except those that are incidental, limited and modest and do not affect the breast or the legs.

The carcasses or cuts that do not fulfil the above mentioned requirements are graded to class B (Figure 11.9).

Conformation   To be graded as class A, the carcasses and cuts of poultry need to have a good shape. The flesh needs to be plump, the breast well developed, broad, long and fleshy, and the legs should be fleshy provided by finely developed muscles. The carcasses or cuts that do not fulfil the above mentioned requirements are graded to class B.

Fat cover   In chickens, young ducks and turkeys there should be a thin regular layer of fat on the breast, back and thighs. In cocks, hens, ducks and young geese a thicker layer of fat is acceptable. In the case of geese, a layer of fat between moderate and thick should be present all over the carcass. The carcasses or cuts that do not fulfil the above mentioned requirements are graded to class B.

Literature and further reading

EUR-Lex. Reg. CEE n° 1208/81, Reg. CEE n° 1186/90, Reg. CEE n° 344/91, Reg. CEE n° 3220/84, Reg. CEE n° 2967/85, Reg. (CE) N. 1249/2008, Regulation (EU) No 1308/2013. Web pages for up to date information on carcass classification: http://eur-lex.europa.eu/en/index.htm (last accessed 22 February 2014).

12

Control, Monitoring and Surveillance of Animal Health and Animal Infectious Diseases at the Slaughterhouse

Ivar Vågsholm

Department of Biomedical Sciences and Veterinary Public Health, Swedish University of Agricultural Sciences, Uppsala, Sweden

12.1 Scope

This chapter deals with the evolution of meat inspection into a tool for control, as well as for monitoring and surveillance systems (MOSS) for animal diseases and zoonoses. The concepts and principles for the design of MOSS activities are presented. The inspiration for this discussion is the European Food Safety Authority (EFSA) reviews on meat inspection.

12.2 Background

12.2.1 An example: UK FMD epidemic 2001 was detected at meat inspection

The usefulness of veterinary alertness at meat inspection is shown by the experience from United Kingdom (UK) foot and mouth disease (FMD) outbreak

Meat Inspection and Control in the Slaughterhouse, First Edition.
Edited by Thimjos Ninios, Janne Lundén, Hannu Korkeala and Maria Fredriksson-Ahomaa.
© 2014 John Wiley & Sons, Ltd. Published 2014 by John Wiley & Sons, Ltd.

in 2001. The first FMD case detected was in a pig slaughterhouse. The epidemiological inquiry indicated that the movement of diseased pigs, or pigs recovering from FMD on 8 and 15 February; resulted in the FMD infection being transferred to a slaughterhouse in Essex, UK, where the first FMD outbreak (FMD/01) was confirmed on 20 February. Recovered pigs in the post-acute phase of the disease could be difficult to identify at *ante-mortem* inspection but possible in the *post-mortem* examination. The subsequent spread of the disease to holdings in Essex was a consequence of mechanical and personnel transmission from the slaughterhouse. It was veterinary students doing their training in meat inspection on pigs that found the suspicious FMD lesions.

12.2.2   Initial purposes of meat inspection

The initial purposes of meat inspection include examining the fitness of the inspected individual carcases for human consumption and measuring the progress of disease control programs in primary production.

Fitness for human consumption   Meat inspection was initially designed for detecting public health risks, such as bovine tuberculosis (TB) in cattle and *Trichinella sp.* in swine, and to examine and verify the fitness of the individual carcase for human consumption. Finding *Trichinella spiralis* larvae in a pig carcase will have a direct impact, as the carcase is unfit for human consumption unless treated (freezing or heat treatment). This remains the primary purpose of meat inspection and each carcase fit for human consumption has to be stamped with an official control stamp identifying the slaughterhouse and its official control. The unsolved problem remains the symptomless carrier animal, being the main source of meat-borne infections. The challenges are how to control the meat-borne transmission of pathogens, for which the most promising idea is a food safety assurance system based on performance objectives. The Performance Objective (PO) is the maximum prevalence and/or concentration of a hazard in a food at a specified step in the food chain before the time of consumption that provides or contributes to an acceptable level of protection. A food safety objective (FSO) is the same at consumption.

Measuring progress of disease control in primary production   Bovine TB is transmitted to humans through by contaminated raw milk and raw milk products and direct contact with infected cattle. Hence, the meat inspection procedure for detecting bovine TB (*Mycobacterium bovis* infections) by palpation and incision of specific (head, respiratory and gastrointestinal) lymph nodes on the carcase is an early example of the successful use of meat inspection as part of animal disease control programmes and as a tool for monitoring progress in zoonoses control. The purpose is also to give feedback to farmers as well as veterinary and public health authorities on the herds' TB status.

Another example of the usefulness of meat inspection is the control of *Echinococcus granulosus* (hydatid disease) in sheep and reindeer. *Echinococcus granulosus* is nowadays a very rare event in reindeer, cattle, goats and sheep. However, in certain areas, such as in Bulgaria, *Echinococcus granulosus* is a common finding at meat inspection, with a reported prevalence of around 5% in cattle and sheep. Portugal also reports frequent findings in farm animal species at slaughter. In this case the results from meat inspection are useful MOSS-tools for assessing the public health risks.

In the 1950s *Echinococcus granulosus* (hydatid) cysts were a frequent finding in reindeer and in humans in the Northern parts of Scandinavia, where the prevalence of hydatid cysts in reindeer carcasses was found to be 10% at meat inspection. The epidemic was maintained by the custom of feeding offal from slaughtered reindeer to shepherd dogs. This zoonosis and public health problem was successfully controlled by compulsory anti-helminthic treatment (deworming) of shepherd dogs, and the prohibition of feeding offal to shepherd dogs. The progress was monitored by the meat inspection findings of reindeer, where currently the finding of hydatid cysts in reindeer is an extremely rare event. This highlights another purpose of meat inspection, which is to monitor the progress of a zoonosis or disease control programme and verify the continued absence of the pathogen after its successful completion.

12.3   Evolution of meat inspection

Nowadays, meat inspection is more than performing applied pathology on individual carcases to assess their fitness for human consumption. The purpose of meat inspection has expanded from fitness of an individual carcase for human consumption to being an integral part of a food safety and quality assurance system as well as national disease MOSS aimed at detecting new diseases, verifying absence of disease and estimating the prevalence of zoonotic and animal pathogens in primary production and at harvest. In other words, meat inspection has evolved into a multipurpose MOSS. In the revisions of meat inspection contemplated in the European Union (EU) this role is further underlined. Official veterinarians became a crucial part of modernized meat inspection that, furthermore, is integrated with animal disease and welfare control strategies. The requirements for official veterinarians outlined in European legislation are a clear illustration of this.

12.3.1   Meat inspection and control

Control embraces all efforts performed to prevent a disease or infectious agent entering a slaughterhouse, to eliminate the agent from the slaughterhouse or to reduce and keep the presence of the agent or disease at acceptable levels. There are usually elements of all three approaches in the practical control strategies that are applied. Therefore, a disease control programme will include elements of monitoring and surveillance, control strategies (prevention of infected animals entering new herds, control of animal movements, containment and treatment of infected herds and, if possible, elimination of the agent from infected herds) and intervention strategies, such as stamping out of infected herds and contact herds, test and slaughter of infected animals, vaccination, test and slaughter, restrictions on movement of animals to other herds and to slaughter, or compulsory vaccination or treatments.

12.3.2   Meat inspection: a part of MOSS and risk management of the food chain

It is important to clarify the purpose(s) the MOSS activities at the slaughterhouse have. The purpose will determine the design of the MOSS activities and the interpretation of findings. The basic tenet is that meat inspection and the monitoring and surveillance at the slaughterhouse are important parts of the overall monitoring and surveillance systems (MOSS) of animal and public health hazards, including infectious diseases. Meat inspection is useful for detecting epizootic diseases like foot and mouth disease, to detect the emergence of diseases in new populations, such as paratuberculosis (Johne's disease) in cattle, to assess the present occurrence of hazards and threats to animal health and welfare, for example the prevalence of pneumonia and pleuritis in pigs and fresh traumatic injuries due to transport and lairage. It is important to put what happens in the slaughterhouse in the national context as, it can be an early warning system for exotic diseases or emerging health and welfare problems, it can give feedback to farmers on their livestock's health performance and also indicate the burden of pathogens entering the food chain.

12.3.3   Meat inspection and disease outbreaks or other disruptive events

It is important that meat inspectors at a slaughterhouse are prepared for at least the following scenarios:

- an epizootic or zoonotic disease is suspected or confirmed at the slaughterhouse, on farms that are nearby or delivering to the slaughterhouse;
- a national standstill, that is the prohibition of all animal movements, being imposed due to outbreaks of epizootic disease;
- similar scenarios like natural disasters (e.g., floods, earthquakes or storms), strikes, civil unrest or war, or blackouts (disrupted electric power supplies).

In all these scenarios, there will be disruptions of the normal slaughter and meat inspection operations, extra work that is not routine will be created, complicated questions on occupational health and waste disposal require rapid answers, and new chains of command will be established. As a consequence of the disrupted slaughterhouse operations, animal welfare and food safety problems might emerge.

The focus here is on the scenario of a suspected or confirmed epizootic and zoonotic disease, but similar considerations will be valid for all disruptive events. Examples of epizootic and/or zoonotic diseases are classical swine fever, foot and mouth disease, highly pathogenic avian influenza or salmonella infections.

In conclusion, there is a need to prepare for outbreaks or disruptive events. The meat inspection veterinarian should be prepared for any disruptive event, including an outbreak of animal diseases or zoonoses. The key principle is making the most critical decisions beforehand when there is calm and quiet and competing interests can be balanced. While it is impossible to foresee disease outbreaks or other disruptive events and all their implications, a good contingency plan enables a prompt and flexible response in situations where urgency is imperative.

A contingency plan should specify what happens in the case of a disruptive event such as an outbreak of epizootic disease. A contingency plan should address the following points if a disease is suspected:

- Whom, for example national competent authorities, diagnostic laboratory and slaughterhouse management, should be notified? If suspected zoonoses or other public health hazards are suspected, who is responsible for notification to public health authorities needs to be clarified.
- Which pieces of information should be collected on the suspected animals?
- When should the transport of animals to the slaughterhouse be stopped?
- What should happen with the animals in lairage?
- What should be done with the carcases from the already slaughtered animals?
- The notification procedures to competent authorities and farmers and food business operators.
- How the collaboration between meat inspection and slaughterhouse management will be organized.
- Who will be the competent supervisors of the meat inspection during an outbreak?
- If the staff at slaughterhouse and meat inspection can contribute to stamping out or culling operations in the event of an outbreak.
- Which samples shall be taken and to which laboratories shall they be sent for analyses, and to whom the laboratory results shall be forwarded?

Moreover, the contingency plan should specify how the clean-up of the slaughterhouse will be done, what kind of disinfectants will be used and for how long thereafter the slaughterhouse will closed before commencing slaughter. The foreseen methods for treatment or disposal of carcases, meat, animal products and waste should be specified. For example, if salmonella is suspected then canning of the meat could inactivate the virus.

A third aspect is the occupational health in the event of disease outbreak in a slaughterhouse. Stress and overwork are concerns, since they increase the likelihood of erroneous decisions. Therefore, if the outbreaks last more than a week, some kind of rotation of responsibilities should be established and specified in the contingency plan. Another issue is the risk to workers and meat inspectors if a zoonotic disease is found or suspected in the slaughterhouse, such as highly pathogenic avian or swine flu or salmonella. The contingency plan should specify additional protective measures foreseen. This could include how the workers protect themselves, access to medical care, special agreements with clinics and doctors.

Sometimes it might be better to have a generic national or industry wide contingency plan that is adapted to the individual slaughterhouses and updated on a regular basis. A contingency plan will often be in the form of checklists. As an example, if suspecting foot and mouth disease (FMD) on cattle at the *ante-mortem* (live animal) inspection, the outbreak the check list could be:

- notify slaughterhouse management;
- stop of all transport to and from the slaughterhouse;
- notify the competent authorities for animal disease control of the suspicion and decide upon which samples need to be taken;
- identify the transport vehicle(s) that delivered the animals and recall them;
- interim stop on the slaughter;
- stop of deliveries of meat, animal products and waste from the slaughterhouse;
- interim trace back investigation including:
  - main clinical symptoms (fever, vesicles, mastitis) and/or *post-mortem* findings (vesicles on the hooves, tongues),
  - age of the cattle, type (dairy, beef),
  - identify transport of the animals to the slaughterhouse,
  - herd(s) or origin, how many, type (dairy, beef), trading patterns;
- the taking of samples (blood, serum) and their transport to a diagnostic laboratory;

Some of these points need to be done in parallel. Unexpected complications can arise if, for example, airlines refuse to transport samples to the laboratory, necessitating transport by car. The next step will be conferencing with the competent authorities (disease control) about what to do next, due to the serious consequences (stop of exports, standstill for animal movements) of a public notification of disease or even suspicion of disease.

12.4 Additional purposes of meat inspection

Soon it became clear that meat inspection could not control all hazards and that the main problem is the presence of clinically healthy animals that are also symptomless carriers of zoonoses. The meat inspection based the current *ante-* and *post-mortem* inspection could not guarantee food safety in regard to these zoonoses (e.g. salmonella, campylobacter) for human consumption. Moreover, the information collected at primary production and meat inspection is not used efficiently. This discussion started many years ago and resulted in several subsequent reviews of meat inspection procedures from public health and animal health perspectives both for biological and chemical hazards. This section is to a large extent based on the discussions and findings of the latest reviews of meat inspection published by the European Food Safety Authority (EFSA). The additional purposes of meat inspection include the following:

- verify the absence of disease or infectious agent in a region, compartment or nationally (compartment means one or more premises in which animals or birds are kept under a common biosecurity management system containing an animal subpopulation with a distinct health status with respect to a specific disease for which required surveillance, control and biosecurity measures have been applied for the purpose of international trade);
- detect the emergence of diseases or infectious agents in a region, compartment or nationally;
- estimate the burden of infectious agents entering the food chain;
- estimate the prevalence of infectious agents in primary production (pre-harvest);
- feed back to farmers, operators for transport and lairage and slaughterhouse operators on animal health, welfare and quality.

These purposes could be summarized as early detection, case finding and prevalence estimation, and will have great influence on the design of meat inspection procedures as well as the design MOSS activities. Some of the points are discussed here. But, firstly, the most important concepts for assessing meat inspection as a tool for MOSS are presented.

12.5 Some useful concepts

In this section some useful concepts are introduced in the slaughterhouse context and illustrating meat inspection as a tool for MOSS activities and disease control are presented and described.

12.5.1 Monitoring and surveillance systems (MOSS)

Monitoring of diseases is the continuing effort aimed at assessing the health, welfare and disease in a population, for example estimating the prevalence of dirty animals at the *ante-mortem* inspection. Monitoring might be either passive or active and no action is foreseen for positive animals.

Passive monitoring is where information is collected as routine. An example could the reporting of necropsy findings at a diagnostic laboratory or number of animals rejected for human consumption at slaughter due to *post-mortem* findings. Meat inspection could be seen as passive monitoring with one great advantage: all animals entering the food chain have to undergo meat inspection – in other words meat inspection has a broad coverage. The disadvantage is that animals sent for slaughter are not representative of the whole animal population; using horses as example following horses are excluded:

- horses that die during rearing, racing and use and sent for cremation or buried in the ground;
- in the European Union horses at end of their useful life are exported to the South of Europe for slaughter;
- age cohorts of horses that are not slaughtered (young horses);
- horses that are showing clinical signs of disease;
- horses undergoing treatments for disease for which there is a withdrawal period or where no such withdrawal period has been established, for example phenylbutazone.

The population of slaughtered horses might also include live horses imported for slaughter. In those Southern European countries that import a lot of horses for slaughter, the population of slaughtered horses is different from the population of horses in that country. Therefore, inferences from the population of slaughtered horses may not be representative of the horse population unless justified in that particular case. Active monitoring is carrying out a specific investigation to assess the prevalence (or incidence) of an agent, such as baseline studies for estimating the campylobacter prevalence in poultry.

Surveillance is monitoring aimed at identifying the infected or diseased individuals or herds with a view to take some action. Surveillance is usually directed at specific diseases or infections that might be present (endemic) or absent (under the design prevalence) in the population surveyed. For exotic diseases the purpose of surveillance is to detect an emergence. A disease surveillance programme has a defined disease monitoring system, design prevalence (threshold) and directed actions. Hence, a disease control programme and disease surveillance are very similar concepts.

There are frequent disagreements on the definitions and differences between monitoring and surveillance and, therefore, it has been suggested to use the terms monitoring and surveillance systems (MOSS). This suggestion is endorsed as it reduces confusion and is used in this chapter. MOSS at meat inspection have three objectives:

1. to work as an early detection system;
2. to detect all cases or diseased animals;
3. to estimate prevalences:
   - for feedback to farmers, for example of pleuritis in pigs or lambs,
   - for estimating the burden of pathogens entering the food chain.

The challenge is that these objectives may require different designs of MOSS at meat inspection.

12.5.2 Population of interest, surveillance population and sample

An issue that requires some thinking is which population to make inferences on – that is to say something about or population of interest – and which population are samples taken from. This is not always obvious and from time to time creates confusion between the experts carrying out the surveillance and the risk managers, including the general public. Examples of different populations of interest include:

- beef and dairy cattle in Finland;
- the ecological pig population of Austria;
- the pigs slaughtered in Denmark during 2012 and intended for export to Russia;
- the broilers slaughtered in the Brittany region of France during 2012;
- the broilers slaughtered at slaughterhouse Y from farm X, in Norway.

The question is then what can the sampling at slaughterhouse really tell us about the disease and health status in these populations. Hence, the population of interest is the population we want to make inferences to, the results from the MOSS should reflect the true status of that population. The surveillance population is the population that samples are taken from, for example animals slaughtered during May 2013. The sample is the animals or carcases that are drawn from the surveillance population and examined. It is important that the sample is representative of the population that is being sampled (surveillance population). If we want to infer about the total population then this needs to be considered carefully, as the slaughter population is a subset of the total population that excludes:

- fallen stock;
- breeding animals;
- young animals;
- diseased animals, as only animals without signs of clinical disease are admitted to slaughter apart from injuries/traumas that are admitted to emergency slaughter;
- animals that are raised for export or as companions;
- animals that are treated with drugs with no maximum residue limit (MRL).

This might introduce a selection bias if the purpose is to measure the level of disease or a welfare problem in the national animal population. On the other hand, if the objective is to study the burden of pathogens entering the food chain, then the slaughter population is suitable as a surveillance population. For example, sampling the population of horses sent for slaughter in Sweden during 2012 may not be representative of the Swedish population of horses.

There are three approaches to slaughterhouse sampling:

- census;
- representative sample;
- risk-based sampling (in particular useful for early detection).

Census Census is if the entire surveillance population is examined. The advantage of a census is that there is no bias, while the disadvantage is usually the costs involved. Meat inspection, being a census of the population of slaughtered animals, is a major advantage of meat inspection as a MOSS tool.

Representative sample A representative sample is a sample is drawn from the slaughter population. The trick is then to obtain an unbiased sample, which means that all animals slaughtered (surveillance population) during, for example, 2012 at slaughterhouse X have an equal chance being sampled. This can be a random sample using a random number table or pulling numbers on pieces of paper from a hat. A sampling frame is needed, such as a list of all animals slaughtered. A sampling frame can be difficult to get prospectively and also to organise practically in the slaughterhouse. One other solution is systematic sampling – that is to sample, for example, every tenth animal slaughtered. However, the assumption is still that the animals being sampled are representative. If slaughtering two animals per day and sampling every tenth animal, this means sampling on the same day (e.g. Friday) and if one farmer sends animals for slaughter only on Fridays there is a problem with bias. One possibility is to use a random number for the first animal sampled and thereafter sample systematically.

Very often the sampling frame used is to sample animals before and after breaks at slaughter. The reason for this approach is the ease of sampling and causing less disturbances for the staff. This is referred to as convenience sampling and the worry then is whether the sample is representative of the surveillance population. The value of such a convenience sampling frame is often questionable. Therefore, a random element should be introduced, say a random selection of the first animal to be sampled and thereafter systematic sampling (every fiftieth animal) or random selection of the days during a year where one may take samples and then one may sample during or at the end and start of the breaks of those days.

Risk-based sampling When the purpose is to detect a disease problem in the population of interest then risk-based sampling is often a more cost effective alternative. That is, sampling those categories of animals thought to be at highest risk for having the disease or agent in question. For example, when designing the MOSS for Aujeszky's Disease, one proposal was for those countries where boars were used to detect heat in sows or for breeding; the surveillance should consist of sampling the boars being sent for slaughter. The boars have been exposed to nearly all sows in the herd and, therefore, should be good sentinels and constitute a high risk population. However, if boars are not used for heat detection or breeding this design would not work. A possibility is to use scenario trees to incorporate the information on risk when assessing the MOSS and the system sensitivity. Risk-based MOSS is a topic where there has been great progress the recent years in quantifying the systems performance – the ability to detect disease.

12.5.3 Measures of disease occurrence

Prevalence and incidence are measures of occurrence of diseases or infections or antibodies to an agent (or that matter any other health or quality attribute of interest).

Prevalence Prevalence is measured as a proportion – the number of infected or diseased animals divided by the population at risk at a point in time. For example, if the lymph nodes from 3000 slaughter pigs were analysed for salmonella bacteria finding 36 positive, then the prevalence of salmonella positive lymph nodes is 36/3000 = 1.2%.

Incidence Incidence is the number of new cases (e.g., newly infected animals) divided by the population at risk during a time period and can be measured as a proportion, that is new cases divided by the number of animals at risk (cumulative incidence) or as a rate number of new cases divided by animal years at risk (incidence rate or density). For example, if 30 000 slaughter pigs were slaughtered during 2011 and 600 pigs had visible acute lesions from traumatic injury at *post-mortem* inspection, then the cumulative incidence was 600/30 000= 2% in 2011.

Incidence and prevalence are linked concepts, as prevalence is the function of cumulative incidence and the duration of a disease or infection. This means that a change in prevalence could be due to changes in incidence and/or changes in the average duration of disease or persistence of antibodies. For example, antibodies to an infectious agent might persist long after the infection has been cleared, which means that the prevalence of antibodies is higher than the observed incidence of a clinical disease. This is very important when results of surveys in the slaughterhouse are interpreted. If the population and the incidence and prevalence are stable and, in addition, the prevalence is small, the relationship can be simplified to the formula:

$$\text{Prevalence} = \text{Incidence} \times \text{duration of disease (or antibody titers)} \quad (12.1)$$

12.5.4 Diagnostic test characteristics

Sensitivity and specificity predictive values of positive and negative tests are important concepts when planning animal health monitoring of a slaughterhouse, as illustrated in Table 12.1.

Table 12.1 Diagnostic test characteristics.

	Diseased	Not diseased	
Test positive	A	B	(A + B) Fraction test positive
Test negative	C	D	(C + D) Fraction test negative
	A + C (prevalence)	B + D (1−prevalence)	N = A + B + C + D

Sensitivity (Se) is the probability of a truly diseased (or infected) animal testing positive or:

$$Se = A/(A + C) \tag{12.2}$$

Specificity (Sp) is the probability of a truly not diseased animal testing negative:

$$Sp = D/(B + D) \tag{12.3}$$

The predictive value of a positive test (PVPT) is the probability that a test positive animal is truly diseased:

$$PVPT = A/(A + B) \tag{12.4}$$

The predictive value of a negative test (PVNT) is the probability that a test negative animal is truly not diseased :

$$PVNT = D/(B + D) \tag{12.5}$$

These concepts might seem academic but understanding them is crucial when interpreting the results from a MOSS programme at the slaughterhouse. For example, if a cow tested positive for a disease (e.g., BSE – bovine spongiform encephalopathy) it would be crucial to assess the predictive value of that rapid test and how likely it is that the cow is a false positive or is truly infected. This means that a confirmative test is often needed. The difficult risk management question is whether the slaughterhouse operations should be suspended while waiting for the results of confirmative tests on the positive sample.

The terms sensitivity and specificity have both an epidemiological and a laboratory interpretation. Therefore, the sensitivity is sometimes referred to as diagnostic sensitivity to distinguish it from the analytical sensitivity, which is the detection limit – a common term in laboratory analyses. In a similar way specificity is referred to as diagnostic specificity to distinguish it from analytical specificity – the ability to detect only the hazard (e.g., infectious agent) of interest without cross-reacting with other hazards. We might also define similar test characteristics for herds or populations that are crucial when designing slaughterhouse MOSS activities and for the interpretation of findings thereof.

12.5.5 Apparent versus true prevalence

An important point is that the true prevalence of a disease might be quite different from the observed or apparent prevalence. If the specificity and sensitivity of a diagnostic testing procedure is know this can be estimated according to the following formula:

$$\text{True prevalence} = \frac{(\text{Observed prevalence} + \text{specificity} - 1)}{(\text{Sensitivity} + \text{specificity} - 1)} \qquad (12.6)$$

A problem with serology is that very often the specificity is not 100% and there is a risk of false positives. If there is an MOSS at meat inspection for PRRS antibodies in oral fluids of pigs and the sensitivity was 0.953 and the specificity was 99.2% at the chosen cut of value (sample-to-positive ratio >0.2), what would an observed prevalence of 2% mean? Using Equation 12.6 we can estimate the true prevalence as:

$$\text{True prevalence} = \frac{(0.02 + 0.992 - 1)}{(0.953 + 0.992 - 1)} = \frac{1.012}{1.945} = 0.00566 \text{ or } 0.6\%$$

The difficulty comes when the observed prevalence is close to 0%. In that case, are the observed positive samples all false positives? In that case one may wish to increase the cut-off, with the result that the specificity is increased to 100% and the sensitivity is decreased to 94.7%. Another alternative would be to add a confirmatory test with a serial interpretation.

12.5.6 Parallel and serial interpretation of tests

If using two tests, a serial interpretation requires that both tests are positive, while a parallel interpretation requires that at least one of the tests is positive, to conclude that an animal is infected. A serial interpretation increases the specificity of the testing procedure (fewer false positive animals) while a parallel increases the sensitivity of the procedure (fewer false negative animals).

For parallel interpretation of two tests A and B (both must be positive), their sensitivity (SeA and SeB) can be calculated as:

$$\text{Se (A and B parallel interpretation)} = 1 - [(1 - \text{SeA}) \times (1 - \text{SeB})] \qquad (12.7)$$

For parallel interpretation of two test results, their specificity (SpA and SpB) can be calculated as:

$$\text{Sp(A and B parallel interpretation)} = \text{SpA} \times \text{SpB} \qquad (12.8)$$

For serial interpretation of two tests A and B (both must be positive), their sensitivity (SeA and SeB) can be calculated as:

$$\text{Se (A and B serial interpretation)} = \text{SeA} \times \text{SeB} \qquad (12.9)$$

For serial interpretation of two test results (both must be negative), their specificity (SpA and SpB) can be calculated as:

$$Sp\ (A\ and\ B\ serial\ interpretation) = 1 - [(1 - SpA) \times (1 - SpB)] \quad (12.10)$$

The practical implication is that if test A is *ante-mortem* examination and Test B is *post-mortem* examination, it will have consequences what kind of interpretation is chosen. If serial interpretation is chosen requiring both clinical signs of disease *ante-mortem* and pathological signs at *post-mortem*, then it is more likely that diseased animals will be overlooked. Implementing a food safety priority could be using a parallel interpretation of *ante-mortem* and *post-mortem* signs indicating public health risks. If designing a MOSS for early warning of animal diseases, a parallel interpretation of *ante-mortem* and *post-mortem* findings could be considered if the negative consequences of a false positive diagnosis are huge, for example mad cow disease. This decision should be made on a case-by-case basis, balancing the consequences of a false positive (false alarm) and a false negative (the failure to detect a risk). Looking forward, the consequences of false positives and negatives should be the basis for the design of output based standards and necessary MOSS activities.

12.6   Quantifying the MOSS of meat inspection

In this section the tools for quantifying MOSS of meat inspection are presented. One of the major steps forward has been to look at meat inspection as an integral part of national MOSS programmes. By intelligent use of the data already collected, a reasonably good overview of the animal health and welfare situation can be obtained.

12.6.1   Detection fraction

If the purpose is to detect all affected animals in a batch then the detection fraction is an important metric – that is how many of the diseased birds, for example, are detected and can be removed. This question could be asked both at individual levels and at batch level – are all diseased birds detected and, if not, what proportion is detected and are all batches with diseased birds detected? The detection fraction in a batch is defined by the formula:

$$Detection\ fraction = \left(\frac{number\ in\ sample}{batch\ size} \right) \times Se\ meat\ inspection \quad (12.11)$$

12.6.2   Input- and output-based standards

In recent years the emergence of output standards in MOSS has given the results from meat inspection new relevance. One conclusion is that the results from meat inspection have been underused for assessing the health and welfare status of food animal populations. A frequent question for slaughterhouse operators is how they can verify that the salmonella prevalence is less than 5% in a batch of poultry meat or that the occurrence of animal welfare problems, for example tail biting in slaughter pigs, is less than 1%.

Previously, the sampling to be done was prescribed in often legally binding terms – input-based standards. For example, the additional guarantees for salmonella within the European Union for fresh poultry meat require that 60 broiler carcases shall be sampled from each batch to be exported to Sweden or Finland. These samples should be analysed without any finding of salmonella before the batch may be exported. This is based on the acceptable level of protection defined as 95% probability that the prevalence is less than 5%. This input-based approach leaves no flexibility on how to attain the required level of safety or protection and might be less efficient.

In recent year a flexible approach has emerged in the food safety and animal health domain. The philosophy is that an acceptable level of protection or detection is defined either locally or nationally and then the food business operators and national competent authorities choose how these standards will be complied with. In food safety once the acceptable level of protection is defined (ALOP), it is possible to derive from the ALOP the associated food safety objectives (FSO) and performance objectives (PO) and performance criteria (PC). The EFSA Biological Hazards panel has given a helpful presentation and discussion in its opinion on risk analysis based targets and criteria. In plain English, once the acceptable risk is defined, say no increase in food-borne salmonella cases (ALOP), one may derive a food safety objective, for example that the prevalence of salmonella in poultry meat at consumption should be less than 0.01% (FSO) or that the prevalence of salmonella on carcases at slaughter should be less than 0.1% (PO) or that a decontamination procedure (hot water rinsing) should achieve a reduction of salmonella numbers by a factor of 1000 or log 3 (PC).

A similar concept, outputbased standards, has been developed for animal health MOSS. The objective is to achieve comparable and harmonized surveillance standards, while allowing flexible approaches that can be adapted to different populations and use all information available. A key point is that the test results are interpreted in a context of history both of the herd and the region/country in which the herd is located, whether it is a risk-based or random sample-based MOSS. One consequence of this development has been the increasing awareness of the usefulness and importance of MOSS at meat inspection.

Several approaches have been proposed to disease freedom questions: (i) population or surveillance sensitivity; (ii) probability of freedom from disease; and (iii) expected cost of error – that is consequences of false positive and false negative results. All approaches show and underline how the value of meat inspection findings will be augmented when interpreted in a broader context to complement other MOSS activities. A useful concept in this discussion is the surveillance (herd, batch, population) sensitivity.

Surveillance sensitivity   The surveillance sensitivity (SSE) is based on diagnostic sensitivity for the individual animal (Se), the design prevalence and the number of animals tested (n); it can be described with the formula:

$$SSE = 1 - (1 - (\text{designprevalence} \times Se))^{\wedge n} \qquad (12.12)$$

The surveillance sensitivity (Equation 12.12) can also be interpreted as batch, herd or population sensitivity. It follows from Equation 12.12 that when the prevalence is high then it is easy to detect infected or disease animals. However, as the prevalence decreases the risk manager has three alternatives, either (i) to use more sensitive tests (or add more tests with a serial interpretation of them), (ii) to increase the number of animals to be tested or (iii) to test in subpopulations at higher risk (risk populations) where the prevalence is expected to be higher. In the latter case, a scenario tree approach can capture the different risks in the different subpopulations and enable an overall conclusion on the national prevalence. One example of focusing on risk animals is BSE monitoring. It was noted that the costs of finding a BSE case at normal slaughter increased from Euro 1.23 million in 2001 to Euro 14.15 million in 2008, while the age of BSE positive cattle also increased. It was therefore suggested to increase the age limit for sampling cattle from 30 to 48 months in EU member states with a favourable situation. One further option was to start testing samples instead of whole populations (censuses) and to focus the sampling on risk populations, that is fallen stock and emergency slaughter.

An example of an output-based standard based on meat inspection is the OIE Terrestrial Animal Health Code on trichinellosis in swine, where the requirement for acknowledging freedom is that the slaughter sow population shall be surveyed with a sample size sufficient to provide 95% confidence of detecting trichinellosis if it was present at a prevalence exceeding 0.02%, or 2 in 10 000 slaughter sows. Some further points might be worth considering before concluding in this case; one assumes a perfect test for trichinellosis (100% diagnostic sensitivity and specificity), that there are no entry or exit to the pig population (or compartment), there is no exposures to wildlife from where trichinellosis can be transmitted to the pigs, and the political decision is made that for example, one trichinella infected sow out of 10 000 sows is an acceptable risk.

Probability of freedom   The probability of freedom is a more intuitive concept and easier to explain to risk managers and lay people. We address the question 'Is the population free from disease based on an evolution of the negative predictive value (PVNT) of the surveillance?' as shown in Equation 12.5. Very often the question when having a series of negative results from meat inspection is how much additional sampling on farms and, for example, serology is needed to validate the claim of disease freedom. The standard could be formulated as – *How many more surveillance activities are needed in addition to meat inspection to achieve a 99% confidence that the prevalence is less than*

1%? This could be addressed using a Bayesian formulation:

Posterior prob freedom

$$= \frac{\{\text{Prior prob of freedom (MI)}\}}{\left\{\begin{array}{c} \text{Prior prob of freedom (MI)} + [(1-\text{Prior prob of freedom (MI))} \\ \times(1-\text{SSe (new activity)}] \end{array}\right\}}$$

(12.13)

Equation 12.13 states the question – given the desired probability of freedom of disease (Posterior prob freedom), what additional evidence is needed from new activities (SSe (new activity)), given the current evidence of freedom based on meat inspection (prior prob freedom (MI)). The advantages include: (i) the ability to deal with great heterogeneity of populations; (ii) multiple sources of surveillance evidence including meat inspection findings; and (iii) historical surveillance can be included explicitly.

In future the output-based standard should be consequence based. One example could be the consequences if there is a failure to detect a diseased animal or the consequences of failing to detect an animal health problem. What would be the optimal allocation of MOSS resources given the current disease panorama and the potential for emerging threats, and costs of early warning, monitoring and surveillance? Hence, it is concluded that consequence-based design of output-based standards seems to include the future meat inspection.

In conclusion, recent work clearly shows that data from meat inspection will be important for determining the risk from both public and animal health hazards. Therefore, the key is how to make the best use of current and historical data from meat inspection and also how to design additional MOSS activities as needed to achieve the necessary level of protection defined as an output-based standard. EFSA has reported on swine that current data from pig health and welfare are greatly underused to support decisions on food safety, animal health and welfare, and points toward a challenge for future veterinary meat inspectors.

12.6.3   A problem with meat inspection MOSS

The difference between diseased animals and infected animals is not always appreciated. The group diseased animals with visible clinical or pathomorphological signs is a subgroup of the larger group – infected animals. Infected animals might have no, mild and typical clinical signs either *ante-mortem* or pathological signs *post-mortem*. This will mean that infected animals are not detected or that the probability of detection is lower. The healthy carrier is one of the major transmission paths for infectious diseases both epizootic and endemic. The practical implications are that the meat inspection will have different diagnostic sensitivities for animals with typical, mild or subclinical signs of a disease. For example, in the evaluation of the *post-mortem* inspection of poultry, the sensitivity of an 'average case' was the average of the sensitivities for typical, mild and subclinical cases. The sensitivity of a mild case and subclinical case was assumed to be 50% and 10% of the sensitivity of a typical case of bird disease, respectively.

12.7   Purposes of MOSS at meat inspection

The purposes for MOSS activities at meat inspection of endemic diseases or welfare conditions include prevalence estimation and/or case identification. The purpose of prevalence estimation is estimating the burden of infectious agents entering the food chain to the processing and consumption stages and prevalence estimation of infectious agents in primary production (pre-harvest) to give feedback farmers on health, welfare and quality on the animals sent for slaughter. This feedback should include those working with transport, lairage and at the slaughterhouse.

12.7.1   Prevalence estimation

The classical examples of prevalence estimation are baseline studies for foodborne pathogens on slaughter carcases or surveys to estimate welfare conditions. For such baseline studies the priority is to be comparable with other baseline studies done in other countries, for example member states in the European Union, and with studies done in earlier years and studies foreseen to be done in the future. Therefore, the methodology for sampling and microbiological analyses must be uniform in these studies. For example for the baseline study of EFSA for salmonella and campylobacter more than 10132 broiler carcases were sampled, from 561 slaughterhouses in 28 countries (EU plus Norway and Switzerland) during 2008. One carcase was sampled from each batch and the neck and breast skin were examined for salmonella and campylobacter. The EU prevalence on broiler carcases were for campylobacter 76%, ranging from 5 to 100% between the countries investigated, and for salmonella 16%, ranging from 0 to 26% between the countries investigated. The total population of slaughtered broilers during this year exceeded five billion broilers. Hence, by carefully designing the baseline survey it is possible to say something about five billion broilers by sampling 10000 broilers. In this case we want to have an acceptable precision (random error due to sampling), sometimes referred to as bounds or error of estimation of the study and that the results should be unbiased (free of systematic errors).

Welfare studies have used, for example, slaughterhouse data for monitoring and surveillance of pig welfare by investigating the prevalence of tail docking and tail lesions and associated carcase condemnations. If investigating the magnitude of health of a welfare problem the study aims will include the estimate of the prevalence with some precision, and to detect changes from, for example, year-to-year or between groups. The sample size (n) assuming large population (>10 000 animals) for an expected prevalence (PV) of 10% with a desired precision (d) of plus/minus 1% with 95% confidence; is estimated by the formula:

$$n = \frac{[1.96^2 \times PV \times (1 - PV)]}{d^2} = \frac{1.96^2 \times 0.09}{0.0001} = 3460 \text{ samples to be taken}$$

$$(12.14)$$

In this case the recommendation would be to take at least 3500 samples. What is the precision if 12 200 animals were sampled and 1200 were found positive (PV = 10%)? This could be estimated by manipulating Equation 12.14:

$$\text{Precision is } 1.96 \times \frac{\sqrt{(PV \times (1 - PV))}}{n} = 1.96 \times \sqrt{(0.09/12000)}$$

$$= 0.0053 \text{ c} : 0.5\% \qquad (12.15)$$

If the diagnostic sensitivity and specificity are available then the true prevalence might be calculated using the Equation 12.6 and then estimating the precision based on the true prevalence.

12.7.2   Case detection

Another purpose of MOSS activities at meat inspection would be to identify each individual case of the diseased animals or batch of animals to implement some risk management action. The key point then is to identify affected animals or batches where the design prevalence is above the accepted level for the disease or welfare problem. In this discussion it is assumed that the diagnostic specificity of meat inspection is 100%, that is, all healthy animals will test negative and moreover, that results of *post-mortem* and *ante-mortem* are not correlated. Looking at the meat inspection there are *ante-mortem* and *post-mortem* diagnostic procedures and with a parallel interpretation (either *ante-mortem* or *post-mortem* positive), the sensitivity of meat inspection is:

$$\text{Se meat inspection is} = 1 - [(1 - SeAM) \times (1 - SePM)] \qquad (12.16)$$

This would be the diagnostic sensitivity for the individual animal at meat inspection including both *ante-mortem* (AM) and *post-mortem* (PM). Equation 12.16 can be extended if the *post-mortem* inspection includes several procedures, such as visual inspection (PM1) and palpation and incision of lymph nodes (PM2) or any further laboratory confirmatory testing. The diagnostic sensitivity of meat inspection can then be formulated as:

$$\text{Se meat inspection is} = 1 - [(1 - SeAM) \times (1 - SePM2)] \qquad (12.17)$$

If looking at samples from batches, for example broilers slaughtered and the batch sensitivity, this would be:

$$Se\ batch = 1 - (1 - Se\ meat\ inspection \times design\ prevalence)^{number\ of\ birds\ sampled}$$

(12.18)

Note that formulas for system sensitivity (Equation 12.12) and batch sensitivity (Equation 12.18) are the same, but this concept is used in two different settings, batch and population levels.

12.7.3   Verify the absence of disease or infectious agent in a region, compartment or nation

The status of having absence of a disease in an animal population in a region or compartment of country is seen as favourable for trade. Meat inspection can be an efficient tool for verifying that the population has been and still is free of a disease. However, this is based on two critical assumptions: (i) the disease in question is found in slaughtered animals and (ii) the other criteria for disease freedom are complied with. In Finland, Norway and Sweden, where bovine TB was eradicated more than 40 years ago, the continued freedom from TB is now substantiated by the continued absence of bovine TB at meat inspection. For bovine spongiform encephalopathy (BSE or Mad Cow disease) this approach is more problematic, as fallen stock has a 10–40 times higher likelihood of being infected than animals sent for normal slaughter. The incubation period for BSE is appears to be more than 50 months and large proportions of the cattle population that are slaughtered are younger than 50 months, creating another bias in the sampling population. Consequently, due to these biases, the absence of findings during normal slaughter does not necessarily prove BSE freedom in the whole cattle population. In this case a MOSS system based on normal slaughter only, would not be sufficient to ascertain absence of BSE in the cattle population. The take home lesson is that the contributions of meat inspection for disease control and MOSS programmes are very context specific and should be assessed on a case-by-case basis.

The recent developments in veterinary MOSS systems to try to quantify the contributions from meat inspection are, therefore, to be welcomed. The gist is to try to use all information available, such as current and historical results from meat inspection, routine necropsies and laboratory investigations, and veterinary notification of clinical diseases both on high risk and low risk populations, to estimate the likelihood of the presence or absence of a disease or its agent at a predetermined prevalence (design prevalence).

12.7.4   Detect the emergence of diseases or infectious agents

The slaughterhouse can be an efficient sampling spot to detect the emergence of public and animal health hazards. The main advantage of meat inspection is that it is a census of animals being slaughtered and, thereby, is a very large sample of the animal population. The quality of the MOSS is influenced by the number of animals included, the sensitivity of the screening tests or procedures and the design prevalence. Design prevalence is the acceptable prevalence of a disease or disease-causing agent in a population. The concept is similar to the Codex Alimentarius concepts of acceptable level of protection. For example, in the additional guarantees granted to Sweden and Finland for fresh beef, pork and poultry meat by the European Union, the design prevalence is defined as −95% probability that it is less than 5%, which means that 60 samples from each batch shall be analysed without finding *Salmonella sp.*

The efficiency can be improved by sampling animals at higher risk. Slaughterhouse surveillance may not be sufficient to detect all emerging diseases, due to the selection biases that fallen stock, exported animals and younger batches are not captured in the meat inspection MOSS. In EFSA's revision of meat inspection in swine it was noted that slaughterhouse surveillance detected a number of epidemic diseases when other MOSS failed, such as the FMD outbreak in the United Kingdom in 2001. The prerequisites for early detection are high coverage and continuous investigation of the susceptible populations, and the diseases should have typical lesions easily detectable at *ante-* or *post-mortem* inspections, for example FMD.

For diseases with less distinct symptoms, such as classical swine fever (CSF), it was of interest to investigate the potential of *ante-* and *post-mortem* inspection by looking at the ability to detect clinical signs and *post-mortem* signs, respectively. The optimal *ante-mortem* diagnostic test, which was based on a combination of clinical signs (unsteady gait/ataxia, not eating, not reacting to antibiotic treatment, conjunctivitis and hard faecal pellets), had a combined sensitivity (Se) of 72.7% with a specificity (Sp) of 52.7%. For *post-mortem* findings the optimally sensitive, specific and efficient test for the detection of a CSF outbreak at *post-mortem* examination was achieved by a combination of pathological findings: the presence of either renal (petechial) haemorrhages or enlarged lymph nodes or dry faecal contents in the colon. This optimally efficient test procedure combined a sensitivity of 69.6% with a specificity of 75.9%. The diagnosis of CSF based on *ante-* and *post-mortem* inspection would be better than by chance only, but not much. There would be a lot of false positives, meaning that meat inspection would not the best approach for early warning. A confirmative test (e.g. serology or polymerase chain reaction) would be required before a definitive diagnosis could be made. Consequently, relying on meat inspection alone would not be appropriate for CSF. However, with a rapid confirmative test available and in a high risk situation, meat inspection will complement the MOSS in place in primary production.

Moreover, meat inspection is an continuous MOSS that provides historical evidence verifying the absence of a disease, and also detecting the emergence of animal diseases. The emergence may not only be of a disease but also of the emergence of animal welfare problems. It is also possible to augment the value of meat inspection as a MOSS activity by targeted sampling of, for example, risk populations. It is important to be able to quantify contributions of these elements for finding the best MOSS approaches.

12.7.5   Quantifying sensitivity for detection

If the wish is to use risk-based MOSS, the formulae becomes more complicated. A brief example is given here, focusing on system sensitivity and the possibilities of risk-based surveillance. The aim is to convince the reader about the usefulness of meat inspection as a population monitoring and surveillance system (MOSS) tool that should be used more intensively.

Plain (representative sample)   The sensitivity of meat inspection can be calculated as:

$$\mathrm{SeMI} = 1 - (1 - \mathrm{SeAM})(1 - \mathrm{SePM})(1 - \mathrm{Se_{further\ tests\ or\ food\ chain\ information}})$$
(12.19)

If it is assumed that, say, 1000 or 10 000 sheep are inspected and the design prevalence of FMD is 10% (PV), the system (batch) sensitivity would be:

$$\text{System Se} = 1 - (1 - \mathrm{SeMI} * \mathrm{PV})^{\char`\^}1000 = 66\%$$
(12.20)

$$\text{System Se} = 1 - (1 - \mathrm{SeMI} * \mathrm{PV})^{\char`\^}10000 = 99.99\%$$
(12.21)

The take home message is that although the meat inspection is lousy on an individual animal level, is it very good on a population level, thereby highlighting the value of meat inspection as a monitoring and surveillance tool. If assuming the design prevalence is 1% and assumed that 50 000 sheep are inspected, the ability to detect an FMD epidemic would still be good:

$$\text{System Se} = 1 - (1 - 0.01 \times 0.01)^{\char`\^}50000 = 99.3\%$$
(12.22)

Risk-based monitoring and surveillance   If the aim is to make monitoring based on meat inspection more efficient, testing of high risk groups can be started. In this case, if a disease is, say, 10 times more likely in the high risk group (RRhr) than in the low risk group (RRlr), and the high risk group is 5% of the population, then it is possible to estimate the new system sensitivity. The adjusted risk in the high risk group is:

$$\mathrm{AR_{hr}} = 10/[(0.05 \times 10) + (0.95 \times 1)] = 10/1.45 = 6.9$$
(12.23)

Adjusted risk in the low risk group is:

$$\mathrm{AR_{lr}} = 1/[(0.05 \times 10) + (0.95 \times 1)] = 1/1.45 = 0.69$$
(12.24)

If it is assumed that 50% of the animals included in the monitoring are in the high risk group and 50% in the low risk group, what would the overall sensitivity be for a sample of 1000 sheep using the sensitivity mentioned above and a design prevalence of 10% (PV)?

$$Se_{\text{riskbased MOSS}} = 1 - (1 - [PV \times Se_{MI}] \times [(0.5 \times AR_{hr})$$

$$+ (AR_{hr} \times 0.5)]\hat{}1000 \qquad (12.25)$$

$$Se_{\text{riskbased MOSS}} = 1 - (1 - [(0.1 \times 0.01) \times [0.5 \times 6.9]$$

$$+ [0.5 \times 0.69])\hat{}1000 = 97.8\% \qquad (12.26)$$

Hence, by using risk-based design the ability to detect an infection improved from 66 to 97.8%. This is a very useful way of saving money and making scarce resources go further.

12.8   EFSA reviews of meat inspection

EFSA has reviewed the meat inspection of several species. These reviews include biological and chemical public health hazards as well as animal health and welfare. Moreover, EFSA has commissioned external reviews (COMISURV) of the animal health part of the meat inspection through structured expert surveys to assess the diagnostic performance of meat inspection. The key question is whether meat inspection reduces the public health risks. On the other hand, would the changes proposed reduce the ability of meat inspection to detect hazards for animal health and welfare? One example is the animal health and welfare consequences from changing to a visual only inspection from an inspection based on visual, palpation and incision of organs.

In Table 12.2 the detection probabilities (diagnostic sensitivities) for typical cases of selected diseases and welfare conditions during *ante-mortem* (live animal) and *post-mortem* inspection of swine are presented. The focus is how likely (in %) a pig with typical symptoms will be detected by *ante-mortem* or *post-mortem* inspection. One caveat is that these probabilities are based on a structured survey of expert opinions, another is that it is based on detection of typical cases. The *post-mortem* inspection is a visual only inspection, which is the proposed future meat inspection procedure.

It appears that the meat inspection procedure based on current *ante-mortem* and visual *post-mortem* inspection should be able to detect animals with typical symptoms of exotic and endemic animal diseases and welfare conditions. One issued raised in the EFSA swine opinion is lower detection probability of non-typical and subclinical cases of animal diseases (or rather infections).

The combined detection probabilities for meat inspection are calculated according to the Equation 12.27, which is derived from Equation 12.16 but in addition includes the correlation between the tests.

$$Se\ \text{meat inspection} = 1 - [(1 - SeAM) \times (1 - SePM)$$
$$+ (\text{correlation between SeAM and SePM}) \qquad (12.27)$$

Table 12.2 Detection probabilities (diagnostic sensitivity of a typical case of *ante-mortem* and *post-mortem* meat inspection of an individual pig for a few selected exotic and endemic diseases and welfare conditions, all assuming the pig shows typical symptoms based on the EFSA review of meat inspection in swine).

Type of disease	Disease	Detection probability ante-mortem (%)	Detection probability post-mortem (visual only) (%)
Exotic	African swine fever (ASF)	70	100
	Classical swine fever (CSF)	74	100
	Foot and Mouth disease (FMD)	70	100
	Swine vesicular disease (SVD)	67	100
	Vesicular stomatitis (VS)	34	89
Endemic diseases	Enzootic pneumonia (*Mycoplasma* spp, *Pasteurella* spp)	59	100
	Pleuropneumonia (*Actinobacillus pleurpneumonia* and *Actinobacillus suis*)	75	100
	Atrophic rhinitis	90	100
	Arcanobacterium pyogenes	75	86
	Ascaris suum	82	98
	Tuberculosis	34	100
Welfare conditions	Arthritis and bursitis	81	68
	Bruising and skin lesions	89	84
	Dark Firm Dry meat (DFD)	2	99
	Lameness	87	71
	Tail biting	86	71

Source: Adapted from EFSA, 2011 with permission from EFSA.

The current *ante-mortem* and visual *post-mortem* inspection for both exotic and endemic animal diseases and welfare conditions has a lower detection rate for sheep and goats, with average symptoms including typical, mild or subclinical symptoms. Therefore, meat inspection may not be that appropriate for case finding on the individual animal, as those with mild and subclinical symptoms will be missed, while it might be useful for finding affected herds. This point is further illustrated in the poultry opinion of EFSA, where the difference in diagnostic sensitivities between a typical and average case for *post-mortem* meat inspection was investigated. In the review of meat inspection of poultry, the sensitivities of the meat inspection for both exotic and endemic diseases as well as welfare conditions were based on expert opinions. In poultry the *ante-mortem* inspection is based on food chain information and visual inspection of crates, while *post-mortem* is based on examination of the individual bird or a sample thereof. One problem is that an infected bird might have typical, mild or subclinical signs and in this study an 'average case' was the average of sensitivities for a typical, mild and subclinical case. The sensitivity of a mild case and subclinical case were assumed to be 50% and 10% of the sensitivity of a typical case of bird disease, respectively. In Table 12.3 some exotic and endemic diseases and welfare conditions for which meat inspection in poultry

Table 12.3 Sensitivity of meat inspection both at *ante-* and *post-mortem* in poultry opinion of EFSA.

Type of disease	Disease	Ante-mortem sensitivity (batch level) (%)	Post-mortem sensitivity (%) Typical case bird level	Post-mortem sensitivity (%) Average case bird level
Exotic	Highly pathogenic avian influenza (HPAI)	98	41	39
	Newcastle disease (ND)	93	41	35
Endemic diseases	Coliform cellulitis	73	100	43
	Mycoplasma gallisepticum infection	92	98	52
	Colisepticaemia	76	100	73
	Botulism	98	0	0
	Necrotis enteritis	93	100	67
	Avian tuberculosis	93	100	49
	Egg periotonitis	62	100	33
	Duck plague	99	100	71
	Infectious Bursal disease	91	99	44
	Aspergillosis	76	100	61
	Histomoniasis	95	100	48
Animal welfare conditions	Thermal discomfort	85	99	61
	Dead on arrival	90	0	0
	Traumatic injuries	99	100	100
	Pododermatitis	70	80	20
	Skin lesions	84	100	100
	Tarsal dermatitis	68	100	100
	Ascites	91	100	100

Source: Adapted from Hardstaff, *et al.*, 2012 with permission from EFSA.

could be helpful for detection and monitoring based on the EFSA opinion are shown. The results are given as detection probability or sensitivity. A slaughter batch is assumed to be between 10 000 and 30 000 birds in the *ante-mortem* evaluation. One caveat is that all estimates are based on a structured survey of expert opinions on the current meat inspection in the European Union, therefore the results should be interpreted with care.

As noted in Table 12.3, the experts' assessment indicates that most batches with diseased birds should be detectable in the live bird (*ante-mortem* inspections) and this is the argument for keeping these inspections. If only *post-mortem* inspections were to be carried out, the experts' assessment suggested that diseases, such as botulism, and welfare conditions, such birds that are dead on arrival, would be missed. For exotic bird diseases, such as highly pathogenic avian flu (HPAI) and Newcastle disease (ND), a single diseased bird could be missed as the diagnostic sensitivity of *post-mortem* was around 40%. Hence, it makes sense to at least examine a sample of birds' from each batch *post-mortem*. It is more realistic is to look at the batch

Table 12.4   Batch level sensitivity of meat inspection based on the EFSA opinion on meat inspection in poultry (EFSA, 2012, Table 4, page 161) for some exotic and endemic diseases, and welfare conditions based on a 1% design prevalence (that is, 1 out of 100 birds are diseased).

Type of condition	Disease of welfare problem	Sensitivity (%)			
		Post-mortem inspection sample size			
		500 birds	100 birds	10 birds	1 bird
Exotic diseases	Newcastle disease	82	29	3	0.3
	Highly pathogenic avian flu (HPAI)	86	32	4	0.4
Endemic diseases	Necrotic enteritis and hepatic disease	97	49	7	0.7
	Egg peritonitis	80	28	3	0.03
	Aspergillosis	95	46	6	0.6
Welfare conditions	Pododermatitis	63	18	2	0.2
	Skin lesions	99	63	10	1
	Thermal discomfort	95	46	6	0.6

Source: EFSA, 2012. reproduced with permission from EFSA.

sensitivity for *post-mortem* meat inspection, too. If a sample of birds is examined *post-mortem*, how big should the sample from each batch be? In this the question of interest is if at least one bird is detected in an infected batch with a design prevalence of 1%, say, which is presented in Table 12.4. A design prevalence of 1% means that in a batch of 10 000 birds, the hypothesis being tested is whether 100 or more birds are diseased. In poultry slaughter the batch size is assumed to be between 10 000 and 30 000 birds. The sample size might vary for *post-mortem* inspection between 1 and 500 birds.

One conclusion from Tables 12.3 and 12.4 is that dropping the *ante-mortem* inspection could reduce the sensitivity of meat inspection dramatically. If the disease is foreseen to have a low prevalence in the batch, then a larger sample size for *post-mortem* inspection is needed; more than 100 preferably closer to 500 birds would be required for a helpful sensitivity.

A difficult risk management question is the relevant design prevalence: Is it is too restrictive at 1%? One way of looking at it would be to ask what design prevalence could the meat inspection detect? The prevalence of diseases that *post-mortem* meat inspection could detect with 95% probability (batch sensitivity is 95%) for different samples sizes per batch, and the expert opinion on the expected prevalence in an infected flock for some exotic and endemic diseases, and welfare conditions are shown in Table 12.5.

On the basis on Table 12.5 it appears that some conclusions on the relevant sample size for *post-mortem* inspection if slaughtering from batches can be made. Given the sometimes harsh economic constraints it is reassuring that even a sample size of 10 birds per batch if representatively (e.g., randomly) selected can give relevant information on the presence of some diseases,

Table 12.5 Sensitivity of *post-mortem* inspection for some exotic and endemic diseases and welfare conditions based on EFSA expert opinion.

Type of condition	Disease of welfare problem	Expected prevalence according to expert opinion (%)	The minimum prevalence that could be detected with 95% probability based on different samples sizes for *post-mortem* inspections (%)			
			500 birds	100 birds	50 birds	10 birds
Exotic diseases	Newcastle disease	100	0.6	3	6	27
	Higly pathogenic avian flu (HPAI)	28	0.6	3	6	26
Endemic diseases	Necrotic enteritis and hepatic disease	65	0.6	3	6	27
	Egg peritonitis	25	0.8	4	8	36
	Aspergillosis	0.5	0.7	3	7	30
Welfare conditions	Pododermatitis	18	0.8	4	8	34
	Skin lesions	80	0.6	3	6	28
	Thermal discomfort	60	0.7	3	6	28

Source: EFSA, 2012. reproduced with permission from EFSA.

and increasing the sample size to 50 birds per batch should enable detection of most affected batches. Only in a situation where the expected prevalence is very low, will a larger sample size be justified.

12.9 Summary and conclusions

Meat inspection has evolved into a key monitoring and surveillance tool for animal health and welfare. The purposes of meat inspection will include:

- early warning system (detection of emerging diseases);
- case finding – identifying all affected animals or carcases;
- prevalence estimation including verifying absence.

By quantitative analyses it is possible to have a reasoned opinion about additional surveillance activities needed to reach the desired confidence levels for disease freedom – output-based standards. The major advantage is that meat inspection is a census of animals slaughtered and entering the food chain. The data collected at meat inspection should be used much more extensively in the future as a strategic tool for monitoring and surveillance.

Literature and further reading

Cameron, A.R. 2012. The consequences of risk-based surveillance: Developing output-based standards for surveillance to demonstrate freedom from disease. *Preventive Veterinary Medicine*, **105**, 280–286. doi: 10.1016/j.prevetmed.2012.01.009.

Cockrill, W.R. 1963. International aspects of meat hygiene. *Royal Society of Health Journal*, **83**, 257–263.

Dupuy, C., Hendrikx, P., Hardstaff, J. and Lindberg, A. 2012. Contribution of meat inspection to animal health surveillance in bovine animals. EFSA Supporting Publications 2012:EN-322.

Elbers, A.R., Vos, J.H., Bouma, A. *et al.* 2003. Assessment of the use of gross lesions at *post-mortem* to detect outbreaks of classical swine fever. *Veterinary Microbiology*, **96**, 345–356 (Erratum: 99 (1), 79).

EFSA (European Food Safety Authority). EFSA opinions on meat inspection. Retrieved from EFSA Journal web pages, swine: http://www.efsa.europa.eu/en/efsajournal/pub/2351.htm; poultry: http://www.efsa.europa.eu/en/efsajournal/pub/2741.htm (last accessed 17 February 2014).

Martin, P.A.J., Cameron, A.R. and Greiner, M. 2007. Demonstrating freedom from disease using multiple complex data sources 1: A new methodology based on scenario threes. *Preventive Veterinary Medicine, B*, **71–97**.

More, S.J., Cameron, A.R., Greiner, M. *et al.* 2009. Defining output-based standards to achieve and maintain tuberculosis freedom in farmed deer, with reference to member states of the European Union. *Preventive Veterinary Medicine*, **90**, 254–267. doi:10.1016/j.prevetmed.2009.03.013.

Prattley, D.J., Morris, R.S., Cannon, R.M. *et al.* 2007. A model (BSurvE) for evaluating national surveillance programs for bovine spongiform encephalopathy. *Preventive Veterinary Medicine*, **81**, 225–235.

Salman, M.D. 2003. *Animal Disease Surveillance and Survey Systems: Method and Applications*. Iowa State Press, Ames, IA.

Scudamore, J. 2002. Origin of the UK Foot and Mouth Disease Epidemic in 2001. Department for Environment, Food and Rural Affairs (DEFRA), UK.

Stärk, K.D., Regula, G., Hernandez, J., *et al.* 2006. Concepts for risk-based surveillance in the field of veterinary medicine and veterinary public health: review of current approaches. *BMC Health Services Research*, **6**,20.

Thrusfield, M. 2005. *Veterinary Epidemiology*, 3rd edn. Blackwell Science, Oxford, UK.

Vågsholm, I. and Smulders, F.J.M. 2012. Symptomless carriers and the rationale for targeted risk management strategies. *WTM – Veterinary Medicine Austria*, **99**, 272–277.

13
Public Health Hazards
A. Biological Hazards

Maria Fredriksson-Ahomaa

Department of Food Hygiene and Environmental Health,
Faculty of Veterinary Medicine, University of Helsinki, Helsinki, Finland

13.1   Scope

Chapter 13 introduces the most important meat-borne public health hazards, both biological and chemical ones. The biological meat-borne public health hazards include bacteria, viruses, parasites and prions (Table 13.1). Occupational hazards, which are not meat borne, are presented in Chapter 21. In Chapter 13A, the most important meat-borne zoonotic hazards, the diseases caused by these hazards in both humans and animals and the epidemiology of the hazards are presented. Antimicrobial resistance in meat-borne bacteria is briefly discussed also in Chapter 13A. Control of the most important biological hazards is discussed in more details in Chapter 13B.

13.2   Bacteria

Animal reservoirs are important for the most common meat-borne bacterial public health hazards like thermotolerant *Campylobacter* spp., *Salmonella* spp., enteropathogenic *Yersinia* spp. and Shiga toxin-producing *Escherishia coli* (STEC). These zoonotic agents enter the slaughterhouse with infected/pathogen-positive animals and may spread during the slaughtering. *Listeria*

Meat Inspection and Control in the Slaughterhouse, First Edition.
Edited by Thimjos Ninios, Janne Lundén, Hannu Korkeala and Maria Fredriksson-Ahomaa.
© 2014 John Wiley & Sons, Ltd. Published 2014 by John Wiley & Sons, Ltd.

Table 13.1   Zoonotic agents causing meat-borne infections in humans.

Zoonotic agent	Animal source	Occupational
Bacteria		
Bacillus anthracis	Ruminants	Yes
Bacillus cereus	Pig, ruminants, poultry	No
Campylobacter jejuni/coli	Pig, ruminants, poultry	Yes
Clostridium botulinum	Pig, ruminants, poultry	No
Clostridium perfringens	Pig, ruminants, poultry	No
Listeria monocytogenes	Pig, ruminants, poultry	Yes
Mycobacterium avium	Pig, poultry	Yes
Salmonella spp.	Pig, ruminants, poultry	Yes
Staphylococcus aureus	Pig, ruminants, poultry	Yes
Shiga toxin-producing Escherichia coli	Ruminants	Yes
Yersinia enterocolitica	Pig	Yes
Virus		
Hepatitis E	Pig	Yes
Parasite		
Sarcocystis suihominis	Pig	No
Taenia saginata	Cattle	No
Taenia solium	Pig	No
Toxoplasma gondii	Pig, cattle, sheep/goat	Yes[a]
Trichinella spp.	Pig	No
Prions		
BSE/TSE	Cattle, sheep/goat	No

[a] Placenta and foetus can contain high numbers of Toxoplasma

monocytogenes and the toxin-producing *Bacillus cereus, Clostridium perfringens* and *Staphylococcus aureus* are ubiquitous bacteria found in animals and the environment. For these pathogens, pre-slaughter contamination plays a minor role. *Bacillus anthracis* and *Mycobacterium bovis/capri* are also described below because the diseases caused by these pathogens can be very severe for humans and animals. Furthermore, meat inspection plays an important role for identification of animals with anthrax and tuberculosis. *Mycobacterium avium* is a common finding in slaughter pigs but whether contaminated meat is a source of human infections is still unclear.

13.2.1   *Bacillus anthracis* and *Bacillus cereus*

Maria Fredriksson-Ahomaa and Miia Lindström

Department of Food Hygiene and Environmental Health,
Faculty of Veterinary Medicine, University of Helsinki, Helsinki, Finland

Aerobic spore-forming bacteria include members of the former genus *Bacillus* and include gram-positive. These bacteria survive a wide variety of environmental conditions because of the resistance of endospores to heat, radiation and disinfectants. This makes them troublesome contaminants in clinical environment, biotechnological processes and in food production. Of medical importance is the *Bacillus cereus* group, which includes the pathogenic *Bacillus anthracis* and *B. cereus* that are very similar both genotypically and phenotypically. However, the differentiation of *B. anthracis* from other aerobic spore-forming bacteria is important because of its clinical significance. *B. anthracis* causes anthrax in mammals and *B. cereus* causes food poisonings and a variety of local and systemic infections in humans.

Bacillus anthracis   *B. anthracis* is a very large spore-forming rod. The virulent bacterial cells are protected by the capsule. *B. anthracis* exists in the environment as a spore which can remain viable in the soil for decades. Spores, usually ingested by grazing herbivores, germinate within the animal to produce virulent vegetative forms that replicate and eventually kill the host. Food products like meat and hides from infected animals can serve as an infection source for humans. *B. anthracis* is genetically and phenotypically extremely homogeneous. Conventional methods like serology, biochemical tests and phage typing are not appropriate to differentiate among strains.

   B. anthracis affects primarily herbivores such as ruminants (especially cattle and sheep, which are more susceptible), although other mammals may also succumb to the disease but less frequently. Anthrax, the disease caused by *B. anthracis*, is often a fatal disease in cattle with the signs not usually observed due to their rapid occurrence. The blood is heavily laden with the *B. anthracis* organism and characteristically the blood does not clot. A vaccine for livestock is commonly used in areas that have anthrax.

   Anthrax continues to be a major problem in several countries worldwide where the disease is endemic. There are around 10 000 human cases and animal outbreaks each year. However, *B. anthracis* infection is rare in developed countries and occurs only sporadically in countries where drastic eradication measures have been established and implemented. Humans generally acquire anthrax from infected animals or as a result of occupational or nutritional exposure to contaminated animal products including meat.

   Anthrax takes one of four forms: (i) cutaneous anthrax, (ii) gastrointestinal anthrax, (iii) pulmonary or inhalation anthrax and (iv) injectional anthrax. Most of the human cases are cutaneous and intestinal anthrax is rare; however, there have been several outbreaks due to intestinal anthrax due to eating meat from a dead cow. There have also been reports of oro-pharyngeal anthrax, which is a rare form of the disease, following the consumption of contaminated beef. Pulmonary anthrax is often referred to as 'wool-sorter's disease' because it is associated with wool processing plants. The disease is usually fatal, although subclinical infections may occur. Injectional anthrax is a soft tissue infection resulting from a subcutaneous drug injection.

   Cutaneous anthrax accounts for over 90% of all human cases and is acquired through a lesion on the skin. It usually develops within 2–3 days. The lesion is painless and is surrounded by oedema. The bacilli remain localized to the lesion in uncomplicated cutaneous anthrax. Without treatment, lymphangitis with subsequent sepsis can develop. The fatality rate in untreated patients can be 5–20%, with therapy it is <1%. Cutaneous anthrax should always be considered when patients who have had contact with animals or animal products present with painless ulcers associated with vesicles and oedema.

Gastrointestinal anthrax results from the ingestion of undercooked meat from animals with anthrax. The incubation time is 2–5 days. It has two forms: abdominal and oro-pharyngeal anthrax. In abdominal anthrax, initial symptoms are nausea, vomiting, anorexia and fever. As the disease progresses, severe abdominal pain and bloody diarrhoea occurs followed by death. The mortality rate is estimated to be 25–60%. Although gastrointestinal anthrax is rare, its incidence may be underestimated due to the non-specific symptoms of the disease. In oro-pharyngeal anthrax, spores settle in the pharyngeal area and produce ulcers. The symptoms include sore throat, dysphagia, fever, cervical lymphadenopathy and oedema. The diagnosis is difficult resulting in a high mortality rate.

Pulmonary anthrax is caused by the inhalation of *B. anthracis* spores. With improved hygiene practices in the food industry and immunization, the number of cases has fallen dramatically. The infection dose is mostly high (around 10 000 spores); however, there are reports that only few spores can cause the disease. The period between exposure and onset of the disease is about 4–6 days, possibly up to six weeks. The illness usually begins with 'flu-like' symptoms of mild fever, fatigue, malaise, myalgia and non-productive cough. As the disease progress, high fever, laboured breathing and meningitis may occur. The mortality rate with adequate therapy is more than 50–75% and without therapy in nearly 100%.

Injectional anthrax, which differs from cutaneous anthrax, is common among heroin users. In this form, spores germinate at the inoculation site. Significant oedema at the injection site is common. Excessive bruising at the injection site may occur early. Severe cases develop thrombocytopenia and coagulopathy.

B. anthracis can be isolated from numerous clinical samples, including blood, skin lesion exudates, cerebrospinal fluid, pleural fluid, sputum and faeces. Culturing with confirmation by immune-histochemical staining and polymerase chain reaction (PCR) are most frequently employed for diagnosis in human infections. A diagnosis of anthrax in animals is most often made by identification of the organism in blood from an infected animal. Blood smears can be directly stained for the organism (usually by using capsule staining) or the organism can be cultured. Anthrax should be considered immediately if Gram stain reveals gram-positive bacilli growing in chains. There are also enzyme-linked immunosorbent assays (ELISA)- and PCR-based laboratory tests available for samples of animal origin. Necrosis should not be performed because of the risk of inducing sporulation and spreading of the organism to the environment.

Bacillus cereus *B. cereus* is a close relative of *B. anthracis*. It is a facultative aerobic endospore-forming bacterium of great importance as a pathogen to humans and animals. *B. cereus* shows a wide range of variation in phenotypes and virulence types. *B. cereus* strains produce a large number of potential virulence factors but their roles in specific infections are mainly unknown. *B. cereus* is widespread in nature and frequently isolated from soil and plants but it is also well adapted for growth in the intestinal tract of mammals. *B. cereus* can easily be transmitted from the intestinal tract to the carcass by faecal contamination during slaughter. Because of the high prevalence and the resistance of the spores, *B. cereus* contaminates nearly all agricultural products and plays a major role in the contamination of food products. A wide variety of foods, including meat, has been implicated in food-borne outbreaks due to *B. cereus*. Under improper storage conditions after cooking, the spores can germinate and the vegetative cells multiply. The optimum temperature for germination of the heat-resistant spores is between 65 and 75°C.

B. cereus causes two types of food poisoning: (i) the emetic and (ii) diarrhoeal syndromes. The incidence of both syndromes has probably been underestimated because the illnesses are usually self-limiting with relatively mild symptoms. *B. cereus* is not always considered in laboratory investigations of food-borne infections, thus the true burden of the illness remains unknown. The combination of the high prevalence, heat resistance and psychrotrophy of many *B. cereus* strains makes it difficult to control this organism in the food processing environment. In many cases, the dishes have been stored at room temperature for some hours before consumption. *B. cereus* concentrations are usually more than 10^4–10^5 cells/g of implicated food.

The emetic illness is caused by the emetic toxin, cereulide, produced by *B. cereus* in the food. At least 10^5 cells/g are usually needed to produce sufficient amount of toxin. The production of emetic toxin is associated with actively growing *B. cereus* cultures under aerobic conditions. Cooking and reheating of the food that may kill the cells will leave the toxin unscathed. The illness is characterized by nausea, vomiting and abdominal cramping, which occur 1–6 h after eating the contaminated food. The symptoms are similar to those of *Staphylococcus aureus* food poisoning. The illness is self-limiting and recovery occurs usually within 6–24 h. Sometimes hospitalization is needed because of excessive vomiting. The foods implicated in emetic food poisonings include meat; however, most outbreaks have been associated with starch-rich foods like cooked rice dishes.

The diarrheal illness is caused by enterotoxins that are produced by vegetative cells in the small intestine after ingestion of spores, or by vegetative cells in contaminated food. The illness is characterized by watery diarrhoea and abdominal pain, occasionally also nausea and vomiting. The onset time ranges from 8 to 16 h and the symptoms usually resolve within 24–48 h. The symptoms are similar to that of food poisoning caused by *Clostridium perfringens*. A large number of cases have been attributed to proteinaceous dishes like meat.

B. cereus has also been associated with illness other than food poisoning, albeit these infections are not common. Systemic and local infections have been reported especially in immuno-compromised patients, neonates, drug addicts and patients with a history of traumatic or surgical wounds. Many of the local infections are mild but also severe deep infections with necrosis and purulence occur. *B. cereus* is also an important microorganism found in severe ocular infections. Bacteraemia is in most cases transient and harmless but can occasionally cause serious infections.

The diagnosis of *B. cereus* food poisoning can be confirmed by the isolation of large number of *B. cereus* organisms ($>10^4$/g) from the epidemiologically implicated food and by determining their toxigenicity using serological (diarrhoeal toxin) or biological (diarrhoeal and emetic toxins) tests. Furthermore, the strain isolated from food should represent the same genotype as the strain isolated from the patient.

13.2.2 *Campylobacter* spp.

Pekka Juntunen and Marja-Liisa Hänninen

*Department of Food Hygiene and Environmental Health,
Faculty of Veterinary Medicine, University of Helsinki, Helsinki, Finland*

Campylobacter spp. are the most common bacterial food-borne pathogens in developed countries. More than 212 000 confirmed human campylobacteriosis cases were reported in the European Union (EU) in 2010, meaning an average notification rate of 49 cases per 100 000 citizens, but variation between countries is remarkable partly due to variable notification systems in different countries. The actual number of cases is estimated to be approximately nine million per year. Most infections are sporadic and outbreaks are rare. The highest number of cases is detected during the summer, mainly between June and August. Outbreaks are mostly caused by drinking of raw milk or contaminated water. The estimation of annual cost for public health systems and decreased productivity is approximately €2.4 billion. Many countries have adopted an integrated strategy from farm to fork to control zoonotic food-borne pathogens including *C. jejuni* and *C. coli*.

Approximately 90–95% of the human *Campylobacter* infections are caused by *C. jejuni*. *C. coli* is isolated in only 5–10% of the cases. Other species, such as *C. lari* and *C. upsaliensis*, are sometimes detected. In these cases, species level determination is usually lacking.

Campylobacter are gram-negative, curved or spiral rods. The shape of cells in old cultures may be coccoid. *Campylobacter* grows in microaerobic atmosphere and the range of growth temperature varies between 30 and 37°C but thermotolerant species like *C. jejuni*, *C. coli* and *C. lari* can grow in temperatures of 41–42°C. The bacteria will survive long periods at refrigeration temperatures but are rapidly killed at room temperature or on dry surfaces.

Clinical symptoms of campylobacteriosis develop after an incubation period of 2–7 days and they include fever, diarrhoea (sometimes bloody), abdominal pain and cramps. Duration of the symptoms is usually 3–4 days and the disease is usually self-limiting. Only in severe cases antimicrobial treatment is needed. *Campylobacter* infection may trigger post-infectious complications. These include reactive arthritis, which may follow the acute infection in few percentage of the patients. In addition, neurologic complications, such as Guillain–Barré syndrome, an ascending paralysis of peripheral nerves, may develop in rare cases a few weeks after the infection.

Antimicrobial resistance of *C. jejuni* and *C. coli* has been an increasing problem among animals and humans in certain countries. Most commonly, resistance against ciprofloxacin has been detected (Table 13.2). Also in broiler chicken, the resistance to ciprofloxacin was very common (Table 13.3). Spain reported the highest resistance (92%) to ciprofloxacin among *C. jejuni* isolates while in Finland only 2% were resistant.

A wide variety of wild and domestic animal species are colonized by *C. jejuni* and *C. coli*. The number of campylobacter in the intestine can be between 10^6 and 10^8 per g, indicating that faecal spills or intestinal ruptures at slaughter can easily contaminate the carcass. *C. jejuni* is the predominant

Table 13.2 Detection rates of resistant *Campylobacter coli* and *C. jejuni* isolates from humans in the EU in 2010.

Antimicrobial agent	Detection rate of resistant isolates (%)	
	C. coli	*C. jejuni*
Ampicillin	25.2	29.4
Ciprofloxacin	66.0	51.6
Erythromycin	11.0	1.7
Gentamicin	0.1	0.3
Tetracycline	32.2	20.6

Source: EFSA 2012b.

Table 13.3 Detection rates of resistant *Campylobacter coli* and *C. jejuni* isolates from broiler chicken in the EU in 2010.

Antimicrobial agent	Detection rates of resistant isolates (%)	
	C. coli	*C. jejuni*
Ciprofloxacin	84	47
Erythromycin	15	0.5
Gentamicin	8	0.8
Tetracycline	73	32

Source: EFSA 2012b.

Campylobacter species in cattle and poultry. Pigs usually carry *C. coli* in their intestine. During 2008 and 2010, the prevalence of *Campylobacter* spp. in slaughter pigs varied between 2 and 67% in nine EU countries. Campylobacter is regularly monitored in chicken flocks/batches in the EU. Most countries monitor *Campylobacter* in the slaughterhouses by using caecal samples, but, for example, Denmark monitors *Campylobacter* at the farm level using boot samples. The prevalence rates of campylobacter in flocks/batches differ highly between countries. In 2010, the lowest prevalence rates were reported in Estonia, Norway and Finland (under 7% of the flocks were campylobacter positive) while the Czech Republic, Slovakia, Slovenia and Spain detected campylobacter in over 70% of the flocks. In cattle, the detection rates varied between 0 and 58% in ten EU countries between 2008 and 2010. *Campylobacter* has also been isolated from turkeys, sheep and goats.

Carcasses from *Campylobacter*-positive poultry flocks are usually contaminated by *Campylobacter* during the slaughter process. Contaminated poultry meat is an important infectious source for humans. Contrary to poultry meat, the contamination levels of beef and pork are very low, usually less than 1% of the meat samples are *Campylobacter* positive.

Several molecular methods have been developed for epidemiological studies of *Campylobacter*. Pulsed-field gel electrophoresis (PFGE) is a widely used method to study sources of human *Campylobacter* infections. Rare-cutting restriction *Sma*I and *Kpn*I enzymes are commonly used in PFGE typing of *Campylobacter* isolates. The high discriminatory power of PFGE is used, for example, by the PulseNet database organized by the Centres for Disease Control and Prevention (CDC) to investigate food-borne outbreaks. Multilocus sequence typing (MLST) is another molecular subtyping method applied for population genetics and source attribution studies. Sequence fragments of several housekeeping genes reveal evolution of *Campylobacter* strains and those can be easily compared between laboratories. Subtyping methods based on polymorphism of single locus, such as flagellin A restriction fragment length polymorphism (*fla*A-RFLP), are less informative but faster and more cost effective to perform.

The main approach to prevent contamination of poultry meat with *Campylobacter* is biosecurity at primary production. The outdoor environment of poultry houses is commonly contaminated by *Campylobacter*, which may be transported into poultry facilities by personnel, visitors or other vectors, such as flies and wild birds. If biosecurity management is unsuccessful, *Campylobacter* can rapidly spread into the flock. Poultry flocks with access to an outdoor environment (usually organic production) are at high risk to be colonized with *Campylobacter*. Thinning is a risk factor for colonization. An all-in, all-out system with a withdrawal period, proper cleaning and disinfection measures is preferable. Highly *Campylobacter*-positive flocks should be slaughtered at the end of the day in order to avoid contamination of other batches during the slaughter. Freezing of contaminated carcasses reduces the *Campylobacter* levels; however, only proper cooking destroys *Campylobacter* from the meat. Cross-contamination is possible before cooking at kitchen, for example via chopping boards and knives to other foodstuffs.

13.2.3 *Clostridium* spp.

Maria Fredriksson-Ahomaa and Miia Lindström

Department of Food Hygiene and Environmental Health,
Faculty of Veterinary Medicine, University of Helsinki, Helsinki, Finland

Clostridia are strictly anaerobic to aerotolerant, gram-positive, spore-forming, rod-shaped bacteria that belong to a large, heterogeneous taxonomic group. They form characteristic spores, the position of which can be useful in species identification. The spores are much more resistant to heat, chemicals, irradiation and desiccation than their vegetative cells. All pathogenic *Clostridium* strains produce toxins that play an important role in the pathogenesis. *Clostridium botulinum* and *C. perfringens* are important food-borne bacteria. However, they are not typical zoonotic bacteria that can be transmitted directly from animals to humans. These pathogens cause diseases through contaminated, improperly handled foods, including meat and meat products. Laboratory testing of anaerobic, spore-forming, toxin-producing bacteria is labour intensive. PCR can be used to detect genes encoding toxin production. For detection of toxins, immunoassays and animal tests are in use. False-negative results might occur due to the toxin degradation.

Clostridium botulinum *C. botulinum* is an obligate anaerobic, rod-shaped bacterium. The spores are located terminally or subterminally in vegetative cells. The species can be divided into four phenotypic groups, of which Groups I (proteolytic) and II (non-proteolytic) are relevant to human disease. The mesophilic Group I strains grow and produce toxin at temperatures above 10°C and produce spores of high heat resistance, while spores of Group II strains are of moderate heat resistance but can germinate and form toxic cultures at temperatures as low as 3°C. *C. botulinum* can produce neurotoxin when the environmental conditions are favourable (warm temperatures, a protein source, moisture and an anaerobic condition). Carbohydrates, such as glucose, induce toxin production. Botulinum neurotoxin is the most powerful natural toxin. The neurotoxins can enter the body (i) by ingestion of toxin with food or drink (food-borne botulism), (ii) by *in vivo* synthesis as a consequence of colonization of the digestive tract by the bacterium (infant botulism, toxico-infectious botulism) or (iii) by *in vivo* synthesis due to contamination of a wound by the bacterium (wound botulism).

Groups I and II *C. botulinum* are involved with food-borne botulism. The disease is rare but extremely severe. Food-borne botulism may occur when *C. botulinum* spores in the food have been allowed to germinate into a toxigenic culture in anaerobic conditions. Such conditions may prevail in hermetically packaged foods or in any foods supporting anaerobic microenvironments. Symptoms typically appear 12–36 h after eating toxic food. The classic symptoms include double vision, difficulty swallowing and muscle weakness. Botulism can result in death due to respiratory failure. *C. botulinum* spores are found in the intestinal tract of animals and in the environment (soils and sediments). Although often present in low numbers, their ubiquitous nature ensures that raw products cannot be guaranteed free of spores. Spore destruction with the mild heat treatments employed in modern food production is unlikely. Raw meat products packed in vacuum may allow *C. botulinum* to grow and must, therefore, be subjected to treatments that destroy spores or stored under conditions that prevent growth and toxin formation from spores. Many cases of food-borne botulism have been associated with meat products such as home-salted ham and home-made pork in glass jars.

Clostridium perfringens   *C. perfringens* can be divided into five types (A–E) according to the major toxins they produce. *C. perfringens* is one of the most common causes of food-borne diseases in the industrialized countries. The food poisoning is caused by the *C. perfringens* enterotoxin (CPE) that is produced by a small population of type A strains. Infection usually develops after the ingestion of foods contaminated with large numbers ($>10^6$ bacteria/g) of CPE-positive vegetative cells, which sporulate in the intestine and produce enterotoxin. CPE produced in the intestine causes diarrhoea and abdominal cramps, typically without fever and vomiting. The symptoms usually occur suddenly within 6–24 h after ingestion of the toxin and last less than 24 h.

C. *perfringens* occurs widely in the environment, such as in soil and dust, and is found in the intestinal tract of humans and animals, including food-producing animals. *C. perfringens* is also commonly found on raw meat, albeit the CPE-positive strains appear to represent a minority. Nevertheless, disease outbreaks are typically associated with temperature-abused meat or poultry dishes. Under ideal conditions *C. perfringens* can grow very rapidly in food. Prevention of human disease is based on appropriate food handling, such as good hygienic practices to minimize initial contamination and proper refrigeration to limit growth. Recent research has suggested humans to be an important source of contamination for *C. perfringens* type A because *cpe*-positive strains have frequently been isolated from human faeces.

Clostridium difficile   *C. difficile* is an anaerobic, spore-forming bacterium that can produce toxins A or B upon colonization of the gut. *C. difficile* infections are usually seen in hospitalized patients who have used antimicrobial therapy. Patients typically develop diarrhoea and, in severe cases, a pseudomembranous colitis. In the community, the bacterium usually causes milder, self-limiting diarrhoea. *C. difficile* is also found in the intestinal tract of animals and in the environment. Toxigenic strains have been found in young production animals, especially in poultry. Pork, beef and chicken meat can also contain this bacterium but the prevalence and the number of bacteria have shown to be low. Despite the low numbers, the spore-forming nature and the heat tolerance of the spores might facilitate food-borne transmission. However, meat-borne transmission of *C. difficile* to humans has not been proven.

13.2.4 *Listeria monocytogenes*

Sanna Hellström

Finnish Food Safety Authority Evira, Helsinki, Finland

The genus *Listeria* contains eight species, of which two are pathogenic, causing disease called listeriosis: *L. monocytogenes* is pathogenic to humans and animals and *L. ivanovii* causes mainly abortions in ruminants. All *L. monocytogenes* strains are considered pathogenic, although only few serotypes cause most listeriosis cases and variation in virulence exists among strains.

L. monocytogenes is a gram-positive, coccoid rod that is motile when cultured <30°C and does not form spores. The optimal growth temperature is 30–37°C but it is able to grow at temperatures from <0 to 45°C and in aerobic and anaerobic conditions. Serotyping divides *L. monocytogenes* into 13 serotypes, based on O and H antigens: 1/2a, 1/2b, 1/2c, 3a, 3b, 3c, 4a, 4ab, 4b, 4c, 4d, 4e and 7. Serotyping has a limited value in subdividing *L. monocytogenes* strains, because of the finite number of serotypes and the fact that most listeriosis cases are caused by serotypes 1/2a, 1/2b and 4b, and food commonly harbours serotypes 1/2a and 1/2c. Nowadays, genotypic methods are mainly used for subtyping of *L. monocytogenes* and pulsed-field gel electrophoresis (PFGE) is still considered to be the golden standard for subtyping *L. monocytogenes*.

Listeriosis appears mainly in two forms: severe invasive listeriosis and non-invasive febrile gastroenteritis. Invasive listerosis manifests as sepsis, meningoencephalitis, perinatal infection and abortion. In cases associated with pregnancy, the mother may have mild flu-like symptoms, while the foetus gets a severe infection or dies *in utero*. Invasive listeriosis has a mean mortality of 20–30% and a hospitalization rate of over 90%, making it one of the most severe food-borne diseases. Invasive listeriosis is mainly a disease affecting susceptible individuals with underlying predisposing conditions. Susceptible individuals include the elderly, pregnant women and their unborn or new-born infants, and patients with severe underlying diseases such as cancer, AIDS or organ transplant. Non-invasive gastroenteritis can manifest in immune-competent adults and it is usually self-resolving. *L. monocytogenes* can also produce a wide range of focal infections and occur in cutaneous form mainly as an occupational disease of veterinarians and farmers.

The majority of listeriosis cases appear to be sporadic. Because of the relative rarity of the disease, long incubation time (1–70 days) and wide spread of contaminated food from a single factory, outbreaks can be unnoticed.

The overall notification rate of confirmed cases of listeriosis in Europe in recent years has been 0.3 cases per 100 000 population, summing up to approximately 1400–1600 ascertained cases annually. The infectious dose of listeriosis remains unclear but according to epidemiological data it is suspected to be high, as the contamination level in foods responsible for listeriosis cases are typically $>10^4$ CFU/g. Consuming foods that contain low levels ($<10^2$ CFU/g) of *L. monocytogenes* is unlikely to cause clinical disease.

The main route of transmission is through the consumption of contaminated food and many listeriosis outbreaks linked to foods have been reported. The majority of listeriosis cases are linked to refrigerated, ready-to-eat (RTE) foods that are consumed without reheating, such as RTE meat products, smoked fish and soft cheese.

In animals, listeriosis is mainly a disease of ruminants. Small ruminants (sheep and goat) seem to be more susceptible to listeriosis than cattle. The source of infection of animals is in most cases feed, silage being a common source in farm animals. In domestic animals, listeriosis manifests mainly as encephalitis, abortion or septicaemia.

L. monocytogenes is an ubiquitous organism that is widely distributed in the environment. The key reservoir for *L. monocytogenes* is soil, and it is frequently found in vegetation, forage, water, sewage and farm environments, but numbers are often low. *L. monocytogenes* can survive in the soil for months and even grow in favourable conditions. Wild and domestic animals commonly are asymptomatic intestinal carriers, frequently shedding the organism and maintaining its populations in the environment.

L. monocytogenes has several characteristics that enable its survival in the food chain. *L. monocytogenes* can survive and grow over a wide range of temperature, pH (4.0–9.6) and water activity (a_w 0.9) limits as well as under aerobic and anaerobic conditions. In addition, *L. monocytogenes* can form biofilms and persist for several months or years in food processing facilities. These characteristics enable the pathogen to survive in food-processing environments and in foods, and make it a concern for the food industry and a threat for public health.

L. monocytogenes has been found in many types of foods but numbers are usually low and seldom above the European legal safety limit of 100 CFU/g during the shelf life of a product. *L. monocytogenes* is frequently found on raw materials and raw products, but these are not likely to be a direct vehicle for listeriosis, because of the usual heat treatment or other listericidal process before consumption. The prevalence of *L. monocytogenes* is often high in products that are minimally processed or have the potential of contamination after heat treatment. Other criteria for risk include support of the growth of *L. monocytogenes* in products, extended storage in chilled temperature and lack of heat treatment before consumption.

Contamination by *L. monocytogenes* can occur at all the steps of the food chain from farm to table (Figure 13.1). The impact of contamination on public health differs at different steps and depends on the type of food. For instance, raw materials can be expected to have small numbers of *L. monocytogenes* and they should be treated accordingly, whereas RTE foods that support the

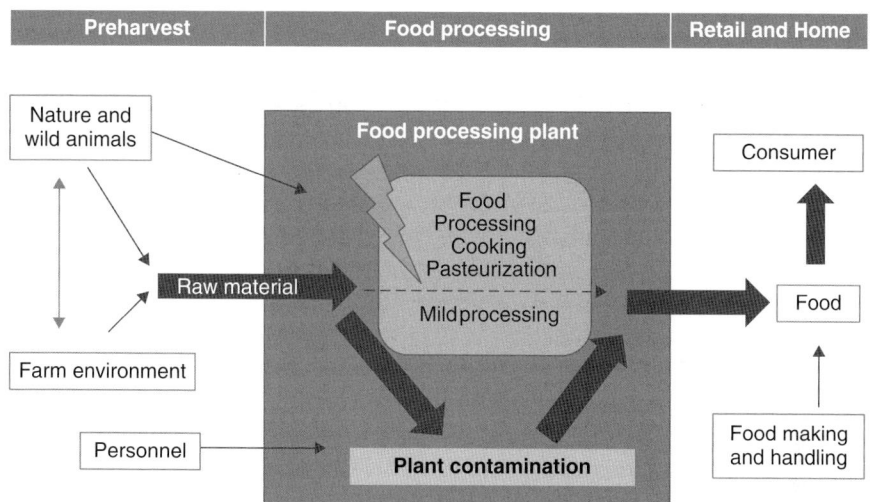

Figure 13.1 Contamination routes of *Listeria monocytogenes* from farm to table. Top bar indicates the processing step. Thick lines indicate the most important contamination routes. Cooking and pasteurization eliminate *L. monocytogenes* but recontamination after heat treatment, especially when foods are handled, is one of the most important sites of contamination. Source: Hellström, 2011. Reproduced with permission.

growth of *L. monocytogenes* should be free of the pathogen at the time of production. Contamination during food processing in a plant can be considered as the major source of contamination foods.

Isolation of *L. monocytogenes* often requires enrichment because of the small numbers of *L. monocytogenes* in food and environmental samples and the presence of sometimes large numbers of competitors. After enrichment *L. monocytogenes* is usually easily grown on selective plates, which also contain indicator substrates to enable recognition of *Listeria* spp. and *L. monocytogenes* among other bacteria. Enrichment, selective plating and identification are time consuming, and several PCR and antibody-based rapid methods have been developed to detect *L. monocytogenes* straight from food and environmental samples or enrichment broth.

13.2.5 *Mycobacterium* spp.

Maria Fredriksson-Ahomaa[1] and Ute Messelhäusser[2]

[1]*Department of Food Hygiene and Environmental Health,
Faculty of Veterinary Medicine, University of Helsinki, Helsinki, Finland*
[2]*Bavarian Health and Food Safety Authority, Oberschleissheim, Germany*

The family *Mycobacteriaceae* contains only one genus, *Mycobacterium*, which includes over 130 species and subspecies, several of which are important

Table 13.4 Pathogenic *Mycobacterium* species belonging to *Mycobacterium tuberculosis* and *avium* complex.

Mycobacterium species	Host	Zoonootic agent
Mycobacterium tuberculosis complex (MTBC)		
Mycobacterium tuberculosis	Human	Yes
Mycobacterium bovis	Cattle	Yes
Mycobacterium caprae	Goat, sheep[a]	Yes
Mycobacterium avium complex (MAC)		
Mycobacterium avium subsp. *avium*	Pigs	Yes[b]
Mycobacterium avium subsp. *hominisuis*	Pigs	Yes[b]
Mycobacterium avium subsp. *paratuberculosis*	Ruminants	Unknown

[a] Wildlife, especially wild ruminants are an important reservoir
[b] Immuno-compromised people

animal and human pathogens. Tuberculosis in both humans and animals is caused by a group of closely related species known as the *Mycobacterium tuberculosis* complex (MTBC) (Table 13.4). Bovine tuberculosis is mainly caused by *M. bovis* and *M. caprae*. *M. caprae* was earlier classified firstly as *M. tuberculosis* (subsp. *caprae*) and then as *M. bovis* (subsp. *caprae*). *M. tuberculosis* can also be transmitted from man to cattle but it happens very rarely. Epidemiology of bovine tuberculosis is very complicated. Both cattle-to-cattle and wildlife-to-cattle transmission routes occur.

Mycobacterium avium complex (MAC) contains *M. avium* subsp. *avium*, *M. avium* subsp. *hominisuis* and *M. avium* subsp. *paratuberculosis*. *M. avium* subsp. *avium* and subsp. *hominisuis* are not able to cause human infection in healthy individuals but they can cause infections in patients with chronic respiratory infections and in immuno-compromised individuals. *M. avium* subsp. *paratuberculosis* (MAP) is the causative agent for Johne's disease, a health problem for ruminant species. There is a possible link, which has not been proven yet, between MAP and Crohn's disease, which is a human inflammatory bowel disease.

Mycobacteria are aerobic, gram-positive, rod-shaped microorganisms that are resistant to acid because of their very complex cell wall envelopes. Pathogenic species can be differentiated from the environmental species, for example by their slow growth in comparison to environmental mycobacteria, which grow more rapidly. Mycobacteria belonging to *M. tuberculosis* complex (MTBC) and *M. avium* complex (MAC) are slow growing. This means that preparing cultures of these mycobacteria can take over six weeks; the culturing of MAP takes up to 12 weeks.

Mycobacterium bovis and M. caprae   *M. bovis* and *M. caprae* are zoonotic members of the MTBC with a significant importance in livestock and a wide range of wild animal species worldwide. During the first half of the twentieth century bovine tuberculosis was considered one of the largest veterinary public health problems in Europe. A significant number of tuberculosis cases in humans were caused by *M. bovis/M. caprae*, especially in children and in rural areas. The risk of getting *M. bovis/M. caprae* infections has become very low in developed countries over the past decades due to the eradication programmes, including mandatory tuberculin testing of livestock, removal of infected animals and pasteurization of milk. Although tuberculosis caused by *M. bovis/M. caprae* is very low compared to *M. tuberculosis*, its potential impact on population groups at the highest risk, such as farmers, veterinarians and slaughterhouse workers, to an increased occurrence should not be underestimated.

Tuberculosis in humans due to *M. bovis/M. caprae* is both clinically and pathologically indistinguishable from cases caused by *M. tuberculosis*. The primary location of lesions depends on the route of infection and also on subsequent dissemination of the bacteria to other organs. Aerosol exposure to *M. bovis* is considered to be the most common route of infection but infection by ingestion of contaminated food is also possible. Characteristic tuberculosis lesions occur most frequently in the lungs and the retropharyngeal, bronchial and mediastinal lymph nodes. Lesion can sometimes also be found in the mesenteric lymph nodes, on serous membranes and in organs, especially in liver and spleen. Transmission of tuberculosis from cattle to humans mostly occurs through consumption of unpasteurized milk and dairy products, and also through close contact to infected animals. Human-to-human transmission of *M. bovis/M. caprae* is rare but transmission of these pathogens from humans back to cattle can occur.

Cattle are considered to be the most important host, even though this pathogen has been isolated from several domestic species, including pigs, sheep and goats, and non-domestic species. The tuberculin testing and culling of positive animals has eliminated tuberculosis in farmed bovine populations in some countries. However, *M. bovis/M. caprae* has shown to exist in wildlife in several countries, and thus acts as a possible source of tuberculosis for livestock.

Bovine tuberculosis, which is often a subclinical infection in cattle, can be diagnosed in the live animals on the basis of delayed hypersensitivity reactions or on the basis of gamma-interferon-release assay, which is a similar to a test used in human medicine. The clinical signs are not specific and may include weakness, anorexia and emaciation. Particularly in advanced tuberculosis, dyspnoea, cough and enlargement of lymph nodes can be present. Lymph nodes of the head and neck may become visibly affected and can rupture. At necropsy, tubercles (typically characterized by small caseous nodules) are frequently seen in retropharyngeal, bronchial and/or mediastinal lymph nodes, which may be the only tissues affected. They can usually be found by palpations and incision during meat inspection. Sometimes lung, liver, spleen, kidneys and the surface of body cavities are also affected. After death, the diagnosis can be made by histopathological, bacteriological and molecular techniques. Bacteriological methods may include microscopic examination to detect acid-fast bacilli. Culture techniques are still considered the gold standard for laboratory-based detection methods.

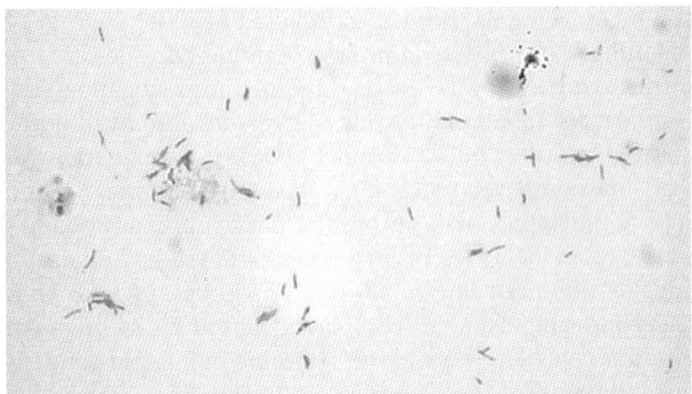

Figure 13.2 *Mycobacterium bovis* visualized using Ziehl–Neelsen stain. Source: Courtesy of Dr Hoermansdorfer. (For colour details, please see colour plate section).

Disease diagnosis and identification of mycobacterial species is not easy. The test mostly used for tuberculosis in animals is still the tuberculin skin test. In human medicine, the tuberculin skin test has been increasingly replaced by gamma-interferon-release assay. *Post-mortem* diagnosis based on examination of gross lesions, followed by histopathology and culturing is mostly used for detection of *M. bovis/M. caprae* in animals. Direct identification of the acid-fast bacilli by microscopy in various samples is also widely used. Ziehl–Neelsen staining followed by light microscopy are most commonly used (Figure 13.2). However, the sensitivity of direct microscopy is very low, thus culturing of the bacteria before microscopy is usually needed to higher the sensitivity. Culturing is very time consuming and, therefore, PCR and serological methods are increasingly used. Using PCR-based techniques, detection and accurate differentiation of mycobacterial species is possible in a short time.

Prevention and control of mycobacteria is difficult. In countries with well-developed control programmes, infected animals rarely show clinical signs being able to remain in a subclinical state for a long period. *M. bovis* infection in cattle rarely presents as clinical disease. Most commonly the animals are apparently healthy and only respond to immunological testing based on tuberculin. Monitoring of bovine tuberculosis in cattle depends on national programmes of herd tuberculin testing and active slaughterhouse surveillance. Limited tuberculin test sensitivity and specificity can contribute to under or overestimation of the disease. Tuberculosis in wildlife can pose serious difficulties for bovine tuberculosis control and eradication. Early diagnosis and intervention to interrupt transmission are the priority for control. Whole-herd depopulation is indicated in some occasions. A number of risk factors has been associated with the bovine tuberculosis, such as purchase of cattle, the occurrence of tuberculosis in the surrounding area, herd type and size, stocking density, spreading of slurry and exposure of cattle to wildlife reservoir when at pasture. *M. bovis/M. caprae* can remain infective up to several months in soil and water.

Mycobacterium avium* subsp. *avium* and subsp. *hominisuis *M. avium* subsp. *avium* and subsp. *hominisuis* are the two zoonotic subspecies of the MAC. They can cause infections in immuno-compromised humans. Bacteria within the MAC are present in the environment (soil and water) as well as in animals such as birds, pigs and cattle. Pigs are the main reservoir for the subspecies *avium* and *hominisuis*. They are primarily infected by the ingestion of mycobacteria from a contaminated surrounding environment. The infection is asymptomatic in pigs, thus diagnosis is based on *post-mortem* findings. Typically, these mycobacteria cause localized lymphadenitis in pigs, particularly affecting submaxillary and mesenteric lymph nodes. However, granulomatous lesions may also be found in the liver. The subspecies *avium* is an obligate pathogen in birds causing mycobacteriosis. Within the exception of backyard poultry, the disease is sporadic and rarely reported in commercial poultry. Classical chronic lesions in the emaciated birds are non-caseated, non-mineralized nodules in different organs. The infection most commonly affects the liver, spleen, intestine and air sacs. Human infection through subspecies *avium* and *hominisuis* is usually acquired through inhalation and is generally limited to lungs without systematic disorders. Whether contaminated meat is an infection source of MAC infections for humans is unclear.

13.2.6 *Salmonella* spp.

Maria Fredriksson-Ahomaa[1] and Sinikka Pelkonen[2]

[1]*Department of Food Hygiene and Environmental Health,*
Faculty of Veterinary Medicine, University of Helsinki, Helsinki, Finland
[2]*Finnish Food Safety Authority Evira, Helsinki, Finland*

Salmonella is an important enteric bacterium of economic significance in animals and humans. Salmonellosis is an important food-borne disease that poses a major threat to human public health in both developed and developing countries. *Salmonella* is a member of the *Enterobacteriaceae* family, which is a large group of gram-negative, facultative anaerobic and non-spore-forming bacilli. The nomenclature and classification of *Salmonella* species have been changed and restructured several times. Currently, the genus *Salmonella* consists of only two species, *S. enterica* and *S. bongori*, based on the DNA similarity. *S. enterica* is further divided into six subspecies: *enterica, salamae, arizonae, diarizonae, houtenae* and *indica*. Strains belonging to subspecies *enterica, arizonae* and *diarizonae* include the serotypes of most interest to veterinary and human medicine. More than 2600 serotypes have been differentiated by agglutinating properties of the somatic O, flagellar H and capsular Vi antigens.

Table 13.5 Host range and disease outcomes of some *Salmonella* serotypes.

Host range		Serotype	Cause
Narrow	Human	Typhi	Septicaemic typhoid fever
	Human	Paratyphi	Septicaemic paratyphoid fever
	Birds[a]	Gallinarum	Systemic disease (fowl typhoid)
	Cattle[b]	Dublin	Severe diarrhoea and septicaemia
	Pigs[c]	Choleraesuis	Pig paratyphoid, septicaemia, pneumonia
	Sheep[d]	Abortus-ovis	Abortion, stillbirth, metritis, septicaemia
Broad	Humans, animals	Typhimurium	Enteritis (diarrhoea, fever, abdominal cramps)
	Humans, animals	Enteritidis	Enteritis (diarrhoea, fever, abdominal cramps)
	Humans, animals	Infantis	Enteritis (diarrhoea, fever, abdominal cramps)

[a] Occasionally infects humans
[b] Occasionally infects small ruminants, pigs and humans
[c] Occasionally infects other species
[d] Occasionally infects goats

Salmonella nomenclature is now based on the name of serotypes. For example, *Salmonella enterica* subspecies (subsp.) *enterica* serotype (ser.) Typhimurium is abbreviated *Salmonella* Typhimurium.

Salmonella serotypes can be grouped on the basis of host range and disease outcomes (Table 13.5). Host-specific serotypes, such as *S.* Typhi and *S.* Gallinarum, are associated with severe systemic disease in adults of a single species. Host-restricted serotypes such as *S.* Choleaesuis are associated with systemic disease in few hosts. Broad-host-range serotypes, such as *S.* Typhimurium and *S.* Enteritidis, cause the majority of human gastrointestinal salmonellosis. They are able to infect humans and many animal species, including domestic livestock and poultry, worldwide. The symptoms range from severe systemic disease to asymptomatic carriage. Mostly, the infection is limited to the gastrointestinal tract and rarely spreads to systemic organs.

Human salmonellosis is common illness with a worldwide distribution. Human-adapted serotypes, *S.* Typhi and *S.* Paratyphi, are important human pathogens of vast concern to public health and with considerable economic impact. They are endemic in regions of the world where drinking water quality and sewage treatment facilities are poor. Typhoidal salmonellosis can lead to typhoid fever, which is a potentially fatal multisystemic illness. The classic symptoms include fever, malaise, diffuse abdominal pain and constipation. Non-typhoidal *Salmonella* are important food-borne pathogens that cause enteritis. They are the second most frequent cause of bacterial intestinal disease in Europe. The typical symptoms are fever, diarrhoea and abdominal pain. The symptoms are often mild and the infection is usually self-limiting, lasting a few days. The elderly, infants and those with a weakened immune system may have a more severe illness. In these patients, the infection can spread from the intestine to the blood stream and cause death if not treated with antimicrobials. Salmonellosis has also been associated with long-term, sometimes chronic sequel like reactive arthritis.

Salmonella infection is typically acquired from the environment by oral ingestion of food or water contaminated with *Salmonella* or by human contact with a *Salmonella* carrier. Each year infections are also acquired through direct animal contact. Following ingestion, *Salmonella* survive the low-pH environment of the stomach to enter the small intestine, where the infection can be established. In *Salmonella* enteritis caused by common broad-host-range serotypes, the infection is usually self-limiting and does not proceed beyond the lamina propria. However, in host-adapted salmonellosis like typhoid fever, *Salmonella* gain access to the lymphatics and bloodstream allowing the bacteria to spread to different organs. *S.* Enteritids and *S.* Typhimurium are common broad-host-range serotypes associated with human salmonellosis. *S.* Enteritidis infections are usually linked to the consumption of contaminated poultry meat and eggs while *S.* Typhimurium infections are linked to a range of food-producing animals like ruminants, pigs and poultry. The global distribution of food and the continuous movement of people around the world enable the introduction of emerging *Salmonella* serotypes into different continents and countries. The prevalence of the different serotypes changes over the time.

A common reservoir of *Salmonella* is the intestinal tract of a wide range domestic and wild animals. A considerable number of serotypes frequently isolated from humans have been isolated from sick or clinically healthy animals. Clinically sick animals are more likely to shed *Salmonella*, and at higher concentration, than apparently healthy animals. However, asymptomatic carriers also can shed *Salmonella* for long periods of time; increased stress can activate shedding. A variety of clinical manifestations has been observed in *Salmonella*-infected animals, ranging from asymptomatic to peracute disease. Infections with host-adapted serotypes generally cause severe systemic disease with high mortality. The economic losses during outbreaks can be very high. In contrast, infections with broad-host-range serotypes usually cause mild or no disease and infected animals may shed *Salmonella* for considerable period of times. The subclinical infections can easily spread between the animals in a herd and flock without detection. Transport and social stress are correlated with increased faecal excretion of animals and reactivation of asymptomatic infection. Stress has the potential to augment *Salmonella* virulence and increase pathogen entry into the food chain and environment.

Salmonellosis has been recognized in all countries and is most prevalent in areas of intensive animal husbandry, especially of poultry and pigs. There are several routes of transmission for *Salmonella*. The main transmission route is through consumption of contaminated food. Although *Salmonella* is a frequent cause of food-borne outbreaks, the majority of the reported cases are sporadic. Contaminated meat and meat products have been implicated in a number of human salmonellosis cases. *S.* Typhimurium is the predominant serotype isolated from humans especially in Europe; cattle, pigs and poultry are an important reservoir of this particular serotype. The majority of human cases attributed to broilers and turkeys have been caused by *S.* Enteritidis and *S.* Typhimurium. The public health risk of *Salmonella* infection from consumption of contaminated meat depends on multiple factors, including the level of infection in the herds, hygiene during carcass processing in the slaughterhouse, meat storage, distribution conditions and, finally, handling of raw meat by consumers.

To reduce *Salmonella* contamination in the meat production chain, it is important to maintain a low or zero *Salmonella* status at the herd level. Measures like cleaning, disinfection and testing are necessary. Sometimes depopulation is needed. Furthermore, restriction on the movement and marketing of animals from infected premises are necessary. Due to efficient *Salmonella* control, both red and white meat produced in Nordic countries (Finland, Norway and Sweden) can be claimed to be virtually free from *Salmonella*. In these countries, infected farms are subjected to restrictions that include a total ban on the movement of animals, except for transport to slaughter. The restrictions remain in place until all animals are declared free from *Salmonella*. In these countries, competitive exclusion gut flora preparation is widely used in commercial broiler production to prevent *Salmonella* infection.

There are several diagnostic methods available. The presence of *Salmonella* using culturing is the most frequently used method. Methods based on the detection of bacterial DNA such as PCR are increasingly being used. Serology is often used for monitoring to indicate exposure of the herd or flock to *Salmonella*. Mixed anti-lipopolysaccharide ELISA is a useful tool for monitoring herd infection *ante-mortem* and *post-mortem* using serum and meat juice samples. Microarrays are also being explored to replace standard serotyping methods. The reasons why some serotypes are highly prevalent in food-producing animals but appear rarely in humans are unclear. There is a need for improved diagnostic tools to predict the zoonotic and epidemic potential of isolates found in animals.

13.2.7 *Staphylococcus aureus*

Maria Fredriksson-Ahomaa and Miia Lindström

Department of Food Hygiene and Environmental Health,
Faculty of Veterinary Medicine, University of Helsinki, Helsinki, Finland

Staphylococcus aureus is a gram-positive and catalase-positive coccus. Growth is possible under both aerobic and anaerobic conditions. There are more than

50 species and subspecies among *Staphylococcus*, of which *S. aureus* is the most pathogenic for humans. *S. aureus* occurs worldwide as part of the normal flora of animals and humans. It is found in the nasopharynx and on the skin of man and animals, including food producing animals. *S. aureus* colonizes persistently approximately one third of population and carriage is generally asymptomatic. Some *S. aureus* strains can produce a wide variety of toxins, including more than 20 different staphylococcus enterotoxins (SE). The most common enterotoxin with emetic activity is SEA, which is related to food poisoning. The enterotoxins are more resistant to environmental conditions, including heat treatment, freezing, drying and low pH, than the cells.

S. aureus food poisoning is common. It is caused by ingestion of enterotoxins that are produced in foods by some strains. Contamination of food occurs through poor handling practices. Foods rich in proteins, such as meat, provide a good growth environment for *S. aureus*. However, raw food is rarely the source of *S. aureus* poisoning, probably because *S. aureus* does not compete well with other bacteria in raw foods. Thus, *S. aureus* grows better when competitive bacteria are removed, for example by cooking. Contamination is most often caused by human contact with hands after the product is cooked. For toxin production, the food must be held under temperature-abused conditions that promote growth of *S. aureus* to 10^6 cells/g of food or more.

Illness results from ingestion of the enterotoxin produced by *S. aureus* and the disease is not transmitted to other humans. The incubation period and severity of symptoms depend on the amount of enterotoxins ingested and the susceptibility of the person. The symptoms appear usually within 1–8 h (3 h on average) after ingesting the contaminated food. The symptoms include nausea, vomiting, abdominal pain and diarrhoea. The illness is commonly confused with that caused by emetic *B. cereus*. In the majority of cases, recovery occurs within 1–2 d without any treatment.

Methicillin-resistant *S. aureus* (MRSA) is an important pathogen in hospitals and the community. It has also recently been isolated worldwide from food production animals, mainly from pigs. The prevalence of MRSA has shown to be high among pigs, pig farmers and veterinarians. MRSA can be transmitted to humans via contact with animals and contaminated environment and by eating or handling contaminated meat. MRSA can cause serious diseases such as endocarditis, pneumonia, and urinary tract, wound and soft tissue infections.

Since *S. aureus* occurs on the skin and hides of animals, it can contaminate the carcass by cross-contamination during slaughter. The spread of *S. aureus*, including MRSA, can already have easily occurred during transport by cross-contamination and in lairage at the slaughterhouse. Dehairing of pigs and defeathering of poultry are critical processing steps for cross-contamination during slaughter. The mechanical treatments of the carcass lead to dissemination of bacteria from mouth, nose, skin and intestinal tract. Chilling the carcasses inhibits the proliferation of *S. aureus* effectively because it does not grow at temperatures under 6°C.

13.2.8   Shiga toxin-producing *Escherichia coli* (STEC)

Maria Fredriksson-Ahomaa[1] and Roger Stephan[2]

[1]*Department of Food Hygiene and Environmental Health,*
Faculty of Veterinary Medicine, University of Helsinki, Helsinki, Finland
[2]*Institute for Food Safety and Hygiene, Vetsuisse Faculty,*
University of Zurich, Zurich, Switzerland

Escherichia coli are facultative anaerobe gram-negative rods within the family *Enterobacteriaceae*. *E. coli* strains are normal inhabitants of the gastrointestinal tract of humans and animals. They are also often used as indicator organisms for faecal contamination and lack of hygiene. *E. coli* bacteria are a genetically heterogeneous group of non-pathogenic and pathogenic strains. Most *E. coli* strains are non-pathogenic but some strains have acquired genes that enable them to cause intestinal or extra-intestinal diseases. Shiga toxin-producing *E. coli* (STEC) were first discovered in 1977 and first associated with the clinical syndrome haemolytic-uremic-syndrome (HUS) in 1983. STEC refers to those *E. coli* strains that produce Shiga toxin, also called verotoxin (VTEC). The name Shiga toxin (Stx) is derived from similarity to a cytotoxin produced by *Shigella dysenteriae*. Those STEC that cause haemorrhagic colitis and haemolytic uremic syndrome are called enterohaemorrhagic *E. coli* (EHEC). EHEC infections have often been traced back to cattle.

Serotyping has been used extensively to characterize *E. coli* strains. The serotype is based on the O antigen determined by the lipopolysaccharide (LPS), the H antigen due to flagellar protein and the capsular K antigen. STEC strains comprise almost 500 serotypes that differ greatly in both their physiological characteristics and their pathogenic potential to humans. At present, it is still not possible to fully define human pathogenic strains. However, some serotypes (O26, O103, O111, O121, O145, O157) are responsible

Table 13.6   STEC serotype associated with human illness.

Serotype	Relative prevalence	Frequency of involvement in outbreaks	Association with HC[a] and HUS[b]
0157	High	Common	Yes
026, 0103, 0111, 0121, 0145	Moderate	Moderate	Yes
063, 091, 0104, 0113, 0128, 0146	Low	Rare	Yes
Multiple (07, 069, 0117, 0119 etc.)	Low	Rare	No
Multiple (06, 08, 039, 046, 076 etc.)	NI[c]	NI	No

[a] haemorrhagic colitis
[b] haemolytic uremic syndrome
[c] not implicated
Source: Adapted from Gyles, 2007.

Table 13.7 Important virulence genes found in STEC strains.

Gene	Location	Virulence factor	Function
stx	Bacteriophages	Cytotoxin called Shiga toxin (Stx)	Responsible for the damages of endothelial cells in the intestine, kidney and brain
eae	Chromosome	Intimin	Responsible for the attachment to the intestinal surface
*ehx*A	Plasmid	Haemolysin (Ehly)	Responsible for enterocyte damage

for the majority of cases (Table 13.6). In many countries, O157 STEC is the serotype associated with the most sporadic cases and outbreaks.

The Shiga toxin genes (*stx*), the *E. coli* attaching and effacing gene (*eae*) and the gene for *E. coli* hemolysin (*ehly* or *ehxA*) are important virulence genes found in *E. coli* strains associated with human disease (Table 13.7). Shiga toxins (Stxs) are the main virulence factors expressed by STEC. There are two types of Stxs, Stx1 and Stx2, and several variants in each type. At present, Stx1 includes three subtypes (Stx1a, Stx1c and Stx1d) and Stx2 seven subtypes (Stx2a-Stx2g). Stx2 is about 1000 times more toxic for human renal microvascular endothelial cells than Stx1. Stx2a, Stx2c and Stx2d are subtypes most commonly associated with severe outcomes of human disease. *stx* genes are located on bacteriophages that are integrated into the bacterial chromosome. STEC strains not only produce Stx but also can produce other virulence factors that may increase the severity of human illnesses. These factors include intimin and enterohaemolysin, which are responsible for the attachment to the intestinal surface and enterocyte damage, respectively. Intimin is encoded by the *eae* gene and enterohaemolysin by *ehx*A gene (Table 13.7). The presence of both *stx2* and *eae* in a STEC strain is considered to be a predictor for highly pathogenic strains. Nevertheless, *ehx*A and *eae* genes are not absolutely required for pathogenicity.

Humans become most frequently infected with STEC by ingestion of contaminated food or water or by direct contact with animals. Also person-to-person transmission occurs. For O157, the infection dose is low, estimated to be less than 50 to a few hundred organisms. Infection by STEC ranges from mild diarrhoea to haemorrhagic colitis (HC) and haemolytic uremic syndrome (HUS), which can lead to death. HC is characterized by bloody diarrhoea, abdominal cramps, sometimes fever and vomiting, while HUS is characterized by thrombocytopenia, microangiopathic haemolytic anaemia and acute renal failure due to the production of toxins that damage the endothelial cells and trigger the clotting mechanism. HUS develops in 5–10% of individuals infected with STEC serotype O157. Infants, children, the elderly and those with compromised immune function are most susceptible to severe complications.

Various STEC strains have been isolated from different animals but they are more prevalent in ruminants than in other animals. Especially, asymptomatic cattle represent the main reservoir host of STEC. In some countries, sheep and goats are also important reservoirs of this pathogen. STEC strains associated with human disease have only sporadically been isolated from pigs and poultry. STEC carrying the *stx*2e gene has been associated with oedema disease in piglets; however, the role of this pig pathogen in the epidemiology of human disease needs further research.

Ruminants transmit STEC to humans by shedding the pathogen in their faeces. A proportion of positive animals, called 'super shedders', excrete more STEC than others. Although the 'super shedders' comprise only a small ratio of ruminants, it has been estimated that they may be responsible for over 95% of all STEC bacteria shed. Once shed into the environment, humans acquire STEC by consuming bovine/sheep-derived products such as meat and milk or contaminated water and vegetables. Direct contact with ruminants at petting zoos or through interaction with infected people represent another transmission route. Cattle and sheep hides have been identified as an important source of STEC contamination of carcasses. The carcass contamination usually occurs during removal of the hide or the gastrointestinal tract.

STEC is an important food-borne pathogen associated with sporadic cases and outbreaks worldwide that poses a serious public health concern. Sources of infection include meat (especially beef and mutton), ready-to-eat sausages, raw milk and milk products, vegetables and water. The first large outbreak of O157 STEC-associated HUS in the United States in 1983 involved consumption of fast-food hamburgers that were not well cooked. Therefore, the disease is also called 'hamburger disease'. Since then, human infections have frequently been linked to food or water contaminated with cattle manure. STEC bacteria grow between 10 and 46°C and can survive for several weeks or months in meat when frozen at −18 to −20°C. Moreover, they are also more resistant to acid conditions, compared to other food-borne pathogens. The food safety practice of cooking minced beef to 72°C or until the meat is no longer pink has been widely adopted by the fast-food industry, contributing to the decrease in STEC infection traced to this food matrix.

The optimal control of STEC needs to involve all stages of food production from farm to fork. To reduce the animal exposure at the farm, it is important to identify the 'super shedders' and to prevent faecal contamination of the feed. Moreover, maintenance of good practices of slaughter hygiene is of central importance to ensure both public health protection in view of STEC and the meat production line.

The strain heterogeneity makes diagnostic testing problematic. Generally, two different approaches are used to detect STEC in foods: (i) serotype-dependent methods, which target a subset of serotypes most frequently involved in human diseases, and (ii) serotype-independent techniques, where the presence of virulence genes are studied. The methods can be culture- or molecular-based methods, or immunological methods. In contrast to cultural methods, which are time consuming, labour intensive and are associated with a low rate of STEC recovery in food, PCR- and immunoassay-based methods are characterized by reduced analysis time and high throughput analysis capabilities. They offer the possibility to quickly rule out negative analyses, which usually represent the majority of the samples. The positive samples can be further studied with different isolation methods if needed.

13.2.9 Enteropathogenic *Yersinia* spp.

Maria Fredriksson-Ahomaa and Suvi Joutsen

Department of Food Hygiene and Environmental Health,
Faculty of Veterinary Medicine, University of Helsinki, Helsinki, Finland

The genus *Yersinia*, which is a member of the family *Enterobacteriaceae*, is a group of gram-negative, oxidase-negative, non-spore-forming rods. *Yersinia enterocolitica* and *Yersinia pseudotuberculosis* are the two species in the genus *Yersinia* that can cause enteric yersiniosis in both humans and animals. *Y. enterocolitica* is primarily a human pathogen that rarely causes disease in animals while *Y. pseudotuberculosis* is a well-known cause of disease in both humans and animals. Asymptomatic finishing pigs are the most important reservoir of human pathogenic *Y. enterocolitica*. *Yersinia* strains are facultative anaerobic and psychrotrophic bacteria which can also grow in anaerobic conditions at refrigeration temperatures.

Y. enterocolitica strains are heterogeneous in their biochemical, antigenic and virulence properties. Six different biotypes (1A, 1B, 2–5) and about 30 serotypes of *Y. enterocolitica* strains have been described. Strains belonging to biotypes 1B and 2–5 have been associated with diseases in humans and animals while strains belonging to biotype 1A are generally regarded as non-pathogenic, due to lack of the virulence plasmid (pYV) and the major chromosomal virulence genes. All pathogenic *Yersinia* strains carry the pYV, which is essential for the replication of the bacteria in the host. Most of the *Y. enterocolitica* strains associated with human disease belong to bioserotypes 1B:O8, 2:O5,27, 2:O9, 3:O3 and 4:O3. These bioserotypes have been shown to

Table 13.8 Ecological and geographical distribution of the most common bioserotypes of *Yersinia enterocolitica* associated with human diseases.

Biotype	Serotype	Host	Geographical distribution
1B	O8	Man, pig, rodents	USA, Japan, Poland
2	O5,27	Man, pig, sheep	America, Australia, Europe, Japan
2	O9	Man, pig, ruminants	Europe
3	O1,2,3	Chinchilla	Europe
	O3	Man, pig, dogs	East Asia
4	O3	Man, pig, rat	All continents
5	O2,3	Hare, goat, sheep	Australia, Europe, New Zealand

Source: Adapted from Fredriksson-Ahomaa, *et al.*, 2010.

have different ecological niches and geographical distributions (Table 13.8). In Europe, bioserotype 4:O3 is the most common type followed by 2:O5,27 and 2:O9 isolated from humans with diarrhoea and pigs at slaughter. Highly pathogenic bioserotype 1B:O8 is mainly restricted to America but has recently been isolated from human clinical samples in Poland also. There is little variation in biochemical reactions among *Y. pseudotuberculosis* strains but they can be classified into 21 serotypes. The most common serotypes are serotypes O1 to O5. All *Y. pseudotuberculosis* strains are considered to be pathogenic. Serotype O3 has mostly been isolated from pigs at slaughter.

Human yersiniosis is the third most common reported food-borne bacterial enteritis after campylobacteriosis and salmonellosis in Europe. The highest incidence of yersiniosis is usually found in young children. Most human cases are caused by *Y. enterocolitica* and only rarely by *Y. pseudotuberculosis*. *Y. pseudotuberculosis* infections are probably under-recognized because it is not routinely tested for in many countries and the clinical diagnosis can be challenging. Even though human yersiniosis cases are typically sporadic, some outbreaks of *Y. enterocolitica* have been reported in Europe, Asia and North America during the past decade and the source of infection has been linked to pork products, dairy products and salad. *Y. pseudotuberculosis* outbreaks have mainly been reported in Finland, Japan and Russia. The infections have been linked to vegetables in Finland and surface water has several times been implicated in Japan.

Infection by enteropathogenic *Yersinia* can cause a variety of symptoms among humans depending on the age of person infected. In children, diarrhoea typically develops 4–7 days after infection and may persist for several weeks. Diarrhoea can be bloody and high fever may occur, especially in infants. In older children and young adults, right-sided abdominal pain and fever can be the predominant symptoms, which are easily confused with appendicitis. The most common post-infectious sequelae are reactive joint infection and skin rash (erythema nodosum), especially among adults. The reactive arthritis is associated with the presence of HLA-B27 antigen. The joint pains are common in the knees, ankles or wrist and usually develop within a month after the initial infection. Uncomplicated cases of yersiniosis usually resolve without any antimicrobial treatment.

Enteropathogenic *Yersinia* enters humans via the oral route. The minimal infectious dose is unknown. After ingestion, *Y. enterocolitica* and *Y. pseudotuberculosis* move through the gastrointestinal tract to small intestine and bind to the intestinal epithelium of terminal ileum. They penetrate the intestinal mucosa through M cells and replicate within the local lymphoid nodes resulting in mesenteric lymphadenitis. They can also disseminate from the intestinal tract to the major lymphatic organs like spleen, liver and lungs. However, usually the host inflammatory response is able to eliminate the dissemination.

The most common transmission route of *Y. enterocolitica* and *Y. pseudotuberculosis* is via contaminated food and water. *Y. enterocolitica* infection has been associated with consumption of raw or undercooked pork while *Y. pseudotuberculosis* infections have been linked to fresh produce and surface water. Direct contact with pigs has shown to be a risk factor for yersiniosis for pig farmers and slaughter workers. Pet animals have also been suspected as being sources of human yersiniosis through close contact with humans, especially children. Direct person-to-person transmission may happen when basic hygiene and hand-washing habits are inadequate. Indirect person-to-person transmission can occur by transfusion of contaminated blood products. In these cases, the sources of yersiniosis are blood donors with subclinical bacteraemia.

Y. enterocolitica strains have been isolated from a variety of animals but clinical manifestations or patho-anatomical changes are rare. Sporadic, small outbreaks of enteritis have been reported in pigs, sheep, goats, chinchillas, hares and monkeys. The infected animals are usually young and in poor condition. Pathogenic *Y. enterocolitica* strains are frequently isolated form asymptomatic pigs at slaughter, especially from the tonsils. At the slaughterhouse, it is impossible to identify asymptomatic carriers in *ante-* or *post-mortem* inspection without microbial sampling. There are several slaughter and meat inspection procedures, such as evisceration and incision of the submaxillary lymph nodes, where contamination can occur. Therefore, strict slaughter hygiene remains important in reducing contamination in slaughterhouses.

Y. pseudotuberculosis has an extremely broad host range. Infections are usually sporadic but outbreaks causing high mortality, especially among wild animals, also occur. In animals, the disease of subacute and chronic forms is characterized by loss of weight, diarrhoea, respiratory distress and muscle weakness. *Y. pseudotuberculosis* can usually be isolated from lungs, kidney, liver, spleen and mesenteric lymph nodes. Yersiniosis is a common infection among deer when they are stressed by weaning, transport, poor feeding, inclement weather or high stocking density. *Y. pseudotuberculosis* is a well-known pathogen of guinea pigs and one of the most important causes of death in hares and in zoo animals, especially among non-human primates. Outbreaks are also common in wild birds due to predisposing factors such as cold weather. Yersiniosis outbreaks among domestic ruminants, pigs and poultry are rare. However, *Y. pseudotuberculosis* has sporadically been isolated from pig tonsil and faeces samples as well as from carcass and offal samples from finishing pigs at slaughter.

Isolating *Y. enterocolitica* and *Y. pseudotuberculosis* from faeces samples or organ abscesses of acutely infected humans and animals is usually easy because these pathogens are often dominant bacteria and can easily be found by direct plating on conventional enteric media. Several difficulties have been associated with isolating pathogenic *Yersinia* from asymptomatic carriers or from food and environmental samples. Culture methods have several limitations, such as long incubation steps taking up to four weeks, low sensitivity and specificity (lack of identification between species, and lack of discrimination between pathogenic and non-pathogenic strains). The low isolation rate is probably due to the low sensitivity of the culture methods. Using polymerase chains reactions (PCR) methods, enteropathogenic *Yersinia* have been detected with higher frequencies in food and environmental samples.

The control of enteropathogenic *Yersinia* in meat chain, especially in the pork chain, is difficult because most of the pigs at slaughter carry pathogenic *Y. enterocolitica* in the tonsils. During the slaughter, it is difficult to prevent the spreading of *Yersinia* from tonsils to carcass and offal. However, it can be reduced by strict slaughter hygiene and changing of slaughter methods. After slaughter, the control is difficult because *Yersinia* can multiply during cold storage and under modified atmosphere. Correct handling of raw meat is important to reduce the cross-contamination to other products. However, *Yersinia* is destroyed during cooking. Consumption of undercooked and raw pork products should be avoided.

13.2.10   Literature and further reading

Barrow, P.A., Jones, M.A., Smith, A.L. and Wigley, P. 2012. The long view: *Salmonella* – the last forty years. *Avian Pathol*, **41**, 413–420.

EFSA (European Food Safety Authority). 2007. Monitoring and identification of human enteropathogenic *Yersinia* spp. Scientific opinion of the panel of biological hazards. *EFSA J*, **595**, 1–30.

EFSA (European Food Safety Authority). 2012. The European Union summary report on trends and sources of zoonoses, zoonotic agents and food-borne outbreaks in 2010. *EFSA J*, **10**, 2597.

EFSA (European Food Safety Authority). 2012. The European Union summary report on antimicrobial resistance in zoonotic and indicator bacteria from humans, animals and food in 2010. *EFSA J*, **10**, 2598.

EFSA (European Food Safety Authority). 2012. Technical specifications on harmonised epidemiological indicators for biological hazards to be covered by meat inspection of poultry. *EFSA J*, **10**, 2764.

Fredriksson-Ahomaa, M., Lindström, M. and Korkeala, H. 2010. *Yersinia enterocolitica* and *Yersinia pseudotuberculosis*. In: *Pathogens and Toxins in Foods: Challenges and Interventions* (eds V.K. Juneja and N.J. Sofos). ASM Press. Washington, DC, pp. 164–180.

Fredriksson, M., Stolle, A.and Korkeala, H. 2006. Molecular epidemiology of *Yersinia enterocolitica* infections. *FEMS Immunol Med Microbiol*, **47**, 315–329.

Gyles, C.L. 2007. Shiga toxin-producing *Escherichia coli*: an overview. *J Anim Sci*, **85**, E45–E62.

Hellström, S. 2011. Contamination routes and control of *Listeria monocytogenes* in food production. Academic dissertation. University of Helsinki, Finland. https://helda.helsinki.fi/bitstream/handle/10138/27420/contamin.pdf?sequence=1 (last accessed 6 February 2014).

Hennekinne, J.A., De Buyser, M.L. and Draqacci, S. 2012. *Staphylococcus aureus* and its food poisoning toxins: characterisation and outbreak investigation. *FEMS Microbiol Rev*, **36**, 815–836.

Hofer, E., Stephan, R., Reist, M. and Zweifel, C. 2012. Application of a real-time PCR-based system for monitoring of O26, O103, O111, O145 and O157 Shiga toxin-producing *Escherichia coli* in cattle at slaughter. *Zoonoses Public Health*, **59**, 408–415.

Huber, H., Koller, S., Giezendanner, N. *et al.* 2010. Prevalence and characteristics of meticillin-resistant *Staphylococcus aureus* in humans in contact with farm animals, in livestock, and in food of animal origin, Switzerland, 2009. *Eurosurveillance*, **15** (16) [online].

Humphrey, T., O'Brien, S. and Madsen, M. 2007. Campylobacter as zoonotic pathogen: a food production perspective. *Int J Food Microbiol*, **117**, 237–257.

ILSI Research Foundation/Risk Science Institute Expert Panel on *Listeria monocytogenes* in Foods. 2005. Achieving continuous improvement in reductions in food-borne listeriosis – a risk-based approach. *J Food Prot*, **68**, 1932-1994.

Karmali, M.A., Gannon, V. Sargeant, J.M. 2010. Verocytotoxin-producing *Escherichia coli* (VTEC). *Vet Microbiol*, **140**, 360–370.

Lassok, B. and Tenhagen, B.A. 2012. From pig to pork: methicillin-resistant *Staphylococcus aureus* in the pork production chain. *J Food Prot*, **76**, 1095–1108.

Laukkanen-Ninios, R. and Fredriksson-Ahomaa, M. 2012. Epidemiology and genetics of enteropathogenic *Yersinia* species. In: *Food-Borne and Waterborne Bacterial Pathogens: Epidemiology and Molecular Biology* (ed. S.M. Faruque). CA Press, Norfolk, UK, pp. 269–287.

Lindström, M., Fredriksson-Ahomaa, M. and Korkeala, H. 2009. Molecular epidemiology of group I and group II *Clostridium botulinum*. In: *Clostridia: Molecular Biology in the Post-Genomic Era* (eds H. Brueggemann and G. Gottschalk). Caiser Academic Press. Norfolk, UK, pp. 103–130.

Lindström, M., Heikinheimo, A., Lahti, P. and Korkeala, H. 2011. Novel insights into the epidemiology of *Clostridium perfringens* type A food poisoning. *Food Microbiol*, **28**, 192–198.

Miche, A.L., Mueller, B. and van Helden, P.V. 2010. *Mycobacteroum bovis* at the animal-human interface: a problem, or not? *Vet Microbiol*, **140**, 371–381.

OIE (World Organisation of Animal Health). 2009. Bovine tuberculosis. OIE Terrestrial Manual, Chapter 2.4.7. Retrieved from http://www.oie.int/fileadmin/Home /eng/Health_standards/tahm/2.04.07_BOVINE_TB.pdf (last accessed 6 February 2014)

OIE (World Organisation of Animal Health). 2010. Salmonellosis. Manual of diagnostic tests and vaccines for terrestrial animals, Chapter 2.9.9. Retrieved from http://www.oie.int/fileadmin/Home/eng/Health_standards/tahm/2.09.09 _SALMONELLOSIS.pdf (last accessed 6 February 2104).

Skuce, R.A., Allen, A.R. and McDowell, S.W.J. 2012. *Herd-level risk factors for bovine tuberculosis: a literature review*. *Vet Med Int*, Article ID 621210, doi: 10.1155/2012/621210.

Songer, J.G. 2010. Clostridia as agents of zoonotic disease. *Vet Microbiol*, **140**, 399–404.

Stenfors Arnesen, L.P., Fagerlund, A. and Granum, P.E. 2008. From soil to gut: *Bacillus cereus* and its food poisoning toxins. *FEMS Microbiol Rev*, **32**, 579–606.

Stevens, M.P., Humphrey, T.J. and Maskell, D.J. 2009. Molecular insights into farm animal and zoonotic *Salmonella* infections. *Philos Trans R Soc Lond B Biol Sci*, **364**, 2709–2723.

Swaminathan, B. and Gerner-Smidt, P. 2007. The epidemiology of human listeriosis. *Microbes Infect*, **9**, 1236-1243.

Sweeney, D.A., Hicks, C.W., Cui, X. *et al.* 2011. Anthrax infection. *Am J Respir Crit Care Med*, **184**, 1333–1341.

Thorpe, C.M. 2004. Shiga toxin-producing *Escherichia coli* infection. *Clin Inf Dis*, **38**, 1298–1303.

Trajman, A., Steffen, R.E. and Menzies, D. 2013. *Interferon-gamma release assays versus tuberculin skin testing for the diagnosis of latent tuberculosis infection: an overview of the evidence. Pulm Med*, Article ID 601737, doi: 10.1155/2013/601737.

13.3 Viruses

Leena Maunula and Tuija Kantala

*Department of Food Hygiene and Environmental Health,
Faculty of Veterinary Medicine, University of Helsinki, Helsinki, Finland*

Unlike most bacteria, all viruses are obligate parasites of cells and cannot multiply outside them. Viruses that propagate in the gastroenteritis tract of the host (enteric viruses) are, however, quite resistant to many physical conditions and can survive for several weeks in a cold environment. Most of these viruses are host species-specific, since the attachment of viruses on the surface of cells is a highly specific event, but some are capable of infecting several hosts.

Viral zoonoses are a concern also at slaughterhouses. The zoonotic viruses that may be relevant in slaughterhouses are listed in Table 13.9. An example of an emerging virus is the highly pathogenic avian influenza (HPAI) virus, which can cause serious disease also in humans. Avian influenza virus poses a threat in poultry slaughterhouses for persons in close contact to infected chickens. Thus far, the zoonotic infections caused by avian influenza virus have been restricted geographically (mainly to Egypt, Cambodia and China).

Globally, the most important virus in this context is the hepatitis E virus (HEV), especially genotype 3 HEV, which seems to be common in pigs worldwide and may cause hepatitis in humans. HEV has probably circulated among healthy pigs for a long time but the potential of HEV to cause zoonotic infections was realized only after the emergence of non-travel-associated human HEV cases in industrialized countries.

Viruses are visible only by electron microscope (EM) (Figure 13.3). HEV has an icosahedral virion with a diameter of about 30 nm. This non-enveloped virus is likely to survive in acidic and mild alkaline conditions while it passes through the stomach. The genome is formed from a positive-sense single-stranded RNA with a length of 7200 bp. Typical for HEV and most other RNA viruses is their capability to evolve rapidly via point mutations, since they lack the proof-reading mechanism for viral RNA polymerase, and also via recombination events.

Influenza viruses have a lipid envelope and are thus more vulnerable in the environment than enteric viruses. Influenza virions are pleomorphic (irregular shape) with a diameter of approximately 120 nm. Since the genome is formed of several RNA segments, the virus can also evolve through re-assortment,

Table 13.9 Viruses that may cause zoonoses at the slaughterhouse.

Virus	Genome, structure	Virus type	Symptoms in humans	Symptoms in animals	Main reservoirs	Geographical regions
HEV	(+)ssRNA, non-enveloped, *Hepeviridae*	Genotype 1	Hepatitis	(infections unknown)		Tropical, subtropical (Asia, Africa)
		Genotype 2	Hepatitis	(infections unknown)		Tropical, subtropical (Mexico)
		Genotype 3	Hepatitis	No symptoms	Pigs, wild boars, game, cattle, small ruminants	Global
		Genotype 4	Hepatitis	No symptoms	Pigs, wild boars	Global (Asia)
Avian influenza virus	Segmented (−)ssRNA, enveloped, *Orthomyxoviridae*	H5N1	Respiratory, high mortality	High mortality (HPAI)	Wild birds, poultry	Asia, Northern Africa
		H7N9	Respiratory, high mortality	Low mortality (LPAI)	Wild birds, poultry	Asia (China)
		H7N7	Respiratory, severe	LPAI and HPAI	Wild birds, poultry	Global (Europe, USA)

Zoonosis suspected Norovirus, rotavirus, sapovirus

Zoonosis rare Cowpox virus, Crimean congovirus, Nipah virus, porcine parvovirus, SARS

Zoonosis unknown Avian HEV, duck hepatitis A virus

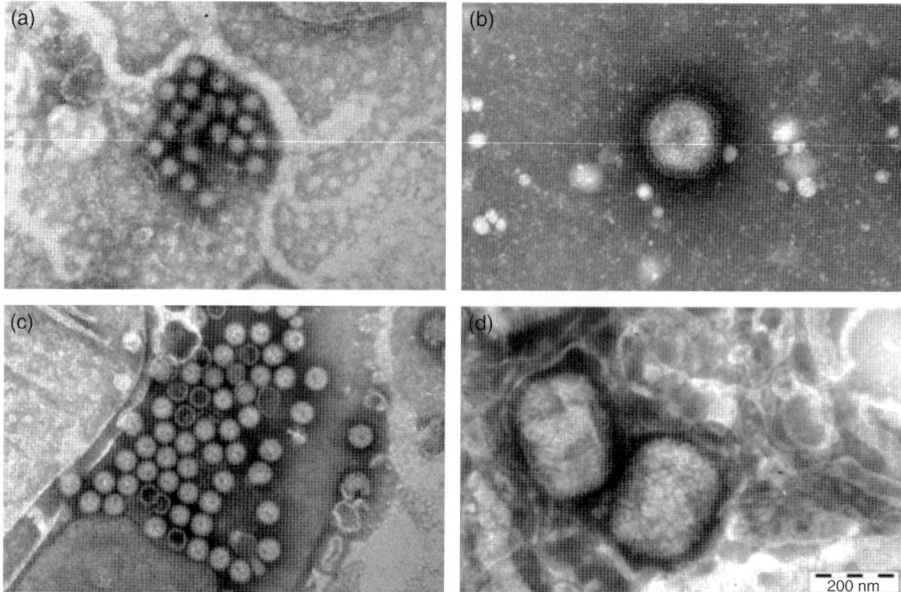

Figure 13.3 Electron microscopic (EM) micrographs showing the morphology of selected viruses: (a) norovirus; (b) influenza A virus; (c) rotavirus; (d) poxvirus. The bar indicates only the scale of Figure 1D (EM pictures from HUSLAB, I. Luoto).

which means an exchange of entire RNA segments and formation of viruses with a 'mosaic' genome in the case of double infections.

Most enteric viruses, for example rotavirus, norovirus and sapovirus, that infect humans also have counterparts in animal hosts and are typically species-specific, non-enveloped viruses resistant to the environment. There is, however, constant discussion concerning the possible zoonotic nature of these viruses. In this section, a detailed description of HEV in humans and pigs is given, while the significance of avian influenza virus and other viruses is discussed briefly in the last paragraphs.

13.3.1 Hepatitis E virus (HEV)

HEV in humans   HEV, currently the only representative of the *Hepevirus* family, was first found in large waterborne outbreaks in tropical and subtropical regions (India, Egypt), and all hepatitis cases caused by HEV in Europe and America were thought to be related to visits to the endemic regions. HE viruses of humans currently belong to four genotypes, of which genotype 1 and 2 viruses infect only humans and cause disease with severe symptoms; genotype 3 and 4 viruses are zoonotic and the symptoms seem to be generally milder; in addition, there is an avian HEV. Large hepatitis waterborne outbreaks due mostly to genotype 1 HEV have occurred in Asia and Northern Africa, and genotype 2 viruses have been reported to infect people in Mexico and Southern Africa. In humans, genotype 3 viruses are encountered globally, while genotype 4 virus infections are generally rare, more common only in Asia. Nowadays it is well established that zoonotic infections caused by genotype 3 viruses occur in Europe. According to a recent report, genotype 4 HEV has also caused some autochthonous infections in Europeans.

HEV usually causes acute, self-limiting hepatitis in humans. The incubation period is relatively long, from two weeks to two months. The most common symptoms are jaundice, anorexia and hepatomegaly; sometimes, patients suffer from abdominal pains, tenderness, nausea, vomiting or fever. Hepatitis E may cause a disease lasting several months; however, increasing evidence suggests that symptomless HEV infections also occur especially caused by genotype 3 HEV. HEV may also cause chronic disease in immune-compromised persons, for example during immunosuppressive treatment. Mortality from HEV in the normal population has been estimated to be 0.5–4%.

Genotype 1 and 2 HEVs cause disease most commonly in 20- to 40-year-old persons. The disease severely attacks pregnant women, with a mortality of 10–20%. Increasing numbers of non-travel-associated HEV cases caused by genotype 3 virus have been reported in England and other European countries. For unknown reasons, symptomatic HEV cases are most commonly encountered in elderly men.

In endemic countries in which genotype 1 and 2 HEVs are circulating, the seroprevalence in the general population is high, for example about 70% has been reported for Egypt. Elsewhere, a seroprevalence of 30% has been described among blood donors in the United States and 5% in the Swedish population. Genotype identification is not possible based on serological tests, since all five HEV genotypes represent only one serotype.

HEV in animals HEV infections of genotype 3 and 4 HEV are common in healthy pigs globally. The zoonotic nature of human genotype 3 and 4 HEV has been verified in experimental conditions in which both viruses of human origin were able to cause infection in pigs. In experimentally intravenously inoculated pigs, the incubation time is 1–2 weeks. In natural conditions, pigs get infected normally at 2–3 months of age and they shed the virus for 3–7 weeks with 1–2 weeks of viraemia. In pig excretions and manure, vast amounts of viruses may be measured.

HEV infection in pigs is usually symptomless and has no significant effect on the fertility of sows or health of offspring. Although most animals lack clinical symptoms, histological changes indicative of hepatitis can often be seen in livers of infected pigs. Liver enzymes and bilirubin levels may be elevated. Mild-to-moderate multifocal and periportal lymphoplasmo-cystic hepatitis and mild focal hepatocellular necrosis can be observed by microscopy. Enlarged hepatic and mesenteric lymph nodes may also occur.

In addition to pigs, genotype 3 HEV, based on genome detection, has been found in the faeces of many wild animal species, such as of wild boars and deer, and genotype 4 HEV in the faeces of deer and birds in China. Antibodies to HEV have also been detected in pets (cats and dogs) and rodents.

Although avian HEV shares the same serotype as human HEV, it is not regarded as likely to pose a health concern for humans. In experimental conditions, it was not possible to infect a monkey with avian HEV. In chickens, avian HEV causes hepatic splenomegaly syndrome, which may lead to a slight increase in mortality and decreased egg production (up to 20%). Other HEV-like viruses have been found in ferrets, rabbits, bats, rats and fish, but the significance of these viruses to human public health is currently unknown.

Transmission routes of HEV   The transmission routes for HEV are shown in Figure 13.4. Humans can contract an HEV infection from other humans or from animals, and the transmission may be direct or indirect through food, water or the environment. HEV, like hepatitis A virus, transmits mainly by the faecal-oral route but, relative to many other viruses using this transmission route, HEV is less easily transmitted from person to person. Transmission of HEV from mother to foetus has also been reported. In blood transfusion, HEV may be transmitted via blood, such as hepatitis B or C viruses, which, unlike HEV, are strictly blood-transmitted.

Humans usually contract HEV infection of human origin indirectly through food, water or the environment contaminated with sewage containing HEV. The viruses may survive in wastewater during treatment or even end up in surface waters, rivers and lakes, without any pre-treatment. Drinking water may serve as a vehicle for HEV infection if individuals consume water that has not been treated properly to inactivate viruses. Fresh vegetables or soft fruit can cause HEV infection if they have been irrigated or washed with

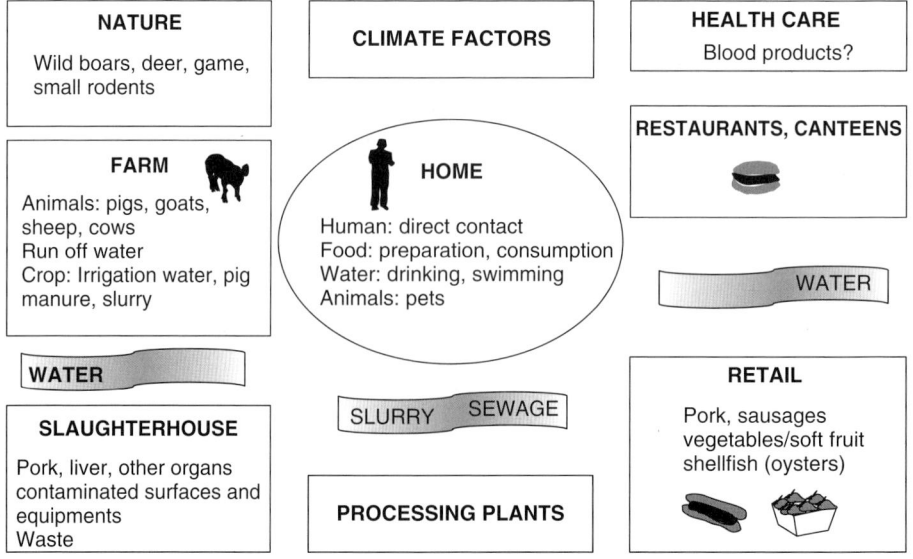

Figure 13.4   Main sources for hepatitis E virus transmission.

HEV-containing water. Humans can become infected also through consumption of raw shellfish containing HEV.

Direct contact with infected animals may lead to HEV infection. In the United States, HEV antibodies were 1.51 times more likely to be found in veterinarians than in normal blood donors, and in veterinarians with needle stick injuries they were 1.9 times more likely to be found. Studies show that persons with frequent contact with pigs, such as farmers, are also more likely to have HEV antibodies. Thus, also, persons working at slaughterhouses are at risk of becoming HEV infected.

Humans can contract an HEV infection indirectly from viremic animals through the consumption of meat or organs. Pork is the most common vehicle, but also uncooked game meat has caused HEV cases, even fatal ones in Japan. HEV infection can be obtained by consuming fresh produce or shellfish or by drinking water containing HEV of animal origin. Pig manure may also contaminate crops with HEV. Increased rainfall and flooding due to climate change in some geographical regions may increase run-off from fields, further contributing to HEV transmission.

Diagnosis of HEV in humans and studies of HEV in pigs   Acute hepatitis in humans caused by HEV cannot be distinguished from the disease caused by other hepatitis viruses. For human HEV diagnostics, serological tests are primarily used for detection. Presence of IgM antibodies or a fourfold rise in IgG antibodies in paired sera reveals an acute infection, and the presence of IgG antibodies indicates that the person has encountered an HEV infection. Direct nucleic acid detection from faeces or serum has become a more common diagnostic method. It is also a prerequisite for genotype determination and may help in source tracking.

Due to an increase in HEV prevalence, some countries have started to screen blood donors for HEV. A vaccination to prevent HEV is currently used only in China. Compounds active against viruses, such as ribavirin, have been used for transplant and other immuno-compromised patients. Determination of viral loads in patient sera to monitor the efficiency of treatment will likely become more common in the future.

Current meat inspection does not allow detection of HEV in pigs at slaughterhouses. Antibody, antigen or genome detection assays have been developed for research studies. In the literature, antibodies against HEV have been measured from pig sera and nucleic acid has been detected from faeces, sera and tissues using reverse transcription-PCR. Vaccine research to achieve protective immunity against HEV in pigs is continuing.

HEV in pig production (farms) and at slaughterhouses   HEV seropositivity of pig herds ranges globally from 46% in Laos to 100% in the United States. HEV affects pigs usually when they are at the production farm. Pig generations at one farm are usually infected with the same HEV strain, although the simultaneous presence of multiple strains has also been described. During infection pigs have been estimated to have a theoretical possibility to infect more than eight other pigs without earlier HEV infections. It is assumed that, ultimately, the environment in farms/fattening farms gets contaminated with HEV, which consequently leads to further transmission of HEV.

Differences in the frequency of HEV RNA positivity of pigs taken to slaughter are reported between countries. Finnish studies show that most pigs become infected while young, and about 90% have anti-HEV antibodies, but usually not acute infections, at slaughter. In Italy and Spain, however, HEV RNA prevalence of 27% in porcine faeces, and 4% and 3% in livers and meat was detected in slaughterhouse samples.

Workers at the slaughterhouse are exposed to HEV while handling pig carcasses (Figure 13.4). HEV replication has been shown to take place in the liver, colon, lymph nodes and spleen. HEV RNA has been found in the stomach, small intestine, spleen, kidney, salivary glands, tonsils and lungs. During viraemia HEV RNA has been demonstrated in muscles in addition to blood. The rinsing water as well as the remnants of the slaughtered pigs must be considered as infectious. In a European multinational study, HEV RNA was found in the slaughterhouse in swabs taken from knives, the floor and conveyor belts as well as on worker's hands, gloves and aprons.

HEV survives well on or in meat when stored cold, which means that the cold chain favours the survival of HEV. Pig liver sausages (Figatellu) containing unheated liver have caused food-borne HEV outbreaks when used for consumption as such. Pork livers sampled at local shops have been shown to contain HEV; in Japan 2% and in the United States 11% of pig livers contained HEV RNA. HEV can withstand heating for 60 minutes at 56°C, but will become inactivated, when boiled or heated for five minutes at 191°C.

13.3.2   Influenza A viruses

Influenza A viruses are among the most varying agents that are spread throughout the animal kingdom. Apparently, every mammalian species carries influenza A virus of its own. Due to the special genome divided into eight RNA segments corresponding to the viral proteins, reassortants can be created whenever two different influenza A viruses infect the same cell. This situation can easily be achieved in passerine bird colonies that function as a reservoir and harbour all known influenza A viruses. The classification is based on two genome segments corresponding to the viral surface proteins H (haemagglutinin) and N (neuraminidase). In birds, these viruses are enteric, in contrast to them being respiratory in mammals. In birds, they show a very low pathogenicity, except for a few virus strains showing particular genome segment combinations.

The role of influenza A viruses as food-borne pathogens became obvious when, in 1996, an outbreak occurred in Hong Kong. Even though only a small number of people contracted the influenza A infection, the mortality was a shocking 60%. The virus was identified as avian (duck) type H5N1 and was defined as highly pathogenic also for birds. Later, H7N7, and more recently, H7N9, have shown similar properties. Human cases by HPAI were registered in birds and man in several locations globally and led to a massive culling, with millions of poultry eradicated to stop the virus (more than 250 million birds globally between the years 2004 and 2007). The economic losses have been massive.

The viruses have been detected on the surface of imported frozen ducks from Asia and also on and in eggs. No human cases have, however, been encountered in connection to food. This is most likely due to the human respiratory tract possessing receptors for these viruses only in the lower part, that is in the alveoli of the lungs. The infection is initiated only when aerosols rich in the virus are inspired deep into the lungs. No human-to-human infections have been registered but the situation warrants alertness, since influenza viruses, being single-stranded RNA viruses, constantly undergo genetic modulations through mutations.

13.3.3 Other viruses

Arthropod-borne viruses infect mammals via blood-sucking insects directly in the blood. From the infected animals, they are spread further only via blood or milk. The latter infection route has been documented in a rare case of tick-borne encephalitis (TBE) virus. Although a slaughtered viremic animal infected with arboviruses carries the virus in organs and muscles, there are no reports of human infections mediated by contaminated meat.

There are numerous, mostly enteric viruses of variable characteristics that can possibly be disseminated through meat products. Associated risks comprise their direct pathogenicity to man (HAV, noro-, rota-, adeno-, astro-, entero-, parvo- and coronaviruses) and the risk of genome recombination (norovirus) or reassortment (rotavirus), with the emergence of new, potentially more pathogenic virus variants.

Porcine rota-, noro- and sapoviruses closely related to human viruses have been reported to exist. However, they usually belong to other genotypes or at least possess different nucleic acid sequences than the corresponding human viruses. In a Canadian study, the norovirus genome, closely related to the human virus genome, was found in pigs and retail pig meat but transmission of norovirus to humans through the consumption of pork has not been shown. There are numerous reports of other closely related animal and human enteric viruses or viruses which contain both human and animal genetic material, but there are no reports about their association especially with meat consumption.

13.3.4 Literature and further reading

Chmielewski, R. and Swayne, D.E. 2011. Avian influenza: Public health and food safety concerns. *Annu Rev Food Sci Technol*, **2**, 37–57.

Di Bartolo, I., Diez-Valcarce, M., Vasickova, P. *et al.* 2012. Hepatitis E virus in pork production chain in Czech Republic, Italy and Spain, 2010. *Emerg Infect Dis*, **18**, 1282–1289.

Kamar, N., Bendall, R., Legrand-Abravanel, F. *et al.* 2012. Hepatitis E. *Lancet*, **379**, 2477–2488.

Kantala, T., Oristo, S., Heinonen, M. *et al*. 2013. A longitudinal study revealing hepatitis E virus infection and transmission at a swine test station. *Res Vet Sci*, **95**, 1255–1261.

Meng, X.J. 2011. From barnyard to food table: the omnipresence of hepatitis E virus and risk for zoonotic infection and food safety. *Virus Res*, **161**, 23–30.

Pavio, N., Meng, X.J. and Renou, C. 2010. Zoonotic hepatitis E: animal reservoirs and emerging risks. *Vet Res*, **41**, 46–66.

13.4   Parasites

Maria Fredriksson-Ahomaa[1] and Anu Näreaho[2]

[1]*Department of Food Hygiene and Environmental Health,*
Faculty of Veterinary Medicine, University of Helsinki, Helsinki, Finland
[2]*Department of Veterinary Biosciences, Faculty of Veterinary Medicine,*
University of Helsinki, Helsinki, Finland

Parasites are organisms that are living at the expense of other organisms (hosts). Their life cycles vary; some need only one host and others go through different developmental phases using different animal or human hosts. The parasites are of different types and range in size from single celled, microscopic organisms (protozoa) to large, even metres long, multicellular worms (helminthes). In mammal muscles, both protozoan and helminthes can be found. Zoonotic parasites can be transmitted from animals to humans through contaminated meat. Control of parasitic diseases is complicated by the often prolonged incubation periods, mild symptoms and unrecognized, chronic sequel. Sometimes the diagnostic methods are not sensitive or practical enough.

Toxoplasma gondii and *Sarcocystis* spp. are important meat-borne zoonotic protozoan that can infect humans. Among helminthes, *Trichinella* spp. nematodes and certain *Taenia* spp. cestodes can be transmitted to humans through raw or undercooked meat and meat products. *Echinococcus* spp. cestodes are important zoonotic parasites found in production animals; however, they are not transmitted to humans via meat. The public health and economic impacts of meat-borne parasitic zoonoses can be considerable, due to high morbidity and even mortality in man and production animals. In addition, condemnation of parasitized meat is still a significant problem in some countries. More effective monitoring of zoonotic parasites in livestock, wildlife and foods is needed to find the transmission routes along the food chain in order to develop effective prevention and control strategies. Thus, sensitive and accurate tools for diagnostic and control of infections are essential. The most important zoonotic meat-borne parasites found in pigs, cattle and sheep are listed in Table 13.10. Zoonotic meat-borne parasites in poultry are rare. *Toxoplasma* has been detected on poultry meat but the public health significance is still unknown.

Table 13.10 The most important meat-borne zoonotic parasites.

Parasite		Disease in humans	Animal reservoirs		
			Pigs	Cattle	Sheep/goats
Protozoa					
	Toxoplasma gondii	Toxoplasmosis	x	x	x
	Sarcocystis hominis	Sarcosporidiosis		x	
	Sarcocystis suihominis	Sarcosporidiosis	x		
Helminthes					
Nematode	*Trichinella spp.*	Trichinellosis	x		
Cestode	*Taenia solium*	Taeniosis	x		
		Cysticercosis	x		
	Taenia saginata	Taeniosis		x	

13.4.1 *Toxoplasma gondii*

Toxoplasma gondii is a unicellular protozoan parasite that is ubiquitous through the world. It causes disease called toxoplasmosis. *T. gondii* has two types of hosts: definitive hosts and intermediate hosts. The definite hosts are domestic cats and other members of the family *Felidae*. Intermediate hosts are many warm-blooded animals, including humans and birds. The life-cycle takes place in two stages: the sexual stage in cats only and the asexual stage in all hosts, also in cats. *T. gondii* occurs in three forms: tachyzoites, bradyzoites and sporozoites; all can cause an infection. Tachyzoites are typical for acute phase of infection; they multiply rapidly and spread over the body. Bradyzoites, which are typical for chronical phase of infection, are located in tissue cysts (clusters of slowly multiplying bradyzoites) and are the source of meat-borne infections. Sporozoites, which are formed within the oocysts, are the end products of the sexual production (Figure 13.5). Oocysts, which are produced in the definitive host (in the *Felidea*), are shed to the environment in faeces where they sporulate in some days (2–5 days), after which the sporozoites are infective to the next host.

Toxoplasmosis is a worldwide infection and it has been estimated that at least one-third of the world's population is exposed to *T. gondii*. However, toxoplasmosis is still an underreported disease. The infection is usually relatively harmless in immune-competent persons. Some may develop 'flu-like' symptoms, such as swollen lymph glands, fever and/or muscle aches. The recovery happens usually without medical treatment in one week to one month after consuming the cysts. However, toxoplasmosis is a serious problem for pregnant women, who can pass the infection to the foetus through placenta. The damage to the foetus can be fatal leading to miscarriage or still birth. *T. gondii* infection can also be a problem in immune-compromised persons, who may develop severe toxoplasmosis, which can even lead to death. Serologic and polymerase chain reaction (PCR) assays are widely used to diagnose congenital infection, encephalitis and ocular toxoplasmosis.

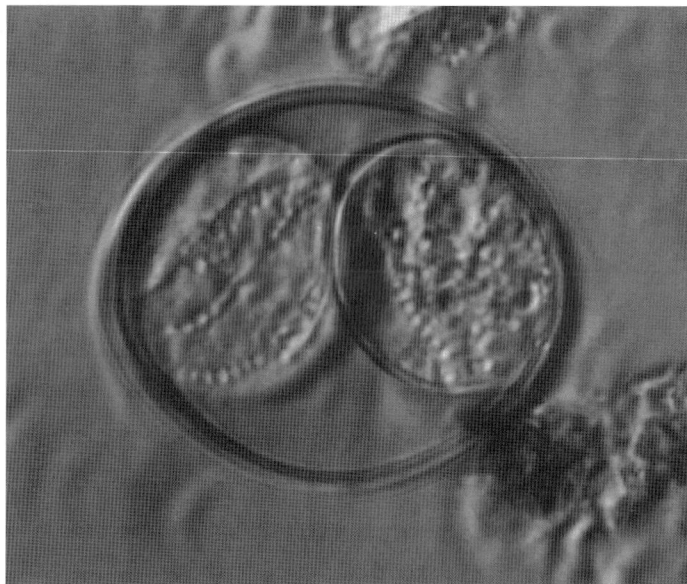

Figure 13.5   A sporulated toxoplasma oocyst (the size is 10–15 μm × 8–12 μm) isolated from cat faeces. Source: Courtesy of Sven Nikande.

T. gondii has a complex life cycle and multiple infection routes are possible. The parasite can be transmitted to humans via undercooked meat containing infective tissue cysts (bradyzoites) and by the oocysts (sporozoites) from the environment shed in the faeces of cats. Also, water contaminated with infective oocysts can be an infection source or transmission vehicle for humans. Consumption of raw or undercooked pork, beef and lamb, especially, has been associated with an increased risk of toxoplasmosis. Meat-borne infections can be prevented through cooking the meat thoroughly, which effectively kills the *T. gondii* tissue cysts. Thus, the risk of toxoplasmosis due to meat, such as poultry, which is usually consumed well cooked, is considered to be low. Raw or undercooked lamb, a delicacy in some countries, is an important source of *Toxoplasma* infection. Outbreaks of acute toxoplasmosis have also been described following the consumption of undercooked pork.

The clinical signs of *T. gondii* infection in animals are usually mild and difficult to recognize; however, noticeable economic losses have been caused to sheep and goats due to abortion and still birth. *Toxoplasma* prevalence in livestock is expected to rise due to an increase in outdoor rearing of animals, which are more likely to come in contact with the sporulated oocysts. Herbivores (ruminants) most likely get *Toxoplasma* infection via pasture, hay, feed or surface water contaminated with sporulated oocysts shed by infected cats. Omnivores (pigs and poultry) can also become infected via uptake of meat carrying *T. gondii* cysts, for example infected rodents. Modern farm production systems, where animals are kept inside and the food is almost sterile, have nearly eliminated *T. gondii* infections in pigs. The prevalence of *T. gondii* cysts in the meat of free-range chickens has shown to be high compared to meat from chickens from commercial indoor farms. Intervention strategies on farms like adequate cat and rodent control should be enforced to reduce the access of *Toxoplasma* into the food chain. Effective post-harvest prevention methods are thorough freezing or heating of the meat.

Toxoplasma cannot be detected macroscopically during meat inspection because the tissue cysts are invisible to naked eye. The testing methods are based on direct detection of *T. gondii* organisms or DNA in tissue or on indirect detection of the infection by demonstrating specific antibodies in serum or meat juice. Currently used molecular and histological methods are rather insensitive to detect *T. gondii* in meat because the density of these parasites is usually low and may vary from site to site. The indirect methods have their disadvantages, too, giving false-negative results at the early stage of the infection before the antibodies appear. More sensitive, robust and reproducible detection methods are needed for efficient monitoring of *Toxoplasma*.

13.4.2 *Sarcocystis* spp.

Sarcocystis spp. are intracellular protozoan parasites with an obligatory two-host life cycle that includes, typically, herbivores as intermediate hosts and carnivores as definitive hosts. *Sarcocystis* forms sexual stages in the definitive host's intestine and, thereafter, infective sporulated oocysts and sporocysts are excreted in the faeces. Intermediate hosts ingest infective sporocysts faecal-orally by contaminated feed or water. Sporozoites from sporocysts invade the intestinal mucosa and the asexual cycle begins in the intermediate hosts. Finally, sarcocysts will be formed within the muscle tissue. Slowly dividing banana-shaped bradyzoites in the sarcocysts are infective for definitive hosts. There are several species of *Sarcocystis* with different hosts. Humans and some primates are definitive hosts for *Sarcocystis hominis* and *Sarcocystis suihominis* and they may get infected after eating tissue cysts (sarcocysts with infective bradyzoites) in raw or undercooked meat from cattle and pigs, respectively.

Sarcosporidiosis (also called sarcystosis) is a disease that occurs worldwide in humans. Eating raw or undercooked beef and pork containing sarcocysts can result in intestinal sarcosporidiosis in humans. *Sarcocystis* may also cause muscular infection in humans. In such cases, humans harbour the sarcocyst stage and, therefore, act as an intermediate host. Human infections are considered rare but it is also possible that they are underdiagnosed, because they are usually asymptomatic. However, symptoms such as nausea, anorexia, diarrhoea and abdominal pain may occur in the intestinal infection and more severe symptoms, such as musculoskeletal pain, fever, rash, cardiomyopathy, bronchospasm and subcutaneous swelling, in the muscular infection. The symptoms are usually transient but in muscular infection they can last even for several years. Diagnosis of human intestinal sarcosporidiosis is based on symptoms and a history of recently having been eating raw or undercooked meat, and detection of oocysts in faeces. Sarcocysts can sometimes be found in muscle sample, especially in skeletal and cardiac muscle, by microscopic examination. Sarcosporidiosis can be prevented by thoroughly cooking or freezing meat to kill bradyzoites in the sarcocysts.

Sarcosporidiosis can be detected in meat by direct observation of macroscopic sarcocysts or microscopic examination of histological sections. Sarcocysts of *S. hominis* are usually microscopic in the muscles of cattle while sarcocysts of *S. suihominis* are mostly macroscopic in the muscles of pigs. The cysts are usually difficult to recognize in meat inspection. When the cysts are visible macroscopically they are seen most frequently in the muscles of abdomen and diaphragm but they may be distributed to heart and skeletal musculature. Mature sows are most commonly affected. Severe cases are comparatively rare, though occasionally carcasses have been condemned due to large numbers of cysts rendering the meat objectionable for food purpose. Eosinophilic myositis, observed as a blue-green tint on the surface of a fresh animal carcass, can be associated with *Sarcocystis* infection. However, sometimes infections occur without eosinophilia. To avoid infection of food animals, they must be prevented from ingesting water, feed and bedding contaminated with oocysts from carnivores.

13.4.3   *Trichinella* spp.

Nematodes (roundworms) in the genus *Trichinella* are characterized by a wide host range and geographical distribution. Among the helminthes, *Trichinella* has a unique life cycle since it completes the whole life cycle in the same host: (i) the enteral phase, (ii) the migratory phase and (iii) the muscle phase of infection. In the enteral phase of infection, the parasite establishes intracellular infection in the enterocytes and reproduces sexually; in the migratory phase, the new-born larvae are widely distributed in muscle tissue by the circulation; and in the muscle phase, the larvae are finally settled in the skeletal striated muscle cells. *Trichinella* spp. have been detected in domestic animals (including meat producing animals, mainly pigs and horses) and wild animals worldwide. Humans may become infected by ingesting larvae, which occur in the striated muscle cells of the infected animals. During digestion, the *Trichinella* larvae are released into the stomach and, subsequently, penetrate the mucosa of the small intestine, where they develop into adult worms. Adult female worms produce larvae starting one week after infection and continuing for up to 4–6 weeks. The larvae penetrate the intestinal epithelium, spread via the blood vessels into the striated muscles, and remain viable there for years.

There are several *Trichinella* species and genotypes infecting a wide range of host species (mammals, birds and reptiles) including humans. All of them are considered pathogenic for humans. The genus *Trichinella* is composed of species in which the host muscle cells they invade become surrounded by a collagen capsule and those species in which no encapsulation occurs. *Trichinella spiralis*, which is encapsulated in the muscle, is considered the most important species in causing human trichinellosis worldwide. In most cases, the larvae of *Trichinella* can be destroyed by freezing, cooking and some curing procedures.

However, certain *Trichinella* species and genotypes (*T. nativa* and T6 genotype) can tolerate freezing, and thus freezing of the meat alone is not recommended for the possible host animals of these species.

Trichinellosis continues to be a public health concern throughout the world. Human trichinellosis may occur even in the countries with negligible risk of *Trichinella* infection due to imported cases following tourism and globalization of trade. The main sources of human trichinellosis are raw or improperly processed pork and pork products, game meat and horse meat. The infection in humans can be divided into two phases: an intestinal (enteral) and a muscular (parenteral) phase. The symptoms correlate with the stage of infection. The enteral phase, which starts within 1–2 days after eating contaminated meat, includes nausea, mild diarrhoea, abdominal pain, vomiting and low-grade fever. These unspecific symptoms are similar to many enteral diseases and are, therefore, easily misdiagnosed. Two to six weeks after infection, the signs due to the parenteral phase (muscular infection) appear. These symptoms include diffuse myalgia, a paralysis-like state, fever, headache, skin rash and conjunctivitis. The duration of the incubation period and the severity of the disease are related to the number of larvae ingested. Host immunity, age and general health are also important factors in the outcome of the disease. Most infections are subclinical or have minor symptoms and do not require any treatment. Sometimes the symptoms are intense and last for months. In these cases, medical treatment is needed. Trichinellosis in humans can be prevented by *Trichinella* testing of the possible hosts (mostly pigs, horses and wild boars) during meat inspection and by avoiding the consumption of raw or undercooked meat, especially pork and pork products but also horse meat and wild game.

The number of human trichinellosis cases is underestimated in many countries due to lack of knowledge and appropriate diagnostic methods. Clinical diagnosis of trichinellosis is difficult because there are no pathognomonic signs or symptoms. The detection of specific antibodies and muscle biopsy may confirm the diagnosis. Many techniques are used for detecting antibodies against *Trichinella* antigens. The most reliable and widely used techniques for diagnosing human trichinellosis are enzyme-linked immunosorbent assays (ELISA) and immunoblotting (western blot). Among commercially available ELISA kits, only those showing no cross-reactions with other nematode antigens should be used. However, usually *Trichinella* antibodies cannot be detected at the onset of clinical signs.

Detection of *Trichinella* spp. larvae in the muscles of host animals, especially in diaphragm, tongue and masseter, is possible by laboratory testing. The sensitivity of the laboratory tests depends on the amount of muscle sample tested. The current testing for the detection of larvae of *Trichinella* spp. is based on the isolation of the larvae by artificial digestion and microscopic identification (Figure 13.6). Trichinelloscopy, which is done by pressing small pieces of meat between glass plates and then using microscope for detection, is not a sensitive method, and thereby it is not allowed in the official meat inspection any more. The species identification of the larvae is done by PCR. In addition to *Trichinella* testing during meat inspection, sylvatic indicator animals should be

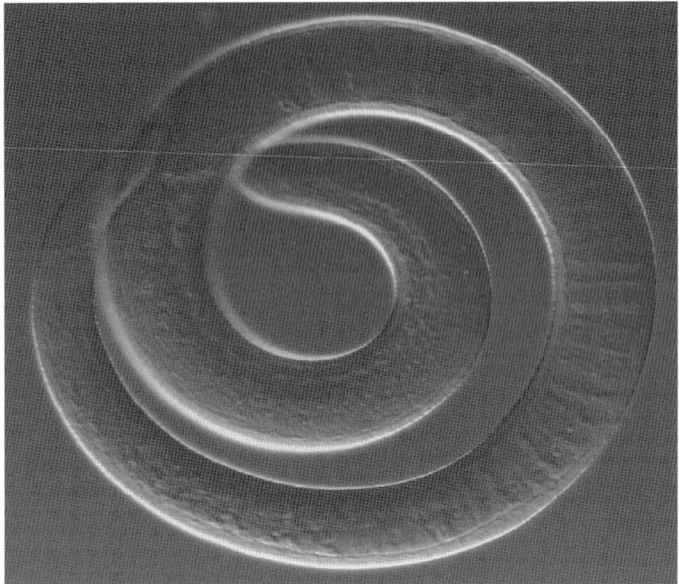

Figure 13.6 *Trichinella* larva (length about one millimetre) freed from the muscle by artificial digestion. Picture has been taken with differential interference contrast (DIC) microscopy. Source: Courtesy of Seppo Saari. (For colour details, please see solour plate section).

monitored for *Trichinella* to follow the infection pressure and to understand the local epidemiology of this parasite.

13.4.4   *Taenia* spp.

Taenia is a genus of cestode (tapeworm) that includes numerous species. The two most important species, which cause infection in humans, are *Taenia solium* (the pork tape worm) and *Taenia saginata* (the beef tapeworm). Humans serve as definitive hosts for these two species, which have a quite similar life cycle. The adult tapeworms, which can be several metres long, live in the small intestine of humans. Proglottids, which contain eggs, are excreted with the faeces. The eggs may remain viable in the environment for months. The pigs become infected by ingesting eggs from *T. solium* and cattle by eating eggs from *T. saginata*. The eggs hatch in the small intestine and the released larvae migrate through the intestine and then spread through the blood stream to skeletal muscles and the heart forming metacestodes and cause cysticercosis, a tissue infection (other than intestine) in the intermediate host. The metacestodes in cattle muscles are called *Cysticercus bovis* and in pig muscles *Cysticercus cellulosae*. Whereas *C. cellulosae* is reported very infrequently in industrialized countries, *C. bovis* has a universal distribution and is common in both industrialized and developing countries.

Taeniasis is the name of the intestinal infection with adult-stage tapeworms. Humans get taeniasis by consuming raw or undercooked beef or pork containing metacestode forms of *T. saginata* or *T. solium*, respectively. The available data on taeniasis in humans are poor. The taeniasis in humans is more common in countries where sanitation practices are substandard and in the areas where pork and beef are consumed raw or undercooked. Most intestinal human cases are without symptoms. Some persons may experience abdominal pain, weight loss, digestive disturbance and possible intestinal obstruction. Irritation of the peri-anal area caused by worms or worms' segments exiting the anus can occur. Taeniasis may last many years without medical treatment. Infections with *Taenia* are diagnosed by recovering eggs or proglottids in the faeces. The eggs of *T. saginata* and *T. solium* are similar but these species can be differentiated based on the morphology of their proglottids and scolex or by PCR.

T. solium can also sometimes cause cysticercosis in humans (the definitive host). Cysticercosis due to *T. solium* typically occurs in pigs (the intermediate host). Humans can get cysticercosis by consuming food or water contaminated with the eggs of *T. solium* or as an autoinfection. Worm eggs hatch in the intestine and the larvae then migrate to various parts of the body and form cysticerci. This can be a serious or fatal disease if it involves organs such as the central nervous system, heart or eyes. Symptoms may vary depending on the organ or organ system involved. Cysticerci in the brain cause neurocysticercosis, one cause of acquired epilepsy. Death is common. Symptoms usually appear from several weeks to several years after becoming infected with the eggs of *T. solium*. Symptoms may last for many years if medical treatment is not received.

T. solium has been eradicated in most industrialized countries as a result of improved sanitation, modern pig production and meat inspection. It is still common in developing countries if pigs have access to human faeces. Control of *T. solium* is based on improvement of sanitation and pig husbandry systems, education and meat inspection. *T. saginata* is found in industrialized countries as well as in developing countries. The ruminants are infected by eating eggs while grazing. Infection of cattle appears to be associated with the effluent from sewage treatment plants, the free access of cattle to surface water and flooding of pastures. Appropriate measures, such as education and treatment of people harbouring *T. saginata*, should be taken to avoid infection of cattle due to exposure to faeces from human carriers. However, it has proved to be difficult to eradicated *T. saginata* due to difficulties detecting animals that are slightly infected and the global habit to consume raw or semi-cooked beef.

Animals usually do not show clinical signs of cysticercosis; infection is usually only found when the meat is inspected. The cysts, which contain the scolex of the worm, are transparent and filled with fluid. They are usually easy to see in pigs (they are about 1 cm in diameter) but they are often smaller and more difficult to detect in cattle (they may be only 2–3 mm in diameter). The cysts can occur in degenerate or calcified stages which have a gaseous or calcified appearance. Cysticerci in pigs (*C. cellulosae*) are found mainly in the brain, liver, heart and skeletal muscle. The predilection sites of *C. bovis* in cattle are the heart (about 15%) and masseter muscles (about 7%). The incisions of the masseter muscles and the heart in the *post-mortem* inspection can reveal the possible infection. The cysts may also be found in the tongue, diaphragm, intercostal and rump muscles, and in organs such as liver and lungs. However, cases with low infection levels or early infections are not diagnosed, which partly explains the persistence of *T. saginata* in industrialized countries. Typically, more than 20 cysts are needed for detection by meat inspection. The serological methods, such as ELISA and western blot, together with meat inspection would improve the efficiency of diagnosis. The cold treatment ($-10°C$ for >14 days or $-7°C$ for >3 weeks) kills the viable cysticerci.

13.4.5   *Echinococcus* spp.

Several species belonging to the genus *Echinococcus* occur worldwide. They are small tapeworms (cestodes) that have only few proglottids. *Echinococcus* requires two mammalian hosts for completion of their life cycle; a carnivore definitive host, in which the adult stage develops in the small intestine, and an intermediate host, in which the cysts develop. The adult tapeworm lives in the intestine of carnivores that excrete eggs in the faeces. The intermediate hosts of *Echinococcus* can be infected after ingestion of eggs, which are spread with definitive host's faeces. The eggs hatch in the digestive system of intermediate hosts and the larvae penetrate the intestinal epithelium and spread along circulation to liver, lungs, brain and other organ where they (metacestodes) form hydatid cysts. The cysts vary in size and they may occur in grape-like clusters.

Human echinococcosis (hydatidosis) is a zoonotic infection that continues to be a substantial cause of morbidity and mortality in many parts of the world, including parts of Europe, North and South America. The infection can be obtained by ingestion of the eggs of the *Echinococcus*, shed in the faeces of carnivores. Humans are not natural hosts of this parasite but in rare cases they can be exposed to the eggs of the tapeworm after close contact with an infected dog or its environment. They can also be infected after eating contaminated food (wild berries, vegetables) or drinking contaminated water. Human echinococcosis occurs usually as two forms: (i) cystic hydatid disease and (ii) alveolar hydatid disease caused by *Echinococcus granulosus* complex and *Echinococcus multilocularis*, respectively. The human cystic echinococcosis due to *E. granulosus* complex infection is the most frequent form globally. *E. granulosus* complex is a complex of rather recently discovered species and genotypes, the most important associated with human cystic echinococcosis being common in sheep. Cysts usually develop in the liver or lungs where they can persist and grow for years. Usually the cysts do not induce symptoms until they have reached a particular size. Sudden onset of symptoms may be due to cyst rupture.

Cystic echinococcosis also constitutes a serious animal health and economic concern in many areas of the world, especially in sheep. In production animals, the *Echinococcus* cysts are usually observed in the liver and occasionally in the lungs. There is high variation in the size of the cysts: in pigs and ruminants, they vary in size from that of a pea to that of a hen's egg. Tissue cysts are not a source of infection for humans but meat inspection of production animals is considered important to estimate the prevalence of echinococcosis in livestock and to prevent infections in carnivores. The cysts in the intermediate hosts' organs are usually big and are, thereby, easily noticed in the meat inspection. The cysts are fluid-filled with concentrically calcified particles (hydatide sand) and protoscolices, preliminary heads of the worms.

E. granulosus complex is mainly transmitted in a cycle between dogs and other carnivores (definitive hosts) that harbour intestinal tapeworm and livestock after the latter ingest the microscopic eggs while grazing pastures that are contaminated with definitive hosts' faeces. The dogs usually acquire infection from hydatid-carrying livestock after eating contaminated offal (liver and lungs). However, the data on reported prevalence of echinococcosis in the dogs and other carnivores are scarce. The presence of carnivores in the enclosures of the meat producing animals can be, however, limited and feeding carnivores with offal should be restricted. Anthelmintic treatment and general hygienic measures (hand washing) are recommended as basic mitigation actions. Monitoring and reporting of echinococcosis in humans should also be improved.

13.4.6 Literature and further reading

Cardona, G.A. Carmena, D. 2013. A review of the global prevalence, molecular epidemiology and economics of cystic echinococcosis in production animals. *Vet Parasitol*, **192**, 10–32.

Dorny, P., Praet, N., Deckers, N. and Gabriel, S. 2009. Emerging food-borne parasites. *Vet Parasitol*, **163**, 196–206.

Fayer, R. 2007. *Sarcocystis* spp. in human infection. *Clin Microbiol Rev*, **17**, 894–902.

Gottstein, B., Pozio, E. and Nöckler, K. 2009. Epidemiology, diagnosis, treatment, and control of trichinellosis. *Clin Microbiol Rev*, **22**, 127–145.

Jones, J.L. and Dubey, J.P. 2012. Food-borne toxoplasmosis. *Clin Infect Dis*, **55**, 845–851.

Kiljstra, A. and Jongert, E. 2008. Control of the risk of human toxoplasmosis transmitted by meat. *Int J Parasitol*, **38**, 1359–1370.

Murrell, K.D. and Pozio, E. 2011. Worldwide occurrence and impact of human trichinellosis, 1998–2009. *Emerg Infect Dis*, **17**, 2194–2202.

Näreaho, A. 2009. Parasites. In: *Handbook of Muscles Food Analysis* (eds L.M.L Nollet and F. Toldra). CRC Press, Boca Raton, FL, pp. 648–662.

Robert-Gangneux, F. and Dardé, M-L. 2012. Epidemiology of and diagnostic strategies for Toxoplasmosis. *Clin Microbiol Rev*, **25**, 264–296.

13.5 Prions

Liisa Sihvonen

Department of Veterinary Biosciences, Faculty of Veterinary Medicine, University of Helsinki, Helsinki, Finland

Prions are 'infectious proteins' causing several neurodegenerative diseases: scrapie of sheep and goats, bovine spongiform encephalopathy (BSE), feline spongiform encephalopathy (FSE), transmissible mink encephalopathy (TME), chronic wasting disease of deer and elk (CWD) and human diseases: kuru, Creutzfeldt–Jakob disease (CJD), variant Creutzfeldt–Jakob disease (vCJD), Gerstmann–Sträussler–Scheinker syndrome and fatal familial insomnia. In each of these prion diseases, the characteristic lesion is spongiform degeneration in the grey matter of the brain, with hypertrophy and proliferation of astrocytes. The term 'transmissible spongiform encephalopathy' (TSE) is also used for these neurodegenerative diseases. The prototype of the prion diseases is scrapie, which was first described already in the fifteenth century in England. BSE ('mad cow disease') was first detected in 1986 in the United Kingdom and became established in cattle through recycling of rendered bovine meat-and-bone meal in the ruminant feed chain. A massive epizootic followed. Humans became infected with the BSE prion through exposure to cattle products. The causative association of the BSE prion and the human vCJD has been shown by epidemiological, pathological and molecular studies. Exclusion of high-risk bovine material from human food chain has decreased the number of human vCJD cases.

13.5.1   Properties of prions

Prions are normal cellular proteins (PrP^c) that have undergone conformational change as a result of post-translational processing of normal cellular protein and become pathogenic disease-specific forms (PrP^{sc}, PrP^{bse}). The normal PrP^c is encoded in the genome of mammals and expressed in many tissues, especially in neurons and lymphoreticular cells. The function of normal PrP^c is unclear. The amino acid sequence of normal PrP^c and the disease-specific protein (PrP^{sc}/PrP^{bse}) are identical in a given host. The conformation of disease-specific PrP is only changed from a structure made up predominantly of α helices to one made up predominantly of β sheets. Disease-specific forms of PrP are very resistant to many chemicals, physical conditions and environmental insults that would destroy any virus.

13.5.2   Human diseases

Kuru, Creutzfeldt–Jakob disease, Gerstmann–Sträussler–Scheinker syndrome and fatal familial insomnia are prion diseases that are manifested predominantly in middle-aged and older humans. Kuru was an acquired fatal neurological disease that occurred only in the Fore tribe in New Guinea, where ritualistic cannibalism was practiced. There are three forms of Creutzfeldt–Jakob disease: (i) sporadic (85%), (ii) inherited (15%) with coding mutations in the PrP gene and (iii) iatrogenic (a few hundred) acquired via contaminated instruments, hormones, corneal grafting. Gerstmann–Sträussler–Scheinker syndrome and fatal familial insomnia are very rare familial prion diseases caused by mutation in the PrP gene.

Variant CJD was first described in the United Kingdom in 1996 affecting unusually young people and having a highly consistent and unique clinicopathological pattern with prominent psychiatric symptoms. All studied patients were homozygotes for methione at polymorphic residue 129 of PrP and no coding mutations were present. vCJD has been linked with exposure to classical BSE. From October 1996 to March 2011, 175 cases of vCJD were reported in the United Kingdom, 25 in France, 5 in Spain, 4 in Ireland, 3 each in the Netherlands and the United States of America, 2 each in Canada, Italy and Portugal, and one each in Japan, Saudi Arabia and Taiwan (WHO). The number of cases of vCJD in the United Kingdom peaked in 2000 with 28 deaths. It has since declined to about two diagnosed cases and two deaths per year in 2008.

13.5.3 Bovine spongiform encephalopathy

Bovine spongiform encephalopathy, BSE ('mad cow disease') was first detected in cattle in England in 1986. The disease was detected in approximately 184 600 cattle in the United Kingdom (1986–2011) and approximately 6033 cattle in the rest of world. The epizootic, measured as numbers of cases reported and confirmed, peaked in 1992. BSE has been detected also in several ungulates kept in zoos and in one goat. The BSE strain has been isolated from cases of feline spongiform encephalopathy (FSE) and FSE is considered to be BSE in felines. The BSE epizootic was caused by contamination of meat-and-bone meal with TSE agent. A single classical BSE strain has been responsible for BSE epizootic and the origin of this food-borne strain is still uncertain. The disease might have originated from rendered carcasses of cattle with a sporadic spongiform encephalopathy, or sheep with scrapie. In 1996, vCJD was described and it is considered to be the human form of BSE. Exclusion of ruminant origin material from the ruminant food chain has resulted in a decrease in new cases of BSE, from 37 301 cases of BSE in cattle in 1992 to eight cases in cattle in 2011 in the United Kingdom. Evolution of the number of BSE positive cases in the European Union from 2001 to 2011 is shown in Figure 13.7.

BSE infectivity can be found in the nervous system, distal ileum and tonsils. BSE titres are relatively low or undetectable outside the central nervous system (CNS). There is no evidence of horizontal transmission of BSE. BSE may show clinically with apprehension, hyper-reactivity and ataxia. BSE cases can only be conclusively confirmed by the detection of PrP^{bse} in the CNS. Cases of BSE have been detected in most European countries and, for example, in North America and in some Asian countries.

Atypical BSE, distinct from the classical BSE, has been identified through the large scale testing for BSE after 2001. Two molecular signatures of atypical BSE prion, H-BSE and L-BSE (or BASE) have been detected. Their PrP^{bse} molecular signature differed from classical BSE in term of protease-resistant fragments size and glycopattern. The origin of the atypical

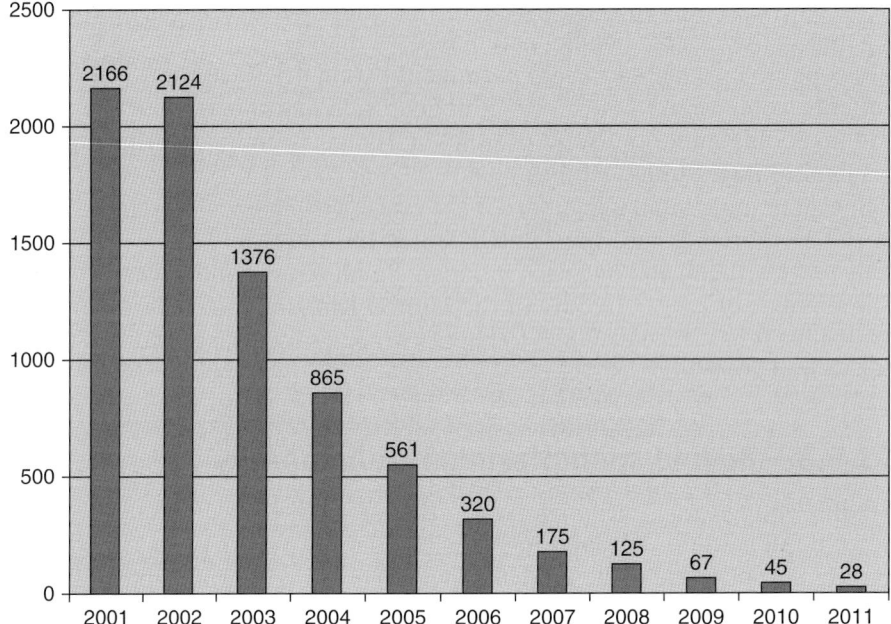

Figure 13.7 Evolution of the number of BSE positive cases in the 27 EU member states since 2001 (data from report on monitoring of ruminants for TSEs in the EU in 2011; European Union, 2012). Source: European Comission (EC). 2012.

BSE cases is unknown. These uncommon cases have been usually detected in aged asymptomatic cattle during systemic testing at slaughterhouse or in the fallen stock surveillance. The unusually old age of atypical cases and their apparent low prevalence in the population could suggest that these atypical BSE forms are arising spontaneously. Atypical BSE cases have been identified in many EU member states and also, for example, in Japan, Canada, the USA and Brazil.

13.5.4 Classical scrapie

Classical scrapie affects both sheep and goats and is widespread. There are multiple strains of scrapie. Many studies have demonstrated that, although classical scrapie is an infectious disease, natural variants of the prion gene are associated with relative resistance to disease and differences in incubation time. Especially valine or alanine at position 136 combined with arginine and glutamine at positions 154 and 171, respectively, are associated with susceptibility to classical scrapie. Of the five most commonly reported PrP alleles, $V_{136} R_{154} Q_{171}$ is generally associated with increased susceptibility, while $A_{136} R_{154} R_{171}$ gives reduced susceptibility for classical scrapie and others are in between. The five most common alleles give 15 different genotypes. Of these,

VRQ/VRQ is considered as the most susceptible to classical scrapie and ARR/ARR the least, with the others in between.

The incubation periods of classical scrapie might last for several years but it is usually 2–5 years. In the early clinical stage, the signs are often vague, such as a change of behaviour. As the disease progresses, the typical clinical signs of classical scrapie develop; such as signs related to the nervous system – pruritus, ataxia, hyperaesthesia and nibbling reflex; secondary signs like wool loss and unspecific signs like weight loss. All signs are not necessarily present in one animal.

The most common entry port for classical scrapie PrP^{sc} is considered to be the alimentary tract. It has been suggested that, following ingestion of PrP^{sc}, infection is initiated in gut lymphoid tissues and, thereafter, PrP^{sc} move to the CNS. PrP^{sc} has been detected in the placenta and placental material has been considered to have important role in the transmission of classical scrapie. The animals might acquire classical scrapie either by eating the placenta, swallowing amniotic fluids or by ingesting material contaminated by these. Sheep milk from clinically affected animals has also been demonstrated to transmit the disease. The occurrence of PrP^{sc} in the lymphoid tissues and other peripheral tissues seems to be dependent on the PrP genotype and has been reported for the most susceptible PrP genotypes.

Diagnosis of classical scrapie has been based on clinical signs, flock history and histopathologic examination of the brain of animal showing clinical signs. Detection of the presence of PrP^{sc} in the CNS is shown to be much more sensitive than histopathologic examination. There are no epidemiological data to suggest that classical scrapie is zoonotic.

13.5.5 Atypical scrapie

In 1998, an atypical PrP^{sc} (Nor98) signature was identified in Norwegian sheep: the partially protein kinase resistant PrP displayed a multiband pattern in Western Blot that contrasted with those normally observed in classical scrapie. Lesions and PrP^{sc} accumulation in atypical scrapie occur in the cerebellum, rather than in the dorsal motor nucleus of the vagus nerve in the medulla as occurs in classical scrapie. Since the implementation in 2002 of an active TSE surveillance, atypical scrapie cases were identified in most EU member states and also, for example, in Canada, the USA and New Zealand. Atypical scrapie cases in goats have also been detected in many EU countries. Atypical scrapie might have been present in small ruminant population for several decades and is spread worldwide. The PrP genetic sensitivity to atypical scrapie is different from classical scrapie. A clearly increased risk for developing atypical scrapie is associated with ARQ and AHQ alleles. Strikingly ARR allele carriers (both homozygous and heterozygous) can develop the disease.

Atypical scrapie isolates appear alike or closely related, suggesting that only one unique TSE strain might be involved in atypical scrapie. The capacity of atypical scrapie cases to transmit disease to other sheep under field conditions is low and possibly nil. In most of the atypical scrapie flocks, only a single case is detected. Atypical PrP^{sc} has not been detected in peripheral tissues but is restricted to the CNS. Both natural and experimental atypical scrapie cases seem to indicate that infectivity levels can be present in skeletal muscle, peripheral nerves and lymphoid tissues of animals incubating or affected with atypical scrapie. Atypical scrapie could be a spontaneous disorder of PrP folding and metabolism, occurring in aged animals without external cause.

Most atypical scrapie cases have been detected through the active surveillance programmes of slaughtered healthy and found-dead animals and only a few animals have been detected as scrapie suspects. Clinical signs most commonly reported have been behavioural changes, emaciation, ataxia and/or circling. Diagnosis of atypical scrapie is based on the detection of PrP^{sc} in the CNS, especially in cerebellum.

13.5.6 Surveillance in animals

The accumulation of disease-specific forms of PrP in the CNS of *post-mortem* animals is used for routine diagnosis of prion diseases. Detection of disease-specific forms of PrP can be done by, for example, immunohistochemical methods, Western immunoblot methods or commercially available rapid test methods. Overall diagnostic sensitivity is influenced by the accuracy of sampling. The passive surveillance consists of testing animals with clinical signs suspicious of a prion disease. After 2001 in cattle and 2002 in small ruminants, a European Union wide active surveillance has been implemented and revealed the existence of unknown prion forms as atypical scrapie in sheep and atypical BSE in cattle. These atypical prion forms have now been identified also in countries outside the European Union.

The intensity of surveillance is variable, as different sampling schemes are applied internationally. Animals are identified within different risk streams, for example as healthy animals at slaughter house or fallen stock, and are submitted to the rapid tests for presence of pathogenic PrP in the brainstem. *Post-mortem* testing can involve a proportion of animals according to a sample-based scheme or all animals above a certain age.

13.5.7 Literature and further reading

Anderson, R.M., Donnelly C.A., Ferguson N.M. *et al.* 1996. Transmission dynamics and epidemiology of BSE in British cattle. *Nature*, **382**, 779–788.

Collinge, J., Sidle, K.C., Meads, J. *et al.* 1996. Molecular analysis of prion strain variation and the aetiology of 'new variant' CJD. *Nature*, **383**, 685–690.

Dewiler, L.A. 1992. Scrapie. *Rev – Off Int Epizoot*, **11**, 491–537.

EFSA (European Food Safety Authority). 2010. Scientific Opinion on BSE/TSE infectivity in small ruminant tissues. *EFSA J*, **8**, 11.

EFSA/ECDC (European Food Safety Authority/European Centre for Disease Prevention and Control), 2011. Joint Scientific Opinion on any possible epidemiological or molecular association between TSEs in animals and humans. *EFSA J*, **9**, 1.

European Comission (EC). 2012. Report on monitoring of ruminants for the presence of transmissible spongiform encephalopathies (TSEs) in the EU in 2011. Retrieved from http://ec.europa.eu/food/food/biosafety/tse_bse/monitoring _annual_reports_en.htm (last accessed 8 February 2014).

Hautaniemi, M., Tapiovaara, H., Korpenfelt, S.L. and Sihvonen, L. 2012. Genotyping and surveillance for scrapie in Finnish sheep. *BMC Vet Res*, **8**, 122.

OIE (World Organisation for Animal Health). 2008. Bovine spongiform encephalopathy. OIE Terrestrial Manual, pp. 671–682. http://www.oie.int/fileadmin/Home /eng/Animal_Health_in_the_World/docs/pdf/2.04.06_BSE.pdf (last accessed 8 February 2014).

OIE (World Organisation for Animal Health). 2012. Scrapie. OIE Animal Health Code, Chapter 14.9. http://www.oie.int/fileadmin/Home/eng/Health_standards /tahc/2010/chapitre_1.14.9.pdf (last accessed 8 February 2014).

Prusiner, S.B. 1998. *Prions. Proc Natl Acad Sci USA*. **95**, 13363–13383.

WHO (World Health Organization), 2012. Variant Creutzfeldt–Jakob disease. http:// www.who.int/mediacentre/factsheets/fs180/en/ (last accessed 8 February 2014).

13.6   Antimicrobial resistance in meat-borne bacteria

Maria Fredriksson-Ahomaa and Annamari Heikinheimo

Department of Food Hygiene and Environmental Health,
Faculty of Veterinary Medicine, University of Helsinki, Helsinki, Finland

Antimicrobial agents (also called antibiotics) are essential for the prevention, control and treatment of bacterial infections in humans and animals. They have dramatically reduced the number of death among humans and animals since their introduction. Antimicrobial resistance (AMR) is resistance of a bacterium to an antimicrobial agent to which it was originally sensitive. The evolution of resistant strains is a natural phenomenon which happens when bacteria are exposed to antimicrobial agents (drugs). However, AMR is a growing public health threat; millions of people fall ill from antimicrobial-resistant bacteria every year. Many bacteria have become resistant to antimicrobials due to overuse, inappropriate use or misuse of them. Critically important antimicrobials for human medicine, which are listed by World Health Organization (WHO), include third and fourth generation cephalosporins and fluoroquinolones. The effectiveness of these antimicrobials for human medicine should not be compromised by overuse, inappropriate use or misuse.

Human infections caused by resistant *Escherichia coli* strains are becoming increasingly common worldwide and they cause a serious health problem for human medicine. Resistance to third generation cephalosporins and fluoroquinolones in *E. coli* has increased consistently over recent years in

Table 13.11 Important antimicrobials for humans which have also been used for meat-producing animals.

Antimicrobial class	Antimicrobial used for	
	Humans	Animals
Fluorochinolones	Ciprofloxacin	Enrofloxacin
	• treatment of severe salmonellosis and campylobacteriosis	• treatment of respiratory and alimentary tract infections in pigs and poultry
Third generation cephalosporins	Cefotaxime, ceftriaxone	Ceftiofur
	• treatment of severe salmonellosis in children	• treatment of bacterial infections in pigs and cattle
Streptogramins	Quinupristin-dalfopristin	Virginiamycin
	• treatment of severe infections caused by gram-positive cocci (including VRE and MRSA)	• growth promoter • banned in EU from1999
Glycopeptides	Vancomycin	Avoparcin
	• treatment of severe *Staphylococcus* infections (including MRSA)	• growth promoter • banned in EU from 1997
Macrolides	Erythromycin	Spiramycin, tylocin
	• treatment of people who are allergic to penicillins • treatment of campylobacteriosis	• banned as growth promoter in EU from 1999 • treatment of infections

Europe. An increasing resistance to antimicrobials used to treat *Salmonella* and *Campylobacter* infections, the two most reported zoonotic infections in Europe, is also a big problem. Ciprofloxacin and third generation cephalosporins are important in the treatment of severe salmonellosis, and ciprofloxacin and erythromycin are important for treating severe campylobacteriosis in humans (Table 13.11). Especially alarming are the increasing numbers of extended spectrum β-lactamase (ESBL)-producing bacteria and methicillin-resistant *Staphylococcus aureus* (MRSA) that have acquired resistance to multiple antimicrobials.

Plasmid-mediated β-lactamases, including ESBLs and AmpC, have emerged among gram-negative bacteria resulting in increased resistance to broad-spectrum β-lactamase antimicrobials. ESBLs have mostly been detected in bacteria belonging to *Enterobacteriaceae* especially in *E. coli*, *Klebsiella pneumonia* and *Salmonella*. ESBLs cause resistance to penicillins, and second, third and fourth generation cephalosporins. ESBL-producing bacteria are frequently multiresistant, showing resistance also to other antimicrobial classes such as fluoroquinolones, aminoglycosides and trimethoprim-sulfamethoxazole. AmpC is found mainly in the chromosome of many gram-negative bacteria but a growing number of AmpC enzymes have recently also been found on the plasmid of gram-negative bacteria. AmpC β-lactamases, which are clinically important cephalosporinases, confer resistance to second and third generation cephalosporins (including cephalotin and cefotaxime) and most penicillins. ESBL/AmpC-producing *E. coli* strains have been found in meat-producing animals, particularly in poultry, which are recognized as important carriers of ESBL-producing *E. coli* and *Salmonella*. There have also been an increasing number of reports of ESBL-producing *E. coli* that have been isolated from meat, especially from poultry meat, but sporadically also from pork and beef. However, person-to-person spread seems to be the main transmission route of ESBL/AmpC –producing *E. coli*.

S. aureus is a common gram-positive bacterium present on skin and mucous membrane of healthy humans and animals. Some strains have developed a resistance to the β-lactam antimicrobials that are used for the treatment of many human infections. Humans mainly get MRSA by contact with infected humans or contaminated devices or equipment in hospital. MRSA strains have also frequently been isolated from intensively reared pigs and chicken. So far, livestock-associated MRSA represents only a small proportion of all MRSA infections in most European countries.

Enterococcus spp., which belong to gram-positive cocci, are natural commensals of human and animal gut but they are also used as acidifying microorganisms in some fermented products. Resistance in *Enterococcus* spp. has received attention due to the rapid increase in the occurrence of vancomycin resistance in *E. faecalis* and *E. faecium* strains (VRE) both among human clinical and animal strains. Avoparcin, which has been used as growth promotion mainly for broilers and pigs, confers cross-resistance to vancomycin. Vancomycin is classified as critically important for human medicine for treatment of severe infections with multidrug resistant *Enterococcus* spp. and MRSA as the main indications. When the connection between avoparcin and VRE in farm animals was confirmed, the use of avoparcin was banned in 1997 in Europe.

Antimicrobials are extensively used in food animal production for a variety of reasons, usually for disease treatment, but they are also applied subtherapeutically for disease prevention and, in some countries, also for growth promotion and to improve feed efficiency. In Europe, subtherapeutic use of antimicrobials for growth promotion in meat-producing animals has been banned 2006. Nevertheless, antimicrobial resistant strains in meat-producing animals are an increasing problem worldwide. The WHO has suggested that national authorities should consider:

- eliminating the use of antimicrobials as growth promoters;
- requiring that antimicrobials be administered to animals only when prescribed by a veterinarian;

- requiring that antimicrobials identified as critically important in human medicine (especially fluoroquinolones and third and fourth generation cephalosporins) only be used in food-producing animals when their use is justified.

Resistant commensal *E. coli* and *Enterococcus* strains present in the meat-producing animals or on meat are considered a potential source of resistant bacteria and resistance genes for humans. High resistance to sulfamethoxazole, tetracycline, streptomycin and ampicillin among *E. coli* strains, which have frequently been isolated from poultry and pigs, has been reported in Europe. Resistance to ciprofloxacin and nalidixic acid among *E. coli* strains isolated, typically, from broilers and chicken meat has been reported in some European countries. Ciprofloxacin is used in poultry production to protect chickens and turkey from *E. coli* infections. Usually, the whole flock is treated by adding the antimicrobial into the drinking water. Alarming is also that a high proportion of *Campylobacter* strains isolated from meat-producing animals and meat are resistant to ciprofloxacin, which is important in antimicrobial treatment of complicated human campylobacteriosis. Furthermore, resistance to at least three different microbial classes in *Salmonella* strains is high in several European countries. *Salmonella* strains from pigs and poultry, especially, have been shown to be frequently resistant to commonly used antimicrobials such as ampicillin, tetracyclines and sulfonamides. In poultry, high resistance to ciprofloxacin has also been observed in *Salmonella* strains.

AMR can spread between meat-producing animals and humans directly by contact and indirectly through consumption and handling of contaminated meat. Farm and slaughterhouse workers, veterinarians and those in close contact with farm animals are directly at risk of being colonized or infected with resistant bacteria through close contact with colonized or infected animals. Humans may be exposed to resistant bacteria via consumption of animal products including meat. There is evidence that meat from different sources and in all stages of processing contains resistant bacteria and their resistance genes. The rise of resistant bacteria among farm animals has been well documented. There is also evidence that consumption of food carrying resistant bacteria has resulted in acquisition of resistant infections.

Meat is susceptible to contamination by food-borne pathogens and commensals at many points from production through to preparation at home. Faecal contamination of carcasses and edible offal during slaughter may result in the transfer of resistant commensal bacteria (including *E. coli*, *E. faecium* and *E. faecalis*) and zoonotic pathogens (including *Salmonella* and *Campylobacter*) to the meat and may lead to meat-borne disease in humans that may not respond to antimicrobial treatment. In the human gut, the resistant strains may transfer resistance also to other gut flora. There is some evidence that (multi)resistant *E. coli* and *Salmonella* strains have been transmitted to humans by consumption of contaminated meat. However, so far there are only a few studies that provide clear evidence of transmission of ESBL/AmpC-producing strains from meat to humans. There is also little evidence of human infections being directly linked to handling or consumption of VRE or MRSA-contaminated meat.

AMR has been recognized as a global health problem for decades but now it is recognized as one of the top health challenges to be faced in the future. The resistant strains spread rapidly and they are difficult to detect. Thus, resistance to commonly used antimicrobials, co-resistance and multidrug resistance should be monitored to prevent treatment failures and increasing health costs. In several countries, AMR in commensals and zoonotic bacteria has been monitored through the whole meat production chain. Monitoring of AMR in *E. coli* and *Salmonella* in farm animals at slaughter is widely used in Europe as an early alert system for tracking emerging resistance in livestock. Guidelines for monitoring of AMR in *Salmonella*, *Campylobacter*, *E. coli* and *Enterococcus* spp. are available on the European Food Safety Authority (EFSA) web site (www.efsa.europa.eu).

Literature and further reading

EFSA (European Food Safety Authority). 2013. European summary report. Antimicrobial resistance in zoonotic and indicator bacteria from humans, animals and food in the European Union in 2011. *EFSA J*, **11**, 3196.

Marschall, B.M. AND Levy, S.B. 2011. Food animals and antimicrobials: impact on human health. *Clin Microbiol Rev*, **24**, 718–733.

WHO (World Health Organisation). 2011. Tackling antibiotic resistance from a food safety perspective in Europe. Retrieved from www.euro.who.int (last accessed 8 February 2014).

B. Control of Biological Meat-Borne Hazards

Sava Buncic

Department of Veterinary Medicine, Faculty of Agriculture, University of Novi Sad, Novi Sad, Serbia

13.7 Scope

Macroscopic meat inspection cannot detect and control the most important bacterial and parasitic meat-borne hazards that may be present on/in carcasses and that presently cause the majority of cases of food-borne illnesses in humans. To improve public health protection via safer meat, a risk-based, comprehensive and coordinated carcass meat safety assurance system targeting the most relevant hazards at the slaughterhouse level has to be developed and implemented in the European Union (EU). In this part of the chapter, the philosophy of and generic framework for such a system and its main elements are discussed and outlined.

13.8 Introduction

The main goals of traditional meat inspection are: (i) to protect public health from meat-borne hazards, and (ii) to control both animal health and (iii) welfare. Although meat inspection can achieve all these goals concurrently in many situations, there are some situations in meat inspection when protection of public health and control of animal health/welfare require opposing actions. For example, some meat inspection procedures, such as palpation and incision, which aim to detect non-zoonotic diseases, are beneficial for animal health but can be detrimental for public health because they may cause cross-contamination of meat with major food-borne pathogens. In such situations, it is assumed that protection of public health is the priority goal of meat inspection and animal health/welfare is important but secondary to public health hazards.

Currently, the most relevant biological hazards causing majority cases of food-borne (including meat-borne) diseases in the European Union are zoonotic bacterial pathogens that are carried and excreted in faeces by clinically healthy animals (e.g. *Salmonella* spp., *Campylobacter jejuni/coli*, *Yersinia enterocolitica*, Shiga toxin-producing *Escherichia* coli (STEC)). However, none of those hazards are detectable by current meat inspection based on macroscopic examination. Consequently, traditional meat inspection is no longer capable of assuring the consumers' health in respect to the most relevant hazards causing alimentary diseases via meat.

Because of the inability of macroscopic meat inspection to detect 'invisible' bacterial hazards being the main public health concern nowadays, and because the impracticality of laboratory examination of those hazards in/on each carcass individually, an effective overall control system for 'invisible hazards' needs to be used. Such a system should be risk-based, meat-chain orientated and comprehensive. It needs to combine a range of preventative and control measures applied both at the farm and slaughterhouse levels in a longitudinally integrated way. This system is more a 'meat safety assurance' system than meat inspection only. The author of this part of the chapter has been heavily involved in the scientific activities of European Food Safety Authority (EFSA) from 2010 to 2013, leading the development of such a generic system by revising and improving the traditional meat inspection. The generic framework and the main elements of the system that are elaborated on and outlined in this part reflect those developed and published by EFSA.

13.9 Hazard identification

There are a large number of public health hazards associated with slaughtered animals and their carcass meat in slaughterhouses that may pose public health risk. Many of them are not present in all parts of the world. In this section, the hazards will be limited to those that are reported in food animal populations and in meat thereof in the European Union (Table 13.12).

The starting point for the development of any meat safety system is to define what is meant by 'hazards posing public health risk' to be controlled. According to its basic definition, 'public health risk' refers to the final product at the time of its consumption; thus, a public health risk assessment requires all food chain stages to be taken into account. The meat inspection is limited to the public health hazards of the carcass at the end of slaughterhouse operation.

Zoonotic hazards of primary interest to be dealt with in carcass meat safety assurance at slaughterhouses are those that are transmitted to humans via the meat-borne route (Table 13.12). The main control of some meat-borne hazards can be during the farm-to-chilled carcass stages or during the post-slaughterhouse stages (i.e. meat processing, storage, retail). Examples of meat-borne hazards primarily controlled during farm-to-chilled carcass stages include *Salmonella, Campylobacter* and *Yersinia* (Figure 13.8). Examples of meat-borne hazards primarily controlled during post-slaughterhouse stages include *Listeria monocytogenes, Clostridia* and *Staphylococcus aureus* (Figure 13.8).

There are also other zoonotic hazards that are transmitted to humans via routes other than meat-borne routes, such as contact-, inhalation- and environment-mediated routes; thus, the control of these hazards differs from control of meat inspection-based hazards. It is based on other legislations and is usually enforced by other governmental agencies/services. Therefore,

Table 13.12 Assessment of zoonotic agents associated with meat-borne hazards in the EU.

Zoonotic biological hazard	Meat-borne hazard associated with livestock				
	Cattle	Sheep/ goats	Pigs	Poultry	Comments
Bacteria:					
Bacillus anthracis	Yes	Yes	–	–	Infection also via contact
Bacillus cereus	Yes	Yes	Yes	Yes	Intoxication: (i) emetic toxin and (ii) diarrhoea toxin
Brucella (abortus/melitensis/suis)	No	No	No	No	Infection via other food (primarily milk) and contact
Campylobacter spp.	Yes	Yes	Yes	Yes	Infection also via milk and water
Clostridium botulinum	Yes	Yes	Yes	Yes	Intoxication also in the intestine (infant botulism)
Clostridium difficile	No	No	No	Yes	Also a hospital infection
Clostridium perfringens	Yes	Yes	Yes	Yes	Intoxication also via other foods
Shiga toxin-producing Escherichia coli	Yes	Yes	–	–	Infection also via other foods, water and contact
ESBL/AmpC E. coli	Yes	Yes	Yes	Yes	Infection also via other foods and contact
Listeria monocytogenes	Yes	Yes	Yes	Yes	Infection also via other foods, also neonatal infection
Mycobacterium avium subsp. avium	–	–	No	No	Infection via meat is not clear
Mycobacterium bovis/caprae	No	No	No	–	Infection via milk and aerosols
Salmonella enterica (non-typhoid)	Yes	Yes	Yes	Yes	Infection also via other foods
ESBL/AmpC S. enterica	Yes	Yes	Yes	Yes	
Staphylococcus aureus	Yes	Yes	Yes	Yes	Infection also via other foods and contact
Methicillin-resistant S. aureus	Yes	Yes	Yes	Yes	
Yersinia enterocolitica	Yes	Yes	Yes	Yes	Infection also via other foods (vegetables)
Yersinia pseudotuberculosis	Yes	Yes	Yes	Yes	
Viruses:					
Avian influenza virus	–	–	–	No	Infection via contact
Hepatitis E virus	No	No	Yes	No	Raw pork and water are associated with the infection
Parasites:					
Ascaris suum	–	–	No	–	Infection due to faeces contamination
Cryptosporidium parvum	No	No	No	No	Infection due to faeces contamination
Echinococcus granulosus	No	No	No	–	Infection due to contamination with dog faeces
Fasciola hepatica	No	No	–	–	Infection via contaminated environment (water)
Sarcocystis hominis	Yes	–	–	–	Infection only via bovine meat
Sarcocystis suihominis	–	–	Yes	–	Infection only via porcine meat
Taenia saginata cysticercus	Yes	–	–	–	Infection only via bovine meat
Taenia solium cysticercus	–	–	Yes	–	Also autoinfection in humans
Toxocara canis/cati	–	–	–	No	Infection via water and soil
Toxoplasma gondii	Yes	Yes	Yes	Yes	Infection via materials contaminated with feline faeces
Trichinella spp.	No	No	Yes	No	Infection only via meat

Yes, Hazard that may cause meat-borne disease.
No, Hazard that is not causing meat-borne disease.
–, Hazard that is not related to the animal species.

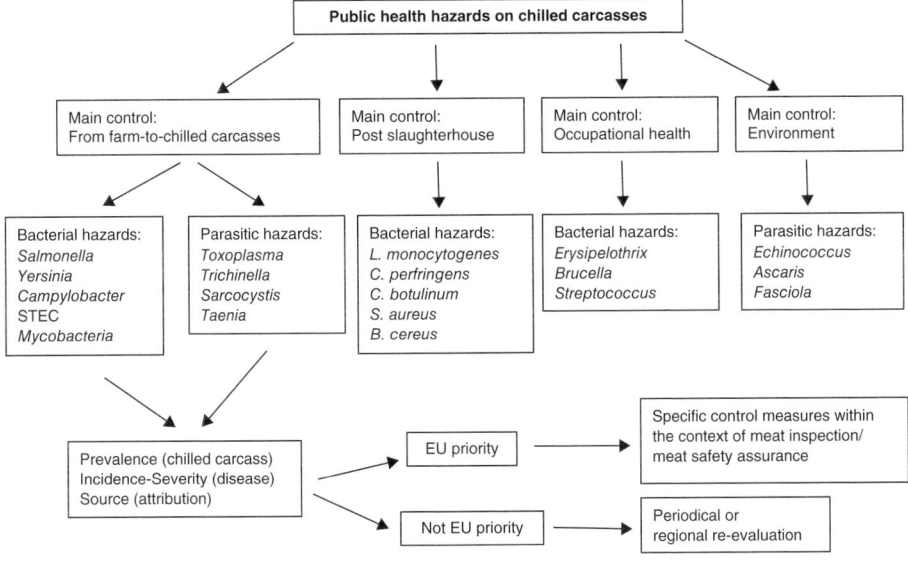

Figure 13.8 Control strategies for main examples of biological hazards that may be present in/on chilled carcasses.

only those hazards, which are both meat borne and controlled along the farm-to-chilled carcass stage (Figure 13.8), are relevant for meat inspection and carcass meat safety assurance at slaughterhouses and will be further elaborated here.

13.10 Prioritization (ranking) of meat-borne hazards

Not all meat-borne hazards associated with carcasses at slaughterhouses have the same importance for meat safety and public health today; thus, equally intensive efforts and resources need not be used to control them. In risk-based meat inspection and meat safety assurance, meat-borne hazards posing a high risk for humans require high level of risk-reduction; hence, specifically designed and sufficiently effective control measures targeting those hazards have to be used. In contrast, for hazards posing negligible or low risk for humans, comparably much lower risk-reduction is needed or even none in some cases. This can usually be achieved by use of only general, good manufacturing/good hygiene practice (GMP/GHP)-based measures. To identify the most relevant (high risk) meat-borne hazards to be targeted by the slaughterhouse meat safety assurance system, several aspects need to be carefully considered for each hazard and combined in an appropriate way (Figure 13.9).

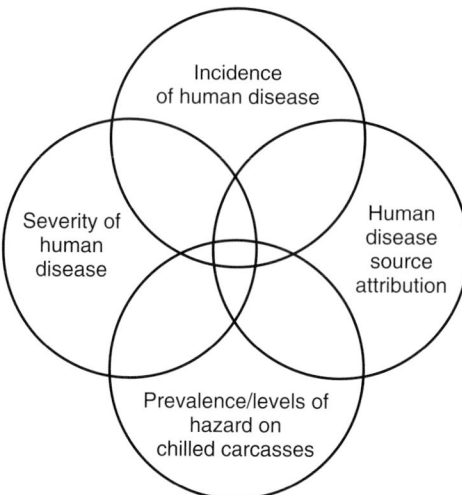

Figure 13.9   Main parameters for prioritization of meat-borne hazards in the context of carcass meat safety assurance.

The incidence data of human zoonotic diseases in the European Union can be obtained from the EU summary report on trends and sources of zoonoses, zoonotic agents and food-borne outbreaks, which is published annually by the EFSA and ECDC (European Centre for Disease Prevention and Control). The report is based on national disease reporting databases obtained from member states. In 2010, campylobacteriosis was the most commonly reported zoonosis (220 209 reported human cases), followed by salmonellosis (95 548 cases). Significantly fewer human cases caused by other bacterial agents were reported: 7 017 yersiniosis cases, 9 485 STEC infections and 1 476 listeriosis cases. One parameter for hazard prioritization is the number of confirmed human cases per 100 000 population. A descriptor is needed to divide the disease in high and low incidence. The value of 10 cases/100 000 EU population is used as such a descriptor by the EFSA.

Regarding the **severity of human diseases**, one parameter, which can be used for prioritization of the disease burden, is proportion (%) of deaths occurring in reported cases. For example, the value of 0.1% deaths has been used for dividing the disease into low and high severity. Sometimes also, a disease severity parameter such as the DALY value (Disability Adjusted Life Years) has been used.

Source attribution of the disease is another parameter that can be used in prioritization (ranking) of the hazard. Because data on total incidence of human food-borne disease are not food-source specific (the same disease can be caused by the hazard originating from different/multiple foods), additional information on the proportion of cases of given disease that is caused via the meat species in question has to be used. For that purpose, epidemiological information (such as information obtained from case-control studies and/or from characterization of isolates from clinical cases and consumed food) from each hazard can be used. Currently, use of source attribution information is particularly hampered by insufficient/lack of data.

The prevalence of the hazard in/on the meat is an important parameter to prioritize the meat-borne hazard. Regarding most hazards, there are only insufficient data available and usually the available data are not comparable due to the different detection methods used. Recently, the values of <0.1%, 1–5% and >5% have been used by the EFSA to define low, medium and high prevalence of a given hazard on pork carcasses, respectively.

By combining the above mentioned hazard prioritization parameters of each meat-borne hazard, the hazards can be ranked into high, medium and low priority hazards in the context of meat inspection and carcass meat safety assurance. The prioritization of the meat-borne hazards (Table 13.13) is based on the information from EFSA documents, published evaluations of food-borne disease burdens in different countries, various published risk assessments and the author's own published research. The meat-borne hazards to be specifically addressed by a carcass meat safety assurance system are those considered as priority hazards, and they include hazards ranked as high or medium. In contrast, meat-borne hazards identified as low priority do not need to be specifically targeted by the system. The priority meat-borne hazards differ between animal species but it should be noted that they comprise mostly bacterial hazards (*Campylobacter, Salmonella*, STEC,

Table 13.13   Ranking of meat-borne hazards in the context of meat inspection in the EU.

Biological hazard	Priority ranking of the hazards			
	Cattle[a]	Sheep/ goats[a]	Pigs[b]	Poultry[b]
Bacillus anthracis	Low[d]	Low	Low	N/A
Thermo-tolerant *Campylobacter* spp.	Low	Low	Low	High[c]
Shiga toxin-producing *Escherichia coli*	High[c]	High[c]	Low	Low
ESBL/AmpC *E. coli*	Low	Low	Low	Medium[e]
Salmonella enterica	High[c]	Low	High[c]	High[c]
ESBL/AmpC *Salmonella enterica*	Low	Low	Low	Low
Sarcocystis hominis	Low	N/A	N/A	N/A
Sarcocystis suihominis	N/A	N/A	Low	N/A
Taenia saginata	Low	N/A	N/A	N/A
Taenia solium	N/A	N/A	Low	N/A
Toxoplasma gondii	Undetermined	High[c]	Medium[e]	Low
Yersinia enterocolitica and *pseudotuberculosis*	Low	Low	Medium[e]	Low
Trichinella spp.	N/A	N/A	Medium[e]	N/A

[a] ranking done on scale: low or high.
[b] ranking done on scale: low or medium or high.
[c,e] high and medium ranked hazard should be addressed by specific risk-reduction measures within carcass meat safety assurance.
[d] low ranked hazards do not need to be addressed by specific risk-reduction measures, but by general GMP/GHP-based preventative measures within carcass meat safety assurance.
N/A, not applicable

Yersinia and ESBL/AmpC *E. coli*) and some parasitic hazards (*Toxoplasma* and *Trichinella*). Among these seven priority hazards, only *Trichinella* is detectable and addressed by current meat inspection systems.

13.11   Carcass meat safety assurance framework

A comprehensive, coordinated and risk-based carcass meat safety assurance system must incorporate several control strategies into a coherent whole. In this way, the ultimate meat-borne risk reduction in respect to priority hazards (both bacterial and parasitic) exceeds the risk reduction achieved by any individual control strategy. This is possible only if: (i) there is a risk manager who coordinates the whole system, (ii) all participants in the system clearly know their responsibility and (iii) there is an efficient flow of all relevant information forward and backward along the farm-to-chilled carcass chain between the participants enabling rapid and reliable identification of the reasons for any meat safety failure if/when it occurs.

The main control strategies and tools brought together and used in a coordinated way in such a system are: (i) identification and traceability of both animals and meat; (ii) food chain information (FCI), including data on risk-reduction performances at the farms and in the slaughterhouses; (iii) risk categorization of slaughterhouses and farms based on their performances; (iv) GMP/GHP- or Hazard Analysis and Critical Control Points (HACCP)-based measures applied at individual points during slaughterhouse operation; and (v) meat inspection *per se*.

Such a system brings together both main players responsible for the main control strategies: the producer (slaughterhouse operator) and the regulator (governmental meat hygiene/safety agencies). The operators hold ultimate responsibility for achieving meat safety targets (slaughterhouses) and animal-related targets (farms) through a range of measures managed through their own system of meat safety/quality control (GMP/GHP- and HACCP-based). The regulator holds responsibility for setting clear meat safety targets (which have to be achieved by the slaughterhouse) and animal-related targets (which have to be achieved by farms), for auditing the operators' systems, as well as for meat inspection based controls. Veterinarians, who have a broad education and skills in both medicine and food chain, are in a central role in the control of meat production chain. A framework for chilled carcass meat safety assurance, controlling specifically all the meat-borne bacterial and parasitic hazards identified as a priority, is illustrated in Figure 13.10. Its main elements are briefly outlined here.

13.11.1   Targets to be achieved by slaughterhouses and farms in respect to priority meat-borne hazards

One of the key prerequisites for a complex, longitudinally integrated meat safety assurance system, based on the main responsibility for meat safety

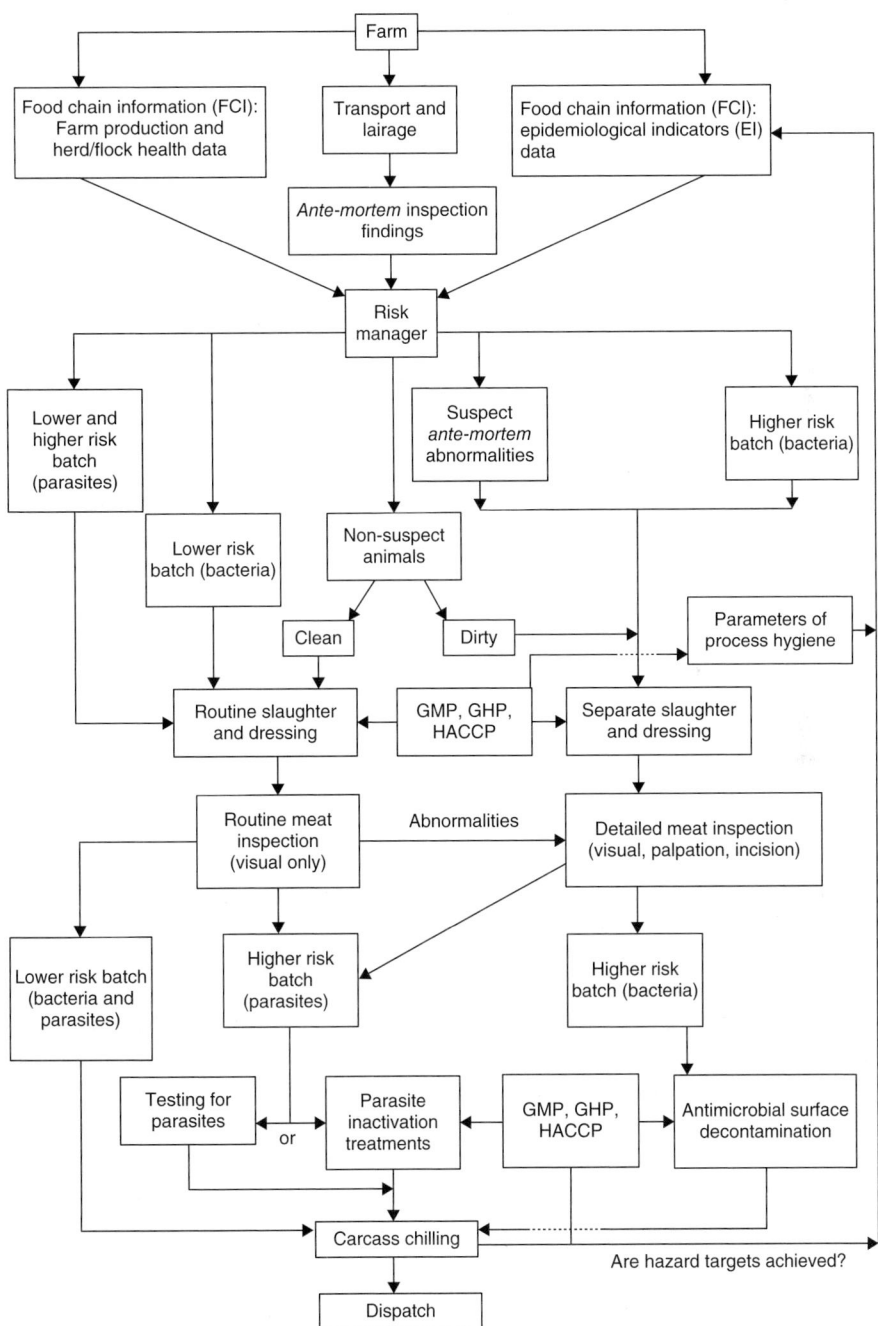

Figure 13.10 Chilled carcass meat safety assurance framework in slaughterhouses.

being allocated to the operators, to be effective is that the main participants in the meat chain are given clear and measurable targets and/or related criteria indicating what they should achieve in respect to specified hazards. In the food safety objectives-driven concept introduced by the Codex Alimentarius Commission (CAC) and International Commission on Microbiological Specifications for Foods (ICMSF), such targets are set by regulators as prevalence/levels of the hazards in the food in question (to be met by operators). The main elements of that concept are defined in following way:

- Appropriate Level of Protection (ALOP): The level of protection deemed appropriate by the Member establishing a sanitary or phytosanitary measure to protect human, animal or plant life or health within its territory.
- Food Safety Objective (FSO): The maximum frequency and/or concentration of a hazard in a food at the time of consumption that provides or contributes to the appropriate level of protection (ALOP).
- Performance Objective (PO): The maximum frequency and/or concentration of a hazard in a food at a specified step in the food chain before the time of consumption that provides, or contributes to, an FSO or ALOP, as appropriate.
- Performance Criterion (PC): The effect in frequency and/or concentration of a hazard in a food that must be achieved by the application of one or more control measures to provide or contribute to a PO or an FSO.

Considerations of ALOP and FSO are governmental responsibility and outside the scope of this chapter. Current EU legislation does not use the terms PO and PC indicated above, but uses two other terms: Process Hygiene Criteria (PHC), applicable to the product at the level of food production, and Food Safety Criteria (FSC), applicable to the product at market level. If PHC are not met, the food safety system of the operator needs to be reviewed and improved. If FSC are not met, the product must be withdrawn from the market. Furthermore, in EU legislation there are no clearly defined targets for chilled carcasses to be achieved by slaughterhouses in respect to each of the priority hazards mentioned above. The legislation states only that carcasses infected with *Trichinella* are unfit for human consumption and that if the occurrence of *Salmonella* on pre-chilled carcasses is above stated value, the process hygiene is unsatisfactory.

Setting and using such priority hazards' targets for chilled carcasses is necessary because they would:

- provide a measurable and transparent focus for a slaughterhouse's meat safety assurance system;
- provide information useful for human exposure assessment for those hazards;
- enable differentiation between slaughterhouses producing end-products (carcasses) of 'acceptable' and 'unacceptable' status in respect to priority hazards, and the information would also contribute towards risk-categorization of slaughterhouses;

- represent a basis for 'backward'-generating of appropriate targets for farms delivering animals to the slaughterhouses and, in turn, differentiation between 'acceptably' and 'unacceptably' performing farms, that is risk-categorization of animals presented for slaughter;
- information as to whether or not the slaughterhouses satisfy their own priority hazard targets would enable judgement whether or not they contribute, at global level, to pre-determined FSOs and ALOP in respect to those hazards.

Furthermore, targets for chilled carcasses in respect to each of the main hazards would serve as a benchmark to derive correlated slaughterhouse process hygiene criteria (PHC). Slaughterhouse PHC should indicate the capacity of the slaughterhouse operation to ensure sufficient risk reduction in respect to priority meat-borne hazards. In defining PHC, the fundamental issue is whether PHC should be linked to individual stages of the process (e.g. reduction of a hazard's occurrence/level at a selected single step along the slaughter line) or linked to two points (e.g. the starting and the end points) or even to multiple points. In other words, the PHC is meant to define the outcome of a step or a combination of steps during the slaughterhouse process, so that the total risk reduction is sufficient to ultimately achieve related target on chilled carcasses.

However, as even the best performing slaughterhouses have only limited hazard-reducing capacity, whether the target set for chilled carcasses for a given hazard will be achieved depends also on the occurrence/level of the hazard in/on incoming animals. Understandably, if the occurrence/level of the hazard in/on incoming animals exceeds the hazard-reduction capacity of the slaughter line operation, the chilled carcass target will not be achieved. Therefore, farms need to be given such animal-related targets, achievement of which would prevent this failure scenario from occurring. This is the reason why the chilled carcass- and the incoming animals-related targets for the same priority hazard must be appropriately correlated, that is the latter has to be derived from the former. Hence, the following two sections of this chapter are presented in the order reflecting that.

13.11.2 Control of meat-borne hazards at the slaughterhouse

The original sources of the priority bacterial meat-borne hazards (*Campylobacter*, *Salmonella*, STEC, *Yersinia*, ESBL/AmpC *E. coli*) are usually asymptomatic animals; hence, they are undetectable by current meat inspection. These bacteria are excreted in the faeces and easily disseminated along the farm–transport–lairage chain directly from animal to animal and indirectly through contaminated environment to animal. This ultimately results in contamination of carcasses with those hazards during slaughter and dressing of animals, particularly during skinning/defeathering and evisceration.

The occurrence and number of priority bacterial hazards on chilled carcasses are highly variable depending on various factors, including particularly: (i) their occurrence in animals before slaughter and the application and the effectiveness of related pre-slaughter control strategies (e.g. biosecurity, vaccination); (ii) the extent of direct and/or indirect faecal cross-contamination during slaughter line operation; (iii) the application and the effectiveness of possible interventions to eliminate/reduce them on carcasses (e.g. decontamination); and (iv) the effects of the chilling technology/regime on the hazards present on the carcass surface. Therefore, the main controls for the priority bacterial hazards are focused on these aspects at the slaughterhouse.

Intramuscular protozoan parasites *Toxoplasma gondii* and *Trichinella*, similarly to above mentioned bacterial hazards, usually do not cause visible abnormalities in animals and can only be detected during meat inspection by laboratory examination. The presence of their viable forms in chilled carcass meat depends on three main factors: (i) the occurrence in animals presented for slaughter, (ii) the application of *post-mortem* testing of carcasses for the parasites, and (iii) the application and the effectiveness of carcass treatments to kill them (e.g. freezing or heat treatment of meat). Therefore, the main controls for the priority parasitic hazards are focused on these aspects at the slaughterhouse.

Bacterial meat-borne hazards of priority at the slaughterhouse Whether carcass meat becomes contaminated with bacterial hazards of priority (*Campylobacter*, *Salmonella*, STEC, *Yersinia*, ESBL/AmpC *E. coli*) is highly dependent on the slaughterhouse process. Technical aspects of individual steps of slaughter line operation vary considerably between slaughterhouses slaughtering different animal species, and even may vary between slaughterhouses slaughtering same animal species. Specifics of technologies and the main sources and routes of carcass contamination during the slaughter line operation for individual animal species are discussed in other chapters.

The presence of priority bacterial hazards on carcasses may also be affected by the hazards residing in the slaughterhouse environment (e.g. surfaces, equipment), which is dependent on the antimicrobial effectiveness of the cleaning-sanitation regimes. More detailed information on sanitation is provided in other chapters.

In addition, manual techniques (palpation and incision) used in current *post-mortem* meat inspection also may mediate microbial cross-contamination of meat through: (i) transfer of pathogens from lymph nodes to surrounding or other tissues due to incisions made to inspect lymph nodes and (ii) transfer of pathogens from contaminated to uncontaminated carcasses/organs via inspector's hands and tools.

The most common abnormalities/lesions found by palpation and incisions during meat inspection are due to: (i) non-zoonotic agents, (ii) zoonotic but not meat-borne hazards or (iii) meat-borne hazards but of low priority. In such situations, public health risk related to cross-contamination of the meat with priority bacterial hazards through manual *post-mortem* inspection is higher than the public health risk related to hazards associated with a given abnormality found by palpation and incision. Therefore, omitting palpation and incision and using visual-only meat inspection for routinely slaughtered (non-suspect) animal batches would provide overall public health benefit. The main benefit from using palpation and incision is related to detection of hazards of relevance to animal health not public health. However, in case that the animals are suspected based on FCI or on relevant abnormalities found at *ante-mortem* or visual *post-mortem* inspection, the use of more detailed meat inspection, including palpation and incisions, and sometimes laboratory examinations is justifiable.

It has been demonstrated that reliable and consistent control of bacterial hazards of priority usually cannot be ensured through a single control measure at a single point within the slaughterhouse process. Rather, this is achieved when several control measures are applied at several points concurrently. For meat-borne exposure of the consumers to the priority bacterial hazards indicated in this chapter, slaughterhouse-based controls are of utmost importance. This has been confirmed by quantitative risk assessment study of *Salmonella* in pigs published by the EFSA. The results demonstrate that a reduction of two logs (99%) of the number of *Salmonella* on contaminated carcasses would result in a 60–80% reduction of the number of human salmonellosis cases attributable to pig meat consumption.

Unfortunately, general design of the individual technological steps, and their order, in industrial high-throughput slaughterhouses have not changed significantly (apart from individual machinery) for decades. The present design and order of slaughterhouse operation is dictated primarily by a desire for ever higher speed and throughput, and cost reduction. The microbiological contamination appears a secondary criterion. However, to effectively carry allocated main responsibility for meat safety, the operators would have to evaluate and adjust the technology so as to maximize its microbiological risk reduction effects.

There is a significant amount of knowledge showing differences between slaughterhouses slaughtering same species in respect to the microbiological status of carcasses and meat safety risk for the consumer. The variation is caused by differences between the slaughterhouse in respect to the hygienic characteristics of the technology and the equipment used, the extent to which the procedures are standardized and documented, the technical knowledge of the operator, the level of food hygiene training and its application, and the motivation of staff and management. These variations individually and their combinations lead to the differences between slaughterhouses in process hygiene performances. The meat safety risks due to bacterial hazards in slaughterhouse operations are strongly influenced by the process hygiene performance of the operators. Thus, the slaughterhouses can be differentiated (risk categorized) in respect to their capacity to control bacterial hazards of priority. The main information required for risk categorization of slaughterhouses are standardized, measurable parameters reflecting their individual process hygiene performances.

Process hygiene criteria (PHC) give guidance on the acceptable functioning of HACCP-based slaughterhouse processes. The maximum values are set for indicator bacteria of (i) overall contamination (total viable count of bacteria, TVC), (ii) enteric origin (*Enterobacteriaceae* counts) and (iii) *Salmonella*. The PHC are applicable to the product at the end of the manufacturing process (final carcass). The PHC are not related to the initial contamination values of the raw materials at the individual operator level. In other words, PHC for slaughterhouses actually do not provide information on ratios between initial contamination associated with incoming animals versus final contamination associated with carcasses, that is on the actual capacity of the process to reduce the incoming contamination, but only on the slaughter process hygiene.

To characterize the slaughterhouse process microbiologically, it would be necessary to determine the microbiological loads at multiple points (or at least at two points: at the beginning and at the end) of the slaughterhouse operation and compare them. It would be best to compare the load of bacterial hazards of high priority, but this is impractical because the pathogens occur in animals and on the carcasses relatively rarely. The bacterial load is also affected by on-farm factors and the detection and quantification of the pathogens is difficult and laborious, needing well equipped laboratories. Pathogen testing is much more valuable for the purposes of consumer exposure assessment and pathogen reduction programmes and is more related to setting of targets for slaughterhouses. Instead, levels of indicator microorganisms (e.g. *Enterobacteriaceae*, and/or *E. coli* and/or TVC) are much better suited for the purpose of process hygiene assessment. The risk categorization of slaughterhouses in respect to their capacities to control bacterial hazards of high priority will be used to decide:

- to which slaughterhouses batches of animals posing higher or lower risk of bacterial hazard will be sent for slaughter (see below);
- whether additional risk-reduction interventions (e.g. carcass decontamination) are to be applied;
- more stringent requirements for monitoring/verification/auditing programmes for higher-risk slaughterhouses;
- which slaughterhouses need improvement of the technology.

Parasitic meat-borne hazards of priority at slaughterhouses *Trichinella* and *Toxoplasma gondii*, which are parasitic hazards of priority, do not cause symptoms in animals and, thus, cannot be macroscopically detected during current meat inspection, either during *ante-* or *post-mortem*, but only through laboratory testing. Direct identification of *Trichinella* larvae in pig and horse muscles is currently based on isolation of the larvae by artificial digestion and microscopic identification with a sensitivity of at least 1–3 larvae/g. Testing of each carcass for *Trichinella* is laborious, expensive and the final decision on the fitness of the meat is delayed until the testing results are obtained. Currently, the slaughtered animals are not tested for *T. gondii*. Currently, there are no sensitive methods available to detect the low number of cysts usually present in the muscle. Furthermore, the methods do not differentiate cysts that are dead from those that are alive.

Alternative approaches to meat safety assurance in respect to *Trichinella* and *T. gondii* should be considered for animal populations at high risk. They are primarily based on meat treatments that inactivate the parasites. Currently, the most reliable inactivation treatments for *Trichinella* larvae and *T. gondii* cysts are based on adequate (i) heating, (ii) freezing or (iii) irradiation of the meat. The former two approaches are much more acceptable to consumers in the European Union. Various temperature–time combinations for heating or freezing can be used but only under controlled conditions (within a HACCP-based system) and the effectiveness of each has to be validated, monitored, documented, verified and audited regularly.

Cross-contamination of *Trichinella* or *T. gondii* between the animals is not an issue, and thus it is not necessary to handle animals from negative and positive herds separately during the transport–lairage–slaughter line period. However, incoming batches of animals could be categorized into *Trichinella*- and/or *T. gondii*-free herds and infected herds (outdoor raised animals are particularly at risk) based on historical testing data and FCI. The slaughtered animals of low-risk farms do not have to be tested for *Trichinella* and *Toxoplasma* and do not have to be subjected to parasite inactivation treatments. In contrast, the slaughtered animals from high-risk farms have to be tested for the parasites or have to be treated by a reliable and validated larvae/cysts inactivating treatment.

13.11.3 Control of meat-borne hazards at the farm level

Bacterial meat-borne hazards of priority Bacterial hazards of priority (*Campylobacter, Salmonella*, STEC, *Yersinia*, ESBL/AmpC *E. coli*) are shed by asymptomatic, carrier animals. These bacterial pathogens can be spread on the farm directly between animals or indirectly via contaminated environment.

A range of preventative measures to control these hazards on the farm exist but the elimination of the hazards is difficult. The hazards may be reduced by combined and coordinated use of multiple control options on the farm. The control measures for bacterial hazards of high priority include:

- preventing the recycling of the pathogens in the environment;
- preventing the introduction and/or spread of pathogens within the farm;
- preventing the ingestion of the pathogens by the animals;
- supressing the pathogen within the animal gastrointestinal tract;
- enhancing the animal host response against the pathogens;
- preventing and/or reducing the pathogen spread during transport.

There are a number of different farming systems for each animal species in the European Union, and within each system there exist significant differences between the farms in respect not only of husbandry practices but also in respect of risk factors associated with, and controls used for, each of the bacterial hazards of high priority. This generates between-farm and/or between-herd/flock differences in the status of animals in respect to priority bacterial hazards. Hence, it is theoretically possible to categorize different slaughter batches into low- and high-risk batches in respect to the priority hazards through the use of FCI, which is based on related historical data for each farm/herd), as well as through application of appropriate epidemiological indicators (EIs), as described in the next section.

Parasitic hazards of priority The most important factors that influence the occurrence of *Toxoplasma gondii* and *Trichinella* in animals on the farm include:

- zoo-sanitary conditions on the farm (e.g. biosecurity);
- size of the farm (i.e. intensive, extensive);
- farming method (e.g. indoor, outdoor);
- age of animals (e.g. fattening, breeding).

Livestock production in developed countries is characterized by marked reductions in the numbers of farms and corresponding increases in herd/flock size. At the same time, there is a rise in numbers of small, outdoor farms.

The occurrence of *T. gondii* and *Trichinella* seems to be lower in animals raised in intensive indoor farming systems than in smaller outdoor farms. This is probably due to effective control measures applied in intensive indoor farming systems, such as (i) biosecurity, including systematic rodent control and exclusion of contacts with wildlife, (ii) hygienic feed handling procedures and (iii) exclusion of cats (for *T. gondii*). On the farms where animals have outdoor access and/or which are poorly managed in respect to other control measures indicated above, the risk of both *Toxoplasma* and *Trichinella* infections are significantly increased. Consequently, the risk of transmission is particularly high with backyard and free-range farm animals. Also, a higher risk is associated with older animals.

It is theoretically possible to categorize the farms/herds into high- and low-risk ones in respect to *T. gondii* and/or *Trichinella* through FCI data based on following information:

- historical data from the parasite testing of slaughtered animals originating from the same farms/herds;
- results from the serological monitoring of the animals originating from the same farms/herds;
- data on farming practices and control measures relevant for the parasites used on the farms;
- epidemiological situation of the parasites in the wider area where those farms are located (geographical risk).

The risk categorization of the farms/herds/batches in respect to *T. gondii* and/or *Trichinella* would facilitate appropriate risk management decision. For example, incoming animal batches with lower risk may not be subjected to any parasite risk-reduction measure at the slaughterhouse, whilst animal batches with higher risk may be subjected to defined risk-reduction measures (e.g. carcass testing for parasites or parasite inactivation treatment of carcasses). Furthermore, the risk categorization would be useful for identification of farms/herds and farming practices where additional or more stringent parasite control measures are needed.

13.11.4 Principles of use of food chain information (FCI) including epidemiological indicators (EIs) in the carcass meat safety assurance framework

Food chain information (FCI) is essential for the control of meat-borne hazards of priority and it should include: (i) hazard- and animal/meat species-specific epidemiological indicators (EIs) for both farms and slaughterhouses; (ii) historical testing data conducted at both farms and slaughterhouses; (iii) production practices and technology used at both farms and slaughterhouses; (iv) risk-reduction interventions applied (e.g. antimicrobial and antiparasitic treatments); (v) data from HACCP verification; and (vi) data whether the animal- and chilled carcasses-related targets are met.

As indicated before, it is envisaged that information obtained at the farm level by use of (i) data on farm practices/production and from herd health plans and (ii) EIs data for the priority hazards at farm would serve to risk-differentiate animal farms/herds/batches. Likewise, process hygiene assessment data and EIs data at the slaughterhouse level would serve to risk-differentiate slaughterhouses in respect to their risk-reduction capacity. All this information on performances and status of each farm and slaughterhouse, including their resultant risk categories, should be fed into the FCI. Once the FCI become able to provide relevant, reliable, up-to-date and sufficient information, it would significantly improve the management of the carcass meat safety and enable informed decision making by the risk manager how to optimally protect public health via safe meat.

Once the use of comprehensive and complete FCI is implemented in a systematic way, optimal meat safety-related decision making by the risk manager will be possible in each specific situation. Examples of possible scenarios and related decisions include:

- Slaughtering lower-risk animal batches (with no or low prevalence of bacterial and/or parasitic hazards of priority) in low-risk slaughterhouses (with good process hygiene and risk reduction capacity); only GMP/GHP- and HACCP-based control measures may be sufficient for achieving pre-set targets on/in chilled carcasses.

- Slaughtering higher-risk animal batches (with certain occurrence of bacterial and/or parasitic hazards of high priority) in slaughterhouses where application of only GMP/GHP- and HACCP-based control measures may not be sufficient; additional risk-reduction interventions have to be applied (e.g. decontamination for the bacterial hazards and inactivation treatment for parasitic hazards) to achieve pre-set targets on/in chilled carcasses.

Development of appropriate EIs and the use of related data are very important elements for the control of biological hazards that should be included in the FCI. For each EI, it needs to be specified: (i) the exact point where the EI is applicable, (ii) the purpose of the EI, (iii) the criteria of the EI to separate acceptable form unacceptable, and (iv) the methodology of the EI. A range of EIs are indicated and principles of their use described in several EFSA reports regarding revision of meat inspection in the European Union.

The main points – as a universal example for all animal species – where information of EIs can be generated are illustrated in Figure 13.11. The data to be collected (descriptors) and the main purpose of the use of the EIs are presented in Table 13.14. All EIs are both hazard- and animal species-specific but in this chapter only the main principles of EIs are highlighted.

There are three types of data provided by these generic EIs: (i) data from structured and standardized auditing of farming and transport–lairage practices (i.e. the risk factors present and related control measures applied); (ii) data from microbiological/parasitological testing of animals and carcasses (i.e. actual presence/absence of the hazards); and (iii) technical data from validation/verification of regimes used for antimicrobial and parasite inactivation treatment of carcasses (i.e. their hazard elimination effectiveness). Deciding on which individual EI and/or combination of EIs will be used for each hazard – within a harmonized regulatory system – as well as implementation requirements for selected EIs (auditing/sampling plans, auditing/testing methodology and numerical criteria/limits) are expected to be a regulatory authority responsibility.

From a practical perspective, FCI analysis and related decision making is a complex task, as, in some situations, the same incoming animal batch may represent either the same or different risk categories relative to different hazards. For example, incoming batches may pose a lower risk of bacterial hazards but a higher risk of parasitic hazards, or the opposite, or the two risks may be similar; a number of other scenarios may occur, too. In each such scenario, the risk manager will have to choose which control options will be applied, so as to ensure that the hazard-based targets for chilled carcasses are achieved and to generate the best overall contribution to public health. This clearly indicates that the risk manager should be playing a central role in future carcass meat safety assurance systems, and thus must have the necessary training, skills and competence. Furthermore, all other participants in such a system also have to be appropriately trained/skilled relative to their roles and responsibilities.

Table 13.14 Descriptors of epidemiological indicators (EI) in the risk-based chilled carcass meat safety assurance.

Epidemiological indicator (EI) of			
bacterial meat-borne risks		parasitic meat-borne risks	
Descriptor	Purpose of EI	Descriptor	Purpose of EI

EI-1:
Audit of animal
  purchase procedures
	Indication of
	  herd/flock-related
	  risk
		EI-9:
		Hazard monitoring in
		  wildlife
			Indication of
			  herd/flock-related
			  risk

EI-2:
Audit of farming
  practices
		EI-10:
		Audit of farming
		  practices (e.g.
		  housing)	

EI-3:
Presence of hazard in
  faeces of animals
  on-farm
		EI-12:
		Verification/audit of
		  parasite inactivation
		  treatment parameters
		  (e.g. temperature)
			Indication of
			  slaughterhouse
			  process hygiene-
			  related risk

EI-4:
Audit of transport and
  lairage conditions
	Indication of
	  batch-related risk
		EI-13:
		Parasite testing of
		  carcasses
			Indication of both
			  carcass- and
			  herd/flock-related
			  risk

EI-5:
Visual animal
  cleanliness scoring
		EI-14:
		Parasitological status of
		  carcasses
		  post-chilling
			Indication whether
			  parasitic hazard-
			  related target for
			  chilled carcasses is
			  achieved

EI-6:
Microbiological status
  of animal coats
  post-slaughter but
  pre-skinning
	Indication of
	  slaughterhouse
	  process hygiene-
	  related risk		

EI-7:
Microbiological status
  of final carcasses
  before chilling			

EI-8:
Microbiological status
  of carcasses
  post-chilling
	Indication whether
	  microbial
	  hazard-related
	  target for chilled
	  carcasses is
	  achieved		

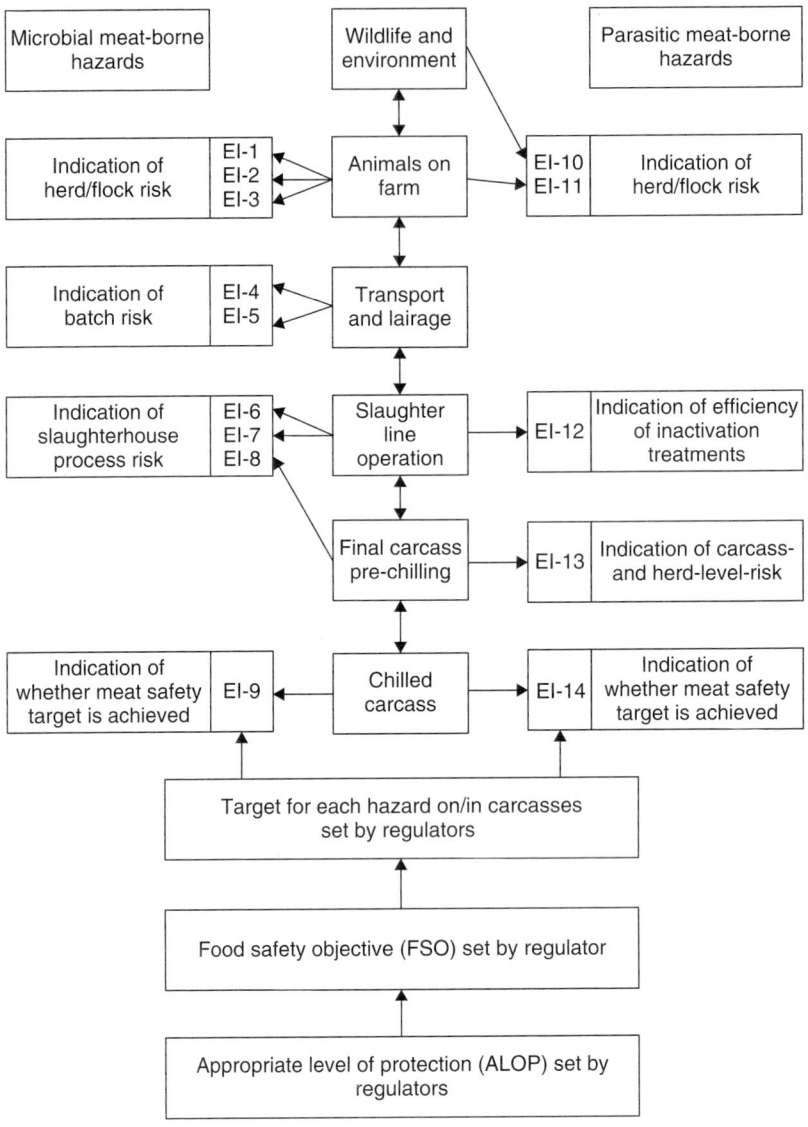

Figure 13.11   Epidemiological indicators (EI) used in risk-based chilled carcass meat safety assurance.

Literature and further reading

Blagojevic, B., Antic, D., Ducic, M. and Buncic, S. 2011. Ratio between carcass and skin microflora as a slaughterhouse process hygiene indicator. *Food Control*, **22**, 186–190.

Buncic, S. and Sofos, J. 2012. Interventions to control *Salmonella* contamination during poultry, cattle and pig slaughtering. *Food Res Int*, **45**, 639–653.

EFSA (European Food Safety Authority). 2007. Opinion of the scientific panel on biological hazards on microbiological criteria and targets based on risk analysis. *EFSA J*, **462**, 1–29.

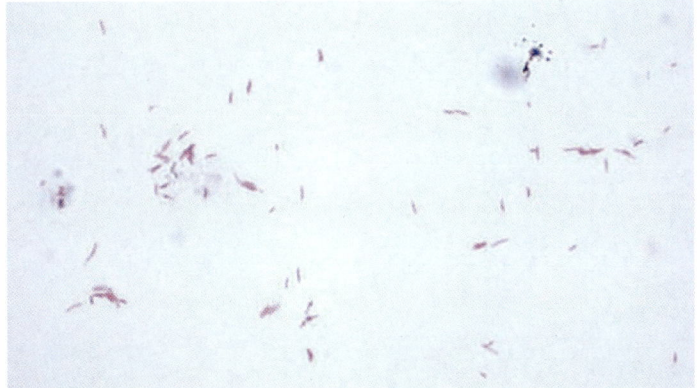

Figure 13.2 *Mycobacterium bovis* visualized using Ziehl–Neelsen stain. Source: Courtesy of Dr Hoermansdorfer.

Figure 13.6 *Trichinella* larva (length about one millimetre) freed from the muscle by artificial digestion. Picture has been taken with differential interference contrast (DIC) microscopy. Source: Courtesy of Seppo Saari.

Meat Inspection and Control in the Slaughterhouse, First Edition.
Edited by Thimjos Ninios, Janne Lundén, Hannu Korkeala and Maria Fredriksson-Ahomaa.
© 2014 John Wiley & Sons, Ltd. Published 2014 by John Wiley & Sons, Ltd.

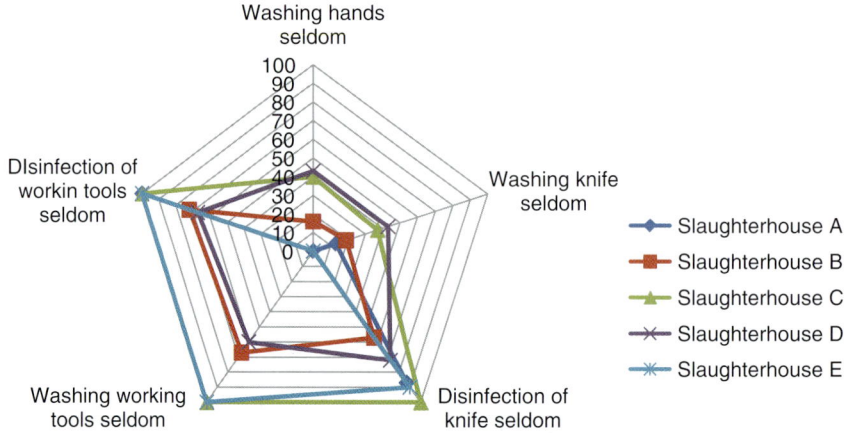

Figure 20.1   Hygiene practices of workers in five slaughterhouses A-E.

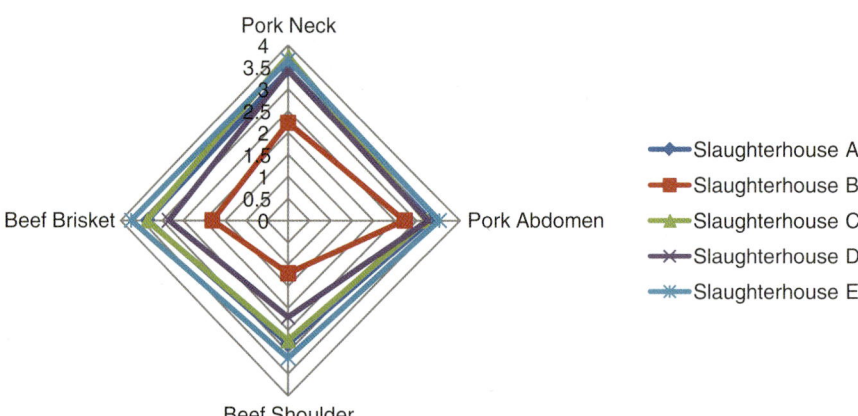

Figure 20.2   Total count (log 10 cfu cm^{-2}) of carcasses in five slaughterhouses A-E.

EFSA (European Food Safety Authority). 2008. Overview of methods for source attribution for human illness from food-borne microbiological hazards. *EFSA J*, **764**, 1–43.

EFSA (European Food Safety Authority). 2012. Scientific opinion on the development of a risk ranking framework on biological hazards. *EFSA J*, **10**, 2724.

[EFSA scientific documents and reports are publicly available at EFSA's web pages: http://www.efsa.europa.eu/en/publications.htm; last accessed 17 February 2014.]

EFSA documents on meat inspection (2011, 2012, 2013, 2014): Scientific opinions on the public health hazards to be covered by inspection of meat (5 documents) and Technical specifications on harmonised epidemiological indicators for public health hazards to be covered by meat inspection (5 documents): swine, poultry, bovines, small ruminants, farmed game.

The European Union Summary Report on Trends and Sources of Zoonoses, Zoonotic Agents and Food-borne Outbreaks in 2012.

Hill, A., Brouwer, A., Donaldson, N. *et al.* 2013. A risk and benefit assessment for visual-only meat inspection of indoor and outdoor pigs in the United Kingdom. *Food Control*, **30**, 255–264.

Hill, A., Horigan, V., Clarke, K.A., *et al.* 2014. A qualitative risk assessment for visual-only post-mortem meat inspection of cattle, sheep, goats and farmed/wild deer. *Food Control*, **38**, 96–103.

Nørrung, B., Andersen, J.K. and Buncic, S. 2009. Main concerns of pathogenic microorganisms in meat. In: *Safety of Meat and Processed Meat (Food Microbiology and Food Safety)* (ed. F. Todra). Springer, New York, pp. 1–30.

C. Chemical Hazards and their Control

Marcello Trevisani[1], Giuseppe Diegoli[2] and Giorgio Fedrizzi[3]

[1]*Department of Veterinary Medical Science, University of Bologna, Bologna, Italy*

[2]*Veterinary Public Health Office of Region Emilia Romagna, Bologna, Italy*

[3]*IZSLER (Public Veterinary Laboratory for Control of Animal Diseases and for Food and Feed Control), Bologna, Italy*

13.12   Scope

This part of the chapter introduces the concepts of risk management in Veterinary Public Health (VPH), specifically the basic tools and the procedures developed and the regulatory requirements that are used by veterinarians for risk management of chemical residues and contaminants in meat production (i.e. primary production and slaughterhouses). This includes information management and the appropriate use of surveillance procedures and analytical methods for detection (and quantification) of residues and toxic contaminants in meat as well as the detection of fraudulent practices. The basics of sampling plans, the organization of monitoring strategies and the tasks of Food Business Operators (FBO) are also discussed.

13.13   Introduction

Chemical residues (drugs) and contaminants in meat are among the most important issues for Veterinary Public Health (VPH). Basic risk management strategies aim at controlling these hazards by developing an effective information system that includes the use of monitoring plans, results of *ante-* and *post-mortem* inspection and investigations in case of substantiated suspicions. Risk may arise from abuse or illegal treatments of food-producing animals with veterinary drugs, anabolic substances and from environmental contaminations. The latter can be due to improper use of pesticides or release of persistent organic pollutants from industries and dump sites. Contaminants can enter the feedstuffs through the use of by-products or ingredients produced with inappropriate methods or derived from contaminated sources. In addition, feedstuffs may be contaminated by mould toxins or through cross-contamination of ingredients that are not controlled for their purity.

Chemical contaminants and drug residues used in veterinary medicine or additives used in animal husbandry can be present in meat. Commonly this cannot be avoided absolutely. Thus, within certain limits the level of risk is considered acceptable. National and international health authorities have to define Food Safety Objectives (FSOs), which are the maximum frequency and/or concentration of hazards in food at the time of consumption that provides or contributes to the Appropriate Level of Protection (ALOP) (i.e. the frequency of related food-borne diseases in a population). For environmental chemical contaminants, FSOs are based on Tolerable Daily Intake (TDI) levels. TDI is an amount that is estimated to be safe for an individual to ingest daily throughout a lifetime. For residues, these are indicated by Maximum Residue Limits (MRLs). MRL is an amount of residue that could remain in the tissues of food-producing animals treated with a veterinary drug because this residue has no adverse health effects if ingested daily over a lifetime by humans. On the basis of risk analysis, the Food Business Operators (FBOs) should define the frequency of the sampling plans that are needed to validate the effectiveness of the management strategies that they have actuated to comply with the FSOs. Health authorities monitor animal tissues, feed and products in the meat production chains and on the market. MRLs for the approved veterinary drugs are indicated by regulation in each country. However, detection of residues is generally not successful on the basis of random sampling, because the prevalence of contamination is usually very low.

FSOs are based on a quantitative risk assessment but the precautionary principle may be used when risk assessment has not given sound and unequivocal evidence that allows the definition of the boundary between safe and unsafe exposure. For the residues that are not tolerated, absence means that a given measure that is possible to achieve with the available analytical methods has been conceptualized. Harmonization of the maximum allowed limits for residues and contaminants is needed in order to facilitate international trades. However, sometimes different opinions have been expressed by the health authorities in different countries and the basis for achieving a harmonized common position is risk analysis. Agreements also involve the use of analytical methods that have comparable performances. Problems in international trades occur with farming practices that are allowed in some countries and forbidden in others, such as the use of growth promoters and antimicrobials used as feed additives or Genetically Modified Organisms (GMOs). Differences are due to the different evaluation and risk management strategies adopted to protect consumers by health authorities in different countries.

Planning of remedial actions is essential to reduce the risks when food products do not comply with FSOs. Withdrawal and recall of food and feedstuffs are remedial actions that apply also when residues or contaminants are detected in products that are on the market. These risk management tools require efficient systems to permit traceability.

The presence of food products that are contaminated at a level above the MRL or the evidence of illegal treatments indicate that control of contaminants at the production level was ineffective and should lead FBOs to improve their control strategies. Effective food safety control measures can be put in practice by adopting good farming practices that fulfil the regulatory requirements. Restrictions and licenses concerning the use of veterinary drugs and pesticides, the importance of 'day to slaughter' and 'day to harvest', information on environmental pollution and the origin of ingredients and additives used for feed production, the liability of suppliers and the use of quality control procedures are needed to prevent or reduce the risk posed by the chemical hazards.

Controls at the slaughterhouse can significantly contribute to the risk assessment and management. Investigations aimed at detecting irregular or fraudulent use of veterinary drugs and the detection of other toxic substances can be prompted by the results of *ante-* and *post-mortem* inspection at the slaughterhouses and by information gathered from FBOs at farm level. At the slaughterhouses, many lines of information converge and they are a core node linking primary production and distribution of meat.

A list of chemical hazards that may occur in meat of slaughtered animals is presented in Table 13.15. These are discussed in more details here.

Table 13.15　Categories of chemical hazards potentially occurring in the meat of slaughtered animals.

Categories of chemical hazards	Subcategory and examples
1.　Residues of veterinary medicine products	a.　Subjected to assessment and approved for use in farm animals (e.g. antimicrobial, antiparasitic, anti-inflammatory and sedative substances)
	b.　Substances for which an acceptable daily intake could not be established (in the EU, chloramphenicol, nitrofurans and their metabolites, medroxyprogesterone acetate, dimetridazol, metronidazole, ronidazol, chlorpromazine dapsone)
2.　Pharmacologically active substances that are used for their anabolic effect	(e.g. steroid hormones, resorcyclic acid lactones, stillbenes, beta-agonists)
3.　Residues of feed additives	
4.　Environmental pollutants	a.　Natural toxins (e.g. mycotoxins and plant toxins)
	b.　Toxic elements (e.g. heavy metals such as Cd, Pb, Cr and Ni)
	c.　Pesticides (e.g. insecticides, acaricides, herbicides, fungicides, rodenticides, plant growth regulators)
	d.　Industrial wastes and by products (e.g. polychlorinated debenzodioxins, dibenzofurans and biphenyls, perfluorooctane sulfonic acid and sulfonate, polybrominated diphenyl ethers)
	e.　Toxicants released from fires and accidental events (e.g. dioxin and dioxin-like compounds, polycyclic aromatic hydrocarbons)

13.14 Residues of veterinary medicine products

Residues of veterinary medicine products (VMPs) can be due to substances that have been subjected (approved) or not to assessment and pre-marketing approval (not approved). MRLs have been set for approved pharmacologically active substances. They are the points of reference for the establishment of withdrawal periods in marketing authorizations for VMPs to be used in food-producing animals, as well as for the control of residues in food of animal origin.

Not-registered substances are prohibited and comprise VMPs for which an acceptable daily intake (ADI) could not be established or anabolic substances and substances that may alter meat quality or affect animal health and welfare. Certain substances are not licensed for use in animals intended for food production because it has not been possible to define their ADI levels (safe levels) and, therefore, also not their MRLs. In the European union (EU), veterinary medicines such chloramphenicol, nitrofurans and their metabolites, medroxyprogesterone acetate, chlorpromazine, dapson, dimetridazol, metronidazole, ronidazol have been included in this category. In order to make results of residue testing reliable and comparable between laboratories in different countries/states, Minimum Required Performance Limits (MRPLs) have been established and these MRPLs have been used in the reporting system.

13.14.1 Antimicrobials

Some antimicrobials have been used in animal feed for growth promotion (bacitracin, carbodox, olaquindox, tylosin, virginiamycin, avilamycin, flavophospholipol, lasalocid sodium, monensin sodium, and salinomycin). Although these antimicrobials are not absorbed and, therefore, cannot be found in meat, liver or kidney tissues, their use was banned in some countries (e.g. the EU) for the potential impact on the emergence of antimicrobial resistant enteric bacteria. Also, bad practices (wrong dosage or administration frequency, prescription not respected) in the use of antimicrobials that are approved can contribute to resistant bacteria. The presence of antimicrobial residues in food products has little effect on the selection of resistant bacteria in humans, instead the resistant bacteria that are spread in the environment and carried by foods pose a risk.

Residues of antimicrobials in meat are normally very low; therefore, the main issues are related to antimicrobials that show direct toxicity at very low doses. Risk of allergenic reactions mainly concerns penicillin and related substances that are both very immunogenic and frequently used, although most often for topical use in milk producing animals. The most frequently quoted case of possible direct toxicity is that of chloramphenicol, known to have caused aplastic anaemia in humans when used in human medicine. This antimicrobial is now prohibited in veterinary medicine in Europe. Some antimicrobials, such as nitrofurans, have been banned because of the concerns about the carcinogenicity of their residues in edible tissue, while some others cannot be used in veterinary medicine because public health authorities have not received a request of approval from producers. The producers must provide the results of toxicological and pharmacokinetic studies, as well as validate the analytical methods used for the detection of residues in foods and animal tissues.

Controls concerning the approved substances are mainly aiming at verifying the appropriateness of their use. Animals must not be sent to slaughterhouses until the withdrawal period has elapsed. This depends on the pharmacokinetics and the known tissue distribution, and information must be reported on the drug label. Treatments have to be prescribed under the care of licensed veterinarians and this information is available at the slaughterhouse. Documentation concerning all antimicrobials supplied and administered to the animals intended for food production should be kept by the prescriber, the supplier and the end user, and their use should be declared in the delivery documents. Information should be delivered to those in charge in the slaughterhouse, who have to decide if the animals can be accepted and make the information available to official veterinarians before the animals are delivered. Farm animals that are sent to the slaughterhouses must be identified. If drugs are used more often than usual on the farm (above average for each specific category of animals) it indicates that hygiene and animal health management may be below the standards achieved in the majority of the farms. Among other factors that can affect an improper management of therapeutic use of drugs is the proper dosage, which is affected by the distribution of the drug in the vehicle used (i.e. water or feed pre-mixes). In some countries, for example, water hardness impairs the solubilization of tetracyclines.

Evaluation of documents together with veterinary inspection can address further controls. Misuse or fraudulent use (not true statements) should be suspected when *post-mortem* inspection shows signs of pathologies that would be treated with medicines not declared, there are signs of injections in farm animals or the information that derives from the farm is suspect (i.e. clinical observations in relation to the prevalence of pathological findings and the results of inspections on other batches delivered by the same farm). Suspect signs and conditions in animals, which indicate that that they might have been treated recently with antimicrobials, are, for example, apathetic animals, animals showing skin disorders, animals with respiratory problems, lameness, dehydration and pain.

Post-mortem findings indicative of a probable recent use of antimicrobials include, for example, acute/subacute pneumonia/pleuropneumonia, arthritis, mastitis, endometritis, nephritis and visible signs of injections. The animal species and information obtained from the pharmacosurveillance reports can be indicative of the type of residues (e.g. macrolides, trimethoprim/sulfa or sulfadimidine or amoxicillin in infectious rhinitis in pigs or aerosacculitis in broilers; quinolones, spectinomycin, ciprofloxacin, amoxicillin, ampicillin, apramicin in enterocolitis; sulfonamides and tetracyclines for protozoan diseases). However, a black market of veterinary medicinal products exists and the best option for safety control is the qualification of farmers through periodic audits, inspections and analytical controls on site. Pharmacosurveillance (or surveillance of the medicaments) is a valid tool to assess the appropriate use of drug in veterinary medicine (i.e. concerning the dose and appropriateness for the specific pathologies).

Non-specific microbiological tests for detection of antimicrobial residues in kidney, liver and muscle tissues can be used, but the identification of these residues is difficult. Therefore, information on substances that are commonly used to treat the pathologies that are detected at meat inspection is useful to address the use of rapid tests, such as enzyme-linked immunosorbent assay (ELISA) tests for groups of suspected antimicrobials and tests for beta-lactamase. However, after a preliminary screening test, confirmatory methods must show that the level of residues is above a threshold level or the MRLs. It must be underlined that microbiological tests do not have a sufficient sensitivity for most antimicrobial residues and the risk of false-negative results is high. Therefore, they have been replaced by use of more specific tests (like ELISA and enzymatic tests for beta-lactamase) or chromatographic screening methods such as liquid chromatography in tandem with mass spectroscopy (LC-MS/MS).

The detection of illegal and fraudulent use of antimicrobials and other categories of veterinary drugs is tricky. There are reports of investigations indicating the illegal commerce of veterinary drugs and evidence of illegal practices. These are almost always due to an economic interest. Examples include both bad practices that may be used by some farmers to solve some issues, like the use of chloramphenicol for treatment of respiratory diseases in young rabbits. In countries, where use of chloramphenicol is allowed in food-producing animals, it is used for the treatment of a variety of infections, particularly those caused by anaerobic bacteria or those that are resistant to other antimicrobial agents. It is well absorbed by both oral and parenteral routes in pigs and poultry but not by ruminants.

The presence of chloramphenicol residues in the urine of pigs and in frozen poultry meat mechanically separated have been shown during routine inspections. Investigation showed that pigs had been fed with milk by-products not intended for animal use. Similar problems might arise from the use of pre-mixes produced with ingredients that are not controlled for their purity. It is not uncommon that some drugs used to formulate approved VMPs and feed pre-mixes are produced in pharmaceutical industries located in countries where other drugs not allowed in the importing country are also produced. Carry-over is not a rare event if quality controls are not strict. Risk management should involve, therefore, the selection of reliable suppliers of feedstuffs and control the purity of drugs used in feedstuffs.

In the European Union, chloramphenicol is not approved for use in animals intended for food production because concerns have been expressed about genotoxicity of chloramphenicol and its metabolites, its embryo- and fetotoxicity, and its carcinogenic potential in humans. There is good evidence that chloramphenicol is haemotoxic in humans, causing a dose-related reversible bone marrow depression and also a severe aplastic anaemia in humans, which is non-dose-related and often irreversible, related to the formation of a metabolite, chloramphenicol aldehyde. Therefore, it was not possible to define an acceptable daily intake. Nevertheless, chloramphenicol is a broad spectrum antimicrobial that is still used in humans and for topical use in companion animals. ELISA kits are available for screening chloramphenicol in muscle tissue and animal urine, but for confirmation LC-MS/MS is the most suitable methods. The European Union has a prescribed minimum required performance limit (MRPL) for chloramphenicol; this represents a minimum concentration of chloramphenicol in a sample which must be detected by the applied methods.

In past years, residues of sulfonamides (such as sulfamethazine) were detected in pig meat. Producers were blamed for not complying with the withdrawal period, but it was later realized that many violations occurred at farms where producers did follow proper withdrawal times. In some cases, it was proved that pigs had no access to medication with sulfonamides. Researchers found that very small amounts of sulfamethazine in the feed (as little as 1 g of sulfamethazine per ton of feed) would cause a residue problem in the tissue (especially in liver). The level of residues is even higher in aged meat products (e.g. cured raw hams and salami) due to the loss of water and, consequently, concentration of sulfonamides in meat. Other causes of residues may include manure or lagoon water recycling, contaminated manure packs, delivery errors and obtaining contaminated ingredients or feed from the feed supplier. The main problem, however, is cross-contamination of non-medicated feed from on-farm mixing and handling. Contamination of the drinking trough was also shown. This strengthens the importance of controls that should be done at the feed producing level.

Nitrofuran, an antimicrobial used for the treatment of bacterial diseases in livestock production, was banned from use in the European Union in 1995 due to concerns about the carcinogenicity of the residues in edible tissue. Nitrofuran is, however, still used in some countries as a growth promoter and prophylactic agent because it is cheap and effective. The presence of nitrofuran residues in meat, aquaculture and other products has mainly been detected in products originating from non-European countries (as documented in the European 'Rapid Alert System for Food and Feed' – RASFF). Due to the instability of nitrofuran, effective monitoring of the illegal use through detection of the parental compounds is difficult, but their metabolites persist. The coupling of high performance liquid chromatography and tandem mass spectrometry with electro-spray ionization (HPLC-ESI MS/MS) has significantly advanced the capabilities of quantitative methods for the detection of nitrofuran. In addition, antibody-based detection methods have also been developed.

13.14.2   Antiparasitic drugs

Many drugs are used to control ticks, mites, lice and other animal pests, and to fight parasitic diseases. This is very important for animal health and welfare and also to ensure good production, but the use of certain drugs that are absorbed into the bloodstream and transported, to different parts of the body may result in the presence of residues in meat. As for other veterinary drugs the rate of metabolism and excretion, and the safety profile, determine the length of the withdrawal time. Variations can exist among species and can also be affected by the route of administration and the dose. Absorption of drugs can be from the GI tract, injection site or skin. Oxidation and cleavage reactions commonly occur in liver. Parasites not only come into contact with drugs that are applied topically, but also with the absorbed fraction in the blood, which is important for the efficacy of many antiparasitic drugs. Depending on the solubility of the drugs and their metabolism, the pharmacokinetics vary and drugs can be recycled, for example, by the biliary route.

Antiparasitic drugs normally do not give evidence of residues in tissues of slaughtered animals and only seldom have been found in organs at concentrations above the maximum limit allowed by regulations, but their proper use has to be monitored because some of them can pose severe toxicological risks, including teratogenicity (benzimidazoles), neurotoxicity (ivermectine) or immunomodulation (levamisol).

Modern antiparasitic veterinary products that are used to control helmints (adult parasites and their larvae) have been synthesized to have a low toxicity and few side effects in farm animals and the environment, which is often due to the specificity of their activity toward the parasites and the metabolisation to inactive compounds that are extracted. A combination of good management and sanitation plus proper use of deworming agents will most effectively control internal parasites but problems are encountered (such as those due to the use of bedding materials, access to intensively used pasture, access to spaces without concrete or solid flooring that cannot be sanitized). The improvement of animal welfare in intensive farming systems also poses new challenges. Therefore, use of antiparasitic drugs is needed in combination with biohazard control programmes when there are not alternatives (such as live coccidiosis vaccines in poultry). The environmental context and the characteristics of animal farming system are important to evaluate the risk due to misuse of antiparasitic drugs. In the control of coccidiosis, antimicrobials like sulfonamides and tetracycline, which have been discussed previously, are also used.

The category of substances that are used to control internal parasites includes antihelmintic and antiprotozoal compounds. Nitroimidazoles used for coccidiosis have already been banned within the European Union, even for therapeutic purposes, since they are mutagens and suspected carcinogens.

13.14.3 Antihelmintics

Substances such as benzimidazoles and probenzimidazole (thiabendazol, flubendazol, fenbendazol, mebendazol, albendazol, oxfendazol and febantel), macrocyclic lactone derivatives (avermectin, ivermectin and moxidectin), imidazothiazoles (levamisole), salicylanilides (closantel, niclosamide, oxyclozanide and rafoxanide), piperazine and derivates (diethylcarbamazine), tetrahydropyrimidines (morantel and pyrantel), organophosphates (haxolon, coumaphos and dichlorvos) are included in the list of drug residues that require monitoring.

Dichlorvos, which has been used to control benzimidazole-resistant nematodes, should no longer be placed on the market in the European Union or United Sates because it is not expected to satisfy the requirements laid down by current regulations. Other organophosphates, which are safer, can be used instead.

Antihelmints that are currently used are well tolerated by both animals and humans. The modifications introduced have made their activities selective to the parasites. For example, avermectins block the transmittance of electrical activity in nerves and muscle cells by stimulating the release and binding of gamma-aminobutyric acid (GABA) at nerve endings with a paralysis of the neuromuscular systems. However, the GABA-ergic receptors are found at the neuromuscular junctions and the central ventral cords in nematodes, whereas in mammals they are found primarily in the brain and ivermectin does not readily cross the blood–brain barrier in mammals at therapeutic doses. A No Adverse Effect Level (NOAEL) was established in dogs, which are the most sensitive species (among those used in toxicological studies). Depletion studies in farm animals have shown that residues in edible tissues are below the ADI a few days after the treatment with high doses.

Cholinergic antihelmintics, such as pyrantel, morantel and levamisole, act as agonists of nicotinic acetylcholine receptors at the nematode neuromuscular junction and cause spastic paralysis. Concerning levamisol, which has been used for years as a broad spectrum antihelmintic, haemotoxic effects (including severe leukopenia and agranulocytosis) have been observed in humans. They appear to be related to immunological activity. Pyrantel and morantel are generally well tolerated by target animal species. In humans, side effects have been observed in the form of central nervous system effects (including dizziness, drowsiness, headache, irritability and trouble in sleeping) and gastrointestinal tract involvements (such as abdominal cramps and pain, diarrhoea, loss of appetite, nausea and vomiting). The reported hypersensitivity (skin rash) seems to be due to hepatic toxicity, which is indicated by increased aspartate-aminotransferase and alanine-aminotransferase activities in the serum.

Salicylanilides include antihielmintics used for fighting cestodes (like niclosamide) and liver flukes (like closantel, rafoxanide, oxyclozanide and nitroxynil). Closantel, rafoxanide, and oxyclozanide have long terminal half-lives in sheep (14.5, 16.6 and 6.4 days, respectively), which are related to the high plasma-protein binding (>99%) of these three drugs. Residues in liver are detectable for weeks after administration, thus needing a longer withholding period.

A wide range of substituted phenols has been used for the treatment of liver flukes and tapeworms in animals. Nitroxynil is an injectable flukicide possessing activity against mature and young flukes. It is used subcutaneously in cattle and sheep. In cows, sheep and rabbits, nitroxynil is highly bound to plasma proteins. Unchanged nitroxynil was the major component of the residues in calf kidney, muscle and fat, accounting for around 56, 69 and 78% of the total residues, respectively. The 4-cyano 2-nitrophenol was the major component of the residues in calf liver with unmetabolized nitroxynil composing only 2% of the residues. In sheep, nitroxynil was the major component of the residues in kidney, muscle and fat, accounting for 45–56, 90–100 and 64–100% of the extractable residues, respectively, at five-day withdrawal. In sheep liver, most of the residues were in the form of 3-iodo-4-hydroxy-aminobenzamide, while 4% was unchanged nitroxynil.

Praziquantel is an isoquinolone analogue that is marketed as a mixture of stereosiomers. Although its action mechanism is still somewhat obscure, praziquantel increases cell membrane permeability in susceptible worms and flukes. This results in intracellular accumulation of calcium, and paralysis and contractions of the parasite musculature. Phagocytes rapidly attach to the affected parasites resulting in cell death. A half-life of 4.2 h was determined and excretion from plasma was rapid, 98% being excreted within 72 h.

Ultra-performance liquid chromatography–tandem mass spectrometry (UPLC-MS/MS) has been developed to simultaneously detect 38 anthelmintic drug residues in bovine tissues.

13.14.4 Antiprotozoals

The most used drugs include benzamides (aklomide, nitromide and dinitolmide), carbanilides (nicarbazin and imidocarb), nitroimidazoles (ronidazole, dimetridazole and metronidazole, which are not approved in the EU, and ipronidazole), polyether ionophore (monensin, narasin, lasalocid, salinomycin and maduramicin), quinolonederivates (buquinolate, decoquinate and methylbenzoquate) and triazines (clazuril, diclazuril and toltrazuril).

Polyether ionophores (such as monensin, narasin, lasalocid and salino-mycin) are drugs that are synthetized by some *Streptomyces* species. They have also been used to manipulate rumen fermentation by increasing propi-onate and decreasing acetate, which improves the efficiency of feed utilization and increases weight gain in growing ruminants. Ionophores also decrease dietary problems from high carbohydrate diets. Ionophore intoxications are often related to errors in feed mixing. Ionophore poisoning is caused by their bioactivity and results in damage to excitable tissues such as cardiac, skeletal and smooth muscles, and the nervous system. Equines are much more susceptible to cardiac muscle damage, but intoxications have also been reported in cattle and ostrich. The major risk for humans is posed, however, by the supplementation of the feed of birds laying eggs. Their broad spectrum of activity on protozoa and bacteria (especially gram-positive bacteria) is due to modification of metal cation and proton transport across the membranes, which leads to an increase in the osmotic pressure inside the cell, causing swelling and vacuolization, and finally death of the bacteria cell. The wide use of these drugs in livestock (among other those that are not absorbed and do not produce residues) has led to selection of resistant bacteria. Thus, several European countries have restricted or banned their use. Probiotics such as *Streptomyces cerevisiae* and *S. ruminantium* are now used in ruminants to manipulate certain biochemical events and the microbial composition of the rumen.

The problem of residues from antihelminthic, anticoccidial and antiprotozoal drugs may be easily controlled by imposing obligatory withdrawal times, generally 7–10 days, but this is, unfortunately, not always respected. On the other hand, given the large number of drugs that may be easily obtained on the market, many producers change one compound for another to avoid development of resistance to drugs.

13.15   Substances having anabolic effects and unauthorized substances

The use of hormones and other drugs has been introduced into animal husbandry to increase the production of meat and to improve the feed conversion efficiency. Controversial opinions on the management options that are needed to prevent risks to consumers and collateral bad effects on animal welfare have led to different legislations worldwide and problems in the food market. In almost all countries, the use of growth promoters is restricted. In countries where anabolic substances are approved for use in farm animals, appropriate administration routes and dosage should prevent risks for consumers.

13.15.1   Sexual steroids

Five steroid sex hormones, including 17β-oestradiol (as such or as benzoate), testosterone (as such or as propionate), progesterone, trenbolone acetate and zeranol, are approved in the United States and some other countries as components in solid ear implants that are used for calves, steers, heifers, bulls, cows, sheep and lambs. Melengestrol Acetate (MGA), which is a progesterone analogue, can be used as a feed additive for feedlot heifers. Ractopamine is a dietary β-agonist that is approved for use in pigs with a 50 μg/kg tolerance in the United States. Zilpaterol-HCl is an active β2-agonist registered for fattening purposes in cattle in Mexico and South Africa. Risk analysis led to different evaluations and control strategies in Europe, where the use of steroid hormones, β-agonists and other anabolic substances in animals intended for food production is forbidden, but a black market exists and many different anabolic substances have been used illegally.

Governments that have approved the use of the anabolic substances have based their decision on following facts: 17β-oestradiol, progesterone and testosterone have low oral bioavailability due to their inactivation in the gastrointestinal tract and by the liver, and have low acute oral toxicity. All adverse effects observed after repeated administration in laboratory animals

are correlated to their hormonal effect, including the occurrence of reproductive disorders, and for 17β-oestradiol and testosterone the carcinogenic
potential is related to their hormonal activity and, therefore, is not below
the NOAEL. Therefore, the ADIs were defined on the basis of exposure to
the possible residual level in food and toxicological studies concerning the
changes in hormone-dependent effects in humans and animals. As for the
dose-response (risk characterization), a safety factor of 10 was applied to
account for normal variation between individuals and an additional tenfold
safety factor added to cover inter-species differences may be included to
account for the use of the Lowest Observable Adverse Effect Level (LOAEL)
in cases where the NOAEL could not be established in animal studies. Consequently, residue levels occurring after approved use generally have to be
considered safe. However, different considerations and other studies have
been used for risk evaluation purpose in the European Union, leading to different control options for the risk management. In particular, 17β-oestradiol
proved to be both mutagenic and carcinogenic. Repeated exposure to hor-
monally active substances in foods has shown that even at very low doses they
can interfere with the synthesis, secretion, transport, binding, action or elimination of natural hormones in the body that are responsible for development,
behaviour, fertility and maintenance of normal cell metabolism. They have
been indicated by the term 'endocrine disruptors'. As for the sex steroid hormones, the adverse biological effects in humans include reduced fertility, male
and female reproductive tract abnormalities, early puberty, impaired immune
functions and various cancers (e.g. breast, vaginal and prostate carcinoma).

Many different steroids with anabolic properties have been detected
during monitoring and control in Europe, among others trenbolone 12α and
12β, 19 nortestosterone 17α and 17β, methyltestosterone, chortestosterone,
chlormadione, stanozolol, 16-OH stanozolol, methandrostenolone, bolde-
none, oestradiol and esters, medroxyprogesterone acetate, melengesterol,
megesterol, methandriol, fluoxymesterone, flugestone, chloroandostedione,
caproxyprogesterone and acetoxyprogesterone. Synthetic drugs of the
stilbene group, such as diethylstilbestrol (DES), were used many years ago,
but abandoned due to the high risk posed to consumers by the orally active
oestrogenic residues persisting in the muscles.

The use of chemical derivatives of the natural hormones (like methyl-,
hydroxyl- and undecylen-esters) is done for the purpose of producing long-
acting injectable products that can be administered by intradermal implants.
Some anabolic products show a low androgenic or oestrogenic potency and
the typical signs that can indicate the use of sexual hormones are therefore
masked. Often they are used in cocktails and combination so each compound,
individually, has a very low or undetectable residue. Endogenous and artificial
steroid hormones are predominantly excreted from the body via urine and
faeces, with differences in animal species. Available data are mainly results
of residue analysis performed in serum or excreta sampled at farm level.
Detection of the parental compounds in urine or faeces at farm level is
very difficult and the metabolism of many substances recently administered

for illegal growth promotion have to be studied in order to define the metabolites that can be used as biological markers for their use. In order to mask the illegal treatments substances that are quickly metabolized, such as 17β-oestradiol, are often used. Natural hormones are frequently used because these compounds lead to the same hormones and metabolites as the endogenously produced hormones. Demonstration of their use therefore requires quantitative measurements and detection of abnormal levels in urine or in blood. The concentration of these endogenous hormones can widely vary according to animal species, sex, age and physiological state of the animal.

Examples of synthetic anabolic hormones with androgenic effects which were used because they are more difficult to be detected include stanozolol (or 5α-androstane-17α-methyl-17β-ol (3,2-C) pyrazole), which most resembles methyltestosterone, norethandrolone and ethylestrenol. Pro-hormones, such as androsta-1,4-diene-3,17-dione (ADD or boldione), which may be converted to the more active synthetic androgenic compound 17β-boldenone in the gut of cattle have also be used.

There is a debate on the capability of controlling the illegal use of anabolic substances and the advantages/disadvantages of allowing the use of some substances with appropriate restrictions. Detection of many different substances used in mixtures with frequently changed chemical composition is not an easy task and there is a debate on when the identification of the banned substances cannot be done but there is evidence of their use. However, the reasons for prohibiting their use, although residual levels can be very low, are due to the fact that authorization would increase their use. With difficulties in ensuring appropriate controls on abuses, this can cause repeated exposure of consumers to these residues that at very low concentration are also endocrine disruptors and their effects may sum or synergize with those of other chemical pollutants. There are also concerns due to the fact that synthetic hormones excreted by animals are present in manure applied as fertilizer and in feedlot retention ponds, and from there they may be retained in soil or transported to ground and surface water. Last, but not least, there are also animal welfare problems.

Some countries (e.g. EU) have, therefore, decided to control/discourage the illegal use with inspection in farms and suspect sampling. Suspect sampling arises as (i) a follow-up the occurrence of a non-compliant result and/or (ii) on suspicion of illegal treatment at any stage of the food chain and/or (iii) on suspicion of non-compliance with the withdrawal periods for authorized veterinary medicinal products.

Suspect signs can be observed at *ante-mortem* inspection, such as heifers showing overdeveloped mammary glands, with leaking milk or males with abnormally small testicles. Signs can be shown in some sex glands (e.g. in calves, steers and heifers) but their existence is not correlated with the detection of residues in tissues in slaughtered animals. Controls are more effective at farm level, although they should include the substances used at that moment in the farms and there is a continuous development of hormone combinations, which are mainly administered as percutaneous pellet implants.

Samples analysed include plasma, serum or urine and muscle, where the hormones and their metabolites can be detected. Relevant levels of natural hormones and the presence of synthetic anabolic in plasma or the detection of their metabolites in urine are indicative of abusive/illegal treatments.

For some hormones the availability of specific antibodies makes possible the use of immunoassays, most often enzyme immunoassays (EIAs), which can be performed in laboratories that do not have mass spectrometry facilities, are highly sensitive and cost effective. Multiresidual methods based on LC-MS/MS are now available for quantitative analyses. They allow detection of all known anabolic steroids such as 17α-methyl, 17α-ethyl, and 19-nortesterone steroids. Samples are generally extracted from serum, plasma or tissues using solid phase extraction (e.g. loaded on C18 mini SPE columns) and quantitatively analysed with LC-MS/MS in presence of internal standards.

Residues of testosterones and related steroids, such as epitestosterone, 4-androsetenedione and dehydroepiandrosterone (4- or 5-ene metabolites), which are normally present in males and females of mammalian species, have to be analysed quantitatively and the obtained levels have to be compared with the normal levels of the males and females of the same age and species. Exogenous administration of testosterone and oestradiol in heifers and steers may be indicated by the presence of significantly increased level of epimerized 17α-metabolites (e.g. 5β-androstane-3β, 17α-diol). The presence of 19-nortestosterones, such as nandrolone, or significant levels of epiboldenone and related metabolites in urine are also indicative of exogenous administration of anabolic hormones in bovine.

Some natural hormones are used for treatment of reproductive problems, but they may also be misused for economic interest due to the anabolic effects. Various analytical methods have mostly been developed to screen the misuse of hormones; however, distinguishing an exogenous administration of hormones from natural origin (i.e. oestradiol-17β or progesterone) still remains a challenging task for authorities. Because the target compounds are always present in the analytical matrix, and because the concentration levels of natural steroids are extremely variable from one animal to another, the establishment of reference thresholds appears very difficult.

The current official approach to detect illegal use of growth promoters relies on the random selection of animals to be sampled, either at the farm or the slaughterhouse, with the subsequent application of analytical techniques based solely on the physicochemical recognition of a given number of molecules, but facts have shown that control on suspicion can be more effective. In order to develop effective low-cost screening methods, many researchers have studied the biological effects of illicit performance-enhancing agents in the target species. The main classes of such agents have well recognized mechanisms at the molecular level, regardless of their chemical structure.

Indirect evidence of hormonal stimulation can be found based on sexual accessory glands, where some histopathological findings are consistent with the use of oestrogens or abnormal changes in the plasma protein profile (metabolomics) that can be related to hormonal treatments, such as serum apolipoproteins in veal treated with the synthetic anabolic androgen 17β-boldenone and its precursor or metabolite. Also, modifications in the binding capacity of sex hormone receptors and a regulation of gene expression, such as genes encoding proteins of receptors *GR-a* and *ER-a*, a number of interleukins, and some apoptosis regulators. Although significant results have been obtained, the measured level of the gene expression needs to be challenged with variability detected in field conditions.

The most consistent histopathological effect of prolonged hormonal stimulation with oestrogen in calves is a metaplasia in the glandular tissue of the disseminated prostate (males) and the Bartolin's gland (females). The glandular tissue replaces the mucinous glandular cells, making them look like excretory ducts, which also have multilayered epithelium. The gland appears to have more ducts than normal glands. The normal histology of the disseminated prostate gland is characterized by alveoli lined with a low epithelium constituted of cylindrical and cubical cells, while the urethra is covered with a transition epithelium. The detection of metaplastic changes in the epithelial cells appears to be facilitated by immunohistochemistry with cytokeratin antibodies. Metaplasia of the bulbo-urethral glands was also described in calves experimentally treated with low-dosages of 17β-oestradiol even after two weeks of suspension. However, animals that are at oestrus normally show some local metaplasia. Hyperplasia of the ducts and local metaplasia without a change in the gland-to-duct ratio in female calves can be due to endogenous hormones and is not always an indication of hormonal treatment. Despite these advantages, its reliability for screening purposes is still questionable due to the lack of accuracy evaluation studies.

13.15.2   Beta-agonists

Anabolic substances illegally used worldwide not only included steroid sex hormones. Clenbuterol is authorized for therapeutic use as a bronchospasmolytic and tocolytic agent in veterinary medicine. In dosages up to 10 times the therapeutic one, clenbuterol has marked anabolic effects, redirecting cellular energy metabolism in favour of protein synthesis. In muscle tissue, β-agonists promote protein synthesis and cell hypertrophy by inhibition of proteolysis. In adipose tissue, β-agonists promote lipolysis. This may result in a reduction of carcass fat up to 40% and an increase of carcass protein up to 40%, an advantage, especially in the market of swine (but not the heavy pigs intended for production of aged hams and long seasoned cured meat products). β-Agonists include analogues of clenbuterol, ractopamine, salbutamol and zilpaterol. Certain modifications on the clenbuterol structure can deceive the tests without substantially affecting the biological activity. Six new clenbuterol congeners (G4, G5, G6, G8, D5 and D9) have been prepared in this way. They cannot be identified by the common tests; however, they were only about one-tenth as effective as clenbuterol. Recently, a new series of clenbuterol-like compounds occurred in black market (trivially named as compound A and clenmeterol). A natural β-agonists such as hordenine has also been used. It is present in 'Zhi Shi', a Chinese herb, which is reported to stimulate the β3-receptors, resulting in breakdown of fat cells. Illegal use of these active substances resulted in several cases of food poisoning with symptoms, such as tremor, tachycardia and nervousness.

The high affinity of β-agonists for melanin makes the retina and the black hairs suitable samples to detect the use of these substances. Relaxant effects of clenbuterol on tracheal smooth muscle can be observed at the *post-mortem* inspection. ELISA tests are available for the phenolic β(2)-adrenoreceptor agonists (salbutamol, fenoterol, and terbutaline) and aryl amine β-agonists (clenbuterol, mabuterol and mapenterol). Spectroscopic methods (LC-MS/MS) are available for screening (multiresidue analysis) and for confirming detection of β-agonists in tissues.

13.15.3 Drugs used to mask signs and avoid collateral effects of sexual steroids and beta-agonists

Synthetic glucocorticoids have been used in cocktails with β-agonists to prevent receptor downregulation, to improve their tolerance in the animal and also to increase weight by incrementing the water content in muscles. It has also been shown that low doses of glucocorticoids improve feed intake and increase live weight gain, reduce feed conversion ratio, reduce nitrogen retention and increase water retention and fat content. Glucocorticoids include dexamethasone, betamethasone, prednisolone, methylprednisolone, prednisone, flumethasone, isofluredone and triamcinolone acetonide, clobetasolpropionate and beclomethasonedipropionate, and esters of cortisol.

Indirect evidence of use of low dosages of dexamethasone in beef cattle can be obtained (although not fully specific) by observing the morphologic changes in the thymus, which shows an abnormal fat infiltration with concurrent cortical atrophy and reduction of the cortex/medulla ratio (C/M). A cut-off for the cortex/medulla ratio 0.93 C/M was used to distinguish control and treated animals. The animals treated with dexamethorphan showed inhibition of cortisol secretion during the treatment period and for up to three days after treatment.

13.15.4 Benzodiazepines

Diazepam can be used as anxiolytic and sedative to prevent stress during sheep transport, thus limiting occurrence of leg injury. Brotizolam can be used to counteract the side effects of β-agonists, in particular to stimulate the feed intake in weak animals and reduce tremors.

13.15.5 Thyreostats

Thyreostatics increase body weight by augmenting water retention. Consumers are misled (water is sold for the price of meat) and the quality of the meat of animals treated with the drugs is inferior. They also lower gastrointestinal motility and, therefore, live weight is also increased by excessive filling of the gastrointestinal tract. Glucosinolates (like oxazolidine-2-thiones) in cattle fattening may originate from feed; 5-vinyl-oxazolidine-2-thione (goitrin or 5-VTO) derivative can be present in large amounts (up to 1%) in the seeds of Cruciferae (e.g. rapeseed). Since rapeseed meal is used as a cheap protein source in animal feed, the highly goitrogenic 5-VTO may appear in milk and constitute a health hazard to infants and children who consume milk in large quantities. Thiouracils and 1-methyl-2-mercaptoimidazole (tapazol) are synthetic drugs, cheap and readily available on the black market. Thyreostatics are drugs capable of inhibiting the production of thyroid hormones. Thyreostatics are generally excreted in urine in the form of both the parent and conjugated compounds. The highest concentrations of thyreostatics occur in the thyroid. Studies in cows have shown that methyl-thiouracil can be detected as a free unchanged compound for several weeks after oral administration. A multiresidue method exists for tapazol, thiouracil, methylthiouracil, propylthiouracil and mercaptobenzimidazol in thyroid tissue using LC–MS/MS after derivation of the compounds with 7-chloro-4-nitrobenzo-2-furazan.

13.15.6   Antibacterial synthetic quinoxaline compounds

The growth promotion activity of carbadox and olaquindox can be due to a stabilization of the intestinal microflora improving the feed conversion and reducing the formation of toxins. Although they do not interfere with the hormonal homeostasis, both compounds are mutagenic, and carbadox also shows carcinogenic effects in animals. The illegal use of these substances in feed premixes has been detected especially in pigs, poultry and rabbits.

13.15.7   Non-steroidal anti-inflammatory drugs (NSAIDs)

Non-steroidal anti-inflammatory drugs (NSAIDs) include enolic acids and pyrazolon derivatives, salicylic acid derivatives, p-aminophenol derivatives, indole indene derivatives, heteroaryl acetic acids, arylpropionic acids, anthranilic acids and alkanones. NSAIDs act by blocking the action of cyclooxygenase (COX), the key enzyme in the synthesis of eicosanoids (prostaglandins and related substances) from arachidonic acid. Salicylates are routinely used to condition animals just after transport to reduce the effects of stress. Administration before slaughtering has been used to give a more efficient bleeding (due to their anti-clotting effect) and to improve the colour in pig and calf meat. At high doses, they can reduce lipogenesis by partially blocking incorporation of acetate into fatty acids. Residues of NSAIDs can be detected in serum, plasma and muscle by using HPLC with diode array detectors (HPLC-DAD). Multiresidue analysis and drug confirmation can be done with spectroscopic methods based on LC-MS/MS.

13.15.8   Arsanylic acid

Arsanylic acid and its sodium salt are most commonly used, particularly in pigs and egg-laying hens. However, their use in animals is generally rather limited and the risk–benefit ratio is questionable because these drugs can produce toxicosis, known as peripheral nerve demyelination.

13.15.9   Somatotropin (or growth hormone, GH)

Porcine growth hormone (GH) has been approved since 1995 for commercial use in growing pigs in Australia. Recombinant bovine somatortopin (rbST) is widely used to stimulate milk production in cows but it is not used commercially for growth promotion. Somatotropin (or GH) is used in cattle, either alone or in association with a β-agonist, since GH will induce upregulation of β-adrenoceptors in some species. GH cannot be administered orally, but by injection. GH can be measured directly by LC–MS/MS (ESI+ ionization); their biological tracers, such as IGFI and IGFBP, can be measured by ELISA.

13.16   Residues of feed additives

Additives are used to improve the quality of feed, for example by providing enhanced digestibility, making them more appetizing, increasing the nutritional value and by preserving the quality. Feed additives are used extensively in livestock nutrition. They can be grouped in different categories: (i) technological additives (e.g. preservatives, antioxidants, emulsifier, stabilizing agents, acidity regulators and silage additives); (ii) sensory additives (e.g. flavouring and colouring agents); (iii) nutritional additives (e.g. vitamins, aminoacids and oligoelements); (iv) zootechnical (e.g. probiotics, regulators of gut flora and digestibility enhancers); (v) coccidiostats and histomonostats. Their use requires authorization and a scientific evaluation demonstrating that the additives have no harmful effects on human and animal health and on the environment. In addition, analytical methods have to be available to determine the presence of an additive in feed and its possible residues in foods.

It is not uncommon that the concentration of additives in feedstuffs is different from those indicated in their label and traces of additive not declared are present. This can be due to human error, carry-over and cross-contamination during production, transport and storage. The following properties of the feed additives and pre-mixes, which have an influence on the cross-contamination, are: adhesive strength, adhesion to walls, particle size and density (carrier, substance) and electrostatic properties. Difficulties in cleaning the machinery used in feed mills or homogeneously mixing the additives in feedstuffs are known problems that industry has minimized using microencapsulated or granular preparations. Relevant evidences of cross-contamination in past years (e.g. in 1996 in Northern Ireland 71 out of 161 complete feeds contained additives that had not been declared) have been reported. Because of their intense use, cross-contamination of not target feedstuffs by sulfonamides and tetracyclines has often been reported. The same phenomenon has not been shown with sulfadiazin or sulfatiazol, which have a plasmatic half-life up to 10 times shorter. Chlortetracycline has also been involved in cross-contamination of not target feedstuffs, mainly concerning the use of the active principle as a powder. Contamination by tetracyclines in Germany was reported 2001 in 87 imported meat and bone meal samples, showing maximum concentrations equal to 2295, 848 and 1274 mg/kg of oxytetracycline, tetracycline and chlortetracycline, respectively. These antimicrobials can bind to calcium and, therefore, persist in bones.

Toxic syndromes can result from overdosage and misuse of antimicrobial ionophores. Monensin, lasalocid, tetronasin, salinomycin, lysocellin, narasin, nigericin, laidlomycin and valynomycin are generally safe in the target animals receiving recommended dosage levels. However, accidental poisoning has been observed due to improper feed mixing, especially for horses and camels, but also for cattle, ostrich, rabbits, poultry and quails. The toxic syndrome is caused by bioactivity and damage to excitable tissues such as cardiac muscle, skeletal muscle, smooth muscle and the nervous system.

13.17   Environmental pollutants

Environmental pollutants may be natural toxins, such as secondary plant metabolites and fungal/mould toxins (mycotoxins) or metals, as well as xenobiotic (synthetic) substances, which are deliberately released into the environment (such as pesticides) or illegal waste of toxic industrial by-products. The long persistence of many chemical compounds facilitates their bioaccumulation and biomagnification along the food chain and even their transport with dust through wind and rain.

13.17.1   Natural toxins (including mycotoxins and plant toxins)

Mycotoxin contamination of crop and the ensuing consumption of contaminated feed ingredients by animals is an inevitable part of the animal production system. Infections by moulds and production of mycotoxins can develop at various stages of crop production: in the field or during harvesting, transport and storage. Approximately 20% of the crops are estimated to be contaminated by significant levels of mycotoxins (like aflatoxins, ochratoxins, fusariotoxins and trichothecenes). Mycotoxins can cause damage to organ systems, reduce production and reproduction, and increase diseases by reducing immunity. Oestrogenic substances derived from moulds of the genus *Fusarium* have caused infertility in cattle and sheep. Synthetic analogues (e.g. α-zearalenol) are used for growth promotion in some countries, such as in the United States and Canada, but they are banned in the European Union. The formulation and implementation of regulatory limits of mycotoxins, regular analysis of animal feed and feed ingredients and employment of proper mycotoxin decontamination and deactivation strategy will help to reduce the economic losses to a great extent.

Ochratoxin A   Ochratoxin A (OTA) is a mycotoxin produced by secondary metabolism of many filamentous species of moulds. In general, *Penicillium verrucosum* is responsible for OTA contamination in cool temperate conditions, whereas *Aspergillus ochraceus* is probably the main ochratoxigenic species in hot tropical regions. These moulds do not invade the crop in the field but mainly in the post-harvest phase. OTA is associated with human nephropathy and it is suspected to be the cause of the fatal human disease known as Balkan Endemic Nephropathy (BEN), an interstitial chronic disease affecting the south-eastern population of Europe (Croatia, Bosnia, Bulgaria and Romania). OTA has been detected in pig's blood, kidney, liver muscle and adipose tissue, with rather high levels found in animals suffering from porcine nephropathy, especially in countries of Balkan Peninsula.

OTA seems to be highly toxic for nerve cells and able to reach at any time the neural tissue (brain and retina). It is also a potent teratogen for laboratory animals. It can cross the placenta and accumulate in foetal tissue, causing various morphological anomalies. Immunosuppression seems to be related to an inhibition of the peripheral T and B lymphocyte proliferation and OTA stops the production of interleukin 2 (IL2) and its receptors. Carcinogenic activity has been shown in liver and kidney in toxicological studies in mice.

OTA toxicity differs between ruminant and monogastric species. In monogastric species (like pigs and poultry), OTA is absorbed from the gastrointestinal tract without, or with little, prior degradation, whereas in ruminants OTA is subjected to microbial degradation in the rumen. It is potentially nephrotoxic in all non-ruminant mammals. Poultry are less sensitive to OTA than pig, probably due to their higher capacity for excreting OTA. Occasionally, high mortality, nephrotoxicity, reduced feed intake and reduced egg production have been observed in poultry. *Post-mortem* findings are pale swollen kidneys and enlarged, yellowish and friable livers. Reduction in size of the bursa of Fabricius can usually been observed. OTA possesses a resistance to acidity and high temperatures; thus, once foodstuffs are contaminated, it is very difficult totally to remove this molecule. A nephropathy occurs in pigs with characteristic macroscopic changes of the type 'mottled or pale enlarged kidneys'. When signs of OTA toxicity are detected in pigs and poultry, which have received feedstuffs made with the same batches of contaminated grains, they should be controlled, in order to detect and measure OTA in their tissues. When pig herds are exposed to OTA through their feed, their kidneys, livers and meat are considered as a possible route of exposure for humans.

Fusariotoxins   Fusariotoxins, which are represented by trichothecenes (diacetoxyscirpenol, deoxynivalenol, fusarenone, monoacetoxyscirpenol, nivalénol), zearalenone and fumonisins, are widely scattered in cereals and their products. Human exposure can be directly via cereals or indirectly via products of animals having eaten contaminated feed. Crop contamination is often inevitable. The main part of absorbed fusariotoxins shows a rapid elimination within 24 h after ingestion, followed by a slower excretion of small amounts. Fusariotoxins are hydrosoluble and, therefore, they are generally weakly accumulated in animal tissues; however, traces of fusariotoxins or their derivates can be found in animal products. More than 95% of the ingested T-2 toxin is eliminated within 72 h with a 0.15 mg/kg BW/day dose, but when the dose increases to 0.6 mg/kg BW, the toxin is more slowly eliminated, probably because of metabolism saturation. The concentration of zearalenone and its metabolites in liver and bile increases with exposition dose. Resistant to fusariotoxins in ruminants is related to a detoxifying role of rumen microbial population. Fusariotoxins globally present a potential danger for animal and human health only when they are absorbed in great amounts or during long term exposure, although sows are highly sensitive to the xenoestrogenic effects of zearalenone.

13.17.2   Cadmium

Cadmium moves readily into the plant. It is present in soil, even in small quantities, and is primarily found in association with zinc. The use of cadmium-containing fertilizers in agriculture increases cadmium concentrations in the crops and derived products. Horse meat, in particular from old horses reared in heavily polluted industrial regions, may often contain substantially high levels of cadmium. Cadmium at a concentration of 368 µg/kg has been found in boneless horse meat imported from Romania and a level exceeding 2 mg/kg has been reported in kidney tissue. Such contents are to be expected as a consequence of the horse's diet and lifespan. As cadmium accumulates in the animal kidney, bound to metallothioneines, this organ may contain rather high levels of cadmium, which is much higher (approximately 10 times) than those observed in the kidneys of cattle reared in the same area. Cadmium accumulates, in particular, in the kidney and liver of animals. Hence, the offal are a source of cadmium in the diet, although vegetables and cereals are generally known as the most significant sources of cadmium in the diet and the atmospheric fall-out contributes to the content of cadmium, especially for leafy vegetables and grain.

Anthropogenic sources of cadmium include industrial emissions (non-ferrous mining and smelting, metal-using industry, industrial and agricultural wastes, coal combustion and phosphate fertilizer manufactures) and urban pollution (incineration of municipal solid waste, road dust and heating). As a by-product of zinc processing, cadmium production has closely followed the demand for zinc. Cadmium has been found to be more soluble and more plant-available in sandy soil than in clay soil. Cadmium is nephrotoxic and may produce renal tubular dysfunction, characterized by increased excretion of proteins. Common analytical methods for measuring cadmium concentrations in biological samples are flame atomic absorption spectroscopy (FAAS), inductively coupled plasma-optical emission spectroscopy (ICP-OES) and inductively coupled plasma-mass spectrometry (ICP-MS). These methods are also applied for measuring levels of cadmium in food.

13.17.3 Pesticides: plant protection products (PPP) and biocides

The term pesticide includes all the substances intended for preventing, destroying, repelling or mitigating any pest (animals, plants or organisms that can provoke damages in agriculture and buildings). It includes herbicides, algicides, fungicides, insecticides (and larvacides), nematicides, rodenticides and molluschicides. The term biocides also comprises disinfectants and products used for sanitation, as well as preservatives (e.g. used for wood, textiles and other manufactures) and any other chemical substance (or organisms) that is used to control harmful organisms. Plant protection products include all chemical (and biological) principles that are used to control or mitigate plant diseases, damaging insects, weeds and fungi.

All residues that can persist in feedstuffs and be adsorbed in the animal body leaving residues in meat have to be controlled at the slaughterhouse. Use of pesticides and plant protection products is regulated in many countries and only licensed products can be used under the restrictions given in the legislation. The amounts of residues found in food must be safe for consumers and must be as low as possible and in all cases below the MRL that is legally tolerated in food and feed. The uses of pesticides and plant protection products that are employed in the food or feed production have to be documented. Legislation covers products used also in countries that are authorized to export food and feed.

The ADI has been developed to assess chronic hazards posed by pesticide residues. It is the assumed amount a human can consume on a daily basis throughout life without causing damages to health. The ADI is assigned on the basis of an examination of available information, including data on the biochemical, metabolic, pharmacological and toxicological properties of the pesticide extracted from studies of experimental animals and observations in humans. Used as the starting point is the NOAEL for the most sensitive toxicological parameter, usually in the most sensitive species of experimental animal. To take into account the type of effect, the severity or reversibility of the effect, and the problems of inter- and intra-species variability, a safety factor (usually 100) is applied to the NOAEL to determine the ADI for humans. The data relative to consumption and contamination levels of the different food commodities are used in the process of risk assessment that lead to definition of maximum levels tolerated (e.g. in animal tissues and vegetables). Limits have been established also for the feed ingredients.

Poisonings of animals reared for food production are rare and have happened as a consequence of accidental contamination of feedstuffs or water.

Severe environmental pollution may derive by accidental events, human mistakes or malfunctioning of the plants producing pesticides or deposits.

The presence of residues consequent to the use of plant protection products is much more frequent in vegetables than in farm animals and poultry. In agriculture, pesticides and other plant protection products can be used only if they exert no harmful effects on consumers, farmers or bystanders and do not provoke unacceptable effects on the environment. Inappropriate use of pesticides is still today a major problem, especially in developing countries, posing there posing there a major health risk. The use of substances that are toxic, bioaccumulative and highly persistent has, therefore, been abandoned. The approved products have to be applied in specific stages of crop cultures and farmers have to adopt measures to avoid persistence in the ground or contamination of water, which can lead to carry-over and the presence of unexpected residues. Systemic pesticides, which can move inside the plant following absorption, should have time to be metabolized, transformed in inactive compounds and depleted, but also the products applied on the surface should be degraded and washed out. Therefore, an adequate period before harvest and usage levels that are indicated on the label of the plant protection products or the constraint given concerning the specific combination (plant–product) must be respected. All this pertinent information has to be made available to farmers or feed producers. In the European Union, the changes in pesticide authorization patterns (which include toxicological and risk assessment for all new effective ingredients and periodical re-evaluation of those that have not led to a definitive assessment), improvement in data reporting systems and the efficient implementation of the general provisions of the food laws have led to reduced levels of pesticides, which very rarely are detected in meat. Under these conditions, risk can only arise when by-products contaminated by some pesticide enter the food chain (e.g. use of biocides for producing these by-products used as a feed component). As an example, propiconazole, a triazole fungicide, has been found on citrus fruits pulp and residues in ruminant meat.

Prominent insecticide families include organochlorines, organophosphates and carbamates. Organochlorine insecticides were commonly used in the past but many have been removed from the market in the developed countries due to their health and environmental effects and their persistence (e.g. DDT and chlordane). They have shown a capability to bioaccumulate in animal (and human) tissues, in particular the fat, thus humans and animals that consume meat deposit these persistent residues in fat. The residues accumulate during their life leading to toxic effects. Some organochlorine pesticides are, however, still used in some countries because there are no alternatives that are considered adequate to protect animals and plants from diseases transmitted by insects. Organophosphate and carbamates are usually not persistent in the environment. Pyrethroid pesticides were developed as a synthetic version of the naturally occurring pesticide pyrethrin to increase their stability in the environment.

Pesticides undergo different modes of action: organophosphorus and N-methyl carbamate pesticides inhibit primarily the acetylcholinesterase and butyrylcholinesterase (plasma cholinesterase) enzymes by phosphorylation and carbamation, respectively. Acetylcholinesterase is responsible for turning off the signal flow ensured by the neurotransmitter acetylcholine between a nerve cell and a target cell. The inhibition of the signal-stopping enzyme leads to an overstimulation. This overstimulation is the reason, usually due to pulmonary secretion and respiratory failure, for the death of the poisoned person and also the activity that kills the insects. However, the anticholinesterase activity is rapidly inactivated by many microorganisms in the environment and by hydrolysis in the food-producing animals; therefore, the risk from residues in meat is insignificant.

13.17.4 Industrial wastes, by-products and toxicants released from fires and accidental events

Feedstuffs can become contaminated as a consequence of radioactive fall-out or the fraudulent use of low cost ingredients, which should not be used for feeding animals. Some striking example have been reported: (i) the use of transformer oil mixed with recoverable vegetable oil, poured onto feed, which led to contamination of swine and poultry meat in several EU countries in 1999; (ii) the use of dry bread, which was produced with polychlorobiphenyl-contaminated oil used in a direct drying system in Ireland in 2009; and (iii) dioxin contaminated non-feed fat mixed with fat for feed in Germany 2010.

The majority of the accidental events due to fraudulent practices and lack of knowledge of the associated risk by some feed producer are related to the presence of persistent organic pollutants (POPs). POPs are chemical substances that possess certain toxic properties and, unlike other pollutants, resist degradation. They accumulate in living organisms, they are transported by air, water and migratory species and they accumulate in terrestrial and aquatic ecosystems. POPs have a tendency to remain in fat-rich tissues.

The Stockholm Convention (in 2001) covers 12 priority POPs produced intentionally or unintentionally: aldrin, chlordane, dichlorodiphenyl-trichlorethane (DDT), dieldrin, endrin, heptachlor, mirex, toxaphene, polychlorobiphenyls (PCBs), hexachlorobenzene, dioxins and furanes. The goal of the convention is to bring about definitive elimination of POPs. Countries that have signed the agreement must develop a national implementation plan detailing how they will fulfil their obligations under the convention.

In 2009, more POPs were included in the list: four types of polybromodiphenyl ethers (PBDEs), α-hexachlorocyclohexane, β-hexachlorocyclohexane, perfluorooctane sulfonic acid (PFOS) and its salts, perfluorooctane sulfonyl fluoride and pentachlorobenzene. The original POPs were mainly pesticides but certain of the new substances have been widely used in consumer products, such as PFOS which is used, for example, in metal plating, firefighting foams and in stain repellents. Their release into the environment can determine their presence, even far from the polluted areas. Important sources of these toxic substances are chlorinated chemical production, such as chloro-aromatics (phenols, benzenes) and oxy-chlorinators, chlorine production using graphite electrodes, oil refining and catalyst regeneration, pulp and paper (elemental chlorine bleaching) processes that involves high temperatures and usually combustion (e.g. drying of feed in the presence of contaminated fumes), thermal manufacturing processes such as cement kilns, asphalt mixing (especially for PAH), production of lime, ceramic, glass, brick and other similar processes carried out on a small-scale, uncontrolled combustion such as fires in forest and bush, agricultural harvest residues and complex industrial combustion of wastes.

Acute effects after high-level exposure have been described for some of the organochlorine pesticides (such as aldrin, dieldrin and toxaphene). PCBs have caused well documented episodes of mass poisoning called 'Yusho' and 'Yu Cheng' disease, which occurred in Japan and Taiwan, respectively. Pregnant women exposed had no or minor symptomatology but their children presented adverse effects and developmental disorders. Chronic toxicity and irreversible damages caused by pesticides include: cancer, mutagenicity, developmental and reproductive toxicity and endocrine disrupting. With some pesticides belonging to the category of organochlorine POPs (DDT, dieldrin, toxaphene and chlordane, mirex, and endosulfan) the issue of being endocrine disruptors has arisen. It means that they may mimic hormones in the body which have not been induced by signals from endocrine glands and which subsequently log on to the receptors and stimulate an effect. These pesticides act as oestrogens and can alter the sex organs and/or induce cancer. Symptoms such as infertility in humans may be apparent only decades later.

Dioxins The term 'dioxins' identifies a family of organic compounds comprising 210 structurally related congeners. Dioxins can be divided into two subgroups: polychlorinated dibenzo-p-dioxins (PCDDs) and polychlorinated dibenzofurans (PCDFs). PCDDs and PCDFs are planar tricyclic aromatic compounds formed by two benzene rings connected by two oxygen atoms in PCDDs and one oxygen atom in PCDFs. The hydrogen atoms can be replaced by up to nine chlorine atoms giving origin to 75 PCDD and 135 PCDF congeners. These compounds have similar chemical properties, they are very stable and resistant to degradation, they are highly fat soluble, and they bioaccumulate through the food chain. Toxicity and persistence are determined by their structure, with the lateral substitutions on the rings resulting in the highest degree of toxicity. The PCDD and PCDF congeners with chlorine atoms in the 2, 3, 7 and 8 positions are the most toxic. This group of dioxins includes seven PCDD congeners and ten PCDF congeners. These compounds induce a common adverse effect and have a common mechanism of action mediated by binding of the PHAh ligand to a specific high-affinity cellular protein. The aryl hydrocarbon receptor (AhR) is a ligand-activated transcription factor that mediates many of the biological and toxic effects of halogenated aromatic hydrocarbons (HAHs), polycyclic aromatic hydrocarbons (PAHs) and other structurally diverse ligands. AhR is a cytosolic transcription factor that is normally inactive and is bound to several co-chaperones. Upon ligand binding to chemicals such as 2,3,7,8-tetrachlorodibenzo-p-dioxin (TCDD), the chaperones dissociate resulting in AhR translocating into the nucleus and dimerizing with ARNT (AhR nuclear translocator), eventually leading to changes in gene transcription after binding of the AhR/ARNT dimer to the DNA.

PCDD/PCDFs are mainly by-products of industrial processes but can also result from natural processes (volcanic eruptions and forest fires). Dioxins are unwanted by-products of a wide range of manufacturing processes, including smelting, chlorine bleaching of paper pulp and the manufacture of some herbicides and pesticides. Often the worst culprits are uncontrolled waste incinerators (solid waste and hospital waste) due to incomplete burning. PCDD/PCDFs are very efficiently absorbed in the digestive tract to the extent of about 95% of the amount ingested. The highest levels of these compounds are found in some soils, sediments and food, especially dairy products, meat, fish and shellfish. Very low levels are found in plants, water and air. The contamination of food products with dioxin-like compounds (PCCD/PCDFs and DL-PCBs) is a well-studied issue because food is generally considered as the major source of dioxin intake for humans. Poultry products are considered to be an important form of intake.

In 1999, high levels of dioxins were found in poultry and eggs from Belgium. Subsequently, dioxin-contaminated animal-based foods (poultry, eggs and pork) were detected in several other countries. The cause was traced to animal feed contaminated with illegally disposed of PCB-based waste industrial oil. In late 2008, Ireland recalled many tonnes of pork meat and pork products when up to 200 times more dioxins than the safe limit were detected in samples of pork. The contamination was traced back to contaminated feed. Another case of dioxin contamination of food occurred in the United States in 1997. Chickens, eggs and catfish were contaminated with dioxins when a tainted ingredient (bentonite clay, sometimes called 'ball clay') was used in the manufacture of animal feed. The contaminated clay was traced to a bentonite mine. As there was no evidence that hazardous waste was buried at the mine, investigators speculate that the source of dioxins may be natural, perhaps due to a prehistoric forest fire. Chickens are mostly farmed for economic reasons, that is the low price of buying and feeding chickens. Chickens are also a means of kitchen waste disposal and provide eggs for consumption.

The contamination of animal feed, pastures and organisms at lower levels of the feed chain leads to bioaccumulation of dioxins in animal fat. Their half-life in the human adult body is estimated to be 7–11 years but depends on age and level of exposure. Short-term exposure of humans to high levels of dioxins may result in skin lesions, such as chloracne and patchy darkening of the skin, and altered liver function. Long-term exposure is linked to impairment of the immune system, the developing nervous system, the endocrine system and reproductive functions. Chronic exposure of animals to dioxins has resulted in several types of cancer. TCDD was evaluated by the WHO's

International Agency for Research on Cancer (IARC) in 1997. Based on animal data and on human epidemiology data, TCDD was classified by IARC as a 'known human carcinogen'. However, TCDD does not affect genetic material and there is a level of exposure below which the cancer risk would be negligible. The complex nature of TCDD, PCDF and PCB mixtures complicates the risk evaluation for humans, fish and wildlife. The concept of toxic equivalency factors (TEFs) was introduced to facilitate risk assessment and regulatory control. At present, there is sufficient evidence available that there is a common mechanism for these compounds, involving binding to the aryl hydrocarbon (Ah) receptor as an initial step. TEFs used have been calculated on the basis of the relative toxicity of each compound. The overall toxicity is expressed by the sum of PCDD, PCDF and DL-PCB TEFs. For the foodstuffs of animal origin the values are calculated as a ratio of fat matter (i.e. pg TEQ/g fat). Many countries have defined MRL for TCDD and PCDFs. The kidney's peripheral fat, liver and muscle samples are analysed by using screening tests (biotests or GC-MS) and confirmatory tests (gas chromatography coupled with high resolution mass spectrometry-GC-HRMS). Determination of naturally occurring (native) compound is made by reference to the same compound in which one or more atoms has been isotopically enriched (internal standard).

Polychlorobiphenyls (PCBs) Polychlorobiphenyls (PCBs) are aromatic compounds formed by substitution of hydrogen with chlorine atoms on the two benzene rings. Theoretically, there are 209 possible congeners, considering the five chlorine binding sites on each ring. In total, 130 of them occur in commercial products. The chemical formula can be presented as $C_{12}H_{10-n}Cl_n$, where n is the number of chlorine atoms. Each congener has been assigned by a number from 1 to 209 in accordance with the rules of the International Union of Pure and Applied Chemistry (IUPAC). In contrast to dioxins, PCBs have widespread use in numerous industrial applications and were produced in large quantities for several decades, with an estimated total world production of 1.2–1.5 million tonnes, until they were banned in most countries by the 1980s. Following the Stockholm agreement, national plans for decontamination and disposal of PCB-containing equipment became mandatory, as well as monitoring that include controls in foods and feedstuffs. From a toxicological point of view, PCBs are divided into three groups: (i) non-ortho, (ii) mono-ortho and (iii) poly-ortho-substituted PCBs. Non-ortho and mono-ortho PCBs can bind to the Ah receptor, with toxicity qualitatively comparable to the dioxins; they are therefore called 'dioxin-like PCBs' (DL-PCBs). This group includes 12 PCB congeners. All PCB congeners are lipophilic, accumulate in fat and enter the food chain. They are eliminated from the body by metabolic degradation and by excretion, which is slow and occurs primarily in faeces, milk and eggs.

The most known event worldwide of PCB and dioxin contamination of meat and products thereof occurred in 1999 in Belgium, The Netherlands and France. In total, 50 kg of PCBs contaminated with 1 g dioxins were accidentally added to a stock of recycled fat used for the production of 500 tonnes of animal feed in Belgium. Thirteen feed producers and approximately 1000 pig farms, 500 poultry farms and 150 cattle farms were involved.

The crisis was resolved by the implementation of a large food monitoring programme for the seven PCB markers (PCBs 28, 52, 101, 118, 138, 153 and 180). When PCB concentrations exceeded the tolerance levels of 100, 200 or 1000 ng/g fat for milk, meat or animal feed, respectively, the 17 toxic poly-chlorinated dibenzodioxins and furans (PCDD/Fs) congeners were also determined. As for the dioxins the toxicity of the dioxin-like PCBs is expressed by Toxic Equivalents (TEQ)/g of fat, whereas for the non-dioxin-like (NDL) PCBs the concentration (ng/g) is calculated on the basis of the sum of the six markers or indicator PCBs (PCB 28, 52, 101, 138, 153 and 180). Analytical methods are similar to those used for TCDD and PCDFs.

Polybrominated diphenyl ethers (PBDEs) Polybrominated dephenyl ethers (PBDE) are organobromine compounds that are used as flame retardant. There are four main classes of brominated flame retardant (BFRs): (i) poly-brominated diphenyl ethers (PBDEs) used in plastics, textiles, electronic castings; (ii) hexabromocyclododecanes (HBCDDs) used for thermal insu-lation in the building industry; (iii) tetrabromobisphenol A (TBBPA) and other phenols used in printed circuit boards and thermoplastics, mainly in televisions; and (iv) polybrominated biphenyls (PBBs) used for consumer appliances, textiles and plastic foams. BFR levels are rapidly increasing in the environment and they are ubiquitous contaminants in the environment because of the high production volume, widespread usage and environmental persistence. Atmospheric transport and deposition has been identified as the predominant pathway for PBDEs present in rural and remote locations. These compounds have similar chemical properties, they are very stable and resistant to degradation, they are highly fat soluble, and they bioaccumulate through the food chain. Detection of PBDEs can be conveniently done in fat tissues by gas chromatography coupled with mass spectrometry (GC-MS, GC-HRMS).

Perfluorinated organic compounds (PFCs) Perfluorooctane sulfonic acid (PFOS), perfluorooctanoic acid (PFOA) and perfluorooctane sulfonyl fluoride (PFOSF) are widely used in industrial and consumer applications, including stain-resistant coatings for fabrics and carpets, oil-resistant coatings for paper products approved for food contact, firefighting foams, mining and oil well surfactants, floor polishes, the photographic industry, photolithog-raphy, semiconductors, paper and packaging, coating additives, cleaning products and insecticide formulations. They are extremely resistant towards thermal, chemical and biological degradation processes and they tend to accumulate in the food chain. They are stable chemicals made of a long carbon chain that is both lipid- and water-repellent. PFOS is persistent, bioaccumulative and toxic to mammals and will, therefore, not accumulate in fatty tissues, as is usually the case with other persistent halogenated compounds. PFOS/F and PFOA are chemically and biologically stable in the environment and resistant to biodegradation, atmospheric photooxidation, direct photolysis, and hydrolysis. Thus, these chemicals are extremely persistent in the environment. Unlike organochlorine compounds, which accumulate in lipid-rich tissues, PFOS/F and PFOA bind to blood proteins and accumulate in liver and gall bladder. The tissue that is analysed to detect residues is, therefore, the liver. PFOS binds to serum proteins such as albumin and can replace a variety of steroid hormones from specific binding proteins in the serum of birds. Several authors reported levels of these substances in the food chain.

PFOS, its salts and PFOSF were added to the list of substances to be controlled under the Stockholm Convention in 2009. The convention calls for the elimination of the production and use of these chemicals except for certain applications for which alternatives still have to be phased-in or are not yet available. PFOS/F and PFOA can be detected by liquid chromatography in tandem with mass spectrometry (LC-MS/MS).

Polycyclic aromatic hydrocarbons (PAHs)　Polycyclic aromatic hydrocarbons (PAHs) are often by-products of petroleum processing or combustion. They include over 100 different chemicals compounds with two or more fused benzene rings in linear, angular or cluster arrangements. The best known are: benzo[k]fluoranthene, anthracene, benzo[b]fluoranthene, benzo[e]pyrene, fluoranthene, naphthalene, phenanthrene, benzo[ghi]perylene and pyrene. Many of them are highly carcinogenic at relatively low levels. Due to their thermal stable structure, PAHs generally exhibit a high melting point, a high boiling poin, and a low vapour pressure. PAHs are typically formed during incomplete combustion of organic matter at high temperatures (e.g. wood and fossil fuels) and can be found in food products as a consequence of certain industrial processing methods, such as smoking, heating (grilling, roasting) and drying, which permit the direct contact between food and combustion products. These are important sources of PAH contamination for seeds, edible oils and meat. In edible oils, the oilseed drying processes by direct combustion can be an important source of contamination. A high amount of PAHs is emitted from processing coal, from motor vehicle exhaust, forest fires and volcanoes or from hydrothermal processes, which are natural emission sources of PAHs. They are ubiquitous in the environment and can accumulate on the waxy surface of many vegetables. Detection of PHAs is conveniently done in fat tissues using liquid chromatography with fluorescence detection (LC–FLD) or gas chromatography with mass spectroscopy (GC–MS, GC–MS/MS).

13.18　Analytical chemical methods and their validation

The illegal use of unauthorized substances is controlled by official inspection and analytical services provided by laboratories that should follow qualitative standards in order to produce reproducible results with the highest sensitivity. Analytical methods aimed at detecting traces of chemical contaminants require a definition of their performances that give confidence in unequivocal identification and reliable quantification of residues monitored.

Screening methods have to permit a high number of analyses with a low rate of false-negative results and low costs. Confirmation methods must give unequivocal identification (e.g. mass spectrometry compulsory for non-authorized substances) and allow accuracy of measurement, which must be compared with the MRLs.

The characteristics to be defined for a validation of analytical methods include the limit of decision for confirmatory analysis (CCα) and the capacity of detection for screening analysis (CCβ). Assessment of CCα and CCβ are based respectively on the determination of signal to noise (S/N) ratios in blank samples and matrix material fortified at the CCα and the determination of the smallest content of the substance that may be detected, identified and/or quantified in a sample with an error probability of β (e.g. 5%). CCα is defined as the limit at which a substance can be concluded as positive with an error probability α. Analysis with reference materials and artificially contaminated (spiked) samples allows the definition of bias (precision) and accuracy (error) within a range of measurements that must include the MRL. Aliquots of the matrix fortified with the target analyte at concentrations equal to or above the MRL are therefore used to determine recovery yield.

For quantitative screening, precision is acquired by the determination of variation coefficients. Inter-assay variation testing gives an indication of the precision of the assay over a longer time. Selectivity or specificity is the ability of a method to distinguish between the analyte being measured and other substances. The stability of the analytes in solution and in the matrix should be included in the validation process. For qualitative screening methods, only CCβ, selectivity and applicability need to be assessed. Any positive findings assessed using screening methods should always be re-analysed by a validated confirmatory method.

For confirmatory methods aimed at detecting forbidden substances (without MRL), the use of spectrometric methods is needed (GC-MS, LC-MS, GC-MS/MS, LC-MS/MS, ICP-MS e HRMS). In order to classify the method as confirmatory, the decision limit (CCα) and trueness/recovery must be determined. An accepted CCα value is normally equal to 1%. Information on the accuracy of a confirmatory method is determined by assessment of trueness (recovery), which refers to the closeness of agreement between the averages recorded for a data set and is determined by the degree of deviation from the mean recovery detection limit (CCß). Precision (bias), selectivity, specificity, applicability, ruggedness and stability must be assessed.

Literature and further reading

Al-Dobaib, S.N. and Mousa, H.M. 2009. Benefits and risks of growth promoters in animal production. *J Food Agric Environ*, **7**, 202–208.

Battacone, G., Nudda, A. and Pulina, G. 2010. Effects of ochratoxin A on livestock production. *Toxins*, **2**, 1796–1824.

Courtheyn, D., Le Bizec, B., Brambilla, G. *et al.* 2002. Recent developments in the use and abuse of growth promoters. *Anal Chim Acta*, **473**, 71–82.

EFSA (European Food Safety Authority). 2008. Perfluorooctane sulfonate (PFOS), perfluorooctanoic acid (PFOA) and their salts. Scientific opinion of the panel on contaminants in the food chain. EFSA, Italy. http://www.efsa.europa.eu/en /efsajournal/doc/653.pdf; last accessed 17 February 2014.

[EFSA scientific documents and reports are publicly available at EFSA's web pages: http://www.efsa.europa.eu/en/publications.htm; last accessed 17 February 2014.]

Lozano, M.C. and Trujillo, M. 2012. Chemical residues in animal food products: an issue of public health. In: *Public Health – Methodology, Environmental and Systems Issues* (ed. J. Maddock). InTech. Retrieved from www.intechopen.com (last accessed 12 February 2012).

Plaza-Bolaños, P., Frenich, A.G. and Vidal, J.L. 2010. Polycyclic aromatic hydrocarbons in food and beverages. *Analytical methods and trends. J Chromtogr A*, **1217**, 6303–6326.

Ramesh, C.G. (ed.). 2012. *Veterinary Toxicology: Basic and Clinical Principles*. Academic Press, New York.

Riviera, J.E. and Papich, M.G. (eds). 2009. *Veterinary Pharmacology and Therapeutics*. John Wiley & Sons, Inc., Hoboken, NJ.

Serratosa, J., Blass, A., Rigau, B. *et al.* 2006. Residues from veterinary medicinal products, growth promoters and performance enhancers in food-producing animals: a European Union perspective. *Rev Sci Tech Off Int Epizoot*, **25**, 637–653.

Stephany, R.W. 2010. Hormonal growth promoting agents in food producing animals. *Handb Exp Pharmacol*, **195**, 355–367.

14
Meat By-Products

Miguel Prieto and María Luisa García-López

Department of Food Hygiene and Technology,
Faculty of Veterinary Science, University of León, León, Spain

14.1   Scope

Routine slaughter activities produce significant quantities of carcass and non-carcass components that are not destined for human consumption. Animal by-products is a commonly used term for this vast group of products. This chapter provides an introduction to the management of animal by-products at the slaughterhouse.

14.2   Introduction

The parts of the animal body that are not part of the dressed carcass are generically named non-carcass components (NCCs). These materials are consumed, disposed of or processed and re-used in many different sectors, including the cosmetic, pharmaceutical, food, feed and fertilizer industry, as well as being used for other technical purposes (e.g. pet food, hides and skins for leather, wool). NCCs include offal, blood, bones, feet, fat, horns, hooves, hide, hair, fleece, feathers, bowels and stomach, intestine and rumen content, manure, and so on. These products should be rapidly and efficiently processed because of their capacity to spoil (short shelf-life), capacity to produce environmentally aggressive liquid, solid and gaseous wastes, possibility to host and spread biotic and abiotic pathogenic agents, and last, but not least, potential economic return.

Meat Inspection and Control in the Slaughterhouse, First Edition.
Edited by Thimjos Ninios, Janne Lundén, Hannu Korkeala and Maria Fredriksson-Ahomaa.
© 2014 John Wiley & Sons, Ltd. Published 2014 by John Wiley & Sons, Ltd.

An *ad hoc* classification of NCCs differentiates between: edible offal or viscera that are hygienically processed and pass satisfactorily the *post-mortem* inspection; edible co-products that, although unsuitable for human consumption when produced at the slaughterhouse, can be incorporated into human food after appropriate transformation; and by-products, which are the parts of an animal not intended for human consumption. This latter group, which is the object of this chapter, is legally defined as 'not intended for human consumption' and is withdrawn from the food chain, either due to a decision by the food business operator based on its low commercial value or for sanitary reasons (i.e. they have been condemned by the veterinary service).

There is a large variety in the characteristics, composition, value and possible destination of the NCCs obtained in the slaughterhouse. Certain offal can either be considered as high-priced delicacies or treated as by-products to be disposed of, depending on economic, social or cultural consumer preferences. Products become animal by-products (ABPs) as soon as the food business operator decides that they will not be used for human consumption, not necessarily because they are unfit for human consumption but because there is no market for them. Once a product has become an ABP, it must not re-enter the food chain. Some viscera and parts (tongue, heart, liver, kidney, brain, thymus, pancreas, testicles, uterus, bladders, tripe and stomachs, tongue, lungs, udder, tail, feet, spleen, head trimmings etc.) can be eaten or incorporated into meat products but sometimes they are considered ABPs due to low commercial value. Offal is usually classified into red offal (e.g. heart, kidney, liver, spleen, tongue, lungs) and white offal (e.g. brain, sweetbreads, feet, testicles) due to their characteristic intensity of colour, which is due to the myoglobin content.

Other carcass or non-carcass components are promptly excluded from the human food chain. Some parts or whole carcasses are condemned during meat inspection by the veterinary service for sanitary reasons and should be adequately transformed or disposed of. Parts of the carcass of ruminants are suspected to contain certain pathogenic agents such as prions. On this basis, they are classified as specified risk materials (SRMs) and transmissible spongiform encephalopathy (TSE) materials, and are handled in such a way as to exclude them from the food and feed chain. Inedible products, such as bones, rendered meat, feathers and hair, can be transformed into compost, feed or disposed of. If cutting and deboning operations are carried out, some products, such as bone, meat and fat trimmings, become more abundant. Wastewater is made up mostly of run-off water with liquids such as urine, blood and gastrointestinal contents; it contains a high load of organic matter and is separately treated. Screening of wastewater helps to retain gross solids intended for composting or fermentation. Certain materials, such as manure and the contents of rumen and intestines (partially digested feed), have limited applications and are also destined for composting or fermentation.

The ABPs industry can process all these materials in multiple ways and for multiple purposes. Some by-products can be transformed into processed animal protein (PAP) and meat-and-bone meal (MBM) in animal feeds, very valuable because of their high digestible protein and fat content.

However, some of these traditional purposes have been banned or restricted as a result of the occurrence of TSE. Where TSE constitutes a public health risk, the materials that can be included in animal feed are regulated by national legislation and restrictions are placed on the use of certain ABPs for human or animal consumption. A ban on the incorporation of mammalian protein in feed to ruminant animals and on the incorporation of mammalian meat-and-bone meal in any farmed livestock feed is currently in place in some parts of the world due to the occurrence of TSE. Regulations usually affect many issues concerning ABPs, such as classification, marking, movement, processing and disposal. Thus, a great part of the protein meals and animal fats are destroyed and used as energy sources for the generation of electricity. Some organs, tissues and glands are used to extract hormones, bioactive compounds and other pharmaceutical products, such as insulin, heparin and cortisone. Hides are cleaned and preserved by washing and salting, and intended for the production of leather goods. Many slaughterhouse products are initially not fit for human consumption but can be transformed and the resulting commodity later incorporated into human food. Examples of these edible co-products are: the tendons, ligaments, hides and skins, which are processed into collagen and gelatine; pig intestines, which are processed into sausage casings; and fatty tissue, which is rendered into edible fats such as lard or tallow.

Approximately 47 million tonnes of animals are slaughtered for meat production in Europe every year. From these, about 17 million tonnes of by-products are produced and 14–15 million tonnes are processed by renderers and fat melters. It has been estimated that NCCs constitute about 30–50% of the live weight of meat animals depending on the species. Table 14.1 shows the proportion corresponding to various carcass meat and NCCs from cattle, pigs, and lamb. The amount of animal waste that is discarded varies between countries because of reasons previously highlighted (economic, social, cultural and eating habits) but this amount increases as society becomes more affluent and food safety becomes more relevant. Some products, especially offal, are considered edible in some countries but are not intended for human consumption in others. Approximate average percentages of major components in rendering raw materials are 60% water, 20% protein and mineral, and 20% fat, although these figures obviously vary depending on the product.

14.3   Advantages of adequate ABP management

Efficient and hygienic handling of by-products from slaughtering and carcass cutting is essential for profitable operation in the meat industry. Proper transformation and use of ABPs is a valuable activity, since it reduces disposal costs, decreases environmental pollution and generates beneficial income for the meat industry. Economic studies have assessed the value of edible and inedible by-products from meat animals at 9–12% of the total value of the carcass, although these numbers have been substantially

Table 14.1 Proportion (%) corresponding to various carcass meat and NCCs from cattle, pigs and lambs.

	Cattle	Pig	Lamb
Carcass meat	34	52	32
Bones	16	17	18
Offal	16	7	10
Skin and attached fat	6	6	15
Blood	3	3	4
Fatty tissues	4	3	3
Horns, hoofs, feet and skull	5	6	7
Abdominal and intestinal contents	16	6	11

Source: Goldstrand, 1992. Reproduced with permission from Springer Science + Business Media.

reduced due to requirements imposed by TSE legislation. Furthermore, the development of synthetic substitutes has decreased the value of some ABPs. On the other hand, there is a growing awareness that products such as pharmaceutical compounds and functional ingredients with a significant added value can be obtained from ABPs, and many components are used in the food, pharmaceutical, cosmetic and rendering industry.

Waste management should minimize the impact of residues in the environment by achieving maximum recovery of ABPs and their adequate transformation or disposal. For this purpose, technological developments in equipment, infrastructure or operational methodology together with advances in scientific knowledge try to prevent or reduce to a minimum the overall impact of emissions on the environment. Some actions that minimize this impact are mostly aimed at reducing the amount and organic load of liquid waste, such as dry cleaning, blood recovery, solid screening and grease trapping. These operations should allow adequate separation, management and handling of liquid and solid wastes.

Another important advantage of adequate waste management is that by carefully handling and disposing of ABPs, the spread of potential biological and chemical hazards to public health is restrained. In this sense slaughterhouses and the ABPs industry can perform as a big filter, interrupting cycles of contamination. All these reasons are further incentives for better management and processing of meat by-products.

14.4 Separation of animal by-products, storage and recommendations on best practices and hygiene requirements

Slaughterhouses should carry out preliminary activities in ABP handling, comprising ABP identification, separation, hygienic operations and record keeping. Similarly to other agrofood industries, HACCP (Hazard Analysis and Critical Control Points), GMP (Good Manufacturing Practices) and general food safety principles apply to the slaughterhouse, the ABP processing and manufacturing plants. As well as including HACCP and GMP, safety and quality programmes implemented in the slaughterhouse and processing industry can also incorporate third-party certification and auditing, as retail and consumer pressures have added to the legislative requirements.

By-products are classified and separated at the slaughterhouse according to the level of risk posed to humans, animals and the environment, and the methods used to dispose or process them. Usually a high-risk category is established to allocate materials with the highest risk for public health, animal health or the environment. In the European Union, it is established that this material (so-called Category 1 material) represents the highest risk and comprises SRMs and animals that are TSE positive or suspected of being infected by a TSE. Not only prions but other infective agents and chemical contaminants also can be present in by-products that can spread animal and human diseases; therefore, ABPs should be properly processed or disposed of. All these products are meant for destruction using methods able to inactivate prions, proteins which show extreme stability and resistance to denaturation by most chemical and physical agents. The bovine spongiform encephalopathy (BSE), commonly known as mad-cow disease, introduced new legislation all over the world, with the aim of excluding risky material from the food and feed chain. SRMs include certain tissues of cattle (the tonsils, the intestines from the duodenum to the rectum, the mesentery of all animals, the skull excluding the mandible but including the brains and eyes, the spinal cord of animals over 12 months, the vertebral column excluding the vertebrae of the tail, the spinous and transverse processes of the cervical, thoracic and lumbar vertebrae, the median sacral crest and the wings of the sacrum, but including the dorsal root ganglia of animals over 30 months) and sheep and goats (ileum, spleen, for animals over 12 months or with a permanent incisor erupted, the skull including the brain and eyes, tonsils, and spinal cord). ABP recovered from wastewater used in the slaughterhouse or plants processing Category 1 material is also considered a SRM.

There is another group of by-products whose risk to human or animal health can be significant although the agents involved are not so difficult to inactivate (Category 2). It includes: any carcass, part of a carcass or offal that has not been presented for *ante-mortem* inspection, or not presented with the necessary food chain information; carcasses or parts of the carcass that contain pathological lesions indicating communicable disease to human or animal (such as cisticercosis or hydatidosis), septicaemia or pyaemia; local conditions (septic lungs, joints; specific conditions and diseases such as tuberculosis etc.), whole bodies either found dead on arrival at the slaughterhouse, in the lairage or animals rejected at *ante-mortem* inspection. Manure, digestive tract content, and solid materials (particle size >6 mm) in wastewater streams of slaughterhouses and plants processing Category 2 materials, fallen animals, or animals containing infectious agents or residues are also included.

A low risk category (3) would include those parts of the carcass which are deemed fit but not intended for human consumption for various reasons, such as the low commercial value. ABPs in this category can be intended for a wider variety of uses, including the manufacture of pet food. Some of the products included are heads, feet, some offal, blood, placenta, wool, feathers, hair, horns, hoof, cuts from animals not showing any signs of communicable disease, parts of carcasses or offal that are rejected as unfit for human consumption because of the presence of lesions with no signs of communicable diseases (lesions in liver and lungs such as melanosis, certain parasitosis etc.) but not permitted by the Hygiene Regulations and considered as unfit for human consumption.

Strict separation of the different categories of ABPs and of ABPs from material intended for human consumption is performed to avoid cross-contamination, mislabelling or misgrading. This separation should be guaranteed at all times, during storage and transport to the processing plant by performing suitable operating procedures. Materials classified as SRMs should not contaminate any other category of ABPs nor should material derived from low risk category by-products be contaminated by other ABPs. If this happen (low risk ABPs are mixed or cross-contaminated with higher risk materials), the ABP in question must be treated as the higher risk category.

To facilitate handling and avoid cross-contamination, storage premises should be situated so that there is a physical separation from the slaughterhouse's main building. The facilities should comply with general hygienic requirements, such as layout, construction materials of walls, roofs and floors, hygienic design of equipment and surfaces, and adequate supply of water and steam. Best practice to prevent cross-contamination involves the use of separate storage rooms for utensils and colour-coded cleaning equipment, tools and vessels. To ensure that the routine work does not result in different categories of ABP contaminating each other, staff should change clothes, footwear and utensils when working in different parts of the premises. Pest prevention plans to avoid infestation by insects, rodents and birds should be in place, since these pests can become a problem during ABP storage. The official veterinarian is in charge of separation and marking of by-products, and for hygiene measures aimed at preventing cross-contamination.

14.5   Identification, transport and marking

The collection, identification, transport and traceability procedures (including identification and marking) of slaughterhouse by-products must comply with hygiene standards that are certified and controlled by competent authorities. Processing industries are highly specialized and able to handle sufficiently large volumes of product in order to ensure and certify the standards requested by the authorities.

Food business operators must apply procedures for the correct identification, separation and marking of ABPs. ABPs should also be labelled and stored correctly, consigned from the premises to permitted approved destinations without delay; records of ABPs consigned from the premises or disposed of on the premises must be properly kept; and a copy of the commercial document or health certificate, when required, created at the time of the consignment is to be kept.

Products from Category 1 and Category 2 material should be permanently marked with glyceroltriheptanoate (GTH) to ensure that they do not enter the feed and food chain. The operators of processing plants should have in place a system of monitoring and recording parameters suitable to demonstrate to the competent authority that the required homogeneous minimum concentration of at least 250 mg GTH per kg fat is achieved.

14.6 Processing of by-products and methods of treatment and disposing of ABPs

On occasion, slaughterhouses carry out animal by-products processing activities. In all cases plants that process ABPs have to be approved or registered. By-products may be processed at the slaughterhouse premises only if proper equipment is available and enough volume exists, although complex processing activities (such as incineration, composting, anaerobic fermentation, rendering) are usually performed in separate facilities outside the slaughterhouse. Hygiene legislation regulates how these activities should be conducted. Slaughterhouses do have some basic equipment that permits preliminary treatments, such as primary treatment of wastewater.

Treatment and disposal of ABPs should be done using the appropriate method for the category and composition of waste. Methods include incineration, rendering, landfilling, composting, anaerobic fermentation and oleochemical processes. Selection of appropriate method of disposal depends mainly on the risk category involved and, secondarily, on the type of wastes and its quantity. Plants processing ABPs of Category 1 and Category 2 must have a layout that ensures the total separation of the two categories from reception of the raw material until despatch of the resulting processed product. Premises for the processing of Category 3 material must not be at the same site as premises processing Category 1 or Category 2 material, unless in a completely separate building.

14.6.1 Incineration

Incineration and co-incineration are procedures approved to dispose of ABPs although they are questioned because of their environmental impact. This is a practical solution in those parts of the world where the feeding of processed animal proteins (MBM) to farmed animals has been banned, since energy can be retrieved as a result of the excellent energy properties of MBM compared to coal or fuel. The design of equipment, hygiene conditions and operation are strictly regulated, as is the disposal of residues and waste material. Regulation of conditions establishes that a temperature of 850 °C for at least two seconds or a temperature of 1100 °C for 0.2 seconds should be attained to accomplish complete inactivation of prions. Incineration and co-incineration of ABPs can be done in authorized hazardous waste incinerators, sewage sludge incinerators, cement plants, power stations, gasification plants or residue incinerators.

14.6.2 Composting

Composting is only applicable to low risk materials, since the conditions and parameters usually employed do not guarantee the complete inactivation of agents posing a risk to animal and human health (e.g. prions, viruses etc.) or its reduction to acceptable risk levels. Particular attention should be paid to conditions, especially to the attained temperature and time combinations required for pathogen inactivation. Composting is based on the microbial activity which decomposes and stabilizes organic substrates. Digestion is carried out in bioreactors and the resulting material consists largely of decayed organic matter that is used for fertilizing and conditioning land. The resulting final product is stable, high in humic substances and can be beneficially applied to land (landfill). Certain ABPs are more appropriate than others for composting; the composition, pretreatments required and possible restrictions on use and application (not to pasture land) may limit widespread use. Composting is very useful for treating slaughterhouse materials such as manure, the contents of the gastrointestinal tract, activated sludge and solids from wastewater treatment.

14.6.3 Anaerobic fermentation

ABPs can be digested and broken down by microorganisms in the absence of oxygen to give a high yield of biogas. The biogas is energy-rich comprising mainly of carbon dioxide and methane. The digestion residues can often be used as organic fertilizers and soil improvers. ABPs, manure and sewage sludge from slaughterhouses can all be treated but, similarly to composting, the process is more effective for gastrointestinal content and manure. Biogas cannot be produced from pure animal material in single state digesters because the nitrogen content is too high. Therefore, animal waste must be mixed with other organic matter to reduce the nitrogen content.

Systems based on fermentation (composting and anaerobic fermentation) have been evaluated to obtain treatment standards based on parameters such as pH, time and temperature. The requirements for such systems are complex, including the necessity for additional protection measures or multiple barriers (i.e. more than one treatment stage) to reduce the possibility that any agent could bypass the system. They usually must be able to meet a minimum time/temperature standard (related to the type of system being used) and characteristically they specify the maximum particle size that may be processed in that system. An example can be a system that reaches a minimum temperature of 70 °C during a minimum time of 1 h, with a maximum particle size of 1.2 cm. Systems must include additional barriers, for example extra storage time. Alternative parameters are also possible when a relevant risk assessment is carried out. A 4–5 log fold reduction in pathogen level is generally required to demonstrate safety.

14.6.4   Rendering

Rendering is used to transform ABPs into fats such as lard or tallow by means of a process involving both physical and chemical transformation. Rendering processes require the application of heat, the extraction of moisture and the separation of fat, but rendering also yields animal proteins. Two rendering processes are available, wet and dry rendering. For the treatment of high risk materials (Category 1), ABPs are homogenized to a particle size of not more than 50 mm and subsequently rendered at 133 °C for a minimum of 20 minutes at a pressure of at least 3 bar (or alternative treatment). Materials are then pressed to separate fat and the leftover is ground to produce meat-and-bone meal.

14.6.5   Oleochemical processes

Fat can be further transformed using several oleochemical processes; these may involve hydrolytic fat splitting, hydrogenation, bleaching, distillation, concentration and refinement steps for purification, modification of properties and deodorizing. These methods are commonly used by the oleochemical industries to transform raw materials into a variety of products, which are then used in soaps, cosmetics, pharmaceuticals, detergents and industrial materials. Oils and fats produced by fat melters may also be used in the food industry, for example in baking and food processing, frying and margarine production. Some of the oleochemical processes have been investigated as an independent method to inactivate hazards (prions) linked to TSE present in ABPs. This would allow the use of such materials in various (non-food) applications, such as in soaps, cosmetic products and plastics, regardless of the category of ABPs that are used as starting materials.

14.6.6   Waste water from slaughterhouses

Slaughterhouses use large amounts of water for premises and equipment cleaning and carcass washing as a means of maintaining proper hygiene standards. Cabinets are sometimes used for showering carcasses with water. Water consumption has become an important environmental and economic operational factor. The use of dry techniques for cleaning procedures and by-product collection, and the limitation of gross carcass washing, reduce the amount of water used and the subsequent treatment at the plant. Treatment of water tends to reduce the organic content (measured as biochemical oxygen demand, BOD) and the concentration of some compounds, such as nitrogen and phosphorous, while some stages also help inactivate pathogens. The main elements of pollution in wastewater from slaughterhouses include faeces, urine, blood, gastrointestinal content, and run-off water from floors and utensils. To reduce BOD, water is pretreated to separate trimmings of meat and fat from carcasses, materials from desanding, grease and oil mixtures, sludge and other material removed from drains and screenings. Additional treatment processes that can be employed are classified in three categories: primary (involving physical and chemical treatments), secondary (i.e. anaerobic or aerobic biological processes) and tertiary treatment (chemical treatments). Slaughterhouses usually carry out primary treatment before releasing water to public watercourses, whereas secondary and tertiary treatments are carried out at wastewater treatment plants.

Primary treatment   Treatment of the wastewater is necessary for slaughterhouses, cutting plants removing SRMs and plants processing material of Categories 1 and 2 (e.g. intermediate and rendering plants) to retain solid materials up to a particle size of 6 mm. Any materials removed from the wastewater by this pretreatment unit are regarded as materials of Category 2 or materials of Category 1 (for establishments processing materials of Category 1 or removing SRMs). To facilitate the objectives of preliminary treatment, screens and racks are commonly used. These stages are also intended to facilitate subsequent treatment processes by removing larger suspended and floating solids and excessive amounts of oils or greases, and to protect pumping equipment.

Primary treatment uses physical and chemical methods to reduce the load of suspended solids and, therefore, the BOD. This stage achieves approximately 40–60% reduction in suspended solids and 25–35% in BOD. This reduction is achieved in a tank (a so-called sedimentation tank, primary clarifier, or primary settling tank) by reducing the velocity and dispersing the flow of wastewater, which causes suspended solids to settle out. Settling can be complemented with chemical processes such as clarification, flocculation and coagulation to remove colloidal material. Other physical phenomena used in treatment consist of dissolved air flotation and skimming (elimination of greases or oils that float to the surface), degasification and flow equalization, to facilitate further treatments. Denser material, such as sand, meat debris, faeces or hair, will settle out and material that is less dense (fats, greases) will float to the surface. The settling rate of a particle depends on the weight of the solid compared to the specific gravity of water, the size and shape of the solid and the temperature of the water.

Secondary treatment   Once the solids in suspension have been partially removed and the BOD reduced in the primary treatment, secondary treatment further reduces the amount of organic matter in the water. The remaining solids in suspension as well as dissolved solids are treated in subsequent processes in secondary settling tanks. Secondary treatment depends primarily upon organisms (bacteria, fungi, protozoa, rotifers, nematodes) that decompose biochemically the organic solids to inorganic or stable organic solids. In general, microbial processes can be carried out aerobically (usually) or anaerobically, converting organic matter into stable forms, such as carbon dioxide, water, nitrates and phosphates, as well as other organic end products. The production of new organic matter is an indirect result of biological treatment processes and it must be removed before the water is discharged into the receiving stream. The devices used in secondary treatment can be trickle filters with secondary settling tanks, intermittent sand filters, activated sludge and stabilization ponds. This stage of wastewater treatment yields wastes such as fat from grease traps, settlings and excess activated sludge as well as flotation tailings, which are taken to composting.

Tertiary treatment   Tertiary treatment is considered an advanced wastewater treatment stage that uses chemical–physical methods based on advanced technologies, such as membrane filtration, reverse osmosis, activated carbon adsorption, ion exchange,and disinfection (chlorination, ozonation). This post-treatment serves to remove chemicals such as phosphorus, sulfides, suspended solids, remaining BOD as well as inactivating pathogens.

14.6.7   Treatment of different categories according to European standards

Very high risk materials (Category 1), such as animal carcasses or parts of carcasses suspected or confirmed as being infected by a TSE, are disposed of by incineration or rendering followed by incineration. ABPs of Category 2 may be processed in a biogas plant only after sterilization with steam pressure, except manure and digestive tract content, which need no pre-treatment. Category 2 materials can also be disposed of by incineration and rendering. Some Category 2 ABPs can be recycled for uses other than feed after appropriate treatment, such as anaerobic fermentation, composting or oleochemical processes. Disposal of Category 3 ABPs can be done in various ways, such as incineration, rendering, authorized landfill (following processing), composting, anaerobic digestion or being used in an approved pet food plant or in a technical plant, where ABPs are used to produce technical products (such as pharmaceuticals, medical devices cosmetics, gelatin and glue, rendered fats etc.).

14.7   Materials obtained from animal by-products at the slaughterhouse

Tallows, lards and greases are obtained from animal fatty tissue by using rendering processes, which separate the fat from the bone and protein using either continuous flow or batch cooking vessels. Grinding precedes cooking, which is generally achieved with steam. Fat is screw-pressed and separated from the cooked material. The leftovers (cracklings), which include protein, minerals and some residual fat, are then further processed by additional moisture removal and grinding. These become a protein source (meat-and-bone meal, poultry by-product meal etc.) and can be used in feed. In the process, the fat cells are broken and the melted fat is released from the tissues. Rendering can produce edible (e.g. shortening, sweets, margarine and chewing gum) or inedible products, mainly depending upon the type of system, technology and process conditions used (approximately 115–145 °C for 40–90 min). Rendering also helps minimize the release of potential biological hazards (bacteria, viruses, protozoa, and parasites) into the environment. Several raw materials can be used, such as trimmings from intestines, mesentery, back fat (in pigs) and other internal organs. However, as mentioned above, the use of meat-and-bone meal made from ruminants is restricted by regulations to reduce the risk of TSE.

Besides the use as an energy source, the most important products from inedible fats are soap and glycerol, which are obtained by saponification of fats with alkali. Glycerol is the base for glycerine, which has numerous industrial uses, such as the manufacture of nitro-glycerine, ointment bases, solvents and plasticizers. Inedible animal fats and oils are also used in the preparation of lubricants and lubricating greases, free fatty acids (e.g. stearic acid and red oil), greases, rubber and miscellaneous other products, such as synthetic detergents.

Organs such as liver, pancreas, ovaries, glands as thymus and so on contain high value compounds that can be extracted, concentrated, dried and further processed. Pharmaceutical compounds are extracted from glands (adrenal, parathyroid, pituitary, thyroid) and organs (thymus, ovaries, pancreas, testes). Table 14.2 includes a list of the glands or organs and compounds extracted from them.

When blood is recovered using open-draining systems it comes into contact with intestinal content, faeces and water, becoming an unhygienic, highly polluting waste that is usually treated together with all the organic material in large anaerobic digesters. Closed-draining systems allow hygienic blood collection using a hollow knife that directly pipes blood by vacuum from the large blood vessels in the throat into a refrigerated container. The product usually contains very low bacteriological load and permits further processing and collection of haemoderivates, which means a high valorization of the product; multiple applications are found in the food industry (additives such as emulsifiers, stabilizers, clarifiers, egg albumin substitute, nutritional additives) and the pharmaceutical sector (Table 14.2). Dried blood (blood meal) is obtained by steam coagulation of the fresh blood, removal of water from blood and then spray-drying to yield the coagulum. The blood may be spray dried as whole blood or after it has been separated into plasma and red albumin.

Hides and skins can be categorized as a by-product for use as fur, leather or leather goods, the manufacture of cosmetic ingredients and medical prosthetics. They can also be used for the production of gelatine and/or collagen for human consumption (edible co-products) with multiple applications, such

Table 14.2 List of glands and organs, and compounds extracted from them.

Glands and organs	Compounds
Adrenal medulla	Adrenaline
Adrenal cortex	Adrenocortical extract
Ovaries	Oestrogens, progesterone
Pancreas	Insulin, trypsin
Parathyroid	Parathyroid hormone
Pituitary	Adrenocorticotropic hormone
Thyroid	Thyroxine and calcitonin
Pituitary gland	Somatotropin
Testicles and ovaries	Testosterone
Serum	Vaccines, antigens, and antitoxins
Blood	Amino acids, foetal serum, thrombin, albumin
Bone	Calcium, phosphorous
Liver	Cortisone
Lungs	Heparin
Spinal cord	Cholesterol
Stomach	Enzymes: rennet (from calves), mucin (from pigs), pepsin (from pigs)

as for confectionery items, ice cream and jellied food products. Their handling must comply with the requirements for fresh meat in the food hygiene legislation, including passing *ante-* and *post-mortem* inspection. Production of gelatine starts by preparing raw materials (hides, connective and cartilaginous tissues, bones, skin, pork snouts) by different curing, acid and alkali processes to extract the dried collagen hydrolysate, although after de-fatting and demineralization of bones and acid treatment of skins, the gelatine extraction steps in some of the processes using bones, hides and skins are very similar. All these chemical and enzymatic processes are followed by refining and other unit operations to obtain material with the required specifications. Cooking in water converts the collagen in these materials to gelatine, which can be later used in confections, jellies and pharmaceuticals. Glue is made from the same raw material as gelatine and the process is very similar but it is extracted by successive boilings in water. The sludge from wastewater treatment from gelatine and glue manufacture is used as fertilizer and soil improver.

Intestines are usually cleaned and washed in the slaughterhouse; the intestine wash water contains valuable mucus that can be recovered, cleaned of pathogens and spray dried. Applications are sausage casings, the strings of musical instruments or tennis rackets and surgical ligatures. To produce the strings, the guts (except from ruminants) are cleaned and cut before being chemically treated to preserve them.

By-products can be transformed into organic fertilizers and soil improvers and are obtained also from multiple sources. Unspecific slaughterhouse waste (manure, digestive tract content, milk, colostrum and blood) is transformed into fertilizers using anaerobic digestion. Blood can be also used as a fertilizer due to its quick action and high nitrogen content.

14.8   Conclusions

Non-carcass components and, specifically, animal by-products are very diverse in regard to their risk, characteristics, composition, value and possible destination. In any case there are important reasons that are further incentives for the better management and processing of ABPs from slaughtering and carcass cutting. An efficient and hygienic handling of ABPs reduces disposal costs, decreases environmental pollution and generates beneficial income for the meat industry. Considering the possible presence of hazards in ABPs, carefully processing and disposal of ABPs is necessary to avoid the spread of biological and chemical hazards that present a risk to human and animal health. Developing and innovative research fields are based on the use of substances with technological or nutritional properties, the characterization of bioactive compounds and the use of animal fats for biodiesel production.

Literature and further reading

Arvanitoyannis, I.S. and Ladas, D. 2008. Meat waste treatment methods and potential uses. *Int J Food Sci Technol*, **43**, 543–559.

European Commission. 2005. Integrated Pollution Prevention and Control. Reference Document on Best Available Techniques in the Slaughterhouses and Animal By-products Industries. European Commission, Joint Research Centre, Seville, Spain.

Toldrá, F., Aristoy, M.C., Mora, L. and Reig, M. 2012. Innovations in value-addition of edible meat by-products. *Meat Sci*, **92**, 290–296.

Liu, D.C. and Ockerman, H.W. 2001. Meat Co-Products. In: *Meat Science and Applications* (eds Y.H. Hui, W.K. Nip and R. Rogers). Marcel Dekker Inc., New York, NY, pp.581–604.

Taylor, D.M. and Woodgate, S.L. 2003. Rendering practices and inactivation of transmissible spongiform encephalopathy agents. *Rev Sci Tech*, **22**, 297–310.

Woodgate, S. and van der Veen, J. 2004. The role of fat processing and rendering in the European Union animal production industry. *Biotechnol Agron Soc Environ*, **8**, 283–294.

15

The Conversion of Muscle to Meat

Frans Smulders[1], Peter Hofbauer[1] and Geert H. Geesink[2]

[1]*Department of Farm Animals and Veterinary Public Health, University of Veterinary Medicine, Vienna, Austria*

[2]*School of Rural Sciences and Agriculture, University of New England, Armidale, NSW, Australia*

15.1   Scope

This chapter reviews the *post-mortem* events in muscles of the main production animal species before, during and after the onset of rigor mortis, and discusses the major biological mechanisms determining the physical–chemical and sensory quality traits of whole tissue meat (colour and water holding of fresh meat, and tenderness and flavour of cooked meat). The effects of various *ante-mortem* and processing factors influencing sensory quality traits of the various major animal meat species are summarized.

15.2   Introduction

The concept of 'meat quality' includes many aspects, such as hygiene and food physiological, technological and sensory properties. Whilst consumers can assume that the meat industry generally adheres to good hygiene and manufacturing practices that promote the safety of meat and allow its storage

Meat Inspection and Control in the Slaughterhouse, First Edition.
Edited by Thimjos Ninios, Janne Lundén, Hannu Korkeala and Maria Fredriksson-Ahomaa.
© 2014 John Wiley & Sons, Ltd. Published 2014 by John Wiley & Sons, Ltd.

for a certain period, they usually are less aware of the many technologies that are primarily directed at improving sensory traits of meat.

The quality of fresh meat, as perceived when buying, preparing and consuming whole tissue cuts, is determined by *ante-* and *post-mortem* factors having interfered with muscle biological events in skeletal muscle. Many of these are common to all animal species and the effects of various processing technologies on meat quality (for instance, that of beef, pork, and poultry) have been studied extensively over the past decades. With few exceptions, game meat species have received less attention and the available information is still scarce and fragmented.

15.3 Muscle structure, composition and function

Lean skeletal muscle (expressed as percentage wet weight) is composed of water (75%), protein (19%), soluble organic compounds (3.5%), variable amounts of lipids (0.5–3%), carbohydrates (1–2%) and small amounts of minerals and vitamins. Of these components, the lipid content is the most variable, with highly marbled beef containing more than 25% lipids.

The visual appearance and the physiological and biochemical characteristics of skeletal muscles vary considerably, reflecting the proportion of fibre types present in the muscle cells. The commonly used fibre type classification is based on the contraction speed (fast or slow) and the energy metabolism (oxidative or glycolytic) of the fibres. The biochemical traits associated with different fibre types (e.g. ATP concentration, calcium, myoglobin, glycogen, lipid, proteinase and their inhibitors contents, and enzyme activities) reflect the diversity of muscle fibres and are related to their physiological function exerted in the living animal. In this respect, meat quality parameters such as colour, flavour, juiciness and tenderness are fibre type dependent.

The coherent structure of muscle fibre bundles is maintained by the intramuscular connective tissue. Three connective tissue structures can be distinguished in muscle, that is the epimysium (a fibrous sheath of connective tissue which surrounds the entire muscle), the perimysium (a three dimensional collagen network surrounding the muscle fibre bundles) and the endomysium (a layer of fine connective tissue fibres encircling individual muscle fibres).

Not more than 5% of the total water of muscle is directly bound to the hydrophilic groups of the proteins, The rest is divided between the so-called 'free water' (i.e. immobilized by the physical configuration of the proteins but not bound to them) and the so-called 'loose water', which is expressed when the water-holding capacity drops.

Fat content varies in quantity and composition between muscles and species, males generally having less than females. Many intracellular lipids are associated with membrane structures. In addition, considerable amounts of lipids are present in the perimysium surrounding myofibre bundles. Macroscopically this is perceived as 'marbling'.

Muscle proteins may be classified as sarcoplasmic, myofibrillar and stroma proteins. Sarcoplasmic proteins are soluble in water or salt solutions of low ionic strength. Many of them are enzymes involved in the breakdown of glycogen. Myofibrillar proteins, which can be subdivided into contractile and cytoskeletal proteins, constitute the major part of the muscle's contractile mechanism and are only soluble in solvents with higher ionic strength than required for the extraction of sarcoplasmic proteins. The main contractile proteins are myosin, actin, and the troponin-tropomyosin complex. Stroma proteins include collagen, elastin and reticulin, all present in connective tissue, and the proteins which are found in the membrane systems of the muscle cell organelles, such as the mitochondria and sarcoplasmic reticulum.

The contractile proteins are arranged in repeating units consisting mainly of actin and myosin. The thin myofilament is composed of two strands of ('filamentous') F-actin, which consist of polymerized ('globular') G-actin. Two strands of F-actin are coiled around each other and have a very precise length. The actin helix is associated with tropomyosin, a double helix of two non-identical peptide chains, and the troponin complex, consisting of three polypeptide subunits – C (binding calcium ions and thus effecting a change of conformation), I (inhibiting the interaction of actin to myosin) and T (the tropomyosin-binding subunit), which cements the tropomyosin to the actin helix. Figure 15.1(a) schematically presents this arrangement.

The troponin complex turns on the contraction process by binding Ca^{2+} ions that are released from the sarcoplasmic reticulum after a nervous stimulus. The thick myofilament consists of packed myosin molecules (Figure 15.1(b)) each of which has a helical tail and a head consisting of two globular units.

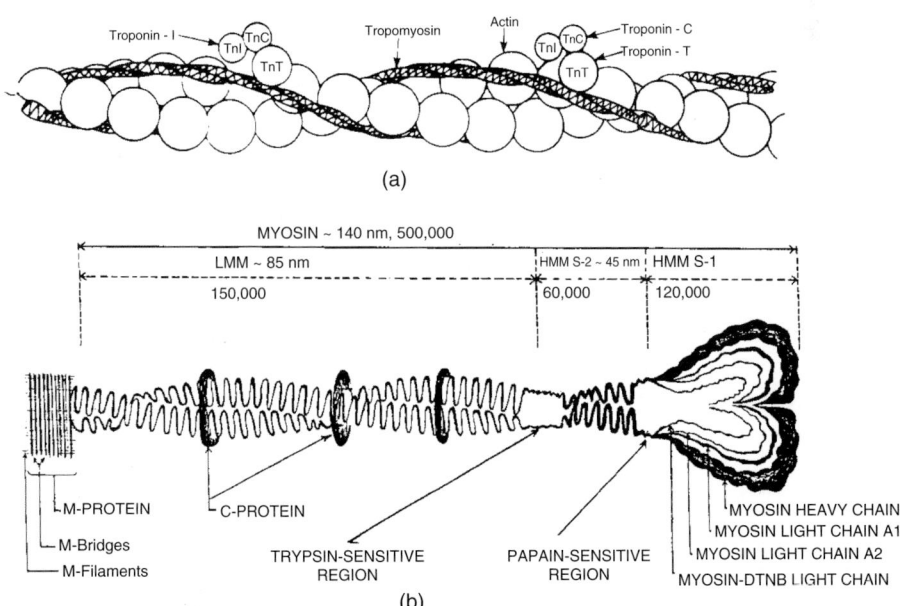

Figure 15.1 The association of actin with the troponin-tropomyosin complex (Figure 15.1(a): Source: Cohen, C., 1975. Reproduced with permission from Nature Publishing.) and the structure of the myosin molecule (Figure 15.1(b): Source: Asghar & Pearson, 1980.). [Note: C protein is not part of the myosin molecule but binds to the myosin tail region to maintain thick filaments in bundles of 200 to 400 molecules.]

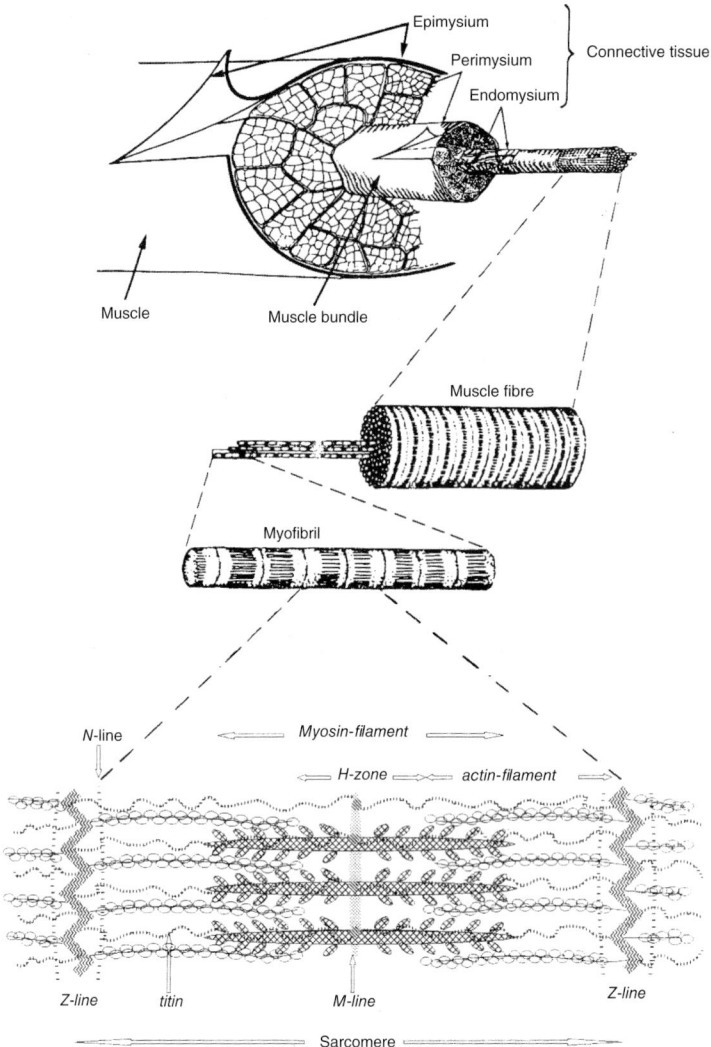

Figure 15.2   A schematic overview of the structure of skeletal muscle (Geesink, 1993, p.15).

The myosin heads contain a site with ATPase activity and a site that forms cross-bridges that interact with actin to form a short-lived actomyosin bond during muscle contraction.

Figure 15.2 shows a diagrammatic representation of the main (ultra)-structural features of muscle. The I-band (an 'isotropic' zone spanning the distance between myosin molecules in two adjacent sarcomeres) mainly consists of actin filaments that are attached to transverse structures called Z-lines. The spacing between Z-lines is referred to as sarcomere length. Myosin fills the entire length of the A-band ('anisotropic' zone). A clear central H-zone contains the M-line representing the cross-bridges between separate myosin filaments.

In developed striated muscle various 'cytoskeletal' proteins serve the role of maintaining cell shape and integrity; they are important for the function of movement. Examples of these are proteins like titin, nebulin, dystrophin and desmin.

15.4 *Post-mortem* muscle physiology; rigor mortis and the conversion of muscle to meat

Figure 15.3 is a diagrammatic representation of the main muscle physiological events that are related to sensory meat quality traits.

In the living animal the blood circulation provides the muscle fibres with oxygen, fatty acids and glucose. Whenever muscle action is required, nervous stimuli effectuate a depolarization of the sarcoplasmic reticulum membrane. This results in a release of Ca^{2+} ions which activate the enzymes that convert glycogen into pyruvate and, eventually, into carbon dioxide and water in the mitochondria (Krebs cycle). In the course of this process adenosine diphosphate (ADP) is phosphorylated to adenosine triphosphate (ATP). Initially, the muscle relies on the muscle specific creatine phosphate (CP) to supply the high energy phosphate. The conversion of ATP to ADP directly supplies the energy needed for muscle contraction and metabolic activities. This is effectuated by Ca^{2+} ions released by the sarcoplasmic reticulum, which stimulate myosin ATPase. Muscle relaxation also requires ATP, not only because the calcium pump transporting Ca^{2+} back into the sarcoplasmic reticulum requires energy, but also because myosin rods will only be allowed to shift out of the sheaths formed by the actin filaments provided ATP is present in its function as 'plasticizer'. Under anaerobic conditions ATP is resynthesized via comparatively inefficient glycolysis leading to the formation of lactate, which may be resynthesized to glycogen in the liver via the Cori cycle (gluconeogenesis).

When an animal is slaughtered it is subjected to a state of shock, which acts upon the muscle as a complex of nervous stimuli. Since carbohydrate and oxygen supply has ceased, the muscles are dependent on glycolysis for their energy synthesis from the moment the muscle specific creatine phosphate (CP) reserves have been depleted. Consequently, comparatively little ATP is resynthesized. Moreover, the enzymes catalysing glycolysis and ATP breakdown are activated and lactate and metabolites such as adenosine monophosphate (AMP) are formed. As a result the muscle pH gradually falls. Rigor mortis sets in when too little ATP is available to keep the actin and myosin filaments apart in a relaxed state. This occurs when the ATP residue is approximately 20% of its initial concentration.

The biochemical reactions in the period before onset of rigor mortis – generally referred to as the period of 'conditioning' – as well as those resulting from the *post-mortem* muscle proteolysis that occurs subsequently – also known as 'aging' – have a great impact on sensory meat quality characteristics.

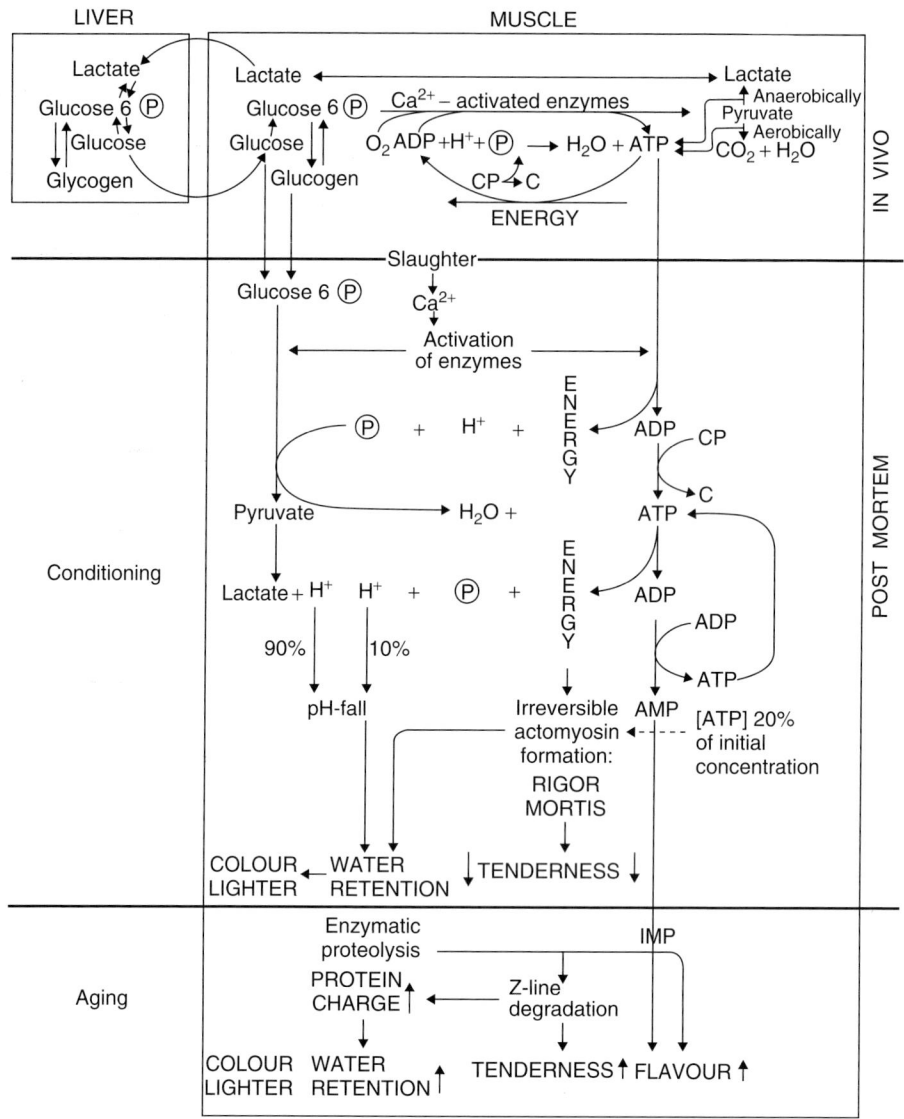

Figure 15.3 Diagrammatic presentation of the physiological events effectuating the conversion of muscle to meat. Source: Smulders, 2007. Reproduced with permission from Wageningen Academic Publishers.

15.4.1 *Post-mortem* muscle pH decline and ultimate pH values of the main meat animal species and major anomalies

The pH decline in *post-mortem* muscle is a result of the breakdown of the energy reserves (particularly glycogen) prevalent in the various muscles at the moment of slaughter. Dependent on their fibre type (glycolytic or oxidative), the course of pH decline differs between muscles and animal species. Also, pH decline is markedly influenced by the intensity of refrigeration. For example, whereas at 37°C beef longissimus would reach pH values <6.0 after about 4 h, this approximately takes 16 h at 7°C. Comparing the major meat species, one often distinguishes between so-called 'slow glycolysing' (e.g. bovine, ovine) and 'fast glycolysing' (e.g. avian, porcine) musculature. In Figures 15.4(a), 15.4(b) and 15.4(c) the pH changes in longissimus (ruminants and pigs) and pectoralis superficialis muscles (poultry) chilled at commercial processing temperatures have been included as examples.

In evaluating the 'normality' of a particular course of pH decline in *post-mortem* muscle, it is essential to consider both the muscle's 'glycolytic rate' (speed of pH decline) and 'glycolytic potential' (the ultimate pH value achieved). The former is dependent on the early *post-mortem* activity of the glycolytic enzymes, the latter primarily on the muscle glycogen reserves present when the animal was slaughtered and the time period during which glycolytic enzymes remain active.

The Porcine Stress Syndrome (PSS) and 'Pale-Soft-Exudative' (PSE) pork In genetically susceptible pigs subjected to acute stress immediately before slaughter, the amount of calcium ions transported out of the sarcoplasmic reticulum may be twice as much as in pigs that are not genetically predisposed. The increased sarcoplasmic calcium concentration leads to the development of malignant hyperthermia, an extremely fast glycolysis (Figure 15.4(b)) and, ultimately, to a meat quality defect commonly known as the 'Pale-Soft-Exudative' (PSE) condition. A detailed description and muscle biological explanation of the specific effects on various sensory quality parameters can be found in Section 15.5. The susceptibility of pigs is partially genetically predetermined, but porcine stress syndrome (PSS) can equally be triggered by incorrect pre-slaughter animal handling.

In the early 1990s it was clarified that the genetic component of PSS is associated with a single nucleotide substitution in the gene encoding for the ryanodine receptor (RyR1) at the halothane (Hal) locus, and since this discovery an accurate diagnosis of the animal's genetic susceptibility has become possible. Three genotypes must be distinguished, that is the normal (nn), the heterozygote (Nn) and the homozygote (NN). Earlier findings relying on halothane challenge testing (i.e. only the homozygote exhibiting muscle rigidity in the limbs rather than the relaxation seen after halothane anaesthesia of normal or heterozygote animals) suggested that the inheritance of PSS is recessive and, hence, that RyR1 heterozygotes were merely carriers of the mutated gene. However, it has meanwhile been shown that these (Nn) carriers exhibit meat quality characteristics intermediate between the NN and nn animals.

It is important to understand that even 'halothane-negative' (Nn, nn) pigs can develop PSS. For example, Dutch studies have shown that a high level of physical and psychological pre-slaughter stress may – particularly in muscles with a 'glycolytic' fibre type (e.g. the longissimus muscle) – lead to a PSE

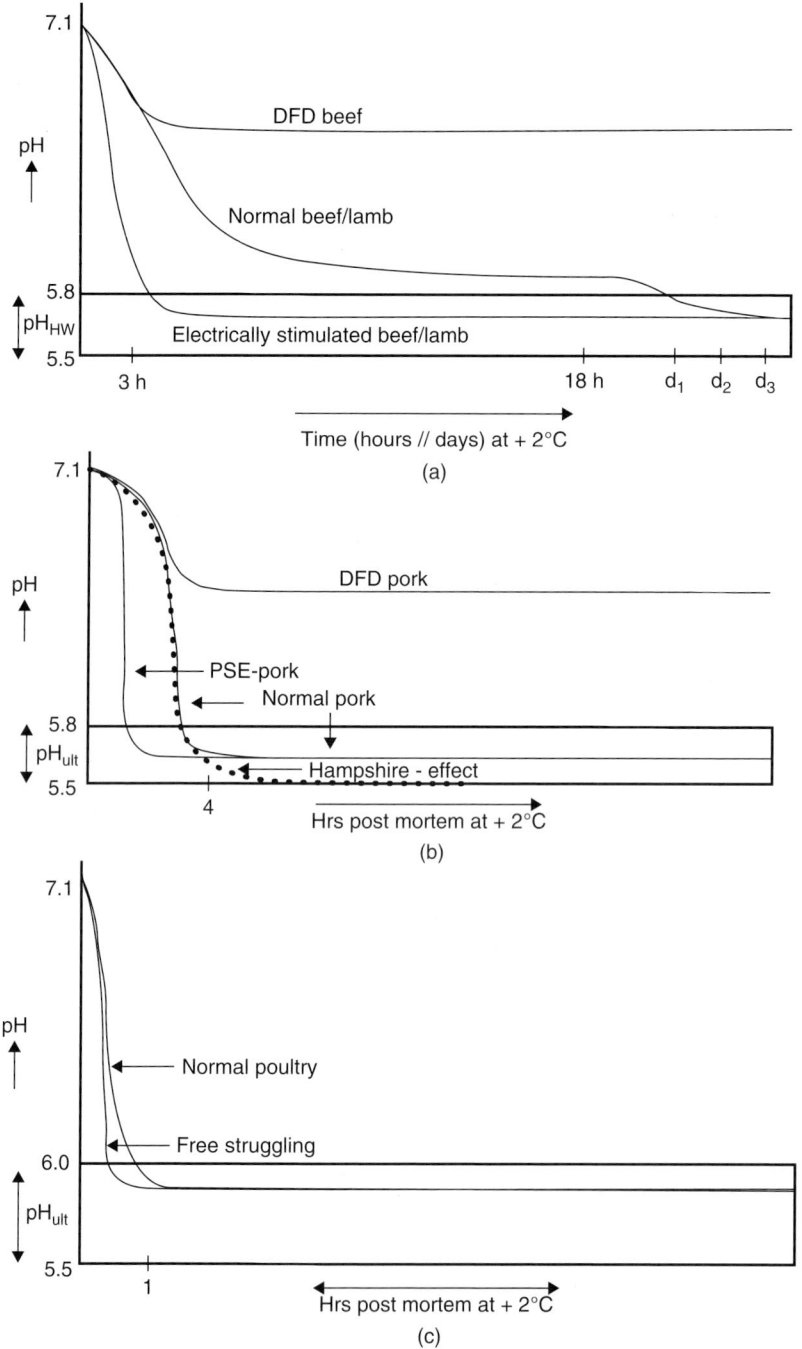

Figure 15.4   Early *post-mortem* pH decline of (a) bovine/ovine and (b) porcine M. longissimus, and of (c) avian M. pectoralis superficialis, and major anomalies observed under common commercial chilling conditions. Source: Smulders, 2007. Reproduced with permission from Wageningen Academic Publishers.

condition. [Note: conversely, muscle with a predominantly 'oxidative' fibre type (e.g. the M. supraspinatus) will react to intense physical activity and/or high psychological stress levels pre-slaughter by exhibiting higher ultimate pH values and DFD (see below)].

Besides proper animal loading, transport and unloading procedures, a most significant preventive measure to reduce the effects of *ante-mortem* stress in pigs is assuring the proper treatment in lairage. Slaughter pigs should be allowed to recuperate for a minimum of 2 h in lairage, during which the animals are generally showered. The latter is particularly suitable at high environmental temperatures. Showering helps to lower the animal's body temperature, reduces body odour (and thereby ranking fights) and as a side effect – favourable when animals are to be electrically stunned – increases the conductivity of the animal's skin.

It is relevant to consider whether or not the PSE condition would also influence the shelf life of such meat. Although it is often assumed that the increased drip loss of PSE meat might favour bacterial growth, Canadian studies have recently demonstrated that the initial microflora is primarily influenced by the dressing conditions at the slaughter plant rather than by the meat quality class (normal or PSE) and that PSE meat has similar microbial loads, even after 35 days of storage at 4°C.

The 'Hampshire effect' in pork Another abnormal condition of pork, exclusively observed in meat of pure or cross-bred Hampshire pigs, is the so-called 'Hampshire effect'. The anomaly has a genetic background: a so-called 'gain of function' mutation (with dominant inheritance) at the PRKAG3 [or RN-('Rendement Napole')] locus results in an increased glucose uptake of skeletal muscle leading to up to 70% higher glycogen levels in muscles of RN-carriers. The course of *post-mortem* pH decline is characterized by a normal rate of glycolysis but an increased glycolytic potential (i.e. lower than normal ultimate pH values; Figure 15.4(b)). Meat with this defect has also a paler colour and lower water-holding properties, most noticeable when manufacturing cooked hams from Hampshire carriers, which have an approximately 5% lower cooking yield.

'PSE-like' conditions in poultry meat The PSE condition in meat results from a combination of a fast glycolytic rate (leading to a rapid muscle acidification) and high muscle temperatures (for details, see Section 15.5). The description 'PSE' has lately been applied to any meat that has a paler than normal appearance and has high drip losses. This includes chickens and turkey although neither the genetic predisposition (in terms of RyR1 or RN- mutations) nor the occurrence of malignant hyperthermia has been demonstrated in poultry.

High temperatures can be generated by wing flapping, free struggling (Figure 15.4(c)), stress and high muscle metabolic rates prior to slaughter, whilst the large muscle mass of the breast muscle is relatively difficult to chill *post-mortem*. Moreover, the (entirely glycolytic) breast muscle of poultry contains large glycogen stores and, therefore, has the potential for a rapid pH decline and/or can lead to a low ultimate pH.

The worldwide growing demand for poultry meat has driven poultry breeders to increase the growth rate, feed efficiency and breast muscle size and to reduce abdominal fatness. Unfortunately, this has been accompanied by an increased occurrence of 'PSE-like' meat in poultry. As long as reliable genetic markers indicative of the predisposition for this condition (which would allow devising targeted breeding strategies) have not been identified, the most efficient short-term preventive measures are based on adhering to the best possible pre- and post-slaughter handling practices.

The DFD condition (beef, pork) Three main mechanisms explain the abnormally high ultimate pH values (>6.0) sometimes observed in beef (Figure 15.4(a)): (i) poor pre-slaughter management practices (e.g. improper penning and animal handling) resulting in substantial depletion of muscle glycogen reserves, whilst animals are not allowed sufficient time for recovery: (ii) adrenergic activation of glycogenolysis by increased adrenaline concentration in the circulation; and (iii) long-term depletion of glycogen reserves as a result of malnutrition or starvation. The phenomenon, generally known as the 'Dark-Firm-Dry' condition (or 'dark cutting' in beef) is particularly observable in the longissimus muscle of young bulls. A seasonal effect has been reported: for example, in the northern hemisphere DFD beef is primarily found in autumn and winter. The specific sensory effects associated with the DFD condition are dealt with in Section 15.5.

DFD is also seen in pork (Figure 15.4(b)). In this context it is relevant to note that in pig production withdrawal of feed prior to transport is common (for reasons of reducing gut contents and, thereby, the odds of cross-contamination in the slaughterhouse). Hence, extending time in lairage beyond 4 h may lead to marked muscle glycogen depletion and, consequently, may result in DFD.

Microbial growth conditions on DFD meat are decidedly more favourable than is the case for meat with normal ultimate pH (ranging from 5.5 to 5.8), firstly because microorganisms generally grow better at high pH, secondly because the lack of carbohydrates causes a shift in the 'bacterial association' towards a proteolytic flora, which may include microbial spoilers, for example *Alteromonas putrefaciens* responsible for the 'greening' of vacuum packaged meat, or even pathogenic bacteria such as *Aeromonas* spp. that can cause disease in immunocompromised individuals.

15.5 Major sensory characteristics of meat

Whilst major interrelated phenomena occurring *post-mortem* and valid across the various meat species have been included in Figure 15.3, the following section discusses in more detail the relevance of the various factors that determine the major quality traits of meat.

15.5.1 Colour of fresh meat

The colour of fresh meat is determined both by light absorption [dependent on the contents of pigments (i.e. apart from the cell pigment cytochrome, particularly haemoglobin and myoglobin)] and light being reflected from water present on the meat surface (largely related to the water-holding properties of muscle proteins and the density of the myofibrillar matrix). Provided an animal is properly exsanguinated, the role of haemoglobin is relatively minor: fresh meat contains no more than 0.3% residual blood.

Myoglobin-related factors The pigment primarily responsible for colour is deoxymyoglobin (DEMb). Like haemoglobin in blood, its physiological function is to store oxygen and deliver it to the muscle whenever necessary. Upon binding of oxygen, its colour changes from purple to an attractive cherry-red, which reflects the presence of oxymyoglobin (MbO). When the haem-iron oxidizes (ferrous iron, Fe^{2+}, turns into ferric iron, Fe^{3+}), metmyoglobin (MMb), which has lost the ability to bind oxygen, is formed. Concurrently, the colour turns into an unattractive greyish-brown. Only through reduction with the aid of reducing enzymes can MMb be converted back to physiologically active (oxygen-binding) DEMb.

The relationship between the partial pressure of oxygen (pO_2) and the chemical form of Mb (DEMb, MbO or MMb) is presented in Figure 15.5. Depending on the presence of oxygen, be it at atmospheric or at higher pressures, as for example prevailing in modified atmosphere packaging, the

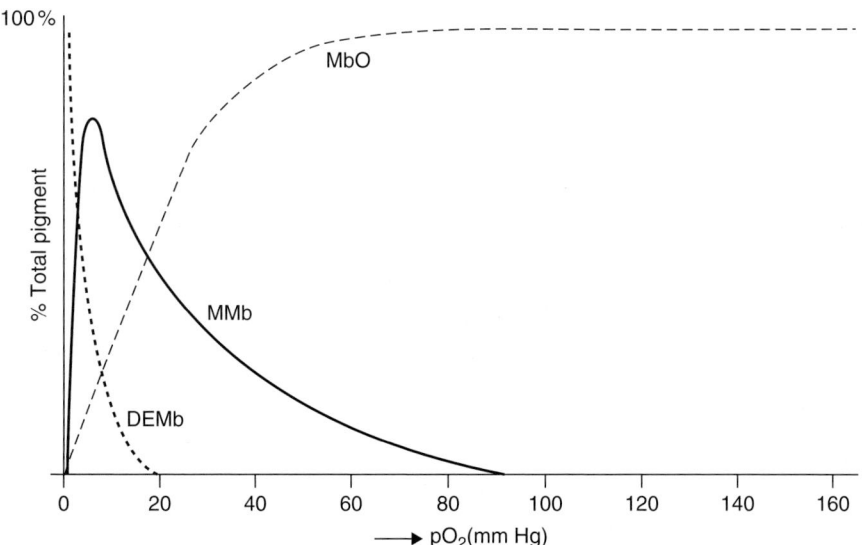

Figure 15.5 The effect of partial pressure of oxygen on the prevalence of various myoglobin forms. Source: Smulders, 2007. Reproduced with permission from Wageningen Academic Publishers.

surface of any piece of meat will have a superficial layer of varying thickness of MbO, which is being replaced by MMb at depths where oxygen penetration is insufficient, and finally by DEMb in the core of the muscle where pO_2 is zero.

Modified atmosphere packaging of fresh meat with, typically, 80% oxygen and 20% carbon dioxide is widely adopted to extend the shelf life. The elevated oxygen content results in a thicker superficial layer of MbO and, as a result, delays browning. The elevated carbon dioxide content has a bacteriostatic effect. A series of recent studies has shown that using elevated oxygen levels to extend the shelf life of fresh meat has a detrimental effect on the organoleptic quality. Due to the elevated oxygen content, oxidative processes are promoted, leading to negative effects on flavour.

An alternative modified atmosphere composition to maintain a bright red colour and limit bacterial growth and oxidation is the use of low levels of carbon monoxide in combination with carbon dioxide and nitrogen (for example: 0.4% CO, 30% CO_2 and 69.6% N_2). Carbon monoxide binds to deoxymyoglobin, forming carboxymyoglobin, which is cherry-red similar to oxymyoglobin. Carbon monoxide binds to deoxymyoglobin more strongly than oxygen, thus making carboxymyoglobin a more stable pigment than oxymyoglobin. Although this method of packaging is approved by the US Food and Drug Administration, a major concern is that products might look fresh even though bacterial levels are high and the product is spoiled.

When interacting with hydrogen peroxide (H_2O_2) or hydrogen sulfide (H_2S), Mb transforms into the green pigment cholemyoglobin (ChMb) or sulfmyoglobin (SMb). The formation of one or both of these pigments is responsible for the green discoloration of meats with severe bacterial contamination.

Fresh meat colour as affected by its water-holding properties   The ability of a muscle matrix to hold on to water has a large impact on meat colour. The lower the water-holding capacity of meat the more water molecules will be released and the more light will be reflected from its surface.

The well-known Pale-Soft-Exudative (PSE) condition predominantly found in pork is largely the result of sarcoplasmic protein and myosin denaturation, caused by a combination of high carcass temperatures associated with stress (malignant hyperthermia- or porcine stress syndrome) and the resulting extremely fast pH decline (see section 3.1.1). As a consequence, PSE meat binds water very poorly, causing exudation (which reduces the muscle cell turgor leading to its soft consistency) whilst the increased amount of water at the meat surface reflects more light causing its appearance to be pale.

Certain stressful *ante-mortem* conditions, for example those associated with physical exhaustion, can cause healthy animals to develop so-called Dark-Firm-Dry (DFD) meat (Section 15.4.1.4), which is characterized by a high ultimate pH, a purplish-black colour, a firm texture and a dry sticky surface. Although the condition is predominantly reported to occur in beef and pork, it is similarly relevant for game species such as deer subjected to stressful conditions, for example regrouping, being chased or being subjected to inappropriate transport conditions. The depletion of muscle glycogen stores caused by these activities results in a lower than normal muscle acidification (higher ultimate pH values) and, consequently, in unusually high electrostatic binding of water by the muscle proteins [i.e. an extremely low release of water (Section 15.5.3)] and, hence, a higher muscle cell turgor (explaining its firmer consistency) and less light reflectance (explaining its darker appearance).

Colour stability of meat In essence, the stability of the colour of fresh meat is characterized by the muscle's ability to keep Mb in the oxygenated form and to prevent the formation of MMb. Although none of them react independently of the other, several biological factors impact on meat colour stability. The major ones are: (a) muscle pH, low values promoting the formation of MMb; (b) temperature, higher muscle temperatures favouring increased formation of MMb and dissociation of oxygen from MbO; (c) relative humidity (rh), lower relative humidity leading to more desiccation and a darker surface colour; (d) exposure to light (particularly the ultraviolet part) resulting in more MMb formation; (e) bacterial contamination, the exponential growth phase of the spoilage population coinciding with the highest oxidation; (f) lipid oxidation; (g) partial pressure of oxygen (pO_2), where zero (not low) pO_2 prevents MMb formation and only partial pressures around or above 80 mm Hg promote the desirable MbO formation (Figure 15.5); (h) the presence of MMb reducing enzyme systems, prolonged refrigerated storage leading to a considerable loss of reducing activity ('fading' of meat); (i) pre-slaughter stress, leading to an aberrantly fast pH decline or high ultimate pH values; and, finally, (j) muscle dependent sensitivity to discoloration, probably related to the muscle's oxidative capacity (fibre type).

A summary of *ante-mortem* and processing effects on meat colour Factors affecting the ultimate appearance (colour) of fresh meat include: (i) species effects or genetic variation within species (responsible for differences in oxygen consumption rate of mitochondria which counteract MbO formation, or determining the vulnerability to stress and the associated PSE condition); (ii) sex and age (e.g. as reflected in the darker colour of bull versus cow meat); (iii) plane and quality of nutrition (e.g. determining the degree of marbling and iron content); (iv) inappropriate *ante-mortem* handling of animals leading to PSE or DFD; (v) accelerated carcass processing methods, such as electrical stimulation, which, when not applied correctly, can lead to increased protein denaturation or hot boning allowing faster cooling of primal cuts and reduced protein denaturation; and, finally, (vi) modified atmosphere packaging (allowing an increased formation of MbO, provided high oxygen concentrations are present in the gas mixture).

15.5.2 Tenderness of meat

The main components thought to determine the tenderness of meat are connective tissue, myofibrillar proteins matrix and fat. The role of fat is primarily

that of a 'dilutant' of the muscle matrix (a high degree of marbling generally being associated with better tenderness and only extremely low contents
<1% being reported to promote toughness). The contribution of connective
tissue and the myofibrillar protein matrix to toughness is dependent on the
time after slaughter, since aging of meat has an important impact on meat
tenderness. From the time of slaughter, tenderness of a given muscle is determined by the background toughness (connective tissue), the toughening phase
(muscle contraction) and the tenderization phase (proteolysis of myofibrillar
proteins). As an example, the evolution of meat tenderness in bovine longissimus muscle during the *post-mortem* period is illustrated in Figure 15.6. The
following sections concentrate on the contribution to toughness of connective
tissue and the myofibrillar protein matrix.

The connective tissue component of tenderness Connective tissue serves an
important physiological function in muscle. It supports muscle fibres and carries over the forces generated by myofibrils to the skeleton. Connective tissue
proteins predominantly consist of collagen (65–95%). In addition, the less
rigid elastin prevails in lower amounts (5–35% of the total connective tissue,
dependent on the muscle), particularly in the epimysium.
  Not only does the collagen content affect the tenderness of meat but so,
too, do its type and the degree of cross-linking between the constituting
tropocollagen molecules. Various isoforms of collagen exist. In muscles, types

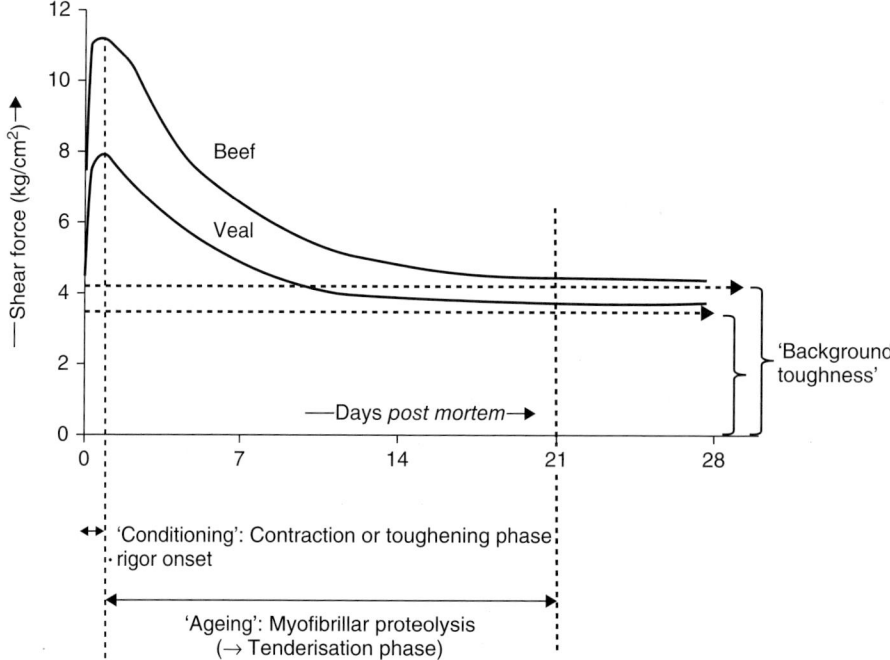

Figure 15.6 Changes in tenderness observed during conditioning and aging of beef and veal
as measured by Warner–Bratzler shear tests (i.e. recording the forces required to shear meat
samples, which are indicative for the tenderness sensation experienced by the consumer).

1–5 are important, each type having distinct properties, For instance, type 3 is markedly more heat stable than is type 1. Consequently, the distribution of collagen isoforms largely determines the solubility of connective tissue. Contrary to the myofibrillar component of tenderness, which is significantly affected during *post-mortem* storage of raw meat, changes in connective tissue are only observed in the course of heating.

Although the ratio of connective tissue:muscle substance decreases during growth, the degree of tropocollagen cross-linking increases, which explains the relative toughness of meat from older versus younger animals (Figure 15.6). It was once thought that connective tissue was the overriding factor explaining tenderness differences between meats. However, research in the past decades has clearly shown that its role is restricted to determining the so-called 'background toughness', which obviously is muscle and species dependent and is particularly related to the content of collagen type 3. Whereas in the course of storage the properties of collagen barely change, heating during meat preparation will affect its solubility markedly, provided temperatures >80°C are reached at which collagen gelatinizes and becomes soft.

The myofibrillar component of tenderness

Density of the myofibrillar matrix

Although a clear relationship between sarcomere length and the tenderness score of meat does not always exist, there appears to be general consensus that long sarcomeres are associated with tenderer meat, short sarcomeres with tougher meat (unless 'super-contraction' has occurred beyond 40% of the muscle's rest length of about 2.5 μm). However, this relationship is not linear and appears to be only valid for muscles with slow glycolysis. Preventing muscles contracting during the onset of rigor mortis (as achieved through the 'tenderstretch' process, e.g. by pelvic suspension) also leads to an end product that is more tender.

The degree of muscle contraction during rigor onset is markedly affected by the muscle's temperature decline. Although at low temperatures ATP breakdown is slower (as the glycolytic enzymes' activity is lower during refrigeration), at critically low temperatures, <6°C, mitochondrial release of Ca^{2+} increases while at the same time the calcium pump responsible for pumping Ca^{2+} ions back into the sarcoplasmic reticulum fails to function properly. The resulting increased concentration of sarcoplasmic Ca^{2+} leads to an intensive contraction known as cold shortening. Cold shortening occurs primarily in meat from animal species exhibiting slow muscle glycolysis (e.g. ruminants). Beef and lamb processors using fast chilling regimes, therefore, generally rely on electrical stimulation, which accelerates glycolysis (Figure 15.4(a)) and thus prevents low temperatures being reached when muscle pH is still high. In this context, combinations of muscle pH > 6.2 while temperatures are already below 12°C should be considered particularly hazardous.

A phenomenon known as 'heat shortening' or 'rigor contracture' may become relevant when slow glycolysing muscle is not chilled immediately after the animal's death, that is when rigor sets in at temperatures >25°C. This situation explains the sometimes extremely tough meat found in PSE

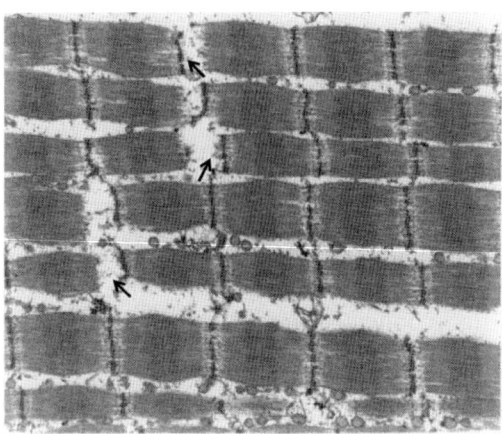

Figure 15.7 Fragmentation of myofibrils adjacent to Z-lines (see arrows) in bovine longissimus muscle after 14 days of storage at 2–4°C. Source: Courtesy of P. Koolmees.

pork, in slowly chilled electrically stimulated meat and also in game animals (ruminants) left unrefrigerated for too long.

Myofibrillar proteolysis

In *post-mortem* tenderization, which results from a loss of the muscle's structural integrity (Figure 15.7), proteolytic enzymes play a dominant role. A number of enzyme systems have been studied in relation to *post-mortem* tenderization, including the calpain system, cathepsins, the multicatalytic proteinase system and caspases. For a variety of reasons the calpain system appears to be the main enzyme system involved in *post-mortem* tenderization, and therefore is discussed in more detail here.

Calpains belong to the cystein group of proteases, which have an optimum pH between 7 and 7.5 and need calcium to become active. In skeletal muscle, this system consists of a least three proteases (μ-calpain, m-calpain and skeletal muscle-specific calpain, or calpain 3), and an inhibitor of μ- and m-calpain, calpastatin. An important characteristic of both μ- and m-calpain is that, once activated by calcium, they not only degrade myofibrillar proteins but also degrade themselves in an autolytic process. In *post-mortem* muscle, only autolysis of μ-calpain is observed but little or no autolysis of m-calpain, indicating its contribution to meat tenderization is minimal. *Post-mortem* proteolysis is not affected in calpain 3 knockout mice, but is largely inhibited in μ-calpain knockout mice (Figure 15.8). Calpastatin is an important inhibitor of μ-calpain activity in *post-mortem* muscle, and therefore tenderization. Differences in the rate of tenderization and proteolysis between species can be explained by differences in the calpastatin content of the muscles.

In summary, μ-calpain and calpastatin are the main determinants of *post-mortem* proteolysis and tenderization during aging of meat. A further understanding of the factors that regulate μ-calpain activity in early *post-mortem* muscle and, consequently, of the developments of methods to maximize the

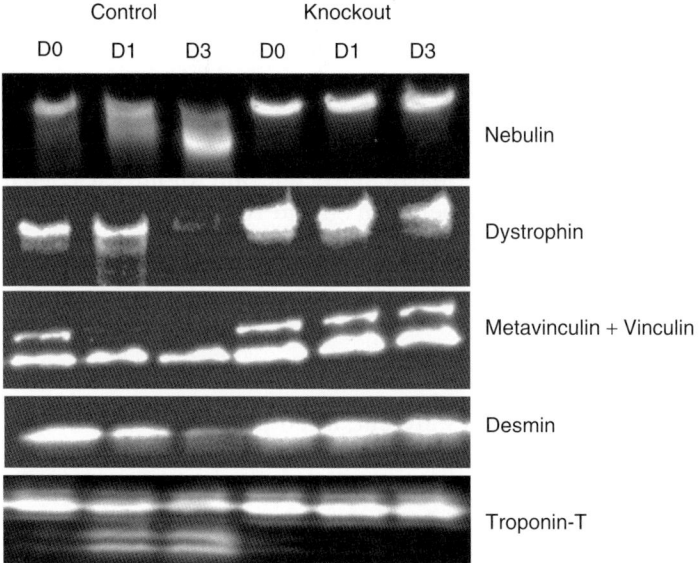

Figure 15.8 Western blot analysis of myofibrillar proteins in muscle extracts of control and μ-calpain knockout mice at death (D0) and after 1 (D1) and 3 days (D3) storage at 4°C. Source: Adapted from Geesink, G., *et al.*, 2006.

potential of μ-calpain to tenderize meat should aid the meat industry in producing consistently tender meat.

A summary of *ante-mortem* and processing effects on meat tenderness Factors affecting the ultimate tenderness of whole tissue meat include: (i) species effects, related to the susceptibility of muscles to phenomena like cold shortening and also to ageing rates (these decrease in the order poultry > pork > lamb > beef); (ii) sex, for example bulls generally rendering tougher meat than cows (a muscle dependent phenomenon also partially explained by differences in fat deposition and the age at which animals are slaughtered); (iii) age, largely based on a decrease in collagen solubility; (iv) nutrition, energy-rich feed rations promoting a higher degree of marbling; (v) pre-slaughter handling, possibly resulting in PSE or DFD meat, the latter condition yielding meat with high tenderness scores, probably because calpain activity is prolonged at high ultimate pH values; (vi) rate of carcass refrigeration, possibly leading to cold (or heat) shortening; (vi) accelerated processing, for example by relying on electrical stimulation possibly combined with hot boning; (vii) tenderness promoting additional processing, such as pelvic suspension or pressure/heat treatment; and, finally, (viii) adhering to optimal ageing times; that is five days for pork and around two weeks for beef, veal, lamb and rabbit.

Not related to processing *per se* but rather to domestic preparation of meat, are various methods for cooking meat, which have a marked influence on tenderness as perceived by the consumer. During heating meat goes through a number of toughening phases, dependent on the temperature reached. In the 40–50°C range, myofibrillar proteins are increasingly denatured, which makes the meat three to four times tougher. Around 65°C, connective tissue and myofibrillar and sarcoplasmic proteins have shrunk maximally, which additionally toughens the meat two times. At 70–80°C meat will become firm and 'well done'. Further heating – particularly relevant for muscles rich in connective tissue – to temperatures between 80 and 100°C for a long enough period leads to solubilization of collagen and disintegration of muscle bundles while the muscle fibres become fairly dry.

15.5.3   The water holding of meat

After the carcass and muscles are cut, a red proteinaceous fluid, called drip, oozes from the cut surfaces. The mechanism of its formation is not completely revealed yet but the water-holding capacity of meat is of great economic relevance, since it directly affects the weight of saleable meat as well as its suitability for further processing. Fluid loss can be partly explained by the decreased water holding by proteins resulting from the *post-mortem* pH fall. However, the amount of water bound firmly to proteins is too small to account for the total fluid loss.

Shrinkage of myofibrils   Since most of the muscle water is present within the myofibrils, a general hypothesis for explaining drip loss is that it originates from the 'lateral shrinkage' of the myofibrils *post-mortem*. Lateral shrinkage is partly due to *post-mortem* pH fall, as with decreasing pH the charges of the filaments and, thus, the negative electrostatic repulsion between the filaments are reduced, causing the space between the filaments to diminish.

The second factor causing ('longitudinal') shrinkage is actomyosin formation during rigor onset. The shorter the sarcomere length, the smaller the myofilament lattice spacing, and a close and linear relationship between drip loss after one week of storage and sarcomere shortening has been observed.

A third factor promoting shrinkage is the denaturation of myosin, which reduces the charge of, and hence diminishes the electrostatic repulsion between, the thick filaments as well as reducing the myosin head length. All these events draw the thick and thin filaments close together and thus result in higher drip loss.

Denaturation of sarcoplasmic proteins   Although it has been well established that in meat with a low water-holding capacity sarcoplasmic proteins are denatured, it is necessary to confront the fact that these proteins only hold about 3% of the water and that upon denaturation these globular proteins lose their compact folded structure and may thus entrap more rather than less water. However, it has been suggested that precipitation of (denatured) sarcoplasmic proteins may contribute to the loss of electrostatic repulsion between filaments and thus to drip loss.

Changes in osmotic pressure   Maximal osmotic pressure reached in post rigor meat is highly muscle dependent and may partly account for the variability in water-holding capacity between muscles within a carcass. As muscle pH drops, extracellular spaces become hyperosmotic, causing migration from water out of the muscle cells until the sarcolemma becomes permeable for proteins and ions.

Fluid accumulation within the muscle   Immediately after the animal's death, the extracellular space of a muscle primarily exists of the vascular system and connective tissue (perimysium and endomysium). 4–6 hours *post-mortem*, large gaps (presumably filled with fluid) appear between the fibre bundles. After rigor onset, these gaps also arise between the fibres and the endomysial network. The separation between fibres is only partial and a fibre may remain in contact with one of its neighbours (Figure 15.9). At this time the sarcolemma loses its integrity and sarcoplasmic proteins can diffuse freely into the extracellular space. The fluid in the extracellular space may be the source of drip. Both perimysial and endomysial gaps form longitudinal channels, which are in connection with the cut surface of the meat. Drip probably arises predominantly by the action of gravity draining the fluid in these channels to the cut surfaces.

A summary of *ante-mortem* and processing effects on the water holding of fresh meat   The following factors affect the water-holding properties of fresh

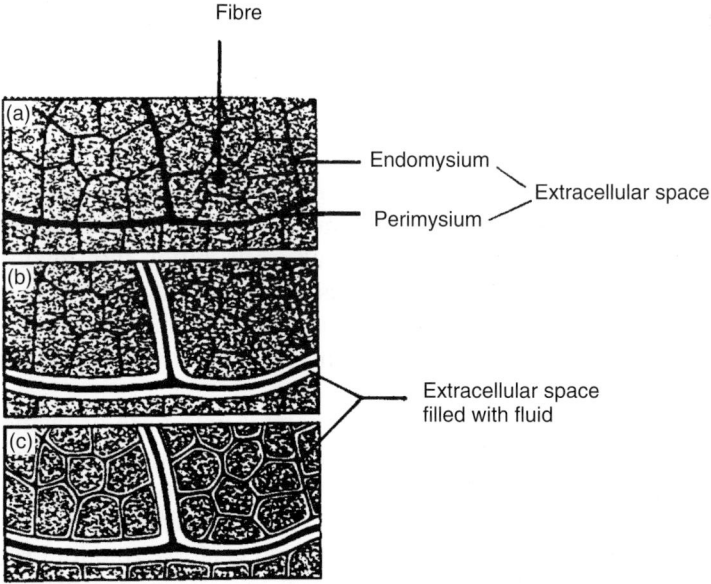

Figure 15.9   Schematic overview of structural changes in beef post mortem: (a) living muscle and immediately after slaughter; (b) 4–6 h *post-mortem*; (c) Rigor. Source: Offer & Cousins, 1992. Reproduced with permission from Wiley.

meat: (i) animal species and breed differences, partly associated with their vulnerability to cold shortening and/or their (largely genetically determined) susceptibility to stress; (ii) pre-slaughter handling, possibly leading to the PSE or DFD condition; (iii) stunning method (e.g. electrical stunning may cause a somewhat accelerated glycolysis); (iv) electrical stimulation, leading to faster *post-mortem* pH decline causing increased protein denaturation; (iv) chilling rate, high chilling rates slowing down protein denaturation but – unless counteracted by appropriate measures such as electrical stimulation – also bearing the risk of inducing the cold shortening phenomenon; (v) hot boning, allowing a faster and more uniform drop in temperature and thus counteracting protein denaturation, but at the same time increasing the risk of cold shortening; (vi) sample characteristics, that is the degree to which muscles are cut and whether the cut is longitudinal (less drip) or transversal (more drip); (vii) method and material of support, suspended muscles losing less than those supported from below, and muscle portions laid on a tray losing less than those placed on absorbent paper; (viii) packaging methods, the effects being largely dependent on the forces applied on the meat and the degree to which cut ends are sealed off; (ix) storage temperature, probably because at higher storage temperatures fluids have a lower viscosity and will migrate more easily; (x) freezing and thawing, fast freezing resulting in the formation of many small ice crystals, which cause less cell damage, and slow thawing allowing remigration of water into the intracellular space; (xi) transport conditions, transport vibrations possibly enhancing drip loss from meats with a low intrinsic water holding capacity; and, finally, (xii) ageing time, extended storage periods possibly allowing re-uptake of drip as a result of pH increase and proteolytic changes.

15.5.4 The flavour of meat

Flavour is determined both by taste and olfactory sensations, although astringency, mouth-feel and juiciness also play a role. Whereas mouth receptors can assess the 4–5 taste sensations (sweet, salt, sour, bitter and 'umami'), hundreds to thousands of odour compounds can be distinguished by epithelial receptors, which are reached either by smelling (via the nose) or via posterior nares at the back of the nose and throat while food is being chewed in the mouth.

The non-volatile or water-soluble compounds with taste or tactile properties can be listed as: inorganic and sodium salts of certain acids (salty), hypoxanthine (an ATP metabolite), peptides and some amino acids (bitter), sugars and some amino acids (sweet) and acids (sour). In addition, there are thousands of low molecular weight compounds that give rise to odour sensations. Although more than 800 volatile compounds have been identified in cooked meat aroma, it is believed that only a relatively small number of these compounds actually play a role in the overall aroma of cooked meat, and whether one of these represents a key odour impact compound depends on both its concentration and its odour threshold, that is how sensitive the human nose is to that particular compound. Two reactions are of particular importance in meat aroma formation: the Maillard reaction (a complex network of reactions that yield both high molecular weight brown coloured products and volatile aroma compounds) and lipid oxidation during heating, which contributes to the formation of desirable flavours. Free amino acids, peptides, sugar and also phospholipids and their fatty acids are particularly significant for flavour forming reactions, and various vitamins and minerals can alter the rate and extent of these reactions. Consequently, changes in composition of nutrients in meat could lead to a change in the balance of flavour forming reactions and, therefore, to a change in the overall aroma and flavour.

A summary of *ante-mortem* and processing effects on the flavour of cooked fresh meat The following factors affect meat flavour: (i) animal species and breed, probably through the genetic control of lipid composition and metabolism; (ii) sex (e.g. 'boar taint' in pork, for which compounds such as androstenon and skatole are thought to be responsible); (iii) nutrition (e.g. 'boiled cabbage' off-flavours of mutton fed with rapeseed or 'fishy' off-flavours of pork fed with polyunsaturated fatty acids, which also makes this pork more prone to oxidation and hence rancidity); (iv) *ante-mortem* stress, for instance, the DFD condition's inherent low acidity and, hence, a flavour that is less 'accentuated', making DFD meat taste rather bland; (v) bacterial contamination, as a result of which, dependent of the bacterial ecology, 'sulfurous' off-flavours (Pseudomonadaceae, Enterobacteriaceae) or 'dairy/cheesy' off-flavours (Lactobacillaceae, *Brochothrix thermosphacta*) may occur; (vi) irradiation at high dosages (causing increased carbonyls or hydrocarbon formation); (vii) lipid (per)oxidation particularly catalysed by haem iron and reported in meat that has been size-reduced (e.g. deboned chicken and turkey); and (viii) phospolipid and fatty acid oxidation ('warmed-over flavour') in cooked meats, probably catalysed by non-haem iron.

15.6 Concluding remarks

Although the meat industry is very keen on being able to routinely assess the sensory properties of whole tissue meat for purposes of targeted marketing, the possibilities are restricted. For instance, whilst the ultimate pH may be routinely monitored, the course of pH decline is only rarely measured, although both are responsible for many desirable (and undesirable) quality changes.

Some (sophisticated) meat classification systems (primarily designed to determine the lean:fat ratio) even allow on-line detection in pork of substances leading to boar taint (e.g. the Danish classification robot). Obvious colour anomalies (e.g. PSE, DFD) can be picked up during carcass classification or are detected during dressing and boning of carcasses and portioning of primal cuts. However, to date, a practicable, all-encompassing and preferably automated system for on-line measurement of the various sensory quality traits of muscle has not yet been developed for application in the commercial abattoir. Therefore, it is important that quality and safety managers (including veterinarians) remain aware of developments in slaughter and processing methods and the associated technologies that may affect meat quality, so as to allow targeted control actions when necessary.

Acknowledgements

This contribution is an update of a book chapter by Hofbauer and Smulders (2010), published by Wageningen Academic Publishers (The Netherlands), which contains most of the references to the original work on which the current review is based.

Literature and further reading

Asghar, A. and Pearson, A.M. 1980. Influence of *ante-* and *post-mortem* treatments upon muscle composition and meat quality. *Adv Food Res*, **26**, 53–213.

Barbut, S., Sosnicki, A.A., Lonergan, S.M. *et al.* 2008. Progress in reducing the pale, soft and exudative (PSE) problem in pork and poultry meat. *Meat Sci*, **79**, 43–63.

Clark, K.A., McElhinny, A.S., Beckerle, M.C. and Gregorio, C.C. 2002. Striated muscle cytoarchitecture: An intricate web of form and function. *Annu Rev Cell Dev Biol*, **18**, 637–706.

Cohen, C. 1975. The protein switch of muscle contraction. *Sci Am*, **233** (5), 36–45.

Den Hertog-Meishke, M.J.A., Van Laack, H.L.J.M. Smulders, F.J.M. 1997. The water-holding capacity of fresh meat. *Vet Q*, **19**, 175–181.

Farmer, L.J. 1994. The role of nutrients in meat flavour formation. *Proc Nutr Soc*, **53**, 327–333.

Faustman, C. and Cassens, R.G. 1991. The biochemical basis for discoloration in fresh meat: a review. *J Muscle Foods*, **1**, 217–243.

Geesink, G.H. 1993. Post mortem muscle proteolysis and beef tenderness, with special reference to the calpain/calpastatin system. PhD Thesis, Utrecht University, The Netherlands, p. 15.

Geesink, G.H., Kuchay, S., Chishti, A.H. and Koohmaraie, M. 2006. μ-Calpain is essential for postmortem proteolysis of muscle proteins. *J Anim Sci*, **84**, 2834–2840.

Hofbauer, P. and Smulders, F.J.M. 2010. The muscle biological background of meat quality including that of game species. In: *Game Meat Hygiene in Focus; Microbiology, Epidemiology, Risk Analysis and Quality Assurance* (eds P. Paulsen, A. Bauer, M. Vodnansky *et al.*), Wageningen Academic Publishers, Wageningen, The Netherlands, pp. 273–295.

Koohmaraie, M. and Geesink, G.H. 2006. Contribution of postmortem muscle biochemistry to the delivery of consistent meat quality with particular focus on the calpain system. *Meat Sci*, **74**, 34–43.

Lawrie, R. and Ledward, D. eds). 2006. *Lawrie's Meat Science*, 7th edn. CRC Press.

Offer, G. and Cousins, T. 1992. The mechanism of drip production: Formation of two compartments of extracellular space in muscle post mortem. *J Sci Food Agric*, **58**, 107–116.

Ouali, M. 2007. Meat tenderization: possible causes and mechanisms – a review. *J Muscle Foods*, **1**, 129–165.

Smulders, F.J.M. 2007. Chapter 9. Substrate properties of foods of animal origin; Section 9.2. Fresh meat – Structure, composition, quality (in German), In: *Tierproduktion und Veterinärmedizinische Lebensmittelhygiene* (ed. F.J.M. Smulders). Wageningen Academic Publishers, Wageningen, The Netherlands, pp 368–392.

16

Microbial Contamination During Slaughter

Claudio Zweifel and Roger Stephan

Institute for Food Safety and Hygiene, Vetsuisse Faculty, University of Zurich, Zurich, Switzerland

16.1   Scope

In the first part of this chapter, the impact of different steps of slaughter on the bacterial contamination of carcasses is discussed. For this purpose, the pig slaughter process is described based on the microbial contamination of carcasses at sequential process steps. In the second part, microbial contamination of carcasses at the end of slaughter is presented by summarizing recent microbial data obtained from carcasses of various animal species. For verification of the slaughter hygiene under practical conditions, the focus is thereby mainly on indicator bacteria such as total viable counts (TVC) and *Enterobacteriaceae*.

16.2   Introduction

To ensure food safety at slaughter, measures additional to the conventional meat inspection are crucial to counter the threats posed by latent zoonoses. Healthy/asymptomatic food-producing animals can be carriers of important bacterial pathogens responsible for human illness. Because no clinical or pathological–anatomical signs are evident in these animals, such carriers are not detected in the traditional meat inspection.

Meat Inspection and Control in the Slaughterhouse, First Edition.
Edited by Thimjos Ninios, Janne Lundén, Hannu Korkeala and Maria Fredriksson-Ahomaa.
© 2014 John Wiley & Sons, Ltd. Published 2014 by John Wiley & Sons, Ltd.

Currently, the focus is therefore on preventive systems instead of endpoint testing. In this context, the hazard analysis and critical control point (HACCP) system is considered as an effective instrument. The gist of HACCP, which is a systematic, science-based approach to control processes, is to prevent or eliminate specific health hazards for consumers or to reduce them to an acceptable level. The HACCP system is based on seven principles contained in the Codex Alimentarius (http://www.codexalimentarius.org). They comprise:

1. conduct a hazard analysis (hazard identification and risk assessment),
2. determine critical control points,
3. establish critical limits,
4. establish monitoring procedures,
5. establish corrective actions,
6. establish verification procedures (e.g. endpoint testing),
7. establish an appropriate documentation.

To fulfil the requirements of a full HACCP system, effective control and reliable monitoring are preconditions for a critical control point.

The HACCP system should always be seen as a part of an integral food safety management system. A suitable model is the 'Food Safety House'. The foundations of this house are the conditions of the premises and the equipment. Its walls are the basic hygiene measures. They include, for example, cleaning and disinfection, pest control, temperature and relative humidity of premises, sufficient separation of production steps to avoid cross-contamination and personnel hygiene. The roof of the house is made up of product- and production-specific preventive measures in accordance with the HACCP principles. This pivotal differentiation between HACCP and good hygiene practices is also emphasized in the food safety legislation of the European Union (EU).

In view of HACCP-based systems applied at slaughter, intervention systems typically used in the United States and Canada and non-intervention systems typically used in Europe must be distinguished. Intervention systems are mainly applied in the slaughter of poultry and cattle. Basically, interventions comprise physical, chemical and biological treatments. For decontamination of carcasses, treatments are often combined following the multiple hurdle approach. The basic principles of different methods used at slaughter and the antibacterial activity of various intervention treatments for carcass decontamination are elucidated in the Chapter 17. Although these technologies reduce microbial loads to some extent, it is difficult to reliably quantify the reductions obtained and to establish an effective monitoring system.

In Europe, carcass interventions with substances other than potable water have been discouraged for a long time because such treatments were perceived to be means of concealing or compensating for poor hygiene practices during slaughter. EU legislation does not categorically ban chemical decontamination of foods of animal origin, but approval of substances is tied to strict prescriptions and can only be authorized after the European Food Safety

Authority (EFSA) has provided a risk assessment. Currently, EU legislation permits the use of lactic acid to reduce microbial surface contamination on bovine carcasses at the slaughterhouse level in compliance with defined framing conditions.

In terms of slaughtering, especially when non-intervention systems are used, a full HACCP system in accordance with Codex Alimentarius can hardly be implemented. Hence, only HACCP principles and the preventive approach of process control can be followed during slaughter. Following the model of the 'Food Safety House', strict and continuous adherence to good manufacturing and hygiene practices, along with risk-based preventive measures, are the basis for preventing microbial contamination of carcasses. It must be emphasized that identification and correction of hygienic weak points in the process and maintenance of good practices of slaughter hygiene are of crucial importance in intervention and non-intervention systems.

16.3   Contamination of carcasses

Despite all advancements in slaughter technologies, significant microbial surface contamination of carcasses takes place during slaughter and there are considerable opportunities for spread of bacteria. Bacteria on carcass surfaces may originate from the animal itself, especially from the hide, fleece or skin, the feet and the gastrointestinal tract. Upon arrival at the slaughter plant, animals have substantial numbers of bacteria associated with them. Although some bacteria are residents of the hide, skin or fleece, much of the external contamination results from direct or indirect faecal contamination. In addition, carcasses can be contaminated or cross-contaminated at multiple steps throughout transport, slaughter and processing. With regard to cross-contamination, environmental sources such as floors, walls, equipment, knives, worker's hands and, eventually, aerosols must also be considered.

The extent of carcass contamination and the relevance of individual sources are dependent on the cleanliness of incoming livestock, the structural design of the slaughterhouse (e.g. the separation of dirty and clean areas), the slaughter technology, the cleaning and disinfection regime, and the process hygiene (e.g. hygienically critical process steps, attitude and behaviour in terms of personnel hygiene). Adherence to good practices of slaughter hygiene, including effective cleaning and disinfection as well as strict hygiene protocols for personnel, is, therefore, of central importance to avoid contaminations. In this context, it must be emphasized that carcasses might be contaminated with bacteria despite the absence of visible debris. Transferred bacteria may affect meat quality and cause spoilage (such as *Pseudomonas, Moraxella, Lactobacillus, Brochothrix thermosphacta*) or act as human pathogens (such as *Campylobacter* spp., *Salmonella* spp., enteropathogenic *Yersinia* spp., Shiga toxin-producing *Escherichia coli*). Amongst pathogens of concern, a good deal originates from healthy animals, in particular from the enteric flora. By contamination and cross-contamination such bacteria might be introduced into the food chain, spread to processing facilities and consumers, and thereby pose a threat for the contamination of other foods at consumer level and for food-borne diseases.

As well as the processing of red meat carcasses, the slaughter of poultry poses special challenges. Poultry meat production and consumption has increased worldwide over the last decades. Healthy broilers may also carry (especially on the feet, feathers, skin and in the intestines) a heterogeneous microflora including both non-pathogenic bacteria and pathogens such as *Campylobacter* and *Salmonella*. In modern large-scale poultry slaughterhouses, nearly all process steps are automated and contact with workers is kept to a minimum. Birds to be slaughtered should be clean and dry. Dependent on the different operation steps, contamination levels change during slaughter. Due to the high automation, for example of the evisceration line, flock uniformity is an important factor in preventing contaminations. Feed removal about 8–12 h before slaughtering is another measure to prevent faecal contaminations. To reduce the amount of dirt carried by the birds, scrubbers and rotating brushes in combination with water washes (before scalding) are sometimes used.

Cross-contaminations must be expected, in particular during scalding, defeathering (plucking) and evisceration. During scalding, carcasses are immersed in hot water to loosen the feathers. Dependent on process parameters (temperature, duration), this process can affect bacterial contamination levels. In commercial poultry slaughterhouses, carcasses are then plucked mechanically. Cross-contaminations among carcasses are a problem due to the plucking equipment or aerosols. The use of an adequate water flow can reduce contaminations during plucking to some extent. During the following evisceration, the risk of carcass contamination, especially of faecal contamination, is linked to the vent cutter, the opener and the eviscerator. After evisceration, carcasses are often briefly washed (rinsed). Typically, an inside–outside washer is used to remove visible dirt. Carcasses are then chilled, which is another important process stage in the slaughter of poultry.

16.4   Microbial contamination during slaughter – pig slaughtering as an example

For implementation of HACCP-based systems, analysis of the slaughter process is of central importance. Such an analysis must also include microbial data, especially because carcasses might be contaminated with bacteria despite the absence of visible contamination. In this section, microbial data from pig carcasses in the slaughter process of two Swiss slaughterhouses (A and B) are presented and discussed. Slaughterhouse A processed 250 pigs per hour and slaughterhouse B 160 pigs per hour. At sequential steps of the slaughter process, 200 pig carcasses were sampled by the wet-dry double swab technique at four sites (neck, belly, back and ham; each of 100 cm^2). Individual samples were examined for total viable counts (TVC), *Enterobacteriaceae*, and coagulase-positive staphylococci (CPS).

The pig slaughter process differed slightly between the two slaughter-houses. After stunning in a restraining conveyor (slaughterhouse A) or by electrical tongs (slaughterhouse B), animals were immediately exsanguinated. The animals were then immersed in a scald tank (59.0–62.0°C) for about 5.0 min (slaughterhouse A) or 8.5 min (slaughterhouse B). At slaughterhouse A, carcasses were dehaired using a rotating drum with scrapers and then passed through a singeing step, whereas a combined dehairing–singeing process was applied at slaughterhouse B. Subsequently, carcasses were wet polished by series of flails and moved into the separated 'clean areas'. After evisceration, carcasses were split along the midline from the hind to the fore using a splitting saw. Carcasses were then trimmed, weighed and washed with potable water (10°C, 15 s). At slaughterhouse A, carcasses were initially blasted with air (–8.0°C, 8.0 m/s, 45 min) before entering the chiller (2.0±1°C, 1.0 m/s), whereas a single-stage chilling process was used at slaughterhouse B (2.0±1°C, 4.0 m/s).

Before scalding, mean log TVCs ranged from 5.0 to 5.6 log cfu/cm^2 and *Enterobacteriaceae* were detected on all carcasses. Scalding clearly reduced (P < 0.05) mean log TVCs and *Enterobacteriaceae* detection rates (Table 16.1). TVCs for the scald tank water remained constant during slaughter. Results are consistent with other studies, in which bacterial reductions were affected by time and temperature conditions. As expected, the dehairing process was an important source of carcass contamination; mean log TVCs and *Enterobacteriaceae* detection rates (Table 16.1) clearly increased (P < 0.05). When considering the effects of dehairing and singeing, differences were evident between the two slaughterhouses (Figure 16.1, Table 16.1). At slaughterhouse A, TVCs after singeing were similar to those after scalding. Differences among sites are probably related to uneven exposure to flames. Although similar reductions have been reported, published data on the effects of singeing differ widely. In contrast, microbial results after the combined dehairing/singeing at slaughterhouse B were increased (P < 0.05) and the values were clearly higher than those after singeing at slaughterhouse A.

Polishing produced opposing effects (Figure 16.1, Table 16.1). At slaughterhouse A, TVC reductions obtained by singeing were almost offset (P < 0.05), whereas changes in *Enterobacteriaceae* results were mainly marginal. Increases may have been due to distribution of residual contamination or to cross-contamination from the equipment. For an effective reduction before the later stages, an additional singeing step could be considered. Based on the higher microbial contamination after dehairing/singeing and probably because of the distribution of contamination, TVC values at slaughterhouse B decreased (P < 0.05) by at least one order of magnitude after polishing. Thus, polished carcasses were typically contaminated with about 2.9 and 3.7 log cfu/cm^2 at slaughterhouses A and B, respectively. *Enterobacteriaceae* accounted only for a small subset of the bacterial load but were more frequently detected at slaughterhouse B.

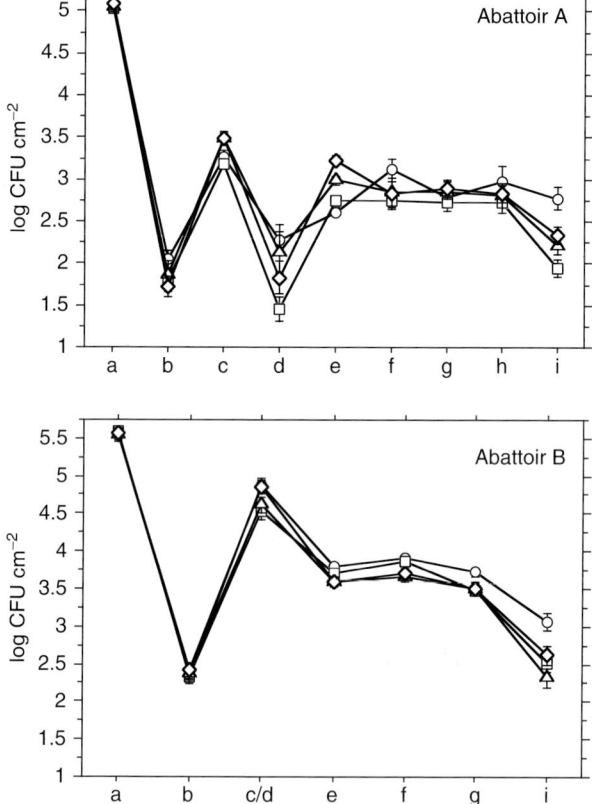

Figure 16.1 TVC results (mean log cfu/cm²) on pig carcasses at the neck (○), belly (□), back (△) and ham (◇) after (a) bleeding, (b) scalding, (c) dehairing, (d) singeing, (e) polishing, (f) trimming, (g) washing, (h) head removal and (i) chilling (n = 100 at each site and process stage, error bars: 95% CI). Source: Intl Assn for Food Protection.

The operations performed manually between polishing and the completion of trimming generally had negligible effects on TVCs (Figure 16.1). However, *Enterobacteriaceae* results provided indications of hygienic weak points at specific sites that were not apparent from the TVC data (Table 16.1), but observed increases were moderate compared with those reported elsewhere. Among sites, the neck, in particular, and the front half, in general, tended to yield higher results, a trend also observed at all later stages (P < 0.05). These observations may be related to the water applied during carcass sawing. Washing might be expected to result in widespread distribution of microbial contamination over carcasses. However, considering the higher overall results at slaughterhouse B, some reductions in *Enterobacteriaceae* were evident at this slaughterhouse. If greater reductions are to be obtained, water temperatures of 85°C or chemical wash components are required.

In the chiller, mean log TVCs and *Enterobacteriaceae* detection rates on sites ranged from 1.9 to 2.8 log cfu/cm² and 0 to 14%, respectively, at slaughterhouse A and from 2.3 to 3.1 log cfu/cm² and 17 to 43%, respectively, at

Table 16.1 Detection of *Enterobacteriaceae* on pig carcasses after different stages of slaughter in slaughterhouses A and B (n = 100 at each site and process stage).

Slaughterhouse	Process stage	*Enterobacteriaceae*-positive (%)				
		Neck	Belly	Back	Ham	Carcass
A	After bleeding	100	100	100	100	100
	After scalding	6	2	4	0	12
	After dehairing	89	82	93	98	100
	After singeing	28	12	42	16	66
	After polishing	27	19	7	20	49
	After trimming	49	21	25	13	66
	After washing	36	26	26	18	62
	Chilling	14	1	2	0	17
B	After bleeding	100	100	99	100	100
	After scalding	22	2	6	6	29
	After dehairing/singeing	69	56	76	86	97
	After polishing	80	87	76	78	97
	After trimming	92	88	83	89	100
	After washing	85	71	69	74	93
	Chilling	43	18	17	21	55

Source: Spescha, Stephan, & Zweifel, 2006. Reproduced with permission from International Association for Food Protection.

slaughterhouse B (Figure 16.1, Table 16.1). The blast of freezing air (slaughterhouse A) or the high speed of air at low temperature (slaughterhouse B) probably led to a fast surface drying and, thereby, to the observed reductions ($P < 0.05$). The slighter reductions on the neck, especially at slaughterhouse A, may be explained by residual water draining across the neck and the accessibility of this location by the air. Literature data indicate that chilling can result in increases, decreases or no changes in microbial contamination, dependent on temperature, air speed, humidity, carcass spacing, and duration.

Although CPS (as indicators for *Staphylococcus aureus*) were constantly found at the beginning of slaughter and scalding considerably reduced detection rates and counts ($P < 0.05$), striking differences between the slaughterhouses were established at later processing stages (Table 16.2). At slaughterhouse A, the low CPS results obtained after scalding remained constant during the slaughter process. In contrast, at slaughterhouse B, CPS reductions obtained by scalding were offset by dehairing/singeing ($P < 0.05$) and results remained at a high level during the slaughter processes. This might indicate that a CPS population has become established on the surfaces of the equipment at this slaughterhouse. In the chiller, CPS were detected on only 6% of carcasses from slaughterhouse A, but on 77% of carcasses from slaughterhouse B.

Although scalding, singeing and chilling did not yield bacteria-free carcasses, these steps can reduce microbial contamination and probably could be integrated into a risk-based system for pig slaughter, when process parameters are standardized. To maintain low contamination levels, proper

Table 16.2   Detection of coagulase-positive staphylococci (CPS) on pig carcasses after different stages of slaughter in slaughterhouses A and B (n = 100 at each site and process stage).

Slaughterhouse	Process stage	CPS-positive (%)				
		Neck	Belly	Back	Ham	Carcass
A	After bleeding	100	100	100	100	100
	After scalding	10	10	2	8	18
	After dehairing	6	4	2	4	15
	After singeing	0	0	0	2	2
	After polishing	1	1	1	1	4
	After trimming	3	0	0	1	4
	After washing	3	2	0	2	6
	Chilling	2	1	1	2	6
B	After bleeding	96	95	93	94	100
	After scalding	9	5	8	1	20
	After dehairing/singeing	93	86	83	86	100
	After polishing	83	76	61	69	96
	After trimming	74	71	52	64	97
	After washing	86	64	79	80	99
	Chilling	63	29	25	36	77

Source: Spescha, Stephan, & Zweifel, 2006. Reproduced with permission from International Association for Food Protection.

cleaning and drying of equipment must be assured to keep bacteria from becoming established on surfaces. Moreover, an additional singeing step may be used after polishing and structural modifications must be considered. Because the effects of certain process stages can be slaughterhouse-specific, each slaughterhouse should develop its own baseline reference data on carcass contamination in the slaughter process and customize HACCP-based systems to match process- and site-specific circumstances.

16.5   Microbial examinations of red meat carcasses at the end of slaughter

To verify slaughter hygiene conditions in a slaughterhouse, feasible and conclusive tools are required. In this section, microbial data obtained from carcasses of various slaughtered animal species at the end of slaughter are presented and discussed (Table 16.3). Samples were mainly obtained by swabbing and examined for TVC and *Enterobacteriaceae*, which have proved to be practical for slaughter hygiene verification. *Enterobacteriaceae* results provided additional indications of hygienic weak points leading to faecal contamination, which is considered as important source of enteric zoonotic pathogens. Alternatively, *Escherichia coli* can be determined for this purpose.

Table 16.3 Average TVC and *Enterobacteriaceae* results on carcasses after slaughter in Switzerland.

Carcass samples	Average TVC (mean log cfu/cm^2)		*Enterobacteriaceae*-positive (%)	
	Different slaughterhouses	Total	Different slaughterhouses	Total
Pig				
Large-scale slaughterhouse	2.18–3.70	2.81	4.0–41.3	20.2
Small-scale slaughterhouse[a]	2.39–4.20	3.28	2.0–56.0	23.9
Cattle				
Large-scale slaughterhouse	2.11–3.10	2.72	12.0–54.0	31.0
Small-scale slaughterhouse[a]	2.65–3.78	3.07	0.0–55.0	20.0
Sheep				
Neck	3.46–3.47[b]	3.47[b]	21.4–29.9	24.3
Brisket	3.34–3.82[b]	3.67[b]	17.4–38.8	27.2
Perineal area	3.05–3.16[b]	3.09[b]	15.4–42.2	22.8

[a] excision samples;
[b] median values.

As well as the sampling method applied and the bacteria determined, knowledge about the pros and cons of evaluation and rating modes is pivotal. Depending on samples at hand, data can be depicted as bar charts, box plots or time trend graphs. Quality control charts thereby offer a biometric-founded and plant-specific concept to categorize microbial results (Section 16.5.1). Comparability of the results from different studies is hampered by the use of varying sampling methods, sampling sites and evaluation modes, as well as the application of carcass decontamination treatments, for example in North America.

In the European Union, carcass samples are usually pooled in daily practice and microbial performance criteria are used for the rating of daily mean values. These criteria are defined for excision samples, but for reasons of practicability the meat industry often prefers swabbing. Based on the data from cattle and pig carcasses, corresponding criteria have also been developed for the wet-dry double swab technique (Section 16.5.1). However, such microbial performance criteria have always to be seen merely as baselines in the context of a standardised hygiene monitoring.

16.5.1   Pig and cattle carcasses

According to the Food and Agriculture Organization (FAO) of the United Nations, world pig and cattle meat production were estimated in the year 2011 to be over 110 million tons and over 62 million tons, respectively (http://faostat.fao.org). Strict maintenance of good practices of slaughter hygiene is of crucial importance for the prevention of microbial contamination of carcasses. For verification of hygiene conditions during slaughter, the microbial status of carcasses at the end of slaughter is often determined. In this section, microbial data from pig and cattle carcasses at the end of slaughter in Switzerland are presented (Table 16.3). Results originated from large-scale and small-scale slaughterhouses.

Large-scale slaughterhouses A total of 650 pig carcasses (slaughterhouse A: n = 200; B: n = 150; C: n = 150; D: n = 150) and 800 cattle carcasses (slaughterhouse A: n = 200; B: n = 150; C: n = 150; D: n = 150, E: n = 150) were investigated at five slaughterhouses with an annually slaughtering capacity of over 14 million kg. Weekly, ten pig and ten cattle carcasses were sampled at four sites (cattle: neck, brisket, flank and rump; pig: back, cheek, ham and belly) by the wet-dry double swab technique. From each carcass, samples were pooled ($4 \times 100\,cm^2$) and examined for TVCs and *Enterobacteriaceae*.

At the different large-scale slaughterhouses, mean log TVCs from pig carcasses ranged from 2.2 to 3.7 log cfu/cm^2 and those from cattle carcasses from 2.1 to 3.1 log cfu/cm^2 (Table 16.3). The finding of mainly significant differences in TVC levels between slaughterhouses ($P < 0.05$) emphasizes that results of microbial examinations of carcasses are slaughterhouse-specific. The results showed also that (i) slaughterhouses showing highest and lowest mean values were identical for cattle and pig carcasses, (ii) on average, with the exception of slaughterhouse D, results from cattle and pig carcasses were on a comparable level, and (iii) the situation at slaughterhouse D (especially with regard to pig slaughtering) indicated slaughterhouse-specific hygienic weak points in the slaughter process.

Time trend graphs showed constant variances of TVCs on relative constant levels within slaughterhouses, but levels varied considerably between slaughterhouses. However, by analysing pooled samples and calculating daily mean values, the results do not consider localized contamination on carcasses. In contrast, the box plot illustration allows a more convenient analysis of variances, median values, ranges and differences between slaughterhouses (Figure 16.2). Furthermore, quality control chart (QCC) methods offer a biometric-founded concept to classify microbial results from carcasses based on slaughterhouse-specific data and boundaries (warn and action levels). Results beyond action levels show a significant deviation from the process mean value, and exceeding the upper action level (UAL) requires corrective interventions. Exceeding the upper warning level (UWL) requires increased process monitoring and the reason has to be evaluated when this happens repeatedly.

Enterobacteriaceae were detected in low counts (or counts below the detection limit) and on 31% of cattle and 20% of pig carcasses (Table 16.3). The more frequent detection and higher counts of *Enterobacteriaceae* on cattle than on pig carcasses ($P < 0.05$) might be associated with faecal contamination on cattle hides being transferred to carcass surfaces during dehiding. At the different slaughterhouses, *Enterobacteriaceae* prevalence on pig and cattle carcasses ranged from 4 to 41% and from 12 to 54%, respectively. Without taking *Enterobacteriaceae* results into consideration, a distorted impression of

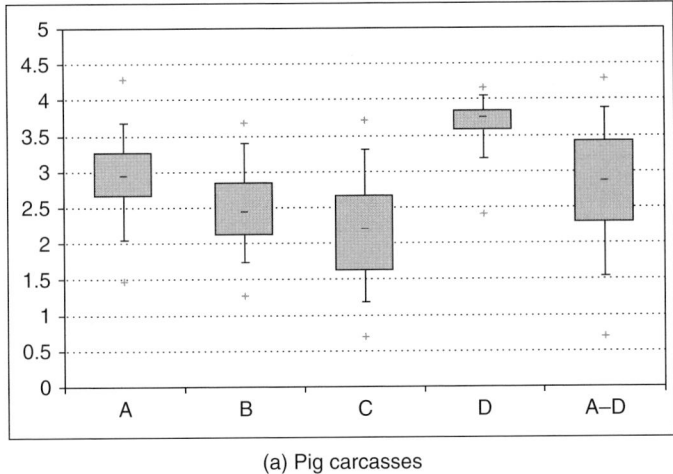

(a) Pig carcasses

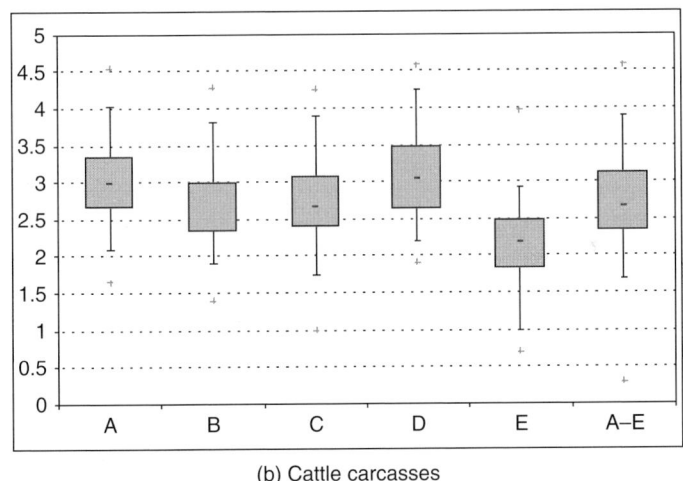

(b) Cattle carcasses

Figure 16.2 TVC results (log cfu/cm^2) on (a) pig carcasses (n = 650) from four (A–D) large-scale slaughterhouses and (b) on cattle carcasses (n = 800) from five (A–E) large-scale slaughterhouses. Source: Zweifel, Baltzer & Stephan, 2005. Reproduced with permission from Elsevier.

the hygienic situation in certain slaughterhouses would have been created (e.g. slaughterhouse C: low–average TVCs, but high *Enterobacteriaceae* results).

Based on the results from the examined pig and cattle carcasses, microbial performance criteria were proposed for the non-destructive wet-dry double swab technique. Hence, results (for daily mean values) are satisfactory, acceptable and unsatisfactory for TVCs when they are <3.0, 3.0–4.0 and >4.0 log cfu/cm^2 and for *Enterobacteriaceae* when they are <1.0, 1.0–2.0 and >2.0 log cfu/cm^2, respectively. However, evaluations in accordance with such specifications include some flaws. Firstly, by analysing pooled samples and daily mean values, only more serious weak points in slaughter hygiene are recognized. Secondly, results are slaughterhouse-specific and mandatory performance criteria can only be seen as baselines. Each slaughter facility should, therefore, implement a monitoring system based on slaughterhouse-specific data and criteria.

Small-scale slaughterhouses   For many small-scale slaughterhouses, the risk-based approach including the microbial verification of hygiene conditions and the implementation of HACCP principles poses a great challenge. Microbial baseline data for carcasses from small-scale slaughterhouses are rare in the literature. Therefore, microbial contamination profiles of pig and cattle carcasses from various small-scale slaughterhouses were determined. A total of 750 pig carcasses and 535 cattle carcasses from 17 small-scale slaughterhouses (median of 600 slaughtered pigs and 220 slaughtered cattle annually) were sampled by excision at four sites (pig: neck, belly, back and ham; cattle: neck, brisket, flank and rump). Individual excision samples (each of $5\,cm^2$) were examined for TVCs and *Enterobacteriaceae*. In addition, values for calculated pooled samples were obtained by calculating log mean from the four sites of each carcass. In the following, mainly the results of these calculated pooled samples are addressed (Figure 16.3).

TVCs on pig and cattle carcasses (calculated pooled samples) averaged out at 3.3 and 3.1 log cfu/cm^2, respectively (Table 16.3). At the different slaughterhouses, mean log TVCs ranged from 2.4 to 4.2 log cfu/cm^2 on pig carcasses and from 2.7 to 3.8 log cfu/cm^2 on cattle carcasses. Amongst sites, the back (pigs) and the neck (cattle) tended to yield higher TVCs than the other sampling sites. Despite limited comparability, results from carcasses of small-scale slaughterhouses did not differ substantially from those of large-scale slaughterhouses. With regard to the European performance criteria, the majority of TVCs from pig and cattle carcasses were rated as satisfactory. Due to the often restricted and inconsistent number of animals slaughtered in small-scale slaughterhouses, not only daily mean values but also each carcass should be rated. Within a long-term trend analysis, such examinations allow to obtain and evaluate reliable microbial data for carcasses from small-scale abattoirs.

Enterobacteriaceae accounted only for a small subset of the bacterial load and were detected on 24% of pig carcasses and 20% of cattle carcasses (Table 16.3). At the different slaughterhouses, *Enterobacteriaceae* prevalence on pig and cattle carcasses ranged from 2 to 56% and from 0 to 55%, respectively. Determination of *Enterobacteriaceae* on carcasses provided indications of additional slaughterhouse-specific hygienic weak points also in small-scale slaughterhouses.

To evaluate the reasons for the increased carcass contamination in certain small-scale slaughterhouses, a detailed analysis of the slaughtering process (including microbial data) is required. Within the slaughtering of pigs, in particular the dehairing and polishing processes must be considered, whereas the contamination risk associated with the hide is of special importance with regard to the slaughtering of cattle. Moreover, practices of evisceration, carcass splitting and other potentially contributing processes must be evaluated.

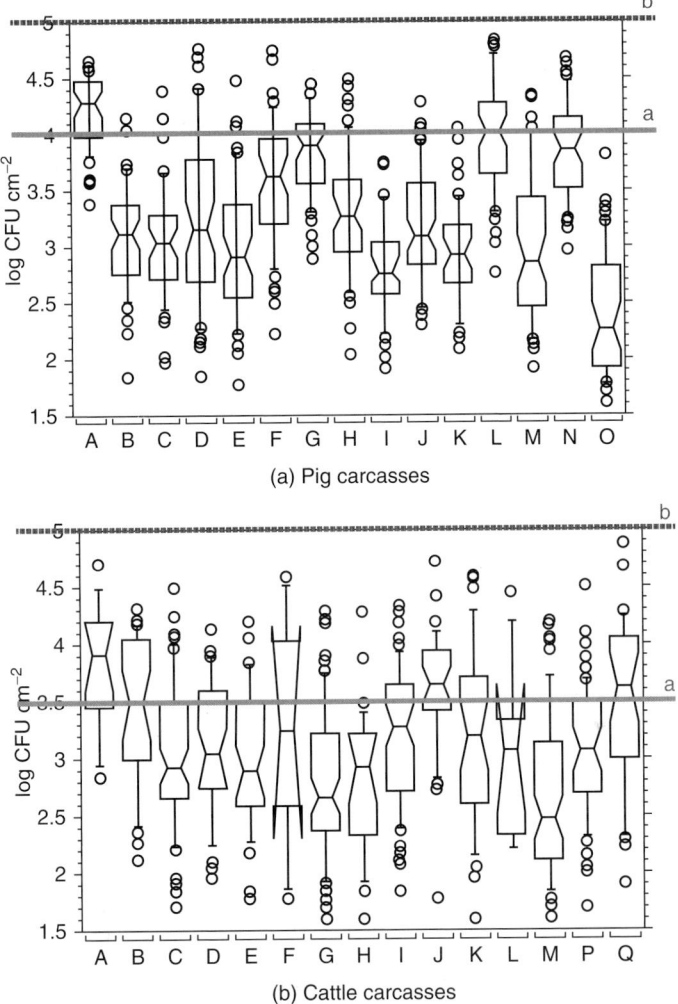

Figure 16.3 TVC results (log cfu/cm²) on pig and cattle carcasses from different small-scale slaughterhouses (calculated pooled samples; a, b: limits for acceptable and unsatisfactory results of Regulation (EC) No 2073/2005; (a): pig carcasses, n = 750; (b): cattle carcasses, n = 535, slaughterhouses C, G, I, M, P: n = 60, slaughterhouses B, D, E, H, J, K, Q: n = 30, slaughterhouses A, F, L: n < 30). Source: Zweifel, Fischer & Stephan, 2008. Reproduced with permission from Elsevier.

16.5.2 Sheep carcasses

World sheep meat production was estimated to be over 7.9 million tons in the year 2011 (FAO, http://faostat.fao.org). To verify hygiene conditions in the slaughter process, the microbial status of carcasses at the end of slaughter is often determined. In addition to analysing carcass pool samples and calculating daily mean values for construction of time trend graphs, examinations for individual carcass sites can yield more convincing data. In the following, microbial data obtained from different sites of sheep carcasses at the end of slaughter in Switzerland are presented. Overall, 580 sheep carcasses from three different slaughterhouses (A–C) were investigated at 10 sites (neck, side of neck, outside foreleg, inside foreleg, shoulder, brisket, back, flank, perineal area and hind leg; each of 40 cm^2) by the wet-dry double swab technique and examined for TVCs and *Enterobacteriaceae*.

At the different sites, TVC median values ranged from 2.5 to 3.8 log cfu/cm^2. Although repeated microbial examinations yielded slaughterhouse-specific result patterns, certain trends for the contamination levels at the various sampling sites emerged (Figure 16.4). The brisket and the neck showed the highest bacterial loads (Table 16.3). Contamination distribution on the sheep carcasses was similar to that of cattle but TVCs of sheep carcasses tended to be higher than those of cattle carcasses. *Enterobacteriaceae* were detected on 68% of the carcasses and in 15% of the samples. Carcasses from one slaughterhouse (A) thereby yielded the highest *Enterobacteriaceae* frequency on the majority of the sampling sites and the relatively high detection rate on the perineal area was striking. Overall, the proportion of positive results ranged from 2.6% (for the hind leg and the flank at slaughterhouse C) to 42% (for the perineal area at slaughterhouse A). In similar examinations of cattle carcasses, lower detection rates of *Enterobacteriaceae* were often found. Based on the higher TVCs and *Enterobacteriaceae* results on sheep carcasses, sheep should be slaughtered on a separate slaughter line or at least after the slaughtering of cattle (and not between sequences of cattle on the same slaughter line).

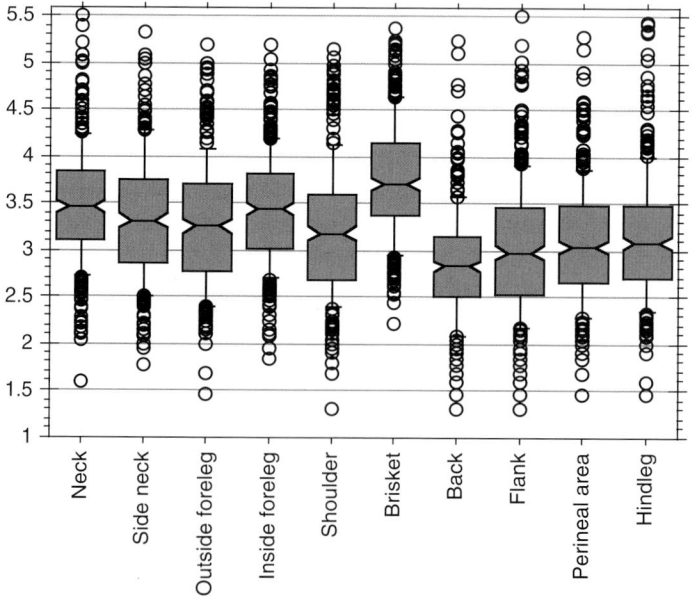

Figure 16.4   TVC results (log cfu/cm^2) on 318 sheep carcasses at 10 different sampling sites obtained from slaughterhouse B (one of the three examined slaughterhouses). Source: Zweifel & Stephan, 2003. Reproduced with permission from International Association for Food Protection.

Examinations for individual sites should include carcass locations with the highest slaughterhouse-specific contamination risk and also comprise a sufficient number of carcasses. Distribution of the microbial contamination on sheep carcasses within a slaughterhouse allows a convenient selection of sampling sites. On the other hand, periodical examinations of individual sites, in addition to the regular examination of pooled carcass samples, enable recognition of localized contamination on carcasses (early-warning system).

16.6 Conclusions

Strict maintenance of good practices of slaughter hygiene is of central importance for the prevention of microbial contamination of carcasses. Adherence to such practices is also an important prerequisite for preventing zoonotic pathogens carried by healthy/asymptomatic slaughtered animals from entering the food chain. Such measures therefore ensure both public health protection and meat quality. To enable the risks involved to be estimated and appropriate measures to be taken, analysis of the slaughter process has to be complemented by collection of slaughterhouse-specific, microbial monitoring data in accordance with HACCP principles.

In daily practice, feasible and conclusive tools are required to verify slaughter hygiene conditions in a slaughterhouse. They must include regular slaughter process analysis, complemented by microbial verification examinations of the environment and carcasses. The present chapter focused on the microbial contamination of carcasses from pigs, cattle and sheep during slaughter and at the end of slaughter. With regard to slaughter process control, it was shown that microbial process analysis (slaughtering of pigs as an example) is of importance for implementation of HACCP-based systems and it identified (slaughterhouse- and site-specific) process steps contributing to increases and others resulting in decreases of the microbial contamination. Furthermore, regular microbial examinations of carcasses by swabbing at the end of slaughter allowed reliable conclusions to be drawn with regard to long-term slaughter hygiene conditions. Such examinations are, therefore, an appropriate tool for verification purposes including the adherence to good hygiene practices and the effectiveness of HACCP-based systems. Trend and long-term aspects must thereby be emphasized.

However, the following framing conditions must be considered for application and appraisal of such tools for slaughter hygiene verification: (i) Testing for the mentioned routine purposes should involve indicator organisms rather than pathogens. Often, pathogens are present too infrequently to be useful for regular investigations and evaluation of process performance. Moreover, detection and characterization of pathogens is often complex and time consuming. (ii) HACCP-based systems must be customized to the specific circumstances of each plant. Checks for verification purposes must be performed and documented systematically to guarantee traceability and plant-specific rating techniques for microbial results are to be encouraged.

Small-scale slaughterhouses and the associated challenges therein deserve special attention. (iii) Should microbial results change for the worse as a trend, review of the system is required. The emphasis is thereby placed on identifying the causative operations. Beside inspection of the slaughter process, additional microbial examinations for individual carcass sites should be performed. Although variation in microbial contamination over sites and between carcasses must be considered, such data are instrumental in recognizing systematic weak points in slaughter hygiene associated with localized carcass contaminations.

Literature and further reading

Gill, C.O. 2004. Visible contamination on animals and carcasses and the microbiological condition of meat. *J Food Prot*, **67**, 413–419.

McEvoy, J.M., Sheridan, J.J., Blair, I.S. and McDowell, D.A. 2004. Microbial contamination on beef in relation to hygiene assessment based on criteria used in EU Decision 2001/471/EC. *Int J Food Microbiol*, **92**, 217–225.

Montgomery, D.C. 2005. Introduction to statistical quality control. John Wiley & Sons, Inc., Hoboken, NJ.

Nørrung, B. and Buncic, S. 2008. Microbial safety of meat in the European Union. *Meat Sci*, **78**, 14–24.

O'Brien, S.B., Lenahan, M., Sweeney, T., Sheridan, J.J. 2007. Assessing the hygiene of pig carcasses using whole-body carcass swabs compared with the four-site method in EC Decision 471. *J Food Prot*, **70**, 432–439.

Pearce, R.A., Bolton, D.J., Sheridan, J.J. *et al.* 2004. Studies to determine the critical control points in pork slaughter hazard analysis and critical control point systems. *Int J Food Microbiol*, **90**, 311–339.

Spescha, C., Stephan, R. and Zweifel, C. 2006. Microbiological contamination of pig carcasses at different stages of slaughter in two European Union-approved slaughterhouses. *J Food Prot*, **69**, 2568–2575.

Zweifel, C. and Stephan, R. 2003. Microbiological monitoring of sheep carcass contamination in three Swiss slaughterhouses. *J Food Prot*, **66**, 946–952.

Zweifel, C., Baltzer, D. and Stephan, R. 2005. Microbiological contamination of cattle and pig carcasses at five slaughterhouses determined by swab sampling in accordance with EU Decision 2001/471/EC. *Meat Sci*, **69**, 559–566.

Zweifel, C., Fischer, R. and Stephan, R, 2008. Microbiological contamination of pig and cattle carcasses in different small-scale Swiss slaughterhouses. *Meat Sci*, **78**, 225–231.

17

Decontamination of Carcasses

Claudio Zweifel and Roger Stephan

Institute for Food Safety and Hygiene, Vetsuisse Faculty, University of Zurich, Zurich, Switzerland

17.1   Scope

This chapter firstly discusses the basic principles of different physical, chemical and biological methods used at slaughter. Basic requirements as well as advantages and disadvantages are described. The second part of the chapter then elucidates the antibacterial activity of various intervention treatments for the decontamination of carcasses at slaughter. The efficacy of physical, chemical and biological methods applied alone or in combination is illustrated by selected bacterial reductions obtained on poultry carcasses, on bovine hides and carcasses, as well as on pig carcasses.

17.2   Introduction

The consumer demands fresh, tasty, healthy and wholesome food products. Food safety is considered self-evident and an absolute prerequisite. However, food-borne diseases are still widespread, thus affecting human health and economies worldwide. In industrialized countries, food-borne diseases are responsible for high levels of morbidity and mortality in the population and particularly in risk groups such as infants, young children, pregnant women, elderly and immunocompromised people.

Meat Inspection and Control in the Slaughterhouse, First Edition.
Edited by Thimjos Ninios, Janne Lundén, Hannu Korkeala and Maria Fredriksson-Ahomaa.
© 2014 John Wiley & Sons, Ltd. Published 2014 by John Wiley & Sons, Ltd.

Zoonotic pathogens are of special concern in food of animal origin, including meat and meat products. Healthy food animals were recognized in recent years as carriers of bacterial pathogens like *Campylobacter* spp., *Salmonella* spp., enteropathogenic *Yersinia* spp. and Shiga toxin-producing *Escherichia coli* (STEC) causing human illness. To counter this threat, the focus is on preventive systems in accordance with the hazard analysis and critical control point (HACCP) principles. Interventions applied at slaughter basically comprise hygiene measures aimed at preventing (faecal) contaminations and a variety of carcass interventions aimed at reducing bacterial contamination, which are discussed here.

17.3   Antibacterial decontamination treatments for carcasses

Despite all efforts targeted on the maintenance of good hygiene practice during meat production, prevention of carcass contamination with meat-borne pathogens during slaughter can hardly be warranted. Antimicrobial intervention technologies are, therefore, gaining interest for reduction of bacterial contaminations or for inhibition and retardation of microbial growth. Interventions applied at slaughter basically comprise physical, chemical and biological treatments. These interventions can also be used in combination or as multiple sequential interventions during slaughter. Various requirements have thereby to be fulfilled; for example, improvement of food safety, no defects in organoleptic and nutritional attributes, no discoloration, no residues on foods, convenient to apply in the production process, inexpensive and no objections from consumers or legislators.

In Europe, carcass interventions with substances other than potable water are subject to strict prescriptions and can only be authorized after the European Food Safety Authority (EFSA) has provided a risk assessment. Currently, EU legislation permits the use of lactic acid to reduce microbial surface contamination on bovine carcasses at the slaughterhouse level in compliance with defined framing conditions. For a long time, interventions were also discouraged because such treatments were perceived to be means of concealing poor hygiene practices. On the other hand, carcass interventions including chemicals are widely used in various other countries such as the United States and Canada. For the United States, the Food Safety and Inspection Service (FSIS) of the US Department of Agriculture lists substances that may be used in the production of poultry and red meat.

17.3.1   Physical decontamination treatments

Physical carcass interventions can be part of the normal slaughter process (e.g. washing steps or scalding and singeing of pig carcasses), whereas other methods are additionally applied with the specific objective of carcass decontamination. Some interventions may be applied at various stages of the slaughter process. General traits of selected physical decontamination treatments are presented below. However, depending on the respective slaughter process, there are other methods that are applied or are under development, for example treatments for the decontamination of bovine hides or ultrasound and freezing for the decontamination of poultry carcasses.

Water and steam   Among the physical methods used for carcass decontamination, water-based treatments predominate. Washing of carcasses with water is routinely used in meat processing plants and has proven to be effective in removing visible contaminants such as soil, hairs, feathers or other debris. For decontamination of carcasses, in particular, the use of hot water has proven to be valuable. An alternative is the application of steam, but more data obtained under commercial conditions are still required. Furthermore, additional investments, costs and potential adverse effects on the appearance and quality of carcasses (in particular, when great heat is used in combination with long exposure times) must be considered. In this regard, the use of hot water or steam followed by ultrasound or air chilling might constitute interesting options, but further investigations are required. On the other hand, probably due to the missing effect of heat inactivation, cold and warm water tend to be less effective for carcass decontamination and may support the release and spread of bacteria on carcass surfaces.

Irradiation   Irradiation of foods (often referred to as 'cold pasteurization') is primarily based on the use of gamma rays, machine generated X-rays (at or below 5 MeV), or accelerated electrons (at or below 10 MeV). The antibacterial activity of ionizing radiation is due to direct DNA damage and the effect of generated free radicals. The efficacy depends on target organisms, the type of food, the presence of oxygen and the water content. Irradiation at adequate dosages seems to be quite effective for carcass decontamination but costs for the infrastructure, the appropriate application during processing and the limited acceptance of this technology by the consumers must be considered.

Chilling   The antibacterial activity of air chilling on red meat carcasses is mainly based on surface desiccation achieved by high air velocity. This desiccation effect seems to be of major importance, in particular for the inactivation of *Campylobacter* on carcasses of pigs and ruminants. With regard to poultry carcasses, a comparable effect is commonly not achieved and not desired because of quality reasons. Conventional air chilling (single-stage chilling process) can be supplemented by blast chilling using freezing air and high air velocity. Moreover, spray chilling or blast-spray chilling are also known chilling procedures for carcasses.

Ultraviolet light   Ultraviolet (UV) light is commonly used in the food industry for decontamination of packaging material or water. For use on carcasses, the restricted penetration depth, the potential impact on fat oxidation and the lack of data obtained under commercial conditions must be considered.

Steam vacuuming Steam vacuum systems, which are applied to relatively small areas, are suited for use on red meat carcasses. The carcass is first treated with steam (hot water) and water and contaminants are then removed by vacuuming. Vacuum cleaning without steam is increasingly used (in the United States and Canada) instead of/in addition to knife trimming to remove localized visible contaminations such as hair from bovine and sheep carcasses.

17.3.2   Chemical decontamination treatments

Chemical compounds used for the decontamination of carcasses comprise a wide variety of substances. Chemicals frequently applied at slaughter are discussed here, although others, such as sodium hydroxide (NaOH), hydrogen peroxide (H_2O_2), peroxyacids and sulfate-based compounds, have also occasionally been tested (mainly under laboratory conditions) for carcass decontamination. A wide selection of chemical compounds is typically used for decontamination of poultry carcasses, especially in the United States and Canada. The antibacterial activity is mainly based on the disruption of cellular membranes, other cellular constituents and/or physiological cellular processes. For appraisal of the applicability in the slaughter process, it must be considered that the activity of some chemicals is counteracted by organic matter or the stability of some chemicals is limited in solution. Additionally, concentrated substances might constitute a health hazard or ecological menace or may have corrosive properties.

Organic acids Organic acids, such as lactic, acetic and citric acid, are frequently used for decontamination of carcasses. Organic acids have a considerable potential in the meat industry because they are quite inexpensive and generally recognized as safe. In addition, organic acids (as certain other chemicals) show some residual bactericidal or bacteriostatic effect. Organic acids (1–3%) are generally without effect on desirable sensory properties when used as carcass decontaminants. However, discoloration of carcass surfaces or respiratory and skin irritation of operators might occur when high acid concentrations are used.

Chlorine-based treatments Chlorine dissolves in water to form hypochlorous acid (HOCl) and hypochlorite ions. The efficacy is greatest in the acid form but hypochlorite is also an effective but slower-acting biocide. It must be considered that the efficacy is reduced by interaction with organic matter. As well as chlorine, the use of chlorine dioxide (ClO_2), acidified sodium chlorite (ASC, $NaClO_2$), monochloramine (NH_2Cl) or cetylpyridinium chloride (CPC) deserves attention as carcass interventions. ClO_2 alters nutrient transport, disrupts protein synthesis after penetrating into cells and seems to be less inactivated by organic matter than chlorine, but the mechanisms are not entirely understood. The antibacterial activity of ASC ($NaClO_2$) is derived from HOCl and ClO_2. NH_2Cl (generated by reaction between ammonia and chlorine) is tasteless, odourless, stable, highly soluble, persistent in water, biocidal and, unlike free chlorine, does not react readily with organic material. CPC (a quaternary ammonium compound) is a cationic surfactant with neutral pH and a non-functional chloride portion. Its antibacterial activity results from interaction with bacterial acidic groups to form weakly ionized compounds inhibiting bacterial metabolism.

Phosphate-based compounds Of the various phosphate-based compounds, trisodium phosphate (TSP, Na_3PO_4) is frequently used for carcass decontamination, in particular for poultry carcasses. Important factors are the high pH and the ionic strength causing bacterial cell autolysis but the mechanisms are not entirely understood.

Electrolysed and ozonized water Electrolysed water (EW) is gaining popularity in the food industry to reduce bacterial populations on foods and processing surfaces. By electrolysis, a dilute NaCl solution dissociates into electrolysed water. Although EW has an antimicrobial activity, the efficacy is counteracted by organic matter. EW has proved to be environmental friendly and seems not to negatively affect the organoleptic properties of various foods.

Ozone, which is a strong oxidant, is effective against microorganisms, including bacteria, viruses, yeasts and moulds. It is recognized as safe and it does not cause a change in the taste and colour of meat. The development of portable ozone generators made the use of ozone recently more practical for the food industry. However, because of its instability in the gaseous and aqueous state, it must be generated on site as needed. Due to its high oxidation potential, ozone reacts with a large number of compounds including organic substrates.

17.3.3   Combinations of decontamination treatments

Different combinations of chemical and physical interventions may be used for the decontamination of carcasses. Compared to single treatments an enhanced decontamination effect can be obtained by the combination of treatments. However, only limited data exist so far for the application of combinations under commercial conditions and results are not always convincing. With regard to practical application, the location in the slaughter process and the effect of sequential interventions ('multiple hurdle approach') must also be considered. Moreover, the sequence of application might influence the outcome when chemicals are used in combination with water sprayings. For example, subsequent water spraying can reduce the efficacy of chemicals by removal or dilution (though water spraying might be of relevance for neutralization of pH conditions after acid treatment). Reversing the application sequence tends to increase the efficacy of some chemicals, probably due to removal of organic matter by the precedent water treatment.

17.3.4   Biological decontamination treatments

Biological interventions such as bacteriophages and bacteriocins (e.g. nisin) show promise as novel carcass interventions. Bacteriophages are generally considered as safe and highly host specific but their practical use on carcasses is still hampered by some challenges, such as sufficient threshold levels, a too narrow host range and the risk for phage-resistant bacterial mutants. Furthermore, as there is increasing interest in 'natural foods' by consumers, certain extracts from plants or fruits and their mixtures will probably gain importance as sources of antimicrobials.

17.4   Antibacterial activity of decontamination treatments for carcasses

The focus of this section is on bacterial reductions on carcasses; topics such as growth inhibition or processed meat products have not been included. The reductions covered refer to the main ranges of reductions. Bacterial reductions obtained on carcasses depend on a variety of framing conditions, such as application modes, temperatures, pressures, exposure times, concentrations, or contamination levels (natural, artificial). Moreover, the effectiveness of some measures is controversial, many interventions have only been tested under laboratory conditions or are still under development, and regulatory issues remain to be solved for certain methods.

17.4.1   Poultry carcasses

The decontamination of poultry carcasses is of worldwide interest. Physical interventions and chemical treatments are frequently used for the removal of surface contamination from poultry carcasses. In the following, the focus is on the antibacterial activity against *Campylobacter*, because consumption of poultry is an important risk factor for human campylobacteriosis and measures that have substantially reduced *Salmonella* in poultry have so far been largely ineffective against *Campylobacter*.

Among the physical interventions used for the decontamination of poultry carcasses, water-based treatments (spraying, immersion, immersion/spray chilling) predominate. Hot water and steam are quite effective for reducing *Campylobacter* on poultry carcasses (Table 17.1). Hot water mainly yielded reductions in the range 1.0–1.7 logs. The use of hot water or steam followed by ultrasound might constitute an alternative (reductions of *Campylobacter* by >2.0 logs) but further investigations are required in practical applications. The same also applies for irradiation, which seems effective at adequate dosages. With regard to ultraviolet (UV) light, the irregular skin surface of poultry carcasses might provide protection for bacteria but the use of pulsed UV is of growing interest. Furthermore, with special regard to *Campylobacter*, freezing

Table 17.1   Selected reductions obtained for *Campylobacter* by different decontamination treatments on the surface of poultry carcasses and carcass parts.

Agent	Reduction (log_{10} cfu)[a]	Application[b]	Contamination
Hot water	1.3–2.8	SP (/IC)	Artificial
	0.6–1.9	IM	Artificial
	0.3–1.6	IM (/SC)	Natural
	0.1–0.4	SP	Natural
Steam	0.5–1.3	Steam	Natural
	1.8–3.3	Steam	Artificial
Acetic acid	1.2–1.4	IM	Artificial
Citric Acid	0.6–1.2	IM or SP	Artificial
Lactic acid	0.3–1.7	IM	Artificial
	0.2–0.9	SP	Artificial
Chlorine	1.2–2.6	SP (/IC, /IM)	Artificial
	0.5–3.0	IM	Artificial
Acidified sodium chlorite (ASC)	2.6	SP	Natural
	1.1–1.9	IM or SP	Artificial
	0.9–1.2	IM	Natural
Cetylpyridinium chloride (CPC)	>4.2	IM	Artificial
	1.4–2.9	SP	Artificial
Trisodium phosphate (TSP)	1.7–2.4	IM	Artificial
	1.6	SP	Artificial
	0.2–1.7	IM	Natural
ASC + TSP	2.4–2.5	IM	Artificial
TSP + ASC	1.1–1.4	IM	Artificial

[a] Values refer to reductions obtained per g, ml, cm² or carcass.
[b] IM, immersion; SP, spraying; IC, immersion chilling; SC, spray chilling.

is promising as reductions of up to 2.2 logs have been reported on naturally contaminated poultry carcasses.

Chemical interventions used for the decontamination of poultry carcasses primarily comprise organic acids (in particular lactic acid), chlorine-based treatments or phosphate-based treatments (in particular TSP). In addition, a wide variety of other substances and combinations has occasionally been tested. It must be emphasized that many studies investigated inoculated samples and were performed under laboratory conditions. Compounds such as organic acids, chlorine, ASC, CPC or TSP are quite effective for reducing *Campylobacter* on poultry carcasses (Table 17.1). Lactic and acetic acid, ASC and TSP mainly yielded reductions in the range 1.0–2.2 logs. Sequential treatments applied during slaughter can further enhance the results. Considerable reductions have been obtained under commercial conditions using ASC. Chlorine-containing compounds are frequently used for pre- and post-chill spray/washing of poultry carcasses or in carcass chillers in several countries. However, reductions obtained for *Campylobacter* vary widely and the maintenance of the residual free chlorine concentration in poultry carcass chillers is difficult due to interaction with organic matter. A promising alternative constitutes the use of EW. It has been shown to reduce *Campylobacter* on inoculated poultry carcasses by up to 3.2 logs. However, the adverse impact of organic matter must not be neglected.

Promising alternatives include biological interventions but studies on the efficacy of bacteriophages or bacteriocins on poultry carcasses are so far very limited, especially in terms of practical application. Depending on the phage titre, reductions of *Campylobacter* by up to 2.0 logs have been reported.

17.4.2 Bovine hides and carcasses

Healthy ruminants, mainly cattle, are regarded as the principal reservoir of STEC, which are responsible for human diseases ranging from diarrhoea to the life-threatening haemolytic-uremic syndrome. The efficacy of water sprayings, organic acids and their combinations has most frequently been investigated for the decontamination of bovine hides and carcasses. Application of interventions at slaughter plants reduced bacterial loads to some extent but the reductions reported on carcasses remained below 2.0 logs using hot water, steam, acetic acid or lactic acid.

Hides Bovine hides often show high bacterial loads and are a major source of carcass contamination during dehiding. Hence, various hide decontamination treatments have been tested to reduce bacterial loads but only restricted data are available for reductions obtained under commercial conditions.

Dehairing can be achieved by hide clipping or the use of chemicals. Probably due to the generation of dust and subsequent spread of bacteria, clipping seems hardly effective in reducing the bacterial load. Chemical dehairing, which often comprises treatment steps using sodium sulphide (Na_2S), hydrogen peroxide and water applied in a washing cabinet, is quite effective on inoculated hides (reductions by several logs) but gives hardly any reduction under commercial conditions.

Interventions used for the decontamination of bovine hides primarily comprise water, steam and organic acids, but other substances such as chlorine, EW, ethanol, isopropyl alcohol, ozone, or sodium hydroxide have also been tested. Water washing of naturally contaminated hides has mainly yielded reductions of <1.0 logs, probably due to the spread of bacteria previously encapsulated in dirt, mud and faeces. However, remarkable reductions (1.9–6.0 logs) have been observed for applications using steam. Acetic and lactic acids have also shown to be quite effective (reductions between 2.0 and 3.0 logs) (Table 17.2). High reductions (>2.0 logs under commercial conditions) were also obtained by CPC. A novel approach constitutes the concept of 'bacterial on-hide immobilization' with a solution of food-grade

Table 17.2 Selected reductions obtained for aerobic bacteria, *Escherichia coli* and *Escherichia coli* 0157:H7 by different decontamination treatments on cattle hides.

Agent	Microorganism	Reduction($\log_{10}$ cfu)[a]	Contamination
Water (20–55°C)	Aerobic bacteria	0–1.0	Natural/Artificial
	Escherichia coli	0–1.0	Natural/Artificial
	Escherichia coli 0157:H7	2.3	Artificial
Steam	Aerobic bacteria	1.9–4.0	Natural
	Escherichia coli 0157:H7	1.9–6.0	Artificial
Acetic acid	Aerobic bacteria	2.4–2.6	Natural
		0.8–1.3	Artificial
	Escherichia coli	2.5–2.8	Natural
	Escherichia coli 0157:H7	0.7–2.1	Artificial
Lactic acid	Aerobic bacteria	1.6–4.1	Artificial
		2.3[b]	Natural
	Escherichia coli	3.3	Artificial
		2.1–2.7[b]	Natural
	Escherichia coli 0157:H7	2.9–4.3	Artificial
Cetylpyridinium chloride (CPC)	Aerobic bacteria	1.9–4.6	Artificial
		3.8[b]	Natural
	Escherichia coli	4.5	Artificial
		3.0[b]	Natural
Acetic acid + water	Aerobic bacteria	0.5–0.9	Artificial
	Escherichia coli 0157:H7	0.6–2.6	Artificial
Lactic acid + water	Aerobic bacteria	0.5–1.0	Artificial
	Escherichia coli 0157:H7	0.8–3.4	Artificial
Water + CPC	Aerobic bacteria	3.0–3.3	Natural
NaOH + water	Aerobic bacteria	0.8	Natural
	Escherichia coli 0157:H7	2.4–3.4	Artificial
NaOH + lactic acid	Aerobic bacteria	2.0–2.4	Natural
	Escherichia coli	2.3–3.0	Natural
NaOH + chlorine	Aerobic bacteria	2.1[b]	Natural
	Escherichia coli 0157:H7	5.0	Artificial

[a] Values refer to reductions obtained per cm^2 or designated area.
[b] Treatment at slaughter plant under commercial conditions.

resin in ethanol in order to reduce transmission of bacteria from hides to carcasses.

Furthermore, various combinations of interventions can be used for the decontamination of bovine hides (Table 17.2). However, they did not consistently enhance the reductions compared to the single treatments. Water and CPC spraying reduced naturally occurring aerobic bacteria by about 3.0 logs but comparison with single CPC treatment is hampered by different modes of application. Combinations of different chemical compounds did not always produce consistent results. For the use of bacteriophages, further investigations are required, in particular to evaluate the eligibility under commercial conditions. It should be mentioned that FSIS approved the use of certain *E. coli* O157:H7 and *Salmonella* targeted bacteriophages for the treatment of bovine hides in 2007.

Carcasses Among the physical interventions used for the decontamination of bovine carcasses, water-based treatments predominate. The antibacterial activity on carcasses thereby increases with increasing water temperature. Hot water spraying mainly yielded bacterial reductions in the range 0.8–1.8 logs under commercial conditions (Table 17.3). Enhanced bacterial reductions were obtained by the combination of steam or dry heat with water sprayings but further investigations are required in terms of practical application. Similarly, irradiation seems effective at adequate dosages. Steam-vacuuming systems are increasingly used to remove localized contaminations, occasionally in combination with water sprayings, but bacterial reductions vary widely (0.2–5.5 logs). Moreover, published data indicate that chilling of bovine carcasses under commercial conditions can result in increases, decreases or no changes in the bacterial load (depending on process parameters such as temperature, air speed, humidity, carcass spacing and duration).

Chemical interventions used for the decontamination of bovine carcasses comprise various substances, although there is less diversity than among substances used for poultry carcasses. Organic acids are primarily used for the decontamination of bovine carcasses and selected reductions of aerobic bacteria, *E. coli* and *E. coli* O157:H7 are shown in Table 17.3. Under commercial conditions, acetic and lactic acid mainly yielded bacterial reductions below 1.6 logs and results seem to be influenced by the point of application during slaughter. Substances such as CPC, TSP or commercially available preparations have also shown promising results on inoculated carcasses but bacterial reductions obtained under commercial conditions have mostly been low (<1.0 log). In addition, other substances and combinations of them, such as chlorine, electrolysed water, hydrogen peroxide, peroxyacids and sodium bicarbonate ($NaHCO_3$) have also been occasionally tested, though mainly under laboratory conditions.

Various combinations of interventions can also be used for the decontamination of bovine carcasses (Table 17.3) but combinations of water/steam and organic acid sprayings predominate. Based on the limited data obtained under commercial conditions, hot water or steam combined with lactic acid mainly yielded bacterial reductions in the range 0.5–2.2 logs. Under laboratory conditions, the combination of water with chemicals such as organic acids, ASC, hydrogen peroxide or TSP, the combination of steam vacuuming with chemicals and the combination of different chemicals also yielded promising results. The application sequence might influence the outcome. In addition, despite the wide ranges of bacterial reductions reported on post-intervention carcasses (aerobic bacteria: 0.4–4.1 logs; *E. coli*: 1.0–4.1 logs), the use of sequential interventions in the slaughter process (pre-evisceration, post-evisceration, end of slaughter) must also be considered.

Table 17.3 Selected reductions obtained for aerobic bacteria, *Escherichia coli* and *Escherichia coli* 0157:H7 by different decontamination treatments on the surface of beef carcasses and carcass parts.

Agent	Microorganism	Reduction(log$_{10}$ cfu)[a]	Contamination
Hot water	Aerobic bacteria	0.3–3.5	Artificial
		<0.3–2.7[b]	Natural
	Escherichia coli	0.8–4.2	Artificial
		1.4–1.8 [b]	Natural
	Escherichia coli 0157:H7	0.8–2.3	Artificial
Steam	Aerobic bacteria	0.1–1.6[b]	Natural
	Escherichia coli	0.1–>1.0[b]	Natural
	Escherichia coli 0157:H7	0.1–4.7	Artificial
Acetic acid	Aerobic bacteria	2.5	Artificial
	Escherichia coli	2.1–2.2	Artificial
		1.4[b]	Natural
	Escherichia coli 0157:H7	0.7–3.2	Artificial
Citric Acid	*Escherichia coli* 0157:H7	1.2–1.8	Artificial
Lactic acid	Aerobic bacteria	0.5–3.3[b]	Natural
	Escherichia coli	2.4–4.8	Artificial
		>0.2–0.6[b]	Natural
	Escherichia coli 0157:H7	1.0–3.0	Artificial
Water + acetic acid	Aerobic bacteria	2.0–3.4	Artificial
	Escherichia coli	1.9–3.7	Artificial
	Escherichia coli 0157:H7	2.4–3.7	Artificial
Water + lactic acid	Aerobic bacteria	4.6	Artificial
		2.2[b]	Natural
	Escherichia coli	1.5–>4.4	Artificial
	Escherichia coli 0157:H7	1.0–5.2	Artificial
Steam vacuuming	Aerobic bacteria	3.5	Artificial
+ lactic acid	*Escherichia coli*	4.4	Artificial

[a] Values refer to reductions obtained per cm^2 or designated area.
[b] Treatment at slaughter plant under commercial conditions.

As mentioned for poultry carcasses, biological interventions constitute interesting alternatives but studies on the efficacy of bacteriophages or bacteriocins on bovine carcasses are so far limited. Yet, the combination of nisin and lactic acid sprayings yielded reductions by up to 2.0 logs on bovine carcasses under commercial conditions.

17.4.3 Pig carcasses

Pig slaughter operations before evisceration commonly involve scalding, mechanical dehairing, singeing and polishing without removal of the skin. The resulting pig carcasses are visibly clean and largely free of hair. Despite this appearance, carcasses can be highly contaminated with bacteria and, thus, constitute a contamination source during slaughtering and processing. Carcass interventions are, therefore, also gaining interest in pig slaughtering. Physical treatments can be part of the normal slaughter process (e.g. scalding or singeing) or they can be additionally applied with the objective of carcass decontamination. Compared to the various interventions and substances used for the decontamination of bovine and poultry carcasses, only a few specific methods have been investigated for pig carcasses. For example, the use of interventions such as irradiation, dry heat or ultrasound has so far not been reported for pig carcass decontamination.

Although it is not their primary intention, slaughter process stages, for example scalding or singeing, might exhibit some bactericidal effect. However, it must be considered that reductions obtained might be offset during subsequent process stages. Dependent on the time and temperature conditions, the bactericidal effect of scalding has been shown in several studies, whereas the effects of singeing or chilling differ widely. For example, scalding (59–62°C, 5.0–8.5 min) reduced aerobic bacteria by 3.1–3.8 logs but, in general, lower reductions have been obtained when a combined scalding and dehairing process was used. Singeing (exposure to flames) mainly yielded reductions in the range 1.5–2.5 logs. With regard to air chilling, the associated surface desiccation seems of major importance for the inactivation of *Campylobacter*. Pigs are a common reservoir for the human pathogen *Campylobacter coli* but consumption of pork is not a predominant risk factor for human campylobacteriosis. However, published data indicate that chilling (air chilling with or without precedent blast chilling, spray chilling, blast-spray chilling) of pig carcasses can result in increases, decreases or no changes in the bacterial load (depending on process parameters such as temperature, air speed, humidity, carcass spacing and duration). Comparisons of published data are thereby often hampered by incomplete information on process parameters.

Among the physical interventions applied with the specific objective of pig carcass decontamination, hot water spraying and steam predominate. The majority of studies have been performed under commercial conditions (Table 17.4). Hot water and steam mainly yielded reductions in the range 1.0–2.1 logs. One laboratory study, which tested UV light on pork skin, reported reductions of *E. coli* by 0.1–3.3 logs (depending on UV intensity applied). Probably, the smooth pig carcass surface facilitates the antibacterial efficacy of UV.

Chemical compounds used for the decontamination of pig carcasses are mainly restricted to organic acids and available data are limited to a few studies. Selected bacterial reductions obtained by acetic and lactic acid are shown in Table 17.4. Under commercial conditions, lactic acid treatment mainly yielded bacterial reductions in the range 0.1–1.0 logs. Occasionally, a few other substances such as EW, potassium sorbate, TSP and their combinations have also been tested, though mainly under laboratory conditions.

Table 17.4 Selected bacterial reductions obtained by different decontamination treatments on the surface of pig carcasses and carcass parts.

Agent	Microorganism	Reduction(log$_{10}$ cfu)[a]	Contamination
Water	Aerobic bacteria	0.2–2.3	Artificial
		0.1–3.3[b]	Natural
	Campylobacter coli	1.3	Artificial
	Coliforms	1.0–2.1	Artificial
		1.7[b]	Natural
	Escherichia coli	0.9–2.1	Artificial
		0.7–2.5[b]	Natural
	Salmonella Typhimurium	1.4	Artificial
Steam	Aerobic bacteria	0.4–2.0[b]	Natural
	Coliforms	1.5[b]	Natural
	Psychrophiles	1.3[b]	Natural
Acetic acid	Aerobic bacteria	1.0	Artificial
	Coliforms	0.3	Artificial
	Escherichia coli	0.5	Artificial
Lactic acid	Aerobic bacteria	0.4-1.7	Artificial
	Aerobic bacteria	0.1–1.0[b]	Natural
	Campylobacter coli	1.7	Artificial
	Coliforms	1.1	Artificial
	Escherichia coli	1.1	Artificial
	Salmonella Typhimurium	1.8	Artificial

[a] Values refer to reductions obtained per g, cm^2 or designated area.
[b] Treatment at slaughter plant under commercial conditions.

17.5 Conclusions

Various foods can serve as sources of food-borne pathogens but meat and meat products are frequently associated with human infections. Many important human pathogens, such as *Campylobacter* spp., *Salmonella* spp., enteropathogenic *Yersinia* spp. or STEC, can be harboured by healthy food-producing animals. Complete prevention of carcass contamination with food-borne pathogens during slaughter can hardly be warranted. Thus, antimicrobial intervention technologies for carcasses are gaining increased interest.

Carcass interventions should be safe, economic and feasible in the production process, widely accepted by the consumers and should not change the organoleptic properties of the carcasses. Moreover, certain treatments might increase the humidity on the surface of carcasses, thus influencing the shelf life of the meat. With regard to practical application, the location in the slaughter process and the effect of sequential interventions must also be considered. Selection and adaptation of decontamination steps have to be customized to plant- and process-specific circumstances.

Interventions used at slaughter basically comprise physical, chemical and biological treatments, which can be applied alone or in combination. Some interventions reduce the bacterial contamination on carcasses to some extent when applied under commercial conditions but complete inactivation cannot be achieved. For example, hot water, steam, acetic acid or lactic acid mainly yielded reductions for several bacterial species by <2.0 logs. Moreover, comparisons of antibacterial activities are often hampered by varying framing conditions, such as modes of application, temperatures, pressures, exposure times, concentrations and contamination levels.

Accurate estimation of the overall effects of various interventions is also difficult because there is a lack of data on industrial-scale processes and extrapolation of experimental results to commercial practices is difficult. Hence, there is a need for more (standardized) commercial data. Similarly, more studies that investigate the antibacterial effects of serial or sequential control strategies under commercial conditions (multiple hurdle approach) are required. In addition, the development and practical application of new or novel processes and technologies has to be carefully monitored.

Carcasses interventions must always be seen as part of an integral food safety system comprising measures at pre-harvest, harvest, processing, storage, distribution and preparation. Decontamination treatments applied at slaughter cannot compensate for poor hygiene practices or replace good manufacturing and slaughter hygiene practices along with risk-based preventive measures. In fact, intervention technologies are most effective when combined with strict hygiene practices.

Literature and further reading

Aymerich, T., Picouet, P.A. and Monfort, J.M. 2008. Decontamination technologies for meat products. *Meat Sci*, **78**, 114–129.

Demerici, A. and Ngadi, M.O. (eds). 2012. *Microbial Decontamination in the Food Industry: Novel Methods and Application*. Woodhead Publishing Ltd, Cambridge, UK.

FAO/WHO. 2008. Benefits and risks of the use of chlorine-containing disinfectants in food production and food processing: Report of a Joint FAO/WHO Expert Meeting, 27–30 May 2008, Ann Arbor, MI. http://apps.who.int/iris/bitstream/10665/44250/1/9789241598941_eng.pdf (last accessed 15 February 2014)

Huffman, R.D. 2002. Current and future technologies for the decontamination of carcasses and fresh meat. *Meat Sci*, **62**, 285–294.

Hugas, M. and Tsigarida, E. 2008. Pros and cons of carcass decontamination: The role of the European Food Safety Authority. *Meat Sci*, **78**, 43–52.

Loretz, M., Stephan, R. and Zweifel C. 2010. Antimicrobial activity of decontamination treatments for poultry carcasses. *Food Control*, **21**, 791–804.

Loretz, M., Stephan, R. and Zweifel, C. 2011. Antimicrobial activity of decontamination treatments for cattle hides and beef carcasses. *Food Control*, **22**, 347–359.

Loretz, M., Stephan, R. and Zweifel, C. 2011. Antimicrobial activity of decontamination treatments for pig carcasses. *Food Control*, **22**, 1121–1125.

18
Cleaning and Disinfection

Gun Wirtanen[1] and Satu Salo[2]

[1]*VTT Expert Services Ltd, Espoo, Finland*

[2]*VTT, Espoo, Finland*

18.1 Scope

Cleaning and disinfection are carried out in order to produce safe products with acceptable shelf life and quality. Fat and protein are the most common types of soil in slaughterhouses. Cutting devices, conveyor belts, eviscerators and slaughtering chains are known as parts that are difficult to clean. The mechanical and chemical power, temperature and contact time in the cleaning regime should be carefully chosen to achieve an adequate cleaning effect. The cleanliness of process surfaces should be detected regularly with suitable methods. Failures in cleaning can be noticed by comparing individually obtained results to results obtained in long-term routine hygiene control.

18.2 Background to cleaning and disinfection

The general aims of cleaning and disinfection are removal of soil and microbes including biofilms, that is microbes embedded in a self-produced extracellular polymeric substance which is also referred to as slime, from process surfaces. This microbial control procedure is used to prevent spoilage of products and to ensure that the quality specifications of the product are met. The most important means for maintaining efficient microbial control include minimizing the microbial load from sources outside of the process, efficient control of growth

Meat Inspection and Control in the Slaughterhouse, First Edition.
Edited by Thimjos Ninios, Janne Lundén, Hannu Korkeala and Maria Fredriksson-Ahomaa.
© 2014 John Wiley & Sons, Ltd. Published 2014 by John Wiley & Sons, Ltd.

at microbiologically vulnerable sites and adequate cleaning and disinfection of the process lines. Correct operation of the entire system under controlled conditions is important and the microbial control programmes, including cleaning and disinfection, should be integrated into the overall processing.

18.3   Cleaning in general

Physical, chemical and microbiological cleanliness is essential in food processing plants. Physical cleanliness means that there is no visible waste or foreign matter on the process and equipment surfaces. Chemically clean surfaces are surfaces from which undesirable chemical residues have been removed, whereas microbiologically clean surfaces imply freedom from spoilage microbes and pathogens. Attached bacteria or bacteria in biofilms can be a problem in food processing because these cells can stick firmly to the surfaces. Once biofilms are formed, cleaning and disinfection becomes much more difficult. Especially in cases where the cleaning is insufficient, the remaining bacteria can start to multiply and grow in humid conditions immediately after the cleaning has been performed. Thus, products treated on these surfaces will be contaminated. The elimination of surface-attached microbes or microbes in biofilms is a very difficult and demanding task in which both detachment and the microbicidal effect of the agents on microbes should be taken into account. Mechanical treatments in cleaning are the most efficient ways of eliminating biofilms but frequently the structure of the equipment makes this difficult.

The key to effective cleaning of a food plant is the understanding of the type and nature of the soil (*e.g.* fat, protein, sugar and mineral salts) and of the microbial growth on the surfaces to be removed. The accessibility and type of equipment and accessories to be cleaned and the availability of suitable cleaning agents are also important. An efficient cleaning procedure consists of a sequence of rinses and detergent and disinfectant applications in various combinations of temperature and concentration (Figure 18.1). This controls the accumulation of soil and development of biofilms on equipment surfaces without corroding the surfaces. The personnel must be properly trained and responsible to maintain a good level of process hygiene in the food plant. In this case, the tools and methods used must also suit the slaughtering process.

18.4   Disinfection in general

Cleaning is mostly carried out in combination with a final disinfection, because it is likely that there will be viable microbes on the surfaces after the cleaning steps have been carried out and these microbes could harm the quality of the meat. The disinfectants have been developed to destroy microbes. Microbes have, nevertheless, been found in disinfectant solutions, which is due to their ability to form resistant strains and the build-up of protective biofilms.

Figure 18.1 Various steps in a sanitation procedure.

This means that microbial contaminants can be spread on the surface instead of being cleaned off. Findings reported as early as 1967 showed that chlorhexidine mixtures were contaminated with *Pseudomonas* sp. Contamination of disinfectants can be avoided through good manufacturing practices, that is keeping the container closed when not in use and not returning leftovers to the container.

Disinfection is required in food processing operations, where wet surfaces provide favourable conditions for the growth of microbes. The aim of the disinfection is to reduce the population of viable microbes left on the surfaces after cleaning, thus preventing microbial growth on surfaces before the restart of production. Disinfectants do not penetrate very well the biofilm matrix left on the surfaces after an ineffective cleaning procedure. The living cells in biofilms are thus not destroyed. Disinfectants are most effective in the absence of organic material, for example fat-, sugar- and protein-based materials. Interfering organic substances, pH, temperature, concentration and contact time affect the efficiency of disinfectants.

18.5   Main soil types and their removal

In slaughterhouses fats and proteins are the most common types of soil. Often the soil is a mixture of protein and fat. Smears of protein and fat can also contain detergent residues as well as mineral salts of calcium and magnesium from hard water. This type of smear can be found where the cleaning procedure is irregular or based on incorrect working routines with the use of the wrong detergents. This waxy mass is not soluble in alkaline or acid solutions. The only way to remove this type of smear is by using mechanical force. Often, manual scrubbing with pads or scrubbing with brushes is the only way to remove this type of residue. It is important to differentiate mineral salt deposits, which are soluble in an acidic environment, from waxy mass. Fat can easily be removed using hot water (50–60°C) to give an optically clean surface and the efficiency can be improved by adding a detergent. Removal of protein depends on the treatment that the soiled surface has been exposed to before the cleaning procedure is started. If the surface has been treated with hot water (≥60°C) for a long time the soil is very difficult to remove. Note that surfaces soiled with protein should be kept wet until the cleaning procedure can start. Protein not dried on a surface can easily be removed with water to obtain a visibly clean surface. Crusts, coatings and adhesive layers are often left behind when the protein has been dried and/or burnt on the surface, even though the dirt has been softened with water and detergent and removed with mechanical force, for example low pressure or manual brushing.

18.6   Cleaning procedure

The cleaning systems in food processing can be divided into open and closed systems. In slaughterhouses most of the systems are open and cleaning is performed using pressure cleaning, brushing and so on. Closed systems are commonly found in dairies and breweries and the most common system is based on cleaning-in-place (CIP) procedures. The ideal cleaning programme can be divided into the following subroutines: clearance of process equipment and its surroundings for cleaning, assembly for cleaning, removal of solid waste from the environment, dismantling of equipment, removal of solid waste from the equipment, pre-rinsing with water, cleaning and rinsing of the environment, application of detergent to the equipment, low pressure rinsing of the equipment, disinfection of the equipment, post-rinsing of the equipment with potable water and post-treatment, for example assembly of equipment, disassembly of cleaning equipment and fog disinfection of rooms with good ventilation possibilities (Figure 18.1).

It is recommended that each person in processing is responsible for clearing a specified area. One person must have the overall responsibility for the whole clearing process. The purpose of clearance after work is to prepare an easy and quick cleaning procedure. If the carcasses and meat have not been removed from the processing area as a part of the processing, the first part of the clearance is to move the carcasses/meat to proper storage. All items that may retard performance or may be damaged by water must be removed. Non-removable equipment should be covered properly in order to protect against penetration of water, because water and electric installations may constitute a safety hazard. Examples of items to be removed from the processing area apart from the meat include packaging materials and mobile equipment, which can be cleaned in a central cleaning area.

Assembly for cleaning is necessary and includes preparing the detergent and disinfectant solutions as well as maintenance and/or preparation/service of cleaning equipment, for example pressure cleaners, hoses and nozzles.

It will always depend on local conditions and may be time consuming. It cannot be neglected, because then the following cleaning procedure will not to be optimal. Solid waste should be removed before using water in the area. It might not be necessary to remove all solid waste in a slaughterhouse but it is recommended, especially in places where the amount of solid waste is large and the water supply is insufficient. The environment should be cleaned thoroughly before the cleaning forces are focused on the equipment. Special equipment should be dismantled so that the cleaning personnel can clean them efficiently. The cleaning of equipment starts with removing most of the solid waste from the equipment before the equipment is pre-rinsed. Removal of waste will decrease the pre-rinsing period. All visible, solid waste should be removed before detergent is applied. Hot water can only be recommended for pre-rinsing of surfaces covered with fat. Pre-rinsing with cold water will be sufficient in most cases and especially when it is followed by an optimal cleaning programme.

If pressure systems are used for rinsing, pressure should not exceed 30 bar, because pressures exceeding 30 bar can harm or destroy equipment surfaces and, furthermore, suspended bioaerosol particles are formed and spread around the facilities being cleaned. The systems for cleaning should be chosen in accordance with the actual conditions in the slaughterhouse. When using pressure cleaners the cleaning results will depend on the water pressure and spreading angle, water temperature, amount of water and detergent as well as time. When an automatic system is chosen, delivery time of spare parts for the system should be considered as well as service possibility when repairs are required. An alternative plan for cleaning and/or online maintenance in the case of failures prevents production delays. Skipping cleaning is not an alternative. The management of the slaughterhouse must set rules for how often the cleaning programme should be carried out.

It is obvious that energy is required for cleaning purposes. Manual cleaning requires considerable input of manpower. Electricity is used for automatic floor cleaners or vacuum cleaners for water suction in food processing areas. Water pressure is used in both low and high pressure systems. Multipressure washing systems, that is systems with both low and high pressures, have limitations because most of them use large amounts of water and the cleaning is thus expensive. Cleaning systems working with low pressure (20–30 bar) and low volume (18–20 1/min) have been developed using special nozzles to obtain the pressures needed. The distance between the nozzle and the surface should be approximately 30 cm and the angle between the surface and the water jet should be approximately 45°. The spreading angle in the nozzle is important to obtain the correct pressure against the surface. An angle of 0° gives a massive water jet. The force of the water jet is less massive when the angle used is larger. Worn nozzles must be replaced. The detergent can be applied using an injector system, in which the water supply for the cleaning system is connected to the detergent and air intakes, when using a pressure cleaning system. The foam formed is spread on the vertical surfaces from below working upwards. Spraying from unclean areas to clean areas must be avoided. High pressure can damage not only the equipment but also wall and floor surfaces; wrongly used it may loosen the wall or floor tiles.

Great care must be taken to ensure that cleaning tools such as brushes, scrubbing pads and squeegees are clean, so that cross-contamination can be avoided. Frequent changes of water and detergent solutions are also essential in prohibiting cross-contamination. Furthermore, the tools and methods used must suit the process and the personnel must be properly trained and responsible in order to maintain a good level of plant hygiene. If the rinsing can be performed with running water this should be preferred. Attention should also be paid to the quality of the processing water, steam and other additives. Post-rinsing with clean potable water is required after chemical disinfection in order to remove chemical residues.

18.6.1   Cleaning of processing environment

The majority of the solid gross soil should be removed by dry methods, for example brushing, scraping or vacuuming, and visible soil residues rinsed off with low pressure water. However, a pure water washing system is not practical due to ineffectiveness and cost limitations. Surfactants, that is detergents containing surface active agents which suspend the adhered particles and microbes from the surfaces in the water, are added to increase the washing effect. After a production run the equipment should be dismantled. Cleaned utensils should be stored on racks and tables, not on the floor.

Emptying of waste containers and cleaning of these is also included in the disassembly. Note that none of the subroutines should be changed without evaluating if the other subroutines should be changed at the same time. It is important to ensure that personnel are aware of individual responsibilities in each area, know who is responsible when irregularities occur, work according to the principle 'from clean towards unclean' and 'clean and unclean procedures are never mixed'. A thorough cleaning programme has to be carried out at least once a day and, eventually, a reduced cleaning programme could be carried out before the type of process is changed during the day.

Fogging can be used as chemical disinfection in specific closed process rooms. Fogging is performed using automatic systems spraying disinfectants. Controlled fogging experiments in industrial scale facilities showed no clear reduction of the microbial load on surfaces below tables and so on. Critical control points in fogging were the amount of fog used, the disinfectant concentration used in the fog, thorough rinsing of equipment and drying of facilities afterwards.

18.6.2   Cleaning of equipment

The cleaning of open process surfaces and surfaces in the processing environment is normally carried out using either foam or gel cleaning. The foam units are constructed to form foam of varying wetness and durability depending on the cleaning to be performed. Surfaces that are not in direct contact with food must also be cleaned in order to avoid accumulation of dirt and cross-contamination. The results of hygiene surveys have shown that neck cutting machines, conveyor belts and cutting devices are difficult to clean. Coliforms and *Eschericia coli* have been found on surfaces of such equipment after cleaning, indicating poor hygiene. The top three places needing special attention in maintenance based on hygiene surveys are conveyor belts, baskets and cutters with modules. These surfaces are all at some point in direct contact with the products. Also, eviscerators, slaughtering chains, scales and trucks (forklifts) in poultry slaughterhouses have been shown to be difficult to clean.

18.6.3 Choosing the cleaning temperature

Temperature is one of the factors affecting the efficacy of the cleaning process besides mechanical and chemical impact and time used in cleaning (Figure 18.2). The optimal water temperature for rinsing depends on detergent used, soil removed, surfaces cleaned, amount of water available and cleaning system used. The water temperature should be thermostatically controlled and a temperature of 40°C will normally be sufficient. Using sufficient water volumes with the correct temperature increases the cleaning effect. Detergents to be used in the cleaning of open systems are formulated to be effective at room temperature or at slightly elevated temperatures in the range 35–50°C. Fats are easily removed at temperatures slightly above their melting point. Sugars and other carbohydrates are water soluble at elevated temperatures but temperatures causing browning should be avoided. Proteins are denatured at elevated temperatures and may adhere strongly to surfaces at too high a temperature. Washing cold surfaces in chilled areas with warm water from a pressure cleaner is challenging, since the temperature of water is decreased on cold surfaces.

18.6.4 Choosing the cleaning agents

The purpose of applying detergent is to decompose and loosen the soil from surfaces and prevent already loosened soil from re-depositing. The choice of detergent depends on the type of soil, the water hardness, the type of surface,

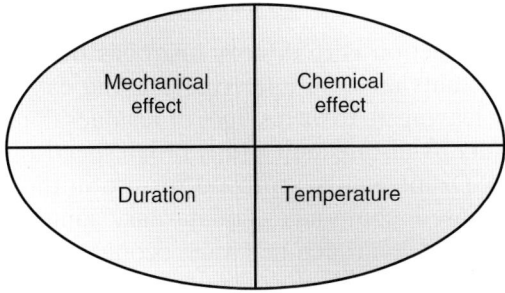

Figure 18.2 Factors affecting the cleaning process.

the system for detergent application and the remaining cleaning programme, including cleaning time, water temperature, chemical impact and mechanical forces (Figure 18.2). A prolonged exposure of the surfaces to cleaning agents enhances the removal. Decomposed and suspended soil and detergent residues should be removed through rinsing with warm water through either a hose or low pressure jet.

Due to safety concerns only mild detergents should be used when the cleaning is carried out manually. Automatic cleaning techniques enable the use of strong detergents. Detergents enhance the mechanical forces applied in a cleaning system. No detergent or cleaning compound can be used as an all-purpose detergent. Mixtures used in detergents combine several properties, making a product effective in cleaning a particular process. Only actual tests with a particular detergent during a specific cleaning operation will give an indication of the agent's efficiency. The detergent costs must not be used as a guide for cleaning efficiency.

In the food industry the selection of detergents and disinfectants depends on the efficacy, safety and rinsability of the agents as well as their corrosive effects and sensory values on products produced. Depending on the application, the ability of the cleaning agent to form foam is either desirable or not. A foam in closed processes and in machine washing decreases the cleaning effect. In pressure cleaning the agent should form a solid foam or gel.

The use of effective cleaning agents and disinfectants on microbes attached to surfaces minimizes contamination of the product, enhances shelf life and reduces the risks of food-borne illness. The basic task of detergents is to reduce the interfacial tensions of soil materials so that it becomes miscible in water. The effect of the surfactants is increased by the mechanical effect of turbulent flow or water pressure, or of abrasives, for example salt crystals. Chelating agents in the cleaning solution enhance the removal of biofilms from processing surfaces. A prolonged exposure of the surfaces to the detergent makes the removal efficient. Attention should also be paid to the quality of additives because poor quality easily spoils the whole process; for example, personnel should be aware of the water quality in all taps and from all hoses, because good quality water coming into the slaughterhouse can be contaminated by dirt in the tap nozzles or by the inner surface in the hoses.

18.6.5 Choosing the disinfectants

The purpose of disinfection is to destroy microbes by chemical or physical means. The disinfectants must be effective, safe and easy to use, and easily rinsed off surfaces, leaving no toxic residues or residues that affect the sensory values of the product. A disinfectant with a broad spectrum is recommended for most purposes. Disinfectants approved for use in the food industry are alcohols, chlorine-based compounds, quaternary ammonium compounds, oxidants (peracetic acid, hydrogen peroxide, ozone), persulfates, surfactants and iodophors.

The number of microbes should be reduced to a level causing no harmful contamination in the food produced. Choosing the correct disinfectant depends on the disinfection method, the surface materials, the type of processing and the remaining part of the cleaning programme. Disinfectants should be chosen based on the factors shown in Figures 18.3 and 18.4. The application of disinfectants can be carried out using a low pressure system. Disinfectants restricted in their effect against microbes may be used for specific purposes. Hypochlorite and peracetic acid can be used for almost all purposes. Quaternary ammonium compounds and amphoteric compounds may be used in areas where heat-treated products are produced. Post-rinsing should be carried out using water of potable quality as in all other process steps. In post-treatment it may be necessary to apply edible oil on surfaces such as cast iron before assembling the equipment after sanitation to avoid corrosion due to the detergent and disinfectant treatment. Good ventilation in the process facilities is needed to enable drying of the process equipment and process lines. Furthermore, cleaning equipment should be stored in a separate room, where detergents and disinfectants can also be stored. The cleaning and disinfection equipment should be maintained in accordance with instructions from the suppliers and the management of the slaughterhouse.

There are several chlorine or chlorine-based compounds that are approved for use in food plants, for example gaseous chlorine, sodium chlorite and calcium hypochlorites. The antibacterial active part is formed when the chlorine

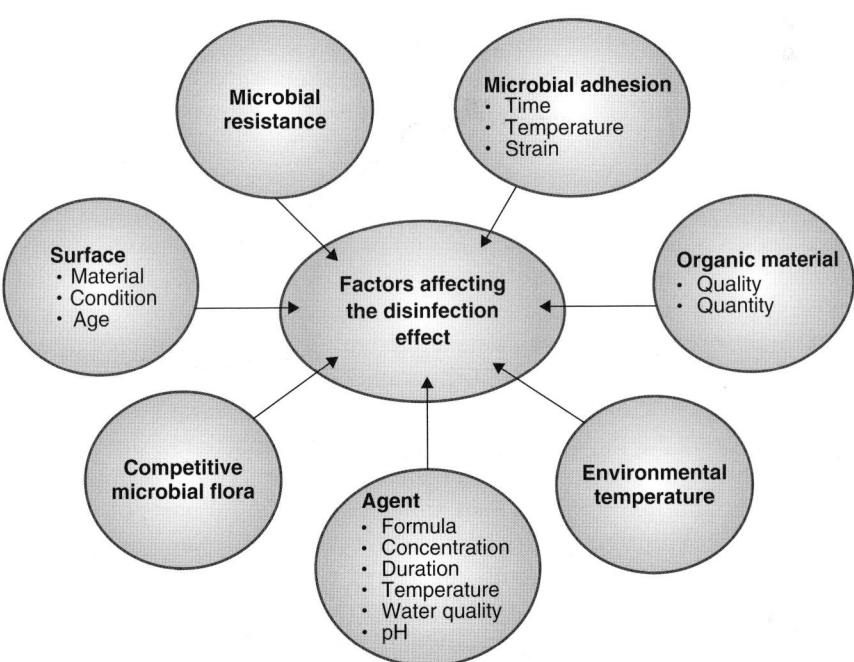

Figure 18.3 Factors affecting the disinfection. Source: Wirtanen, 2002. Reproduced with permission from VTT.

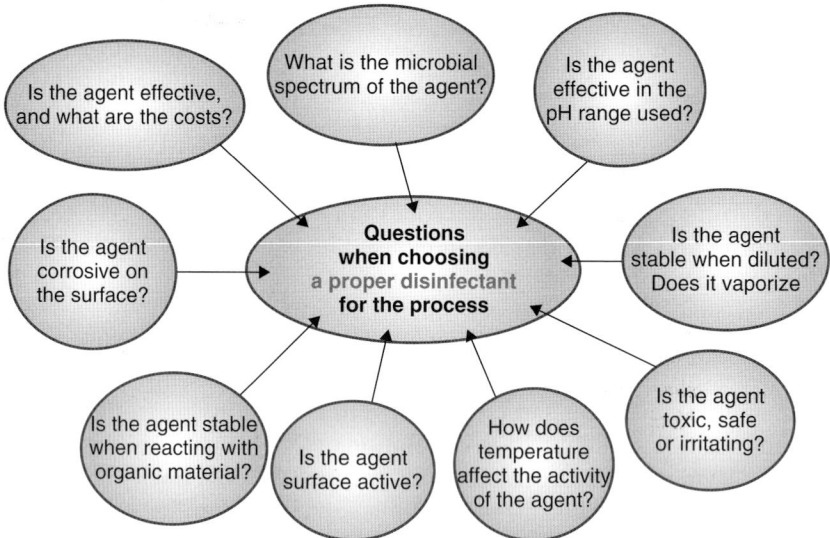

Figure 18.4 Factors affecting the choice of disinfectant.

compound is added to water. Hypochlorous acid is formed; this dissociates further into protons and hypochlorite anions. Stabilized hypochlorites are used when long duration disinfection is required. The range of microbes killed or inhibited by chlorine-based compounds is probably broader than by any other approved sanitizer.

Hydrogen peroxide has been found to be effective in removing biofilms from equipment surfaces. The effect of hydrogen peroxide is based on the production of free radicals, which affect the biofilm matrix. The microbicidal effect of peracetic acid on microbes embedded in biofilms varies and aldehydes do not break the biofilm. The biofilm must be disrupted in some way before agents such as peracetic acid and aldehydes can be used effectively, preferably by cleaning. The effect of ozone treatment has been found to vary depending on the processing circumstances and the microbes tested, for example ozonation proved very effective in treatment of cooling water systems and work wear. Note that the ozonation must be carried out in tightly closed areas because of its toxicity to humans, that is in water system containers and in work wear cupboards with ventilation connected to outdoor air.

Iodophors are used extensively in the food industry. In the disinfection the iodine compound takes part in the oxidation of essential parts of the microbial cells. Like chlorine-containing products, iodophors are active against gram-positive and gram-negative bacteria, yeasts and moulds. Bacterial spores, however, are resistant to iodophors. Iodophors should not be used in food plants producing starch-containing products because iodine forms a purple complex with starch. Quaternary ammonium compounds are used as sanitizers in the food industry because they have good wetting properties. The greatest effect of quaternary ammonium compounds is observed against gram-positive bacteria; many of the gram-negative bacteria, significant in the contamination of slaughtering products, are affected only a little or not at all.

18.6.6 Surface materials with limitations in cleaning and disinfection

The metals, plastics, concrete and paints used in these areas limit the choice of detergent. Detergents can be formulated so that they do not extensively corrode these surface materials but this also means that the cleaning effectiveness is reduced. In most instances the choice of a suitable detergent must be a compromise between the efficiency of the detergent, the need to protect metals, building materials and personnel, the cleaning methods, the frequency of cleaning and economics. The cleaning chemicals can contain: alkaline compounds, for example sodium hydroxide (also called caustic soda), which suspend protein and convert fats into soap; surface active agents, which lower the surface tension, for example wetting, emulsifying and penetrating agents; sequestering agents, which bind calcium and magnesium to prevent the formation of insoluble salts on surfaces cleaned with hard water, for example ethylenediamintetraacetate (EDTA) and nitrile triacetic acid (NTA); inhibitors, which neutralize the corrosive effect of some chemicals, for example silicates; inorganic or organic acids, which remove mineral deposits; and fillers, which convert detergent powders into detergent fluids or detergent fluids into powders.

Materials used in slaughterhouses should be chosen to facilitate cleaning and disinfection, as in other food processing plants. The materials in the equipment should be resistant to both detergents and disinfectants generally in use. Most commonly used materials are stainless steel, plastics and rubber. Note that even stainless steel can corrode when exposed to strong alkaline, acid and chlorine solutions for extended periods. In the case that the equipment contains galvanized iron parts, these parts will corrode when exposed to either alkaline or acidic detergents. Wooden material, if used, is susceptible for deep splits and swelling due to moisture, making the surface difficult to clean. To keep up a good hygiene standard the material in cutting boards should be plastic instead of wood, because theoretically plastic is easier to clean than wood. Cutting boards and chopping blocks must be replaced regularly or at least when needed, because the score lines in the surface can harbour microbes. Plastic material may be damaged when exposed to organic solvents, as may rubber surfaces when exposed to acidic detergents or detergents containing organic solvents. Glass can frost when exposed to alkaline detergents. Moreover, glass is also a safety risk because it can break in the process. Painted surfaces will deteriorate when strong alkaline detergents are used. Alkaline detergents containing metasilicate are recommended for cleaning surfaces of concrete and cement, because these surfaces will deteriorate when exposed to acidic detergents.

18.6.7   Efficacy testing of disinfectants against microbes

It is important to know that the disinfectants used are effective in the environment in question. The efficacy testing of disinfectants can be performed in suspension and on surfaces. The European biocide directive requires that the agents on the market are proven to be effective, and thus the manufactures of cleaning agents and disinfectants are obliged to give information on efficacy tests. Determination of disinfectant efficiency is often performed using suspension tests with free cell and ready-to-use dilutions. These circumstances do not mimic the growth conditions on surfaces, where the agents are required to inactivate microbes in food processing. Several of the efficacy tests launched by the European Committee for Standardization (CEN) are based on the Dutch 555-suspension test protocol for finding out bactericidal, fungicidal and sporicidal activity of the disinfectant. The standard suspension tests have proved sufficiently reliable because the variations of results are within acceptable limits when using adequate amounts of replicates. There can, however, be problems with the repeatability and reproducibility of suspension tests performed with an organic load. In the Dutch 555-suspension test the activity is measured after a challenging time of five minutes. In order to be considered effective the agent must reduce the vegetative cells with at least 5 log units/ml and bacterial spores with at least 1 log unit/ml. The standard test microbes used represent spoilage bacteria, pathogens and bacterial spores of concern in hygiene, that is *Salmonella* Choleraesuis, *P.aeruginosa*, *Staphylococcus aureus*, *Bacillus cereus* (spores) and *Saccharomyces cerevisiae*. The test is carried out using bovine albumin as the organic load. In a modified 555-suspension test the disinfectant is tested against a chosen panel of process contaminants in bovine albumin as organic soil.

It is important not to use concentrations lower than the recommended in-use concentrations of disinfectants because resistant microbial strains can develop when the microbes are not killed properly and they are adapting to low concentrations of disinfectants. Floors, especially, have plenty of sites that may harbour microbes and, therefore, the cleaning order is firstly the floors followed by cleaning of equipment. It is important that floors as well as all other process surfaces are properly dried after the cleaning and disinfection procedure.

In developing the testing procedure for disinfectants on surfaces it is important to identify the major sources of variation in the procedure. Surface tests show how the agent acts in practice and are, therefore, important. Various surface tests have shown that surface-attached cells are more resistant to disinfectant treatment than are cells in suspension. It is thus obvious that the surface tests are even more difficult to perform than suspension tests, because of the carrier material used and the viability of dried cells on the surfaces. Microbes growing or dried on surfaces are not susceptible to disinfectants from all sides as they are in suspensions and, due to the requirement for penetration, disinfectants are used in higher concentrations on surfaces than in suspensions. The goal for reduction of surface-attached bacteria is at least 3 log units/cm^2.

18.6.8 Chemical residue tested with microbes

As mentioned above, cleanliness is essential in food plants. Chemically clean surfaces are surfaces that are free from undesirable chemical residues. However, the total cleanliness of the process facilities prior to initiating production is based mainly on measuring either the microbial load using culturing techniques or the organic load using protein or adenosine triphosphate (ATP) kits. It is possible that these results do not indicate the total cleanliness, because these methods show only that there are no viable microbes or organic soil on the surface. Thus, it is still possible that there are cleaning agent and disinfectant residues left on the surface. This is not allowed but usually no tests are run to ensure this. The toxicity of samples, that is low amounts of detergent and disinfectant residues, can be measured based on the reduction in light output due to interactions between the bioluminescent photobacterial (*Vibrio fischeri*) cells and the toxic compounds in cleaning agents and disinfectants. In addition to the fact that chemical residues must not be present on food contact surfaces, chemical residues may also interrupt detection of microbes. Some chemicals prevent the growth of microbes on nutrient agars but they still do not kill the microbes. In some culture-based detection methods, inactivation of active compounds in cleaning agents and disinfectants is taken care of. The enzymes used in the ATP-method can also be disturbed by chemical residues.

18.6.9 Ultrasound cleaning – an alternative method for utensils and open process lines

The removal of microbial cells by ultrasonication can be achieved by choosing combinations of ultrasonic power, exposure time, temperature and distance from the contaminated surface. Ultrasonic cleaning is a procedure in which ultrasonic energy is transmitted by water. Ultrasonication creates microscopic bubbles to produce a scrubbing action for cleaning. Bubble collapse during cavitation is believed to induce cellular stresses that are sufficient to disturb cell wall structures, leading to cell leakage. The frequencies used in ultrasonic cleaning applications are normally between 20 and 100 kHz. The cleaning of microbes from surfaces is mainly based on the mechanical detaching force caused by cavitation. Localized high temperatures due to cavitation are also likely to contribute to microbial inactivation. The main advantages of ultrasonic cleaning are the effective cleaning of holes, narrow channels, cavities, inner surfaces and other complicated forms through a process of short treatments, which can be performed with a wide range of chemicals. An example of ultrasonic application for an open process in a food plant is shown in Figure 18.5.

18.6.10 Corrective action – power cleaning

It is important to address the economic, efficacy and safety issues of the cleaning procedures in a slaughterhouse, but inadequate cleaning and sanitation

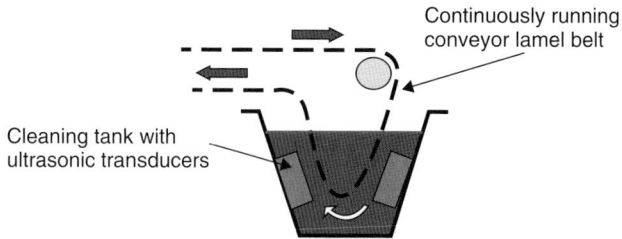

Figure 18.5   Ultrasonic cleaning of a conveyor belt in food factory.

can lead to contaminated products. The cleaning and disinfection may occasionally need to be improved, for example when biofilm formation is found on process surfaces, when quality and safety of the product must be enhanced or when pathogens have been detected on surfaces in the process. Moreover, increased resistance to chemicals will also occur in surface-attached microbes due to changes in the cell physiology. In this type of case routine cleaning and disinfection procedures are not sufficient to kill the microbes. The remaining organic matrix must be removed because it provides excellent opportunity for regrowth of microbes. A power cleaning procedure with enhanced mechanical and chemical forces, for example through manual scrubbing and by using strong agents based on other chemical reactions, and prolonged time at an appropriate temperature can improve process hygiene.

18.6.11   Controlling the cleaning results

The results of the cleaning and disinfection programmes used can be controlled by visual, chemical and microbial methods or by using a combination of these measures (Table 18.1). When visual control is used, it must be done in a systematic way and the personnel performing the control must be well trained. This type of evaluation gives information on the efficiency of the cleaning and disinfection programme. The visual control shows at which points cleaning must be improved. The detection and enumeration of indicator organisms is used to assess the efficacy of sanitation procedures. A collaborative study using Hygicult®dipslides, contact agar dishes and swabbing in measuring total counts of aerobic bacteria and enterobacteria has shown that about 15–20% of the theoretical yield of the artificially soiled stainless steel surfaces was obtained using these three methods. Based on these results, it can be concluded that the methods presented do not differ in practical terms either in yield or in precision.

The microbial analysis will reveal the points in which microbial hygiene is appropriate and where it must be improved. The risk of the presence of pathogens is elevated in cases where *Enterobacteriaceae* is found because these microbes are indicators for low hygiene and their appearance proves that the environment is suitable for growth of pathogens. *E. coli* counts can

Table 18.1 Practical methods for analysing microbial residues, including biofilms, on food contact surfaces.

Method	Application with comments
Contact plate (culturing)	Used for hygiene testing *Advantage*: identification, selective growth media, easy to use *Limitation*: sensitivity, time consuming
Swab method (culturing)	Used for hygiene testing *Advantage*: identification, selective growth media *Limitation*: sensitivity, time consuming
Impedance	Used for hygiene testing, informative also in laboratory studies *Advantage*: more rapid than cultivation, selective growth media *Limitation*: sensitivity
ATP	Used for hygiene testing *Advantage*: on-site ATP-based biofilm detectors *Limitation*: sensitivity, large scatter when analysing multilayer biofilms due to different metabolic status of cell layers
Protein test	Used for hygiene testing *Advantage*: rapid, easy to use *Limitation*: only protein-based dirt

Source: Adapted from Wirtanen, *et al.*, 1999.

be used as an indirect measure of faecal contamination. Hygiene assessment can also include checking of the amount of surface-attached soil, including protein, polysaccharide, other organic and inorganic residues, biofilm, dead and/or living microbes in general, or specific pathogens and other harmful microbes. In assessing hygiene, it is essential to detect the microbes remaining on surfaces after cleaning, even though the results are available after the products have been manufactured. Hygiene assessment also helps with tracing contamination sources and in optimizing cleaning systems.

Preventive Hazard Analysis and Critical Control Points (HACCP)-based food safety management systems require that hygiene monitoring results are available rapidly. For example, ATP and protein detection kits can provide a real time estimation of the overall cleaning efficacy. However, these methods are unable to detect the presence of low numbers of bacteria on surfaces. The ATP method can differentiate between bacterial and other organic soil if non-bacterial ATP is removed enzymatically before the hygiene assay is measured. However, hygiene monitoring in the food industry normally includes cultivation-based methods to enable detection of low amounts of and/or specific microbes present. The methods used must be tested for each case, since the methods may not be sensitive enough in all applications. A visual overall hygiene assessment can be done using a UV-lamp, because organic soil itself is fluorescent. For some product soils, residues can be observed visually by wiping the surface with white gauze or paper tissues. Chemical residues remaining after cleaning and disinfection can be detected using a rapid bioluminescence method (Section 18.6.8). The light output decreases when there are chemical residues in the sample(s). The goal to achieve a clean food plant must also be desired by the plant management, which has to invest the necessary time and money to accomplish it.

18.6.12   Interpreting the microbial results – limits for microbes on cleaned surfaces

Interpretation of the results from hygiene monitoring is often carried out case by case, since there are many factors that affect an acceptable level of cleanliness. The acceptable level depends on the surfaces sampled. The surfaces in contact with ready-to-eat food products must be much cleaner than surfaces with products that will undergo heat-treatment or surfaces not in direct contact with foods. Special attention should also be paid to the surfaces next to surfaces with food contact, since there is a high risk of spreading contaminants to food products. The shelf life of the (processed) product also affects to the desired cleanliness level. The threshold limit for a clean surface must be based upon a perception of a specific risk and a decided, acceptable level. Alternatively, microbial yields obtained from a surface on which a well-designed cleaning programme has been applied can be used as guideline values. The available guidelines and standards for aerobic colony counts on clean surfaces vary widely from <1 to 80 cfu/cm^2. Comprehensive studies have indicated that in many cases levels of <2.5 cfu/cm^2 on a clean surface are attainable.

18.6.13   Optimization of cleaning procedures

Cost effectiveness is becoming more and more important in the food industry. This means that the sanitation procedure is run as infrequently as possible, in the shortest possible time, with low chemical, energy and labour costs, producing as little waste as possible and with no damage to the equipment. Nevertheless, the cleaning programmes should still be performed so that the surfaces are thoroughly cleaned, so securing meat safety.

In the food industry there is a trend towards longer production runs with short intervals for sanitation. Optimization of cleaning procedures offers significant savings in many food factories, since the cleaning procedures in many cases are based on empirical knowledge and old habits. Thus, the cleaning procedure can be oversized, which leads to excessive use of time, chemicals, water or energy, which is costly for the company. An excessive cleaning procedure also increases waste water volume and excessive use of cleaning chemicals affects the surface quality. On the other hand, inadequate cleaning increases the risk of both microbial and chemical contamination of food products, which may harm the company image and cause economic problems.

Simulation can be used in improving the hygienic design and process flows. It provides useful and visual information about parameters related to cleaning liquid patterns. Simulation is a potential tool to be used for cleanability studies of complicated equipment, because equipment is getting bigger all the time and it makes practical trials hard to perform. It is a quicker and cheaper way to gain the knowledge needed for improvements than relying solely on laborious pilot scale studies. Comparisons with information obtained from cases in either pilot or process scale studies are needed in order to validate the simulations, because computer simulations must be related to empirical cleanability knowledge. Optimization of cleaning procedure depends on knowledge about whole process system.

18.7   Improved cleaning possibilities through hygienic design

Hygienic requirements should be adopted at the initial stage in developing process equipment and components, because upgrading of already existing designs to meet hygienic requirements is often both expensive and unsuccessful. Correct design of process equipment has a tremendous impact on diminishing the risks of food contamination, because poor equipment design complicates effective cleaning. With a good hygienic design the lifetime of the equipment will be increased. The maintenance and the manufacturing costs will also be reduced. The hygienic design of process equipment and components should be based on a sound combination of process engineering, mechanical engineering and process microbiology. The choice of materials and surface treatments, for example grinding and polishing, are important factors in inhibiting the formation of biofilm and in promoting the cleanability of surfaces. The process equipment is easier to clean when the surface materials are smooth and in good condition. Dead ends, corners, cracks, crevices, gaskets, valves and joints are vulnerable points for microbial growth and biofilm accumulation. Poorly designed sampling valves can destroy an entire process or give rise to incorrect information due to biofilm effects at the measuring points.

There is a need for criteria and procedures for testing, assessment and certification of equipment through which conformity with predefined requirements can be shown. The European Hygienic Engineering and Design Group (EHEDG) has developed and published various guidelines on the hygienic design of equipment, including one on system integration. The latter EHEDG guideline has the task of linking and supporting current guidelines on hygienic design regarding specific equipment and hygienic tests. It can be viewed as both vertical and horizontal guidelines. The most fundamental EHEDG guidelines in hygienic integration are: Document 8 'Hygienic equipment design criteria', Document 10 'Hygienic design of closed equipment for the processing of liquid food' and Document 13 'Hygienic design of equipment for open processing'.

18.8   Concluding remarks

The factors affecting the outcome of cleaning procedures are mechanical and chemical forces as well as temperature and time. Assessment of process hygiene should be based on observation of technical solutions from a hygienic design point of view, on soil accumulation in processing lines or on detecting microbes. A clean surface has to be free of microbes, physical particles and chemicals. New cleaning results can be compared to results from long-term routine hygiene control. The interpretation of results obtained in liquid flow and on surfaces needs to be based on holistic risk assessment, since official recommendations are scarce. The basic rule in hygienic design is to use simple structure that is stated in common guidelines. The procedures for microbial sampling from surfaces are at the moment the weakest link in the detection of microbes on process surfaces after cleaning. All available methods have limitations and the suitability of the method strongly depends on the type of microbes and the microbial load present. Visual observation of the cleaning results is an important practical method, which can be improved by using UV light.

Several standards exist for evaluating disinfectant efficacy. These suspension tests function as screening methods, because testing based on microbes in suspensions does not reveal the true effect of the disinfectants on microbes attached to a surface. Choosing a suitable disinfectant should be based on efficacy tests modified with process isolates in suspension and on surfaces, because different strains can have different tolerance against disinfectants. Note that the concentrations of agents used as well as the exposure times are significant factors in disinfection. The tolerance of microbes towards disinfectants can increase over time and, therefore, rotation of disinfectants based on different reactions can be recommended in certain cleaning procedures. However, many microbes are cross-resistant, that is resistant to several mode of action of disinfectants, and thus rotation of disinfectants must be based on precise studies such that cause multiple resistant microbe strains are not created.

Hygienic requirements must be included in the design of processes and the hygienic integration of process equipment in process lines and in performing the cleaning procedure. Many suppliers can produce customized systems that have an excellent cleaning efficiency. The challenge is still to optimize the design of cleaning systems with respect to efficiency and economy. Minimizing cost implies that consumption of water and cleaning agents as well as the cleaning time must be reduced. Producers of cleaning agents can today deliver effective cleaning programmes applicable in most practical situations. On the other hand, the tendency is to develop new environmentally less harmful cleaning agents.

Literature and further reading

Lelieveld, H.L.M., Mostert, M.A. and Holah, J. (eds). 2005. *Handbook of Hygiene Control in the Food Industry*. Woodhead Publishing Ltd, Cambridge, UK.

Wirtanen, G. and Salo, S. 2003. Disinfection in food processing – efficacy testing of disinfectants. *Rev Environ Sci Biotechnol*, **2**, 293–306.

Wirtanen, G. and Salo, S. 2009. Risk Management by Hygienic Design and Efficient Sanitation Programs. VTT Symposium 261. VTT, Espoo, Finland. http://www.vtt.fi/inf/pdf/symposiums/2009/S261.pdf (last accessed 15 February 2014).

Wirtanen, G. and Salo, S. 2012. Microbiological limits used for various types of food process surfaces based on case study evaluations. *New Food*, **15** (1) (Suppl), 9–13.

19
Pest Control

Mirko Rossi[1] and Francesco Andreucci[2]

[1]*Department of Food Hygiene and Environmental Health,
Faculty of Veterinary Medicine, University of Helsinki, Helsinki, Finland*

[2]*AULSS 18, Rovigo, Italy*

19.1   Scope

This chapter describes the main pest species relevant in slaughterhouse environment and the different strategies that can be applied for controlling the infestations. It also describes methodologies that may be used in routine monitoring and periodic inspection related to pest control.

19.2   Introduction

Pest control refers to the prevention, monitoring and eradication of certain hazardous animal species in slaughterhouses. It is fundamental that each slaughterhouse implements a regular control plan, which needs to include inspection of the premises for detecting the presence of pests as well as a list of actions for the eradication of the infestation(s).

19.3   Control plan

The hazardous nature of pest animal species is due to the facts that they are incompatible with the most basic concepts of environmental hygiene, can behave as sources of contamination and may be a physical threat for carcass

Meat Inspection and Control in the Slaughterhouse, First Edition.
Edited by Thimjos Ninios, Janne Lundén, Hannu Korkeala and Maria Fredriksson-Ahomaa.
© 2014 John Wiley & Sons, Ltd. Published 2014 by John Wiley & Sons, Ltd.

and meat quality as well as structure. Therefore, pest control in the meat processing plant is an essential prerequisite of the food safety programme. Pests can occur for several reasons; they are even more relevant if the facility provides a breeding site and food supply. It is challenging to have a totally secure plant in spite of an optimal design that impedes the access of pests. Therefore, it is essential to establish a specific regular pest control plan to prevent adulteration or contamination of the products. All meat processing facilities, regardless of size, must have a written pest control programme; this should be part of the plant's monitoring system, including the food manufacturer's sanitation programme, food safety and good manufacturing practices (GMPs). Depending on the level of control required, specific requirements are available and they dictate strategies and products that can be used. Although these requirements may differ from country to country, all are based on a simple concept: precautions must be adopted to keep pest control treatments from compromising food safety and quality.

In order to respect the requirements of a food plant's environment, pests should be controlled following the principles of integrated pest management (IPM). An IPM programme considers the management of pest populations as a process not as a one-time event. It combines control practices to avoid and solve pest problems by considering all the information accounting for multiple points, and all available preventative and curative options. In general terms the goal of IPM is to provide a safe, effective and economical outcome by addressing the underlying causes of pest infestations (e.g. access to food, water and shelter) before considering the use of pesticides, which may threaten food safety. The control plans include several phases, which can be summarized in the following steps:

- detection and identification of pests and inspection of the facility;
- application of adequate control techniques ;
- implementation of a specific regular monitoring programme.

19.4   Identification of the pest and inspection

Since different pest animals have different lifestyles, the primary action that should be taken into consideration in order to set up a correct plan is to identify the problematic species that infest the slaughterhouse. Although there are several pest animals, which can vary depending on the geographical area, this chapter considers only the most common ones. These include:

- flying arthropods (i.e. flies, order *Diptera*);
- crawling arthropods (i.e. cockroaches and ants, orders *Blattaria* and *Hymenoptera*);
- rodents (i.e. rat and mice);
- birds (i.e. pigeons, sparrows, starlings, gulls);
- others (i.e. dogs, cats, racoons, wild animals).

Certain other animal species, including snakes, wasps, bats, spiders, racoons and other small vertebrates, can also be considered pests. In addition, it is quite frequent that domestic animals (e.g. dogs and cats) usually move around the premises, increasing the risk of contamination of the slaughterhouse environment. Nevertheless, the methods available for the control of these pests are similar to those applied in the management of the most common ones. A detailed list of the most relevant species considered as pests is given in Table 19.1, along with inspection criteria useful for detection.

The basis of an effective IPM programme is a schedule of regular and detailed inspections of both the interior and exterior of the facility. Pest animals will behave differently during different times of the year, according to their lifestyles, and therefore inspections need to be undertaken at frequent intervals. For slaughterhouses weekly inspections are common and in some plants occur even more frequently. The inspection, by its nature, relies on the ability of the inspector (an employee of the slaughterhouse or a specialist from an external company), thus it should be performed by well-trained personal. In the absence of a specialist or due to general lack of staff, external companies may be contacted to carry out the inspections and, in general, to manage and implement the control plan. The inspections should focus on critical areas of the plant where pests are most expected to appear, such as receiving docks, storage areas, employee break rooms, sites of recent ingredient spills and so on. Once the pests are recognized, it is then necessary to identify the critical points leading to such a condition in the facility. There are some hints that can help to ascertain the best control technique to adopt. Indications of food debris, moisture accumulation, odours, wall/ floor/ceiling cracks and broken windows are just some aspects that should be taken in consideration by the inspector. Any potential entry points, food and water sources or harbourage zones that might encourage pest problems should be identified. The inspector must also verify that, in general, the facilities, the equipment and procedures conform to the slaughterhouse's own-check plan or, alternatively, to the Sanitation Standard Operating Procedures (SSOP) or Sanitation Performance Standards (SPS) plan. In addition, the inspector should also monitor employees to see if they are following the slaughterhouse's Hazard Analysis and Critical Control Point (HACCP) plan. In fact, good hygiene practices are essential to avoid creating an environment conducive to pests. The inspector will produce reports that are critical for deciding which control techniques should be applied and assist in the implementation of an efficient monitoring programme.

19.5 Control techniques

The control techniques are based on the principles of exclusion, restriction and destruction. Exclusion refers to the methods adopted to prevent pest entry into a building. Restriction refers to the methods used to create unfavourable

Table 19.1   List of the most common animal pests, associated hazard and respective inspection criteria.

Group	Common name	Scientific name	Hazard	To inspect /To search
Flying arthro-pods	Common house fly	*Musca domestica*	Potential vectors of pathogens, such as *Salmonella* spp., *Listeria monocytogenes*, *Campylobacter* spp., *Escherichia coli* 0:157 and the genus *Cronobacter* (formerly *Enterobacter sakazakii*) and helminths.	• All drainage channels and interceptor wells. • Areas where debris can accumulate (e.g. under refrigerators etc.). • Under or behind equipment, pallets, at the bottom of the wells and bins, and areas of elimination of waste.
	Lesser house fly	*Fannia canicularis*	Potential vectors of pathogens (similar to house fly).	
	Blow flies	*Calliphora* spp.	They are attracted by rotting animal remains on which they lay their eggs. Potential vectors of pathogens (similar to house fly).	
	Flesh fly	*Sarcophaga carnaria*	Flesh flies exploit decaying organic matter for larval feeding sites (e.g. rotten meat) and they can also utilise stored meat as a larval deposition site. Potential vectors of pathogens (similar to house fly).	
Crawling arthro-pods	Common cockroach	*Blatta orientalis*	Pollute the environment and food with their faeces, regurgitated food and residues of their exoskeleton. They are passive carriers of different microorganisms and helminths.	• Tracks and signs on the surfaces. • Presence of brown stripes and irregular droppings. • At night with the help of sticky traps (or glue traps).
	German cockroach	*Blattella germanica*		
	American cockroach	*Periplaneta americana*		

Table 19.1 (*Continued*)

Group	Common name	Scientific name	Hazard	To inspect /To search
	Ants	*Formicidae*	The presence of ants can have an impact on the safety and marketability of food, if found in the areas of production.	• Around door thresholds, sinks and sideboards. • Under cutting board and under the floor.
Rodents	House mouse	*Mus musculus*	Mice and rats damage food and property. They can spread pathogens.	• Marks: faecal residues, gnaw marks, droppings, grease marks, foot prints and tail marks.
	Black rat	*Rattus rattus*		• Use of Kaolin fluorescent dust for detection of rodent marks.
	Brown rat	*Rattus norvegicus*		• Damage to food, food packaging and property. • Holes in ground outside property. • Vents or ventilation stacks and windows. • Walls holes that provide access.
Birds	Pigeons, sparrows, starlings, gulls etc.		They are vectors of several pathogens. In addition, the corrosive effects of bird droppings can cause irreversible damage. These acidic droppings eat away at paint, concrete and metal, and can eventually cause structural failure.	• Vegetation or other potential attractants around the buildings. • Windows and ceilings.

conditions for pests. Destruction covers all the physical and chemical methods applied for controlling the pest population.

To minimize the likelihood of infestation and, thereby, limit the need for pest control measures, such as the use of pesticides, good hygiene and sanitation, stock management and exclusion practices must be applied. A detailed list of preventive measures should be taken into consideration in slaughterhouse is shown in Table 19.2. The food processing facility and its equipment should be designed and purchased with consideration of the IPM programme and regular and preventative building maintenance should be performed. The facility should be operated to minimize the opportunity for pests to become established by: reducing or eliminating infestations in incoming materials; cleaning thoroughly, regularly and frequently, premises and equipment (with written procedures identifying cleaning methods); keeping dust, flour, insects or other material from reaching areas that are inaccessible for cleaning, treating and inspecting; and removing potential food sources and pest harbourages.

Staff should be adequately trained, informed and made aware of the importance of using behaviour that promote hygiene. In addition to general preventive measures, a variety of control actions can be targeted at pests. Listed in Table 19.3 are the most common physical and chemical control techniques that could be applied in the food industry in general, including a slaughterhouse.

19.6   Monitoring programme

A monitoring programme (MP) is a planned set of operations required to detect and report the level of infestation present in the slaughterhouse. In order to be effective, the monitoring programme must be planned in advance, carried out systematically and tested at appropriate intervals, or whenever there are modified objectives. An effective monitoring programme minimizes the amount of pesticides applied by focusing treatments in specific harbourages with active signs of infestation, resulting in a reduction in risk of contamination of the products. The monitoring programme should include the definition of the targets and actions such as calendars for placing and checking traps, physical inspection of the plant and process equipment. During monitoring, detailed collection of any data capture and technical comments about the lack of sanitation should be performed.

Trapping is an efficient method of monitoring and should be correctly implemented in the monitoring programme. Traps should be located throughout the facility (for example besides the food waste containers or besides the backdoor etc.) and their positions should be clearly registered on the map of the plant. For correct placement of the traps, the operator should take in account the biology of the pest to be controlled (different positions if the traps are for rodents or for flies). The traps should be checked on a regular basis and replaced on a regular interval. Insects could be monitored using pheromone traps, adhesive traps, bait bags or pitfall traps, while for

Table 19.2 Examples of preventive measures to be taken in the plant to decrease the likelihood of pest infestation.

- To avoid standing water, the areas outside the plant should be paved and adequately drained.
- The external areas, for a range of 8–10 m from the walls, must be kept clean and free from objects and obsolete materials and kept in good, clean condition.
- For more distant areas that are not paved, there must be an effective control of vegetation.
- The walls must be as smooth as possible and the fixtures installed on the outer edge with the sill inclined by 45° towards the inside, or on the inner edge of the wall but with the sill inclined outwards to impede climbing of rodents and stop birds landing.
- Materials, construction and layout of the building and equipment should avoid creating surfaces and cavities where dust and food material can collect.
- Floors, walls, ceilings and equipment surfaces must be made from a material that is appropriate to the use of the area and is easy to clean.
- All windows and other openings used for ventilation (including ports), should be equipped with anti-insect nets.
- Service doors, when it is possible, should be automatic in order to reduce the time of opening; all other doors must remain closed, including those of communication between the departments.
- Door jambs and shutters must be intact and tight.
- All manholes of underground facilities should be closed.
- All cable ducts, which represent corridors for rodents and receptacles for insects, must be closed.
- All exhaust pipes must have siphons to prevent climbing of rodents and they should be periodically sanitized to prevent the proliferation of insects.
- All the holes must be closed with mortar or metal plates.
- Leaks in pipes should be avoided, and when they occur quickly identified and repaired.
- Waste must be evacuated from the areas at least once daily and must be stored in appropriate bins. The bins should be located far from the premises and placed in a paved area to facilitate cleaning.
- To limit the possibility of infestation and to allow periodic cleaning and inspections, it is recommended that facilities and goods have to be placed 50 cm from walls and lifted off the ground;
- The pallets (especially wood pallets) should not be stored outside but inside in adequate dry areas; regular sanitation is required.
- Departments and machinery temporarily out of production should be kept clean and inspected regularly.
- The incoming material should be inspected before unloading and, if it is possible, should be received in a building that is separate from the main processing facility, or in an isolated area of the facility. If necessary, materials should be treated before entering the main process flow.
- Employees must be trained with an awareness of the IPM programme and its elements, particularly sanitation.
- Written procedures should be available for all aspects of an IPM programme, including cleaning, monitoring, identifying and correcting of problems, and treatments.

Table 19.3   List of the most common physical and chemical control treatments.

Pest	Method	Characteristics	Risk	Precaution	Remarks
Arthropods	Electric Fly Control Units	• UV light lamps (attractants) should be replaced every 6–12 months (approximately 6000–8000 working hours). • Avoid placement of the UV light lamps on top of the food processing area, as the remains of the pests will be aerosolized (in these areas, use adhesive film traps instead).	• None if used inside the slaughterhouse.	• Placement and type of unit should be carefully considered: UV light lamps should not be placed near normal light bulbs – as it diminishes the attractiveness.	• Suitable for flying arthropods. • Should not be used outside the facility.
	Pheromone traps	• Avoid placement of the units near door openings, as they can attract insects into the facility. • The baits must be replaced regularly (intervals of 2 or 12 weeks) depending on the charge and on the lure.	• None if used inside the abattoir.	• Carefully chose the trap.	• Suitable for flying (funnel traps) and crawling arthropods (glue traps).
	Pesticide spray	• There are a number of different ways to formulate insecticides and acaricides. • Chemical control of arthropods involves the use of insecticides or acaricides. • These are chemicals that kill insects and mites or prevent their development, thus preventing the production of the next generation.	• Possible contamination of goods. • Depends on the product.	• Only when abattoir is non-operational, and followed by strict cleaning measures.	• Most suitable for crawling arthropods.

		Description	Risks	Notes	Suitability
	Pesticide baiting		• None.	• Application should be effective to avoid acquired resistance.	• Most suitable for crawling arthropods. • High efficiency for certain applications.
	Pesticide Fumigation		• Locations with unwished deaths.	• Only when abattoir is non-operational, and followed by strict cleaning measures.	• Suitable for crawling and flying arthropods. • Not applied outside the facility.
Rodents	Traps	• Spring traps (kill the rodent), live traps (collect the rodent) and sticky or glue board traps (collect the pest). • Can be single or multicatch. • Efficiency will depend on the trap model.	• The presence of a bait attractant may pose a contamination risk. • If used outside, risk of capture of non-target species.	• Frequent control to remove the dead rodents in the case of spring traps.	• Extremely useful in areas where it is not possible to use rodenticides. • Live traps are particularly useful when there is a risk of catching non-target species with other methods.
	Baits	• A non-toxic pre-bait can be used and later replaced by a rodenticide bait – to attract 'shy' rodents. • The bait can only be consumed within the dispenser and cannot be dispersed.	• Environmental contamination and dispersion. • Can be ingested by non-target species.	• Dispensers should be specific for the rodenticide used.	• Dispensers of sturdy plastic bait in non-productive (outside areas, waste areas, loading/unloading, warehouses etc.).

(continued overleaf)

Table 19.3 (*Continued*)

Pest	Method	Characteristics	Risk	Precaution	Remarks
		• The rodenticide bait can be either from chronic (usually anti-coagulants) or acute categories.			
Birds	Traps	• Made of wire mesh.	• None.	• None.	• These are fundamental for release species that are not pests.
	Anti-perching systems	• Consist of sprung wire or spike systems and are designed to prevent birds from alighting on ledges or similar surfaces. • Electric wire systems are also available.	• None.	• None.	–

rodents non-toxic bait clocks and the application of tracking dust could be used (Table 19.3). Traps and attractants should be kept in such a condition that they operate with maximum efficiency; surface adhesives should be restored or pheromones and attractants replaced in accordance with specified intervals. Because the trap provides a relative measure of pest population, the trends in trap catches should be considered for a more accurate interpretation of the data. Low or no trap catch followed by a sudden increase in numbers is an indication of a developing infestation.

A monitoring programme must be a continuing activity, not a periodic exercise, and requires accurate record keeping. The documentation relating to the monitoring programme should include the following elements: the pests monitored (species and number of individuals observed) within the environment and the description of each intended station; the number and type of stations installed and the type of attractant used; the date of placement of the trap in the environment and the replacement date of the attractant and data collected during each inspection regarding bait consumption; a description of the non-compliance; the corrective action taken to manage the non-compliance in relation to the implementation of the monitoring programme.

Literature and further reading

CIEH (Chartered Institute of Environmental Health). 2009. Pest Control Procedures in the Food Industry. Chartered Institute of Environmental Health, London.

Keener, K.M. 2012. Safe food guidelines for small meat and poultry processors: a Pest Control Program. Purdue Extension FS-22-W, Purdue University, West Lafayette, IN, USA.

Pava-Ripoll, M., Pearson, R.E., Miller, A.K. and Ziobro, G.C. 2012. Prevalence and relative risk of *Cronobacter* spp., *Salmonella* spp., and Listeria monocytogenes associated with the body surfaces and guts of individual filth flies. *Appl Environ Microbiol*, **78**, 7891–7902.

Bennett, G.W., Owens, J.M. Corrigan, R.M. (eds). 2010. *Truman's Scientific Guide to Pest Management Operations*, 7th edn. Questex Media Group LLC & Purdue University Press, West Lafayette, IN.

20
Working Hygiene

Marjatta Rahkio

Finnish Milk Hygiene Association, Helsinki, Finland

20.1 Scope

This chapter deals with the hygienic working and motivation of slaughter-house workers. The knowledge of hygiene, hygiene facilities and social support has been found to motivate slaughterhouse workers to practice good hygiene. Studies on the effect of hygiene facilities on hygienic working practices in slaughterhouses are reviewed.

20.2 Introduction

There are three hurdles that harmful microbes must overcome before they cause problems in the food items. Firstly, these microbes must enter the food chain; secondly, they must multiply in the food chain; and, thirdly, the risk management operations for their destruction do not work.

Hygiene practices are preventative measures that effect following steps:

- restrict and prevent the access of microbes into the food chain;
- limit the multiplying of microbes;
- ensure that procedures for destruction of microbes are functioning.

Hygiene practices reduce the spreading of spoilage and pathogenic bacteria and, thereby, ensure that the product is appropriate for human consumption. Hygienic working methods are part of hygiene practices. There used to be detailed requirements on hygienic working in legislation, particularly for

Meat Inspection and Control in the Slaughterhouse, First Edition.
Edited by Thimjos Ninios, Janne Lundén, Hannu Korkeala and Maria Fredriksson-Ahomaa.
© 2014 John Wiley & Sons, Ltd. Published 2014 by John Wiley & Sons, Ltd.

meat chain and slaughtering work. However, nowadays the legislation is more general in nature and food business operators are obliged to plan and tailor necessary and specific instructions. The requirement to consider hygiene practices is incorporated in quality control and food safety management standards, too. Hygienic working methods are intended to be followed through the meat chain starting from the primary level.

The slaughter animals, especially their skin, hide and intestines, are the main sources for many harmful bacteria entering into the food chain. The meat of healthy animals is sterile. Avoiding contact between the hide and the meat surface and between the contents of the intestine and the meat surface are crucial in the slaughter process. The slaughterhouse worker has a key role in this preventative work. The common way for microbes to spread from the hide and intestine to the meat is cross-contamination via the hands or tools of the workers. The workers themselves can also be the direct source of bacteria and viruses; however, the workers' role in cross-contamination is more relevant.

The effective way to prevent bacteria from multiplying is to chill carcasses immediately after slaughtering. Microbes can be killed by heating and processing, which can be done later in the food chain after slaughtering. Proper disinfection of the working tools kills microbes on the tool surfaces and diminishes cross-contamination. Decontamination of carcasses by lactic acid in a slaughter line is possible and allowed in some countries and for certain species of carcasses.

20.3   Hygienic slaughtering

Before animals enter the slaughterhouse, there are many possible preventive measures that can be carried out at the primary level in order to decrease the incidence of harmful microbes, such as *Salmonella* and STEC (Shiga toxin-producing *Escherichia coli*), in slaughter animals. However, the hide as well as the intestines of the slaughter animal must always be considered as a possible source of harmful bacteria.

Since the hide and intestines are the main sources of bacteria, the dehiding and evisceration are the critical points at slaughter. Although the animals are visually clean when slaughtered, there are bacteria on the hide and hair. When skinning, direct contact between the hide and meat should be avoided or minimized. However, the knife will always be contaminated when it cuts through the hide and if the same knife is used to free the skin, the microbes are transferred to the carcass surface via the contaminated knife.

The spreading of bacteria from the intestine to the meat surface can be prevented by avoiding contact between the content of the intestine and the meat surface. Thus, the most crucial aspect during slaughter is to avoid damaging the intestine. Although the evisceration is done as carefully as possible and the intestine removed from the carcass cavity as a whole, for example by sealing off the rectum with a plastic bag and tying the oesophagus, the hands and knives of the slaughterhouse worker are forced into some kind of contact with the content of intestine and its microbes.

From these examples it can be seen that the hands, knives and tools of slaughterhouse workers working in critical points are possible, and even probable, cross-contaminators between the clean and unclean parts of the individual carcass and between carcasses. In hygienic slaughter, this contamination is minimized by washing hands and tools and by disinfection of the tools. Since the meat surface of an individual carcass can be contaminated from intestinal content or the hide, this contaminated carcass surface can further contaminate other carcasses; this cross-contamination takes place via the hands and tools of the workers. This is why the washing of hands and washing and disinfection of tools between carcasses is also so important in those non-critical working points, where the workers handle only the meat surface of the carcass.

The role of working hygiene is higher in slaughter lines that are less automatized and have numerous workers engaged in the actual slaughter work. In more automatized lines, the workers do not necessarily have as many contacts with carcasses and their main work is to control the machinery. However, hygienic working methods and hygienic behaviour must not be forgotten in these slaughterhouses either. Washing hands and tools is only one part of working hygiene. Working hygiene includes all the guidelines and orders to be followed in order to get carcasses that are as clean as possible. The separation of hygienic levels and movements of workers between the different hygienic levels is extremely important in bigger establishments.

Slaughterhouses can have instructions for workers on how to get to the working place and slaughter line from the dressing room and how to handle working clothes. Following these instructions is part of hygienic working. Even the instructions dealing with chilling procedures at slaughterhouses can be considered as part of hygienic working.

20.4 Motivation of workers

Since the slaughterhouse workers and their working habits do have a key role in the hygiene of meat, their working motivation is important. Hygiene practices used to be considered simply a matter of motivation, in which very little can be changed. When analysing motivation of workers profoundly, three factors were found to have an effect on hygiene practices. These main factors are:

- actual knowledge of the relationship between hygiene and safety and hygienic working methods is given to workers and this education is continuous;
- facilities for hygienic working are available and reachable;
- social support.

The first two of these factors, knowledge and facilities, can be organized by management, the third one is more complex one. Social support means that the behaviour of other workers and superiors is exemplary. They practice hygiene themselves and their attitude to hygiene practice is positive. The atmosphere of the working place is hygiene orientated. When there is no social support, the behaviour of others is dismissive. They do not practice hygiene and they can make negative remarks on hygienic working. The atmosphere is non-hygiene orientated.

The obvious result of washing hands is that hands are clean and represent a reduced microbiological risk. In addition to this hygiene-related result, there are many other consequences. When the positive consequences due to a certain action are numerous compared to negative consequences, an individual is more anxious to take certain action.

The possible positive consequences of hand washing are:

- hands become clean;
- dangerous bacteria are removed;
- reduced transmission of bacteria;
- reduced chance of skin complaints;
- active contribution to a good product;
- superiors are satisfied;
- colleagues are satisfied.

The possible negative consequences of hand washing are:

- inability to keep up with the working pace;
- wet and irritated hands;
- negative or no reaction on the part of the superior;
- negative or no reaction from colleagues;
- getting the feeling that one is behaving differently from other workers.

Although the number of positive consequences is more numerous, the negative consequences are more immediate. In addition to this, the positive consequences are partly dependent on the knowledge of the individual worker and partly dependent on the behaviour of superiors and other workers.

A list of positive and negative consequences from following instructions can be made for other procedures as well. The example of factors influencing the decision whether to change clothes or not when visiting outdoor warehouse is used here.

The possible positive consequences from following the instructions to change clothes when visiting an outdoor warehouse are:

- indirect contribution to a good product;
- superiors are satisfied;
- colleagues are satisfied.

The possible negative consequences from changing clothes are:

- it is time consuming;
- it requires effort;
- negative or no reactions on the part of superiors;

- negative or no reactions from colleagues;
- getting the feeling that one is behaving differently than other workers.

The number of negative consequences is more numerous. Since there are no immediate positive consequences, the importance of the behaviour of superiors and colleagues is even more important in this example than in the example of factors influencing on the decision to wash or not to wash ones hands.

20.5   Hygiene practice at the slaughter line

20.5.1   Methods

Hygienic working of 73 slaughter line workers in five Finnish slaughterhouses was observed. The study was made in non-automated slaughterhouses, which all had both a beef and a pork line. The workers were observed over several periods of about 15 minutes each. At least 10 carcasses were handled during each observation period. The workers were observed by the same person. The hygienic practice of the workers was estimated as the frequency of washing hands, knives and other working tools, and the frequency of disinfection of knives and other working tools with hot water. These practices were recorded using a three-point scale: (1) after every carcass, (2) after every other carcass and (3) more seldom than after every other carcass. Incorrect slaughter technique, such as damaging the intestinal canal, unskilful use of a knife and unhygienic behaviour like touching the carcasses unnecessarily, was also recorded.

Hygiene facilities for washing and disinfection were recorded using a two-point scale. The placement of the facilities was recorded as being placed near the working point when the worker did not have to move in order to reach them, or as being remote when one or more steps were necessary to reach the facilities. In practice facilities meant wash basin and sterilizers.

The contamination level of carcasses in these five slaughterhouses was studied with the same kind of method and swab samples were taken by the same person. Carcasses were sampled during the observation periods.

20.5.2   Results

Two out of 73 workers in the beef and pork line were inexperienced in using the knife. Four out of these 73 workers behaved in an unhygienic manner, touching the carcasses unnecessarily with the knife. Actual slaughter mistakes were not found. The results from observing the work are shown in Table 20.1. Of the observed workers, 81% washed their hands at least after every other carcass and 78% washed their knife at least after every other carcass. Only 24% disinfected their knives and 15% disinfected other working tools at least after every other carcass. And only 21% washed other working tools at least after every other carcass. The more tools the worker had to operate with, the less seldom was the disinfection of the tools.

Table 20.1 Hygiene practices of workers on pork and beef lines in five slaughterhouses (numbers and percentages of workers).

Hygiene practice	Frequency at slaughterhouse A				Frequency at slaughterhouse B				Frequency at slaughterhouse C				Frequency at slaughterhouse D				Frequency at slaughterhouse E			
	Frequent[a]		Seldom[b]		Frequent[a]		Seldom[b]		Frequent[a]		Seldom[b]		Frequent[a]		Seldom[b]		Frequent[a]		Seldom[b]	
Washing hands	15	100%	0	0%	20	84%	4	16%	6	60%	4	40%	8	57%	6	43%	10	100%	0	0%
Washing knife	13	87%	2	13%	17	81%	4	19%	5	63%	3	37%	8	57%	6	43%	10	100%	0	0%
Disinfection of knife	2	13%	13	87%	9	43%	12	57%	0	0%	8	100%	4	28%	10	72%	1	10%	9	90%
Washing other working tools	1	100%	0	0%	3	30%	6	67%	0	0%	7	100%	2	40%	3	60%	0	0%	7	100%
Disinfection of other working tools	0	0%	5	100%	3	27%	8	72%	0	0%	8	100%	3	33%	6	67%	0	0%	8	100%

a washing/disinfection at least after ever other carcass
b washing/disinfection more seldom than after every other carcass

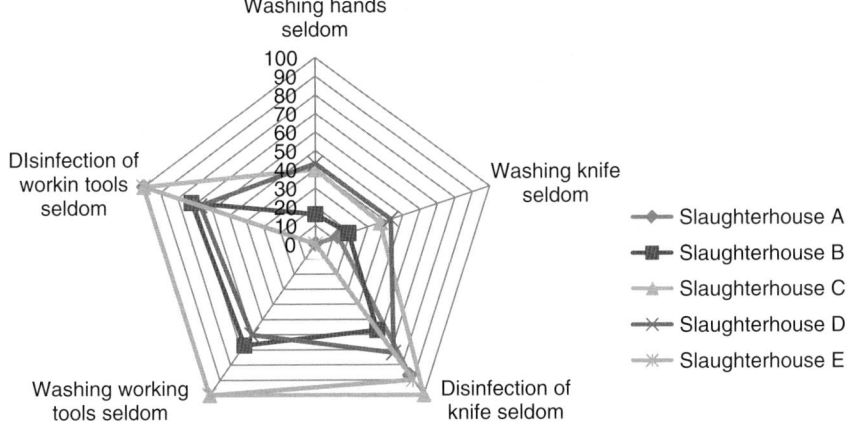

Figure 20.1 Hygiene practices of workers in five slaughterhouses A-E. (For colour details, please see color plate section).

These results contribute to the theory of factors having influence on the decision to practice hygiene or not. Hands are washed because it is inconvenient to work with dirty hands and the knife is washed at the same time as the hands. The positive consequence because of hand washing is relevant. The disinfection of the knife or the washing of the other tools do not have this kind of positive consequence and these hygiene practices are performed less frequently.

An association between daily routine practices and microbiological contamination of the carcass was found in the slaughterhouses studied. The cleanest carcasses were encountered in slaughterhouses and lines where tools were frequently washed and disinfected. In Figure 20.1, the hygienic practices of workers in the five slaughterhouses studied are described by the percentage of workers that seldom practice hygiene. Seldom means that washing or disinfection is done less frequently than after every other carcass. The nearer the line in the radar figure is to the beginning of the axis and centrum of the radars, the more hygienic are the working methods in the slaughterhouse the line represents. In Figure 20.2, the hygiene levels of carcasses in the same five slaughterhouses are shown as the mean of the bacterial count of pork neck, pork abdomen, beef shoulder and beef brisket. The nearer that the line in the radar figure is to the beginning of axis and centrum of the radars, the cleaner are the carcasses in the slaughterhouse the line represents. The carcass contamination and hygiene practices of workers rank the slaughterhouses in the same way.

20.5.3 Effect of facilities

The placement of facilities, sufficiency of working space and line speed did have effect on the hygiene practice of the slaughterhouse workers. The effect

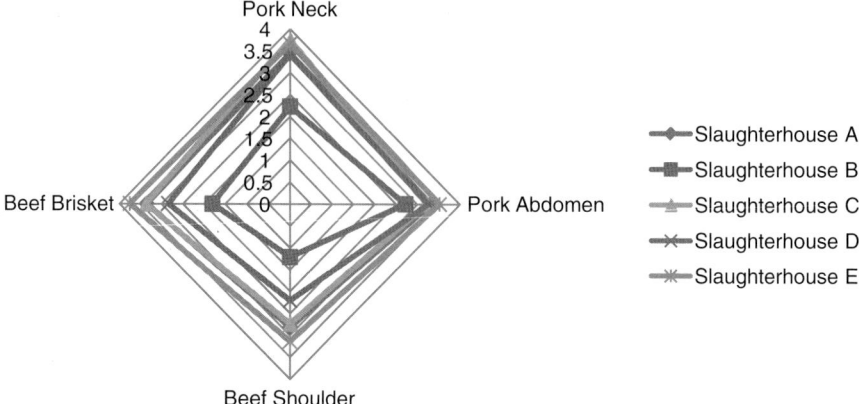

Figure 20.2   Total count (log 10 cfu cm^{-2}) of carcasses in five slaughterhouses A-E. (For colour details, please see color plate section).

Table 20.2   Effect of accessibility of hygiene practice facilities on hygiene practice frequency of slaughterhouse workers (numbers and percentages of workers).

Location of facilities	Hygiene practice frequency					
	Frequent[a]		Seldom[b]		Total	
Knife disinfection facilities						
Next to working place	14	19%	27	37%	41	56%
Distant from working place	2	3%	30	41%	32	44%
Tool[c] washing facilities						
Next to working place	6	22%	5	19%	11	41%
Distant from working place	1	4%	15	56%	16	59%
Tool disinfection facilities						
Next to working place	5	16%	7	22%	12	38%
Distant from working place	1	3%	19	59%	20	62%

[a] washing/disinfection at least after ever other carcass
[b] washing/disinfection more seldom than after every other carcass
[c] workers with one tool other than knife

of placement of facilities on washing and disinfection practice is shown in Table 20.2.

Those workers who had easy access to the washing point, washed their tools more often than those whose access was more difficult. Sufficiency of working space had the same kind of effect. However, if a worker had more tools to operate with, this kind of effect was not found.

The accessibility of disinfection facilities as well as the sufficiency of working space and working time did have a positive effect on the frequency of knife disinfection and also on the disinfection frequency of tools other than a knife. The effect of location of disinfection facilities was not dependent on the number of tools. In the Figure 20.3, the effect of placement of facilities is illustrated

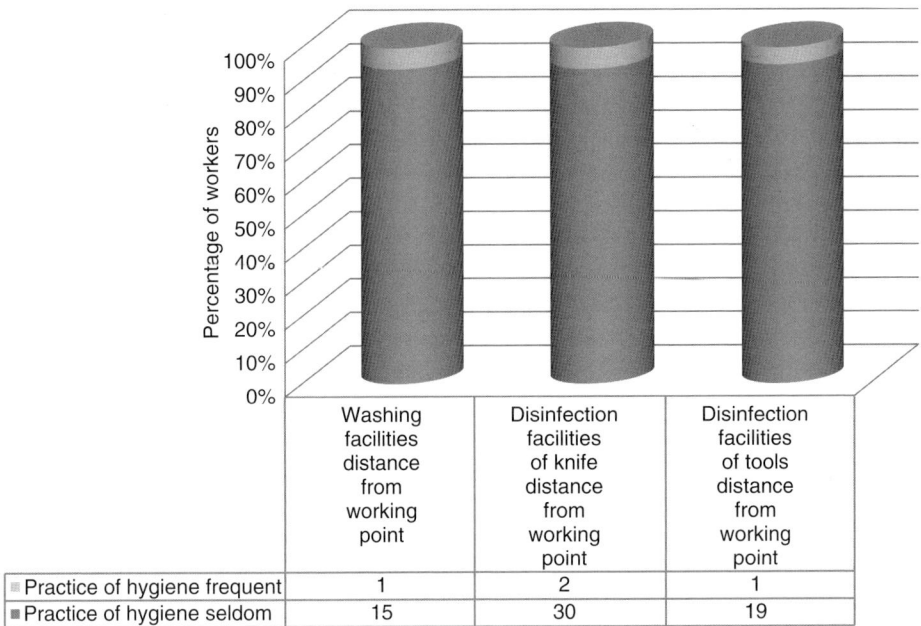

	Washing facilities distance from working point	Disinfection facilities of knife distance from working point	Disinfection facilities of tools distance from working point
■ Practice of hygiene frequent	1	2	1
■ Practice of hygiene seldom	15	30	19

Figure 20.3 Washing and disinfection practice of slaughterhouse workers distant from facilities.

by the hygiene practice of workers some distance from the washing or disinfection facilities. The percentage of workers who disinfected their knife and tools and washed their tools was very low when the facilities were not within immediate reach.

20.6 Conclusions

These results show how workers can be motivated to hygienic working and how important small details can be. The slaughter lines are typically highly automated, especially in big slaughterhouses. However, there are still working points with workers and knives in modern slaughter lines. Meat inspection working points are among these handwork points.

Literature and further reading

European Commission. 2001. The Cleaning and Disinfection of Knives in the Meat and Poultry Industry. Opinion of the Scientific Committee on Veterinary Measures Related to Public Health. Retrieved from http://ec.europa.eu/food/fs/sc/scv/out43_en.pdf (last accessed 16 February 2014).

Gerats, G. 1987. What hygiene can achieve – how to achieve hygiene. In: *Elimination of Pathogenic Organisms from Meat and Poultry* (ed. F.J.M. Smulders). Elsevier, Amsterdam, pp. 269–283.

Heuvelink, A.E., Roessink, G.L., Bosboom, K. and de Boer, K. 2001. Zero-tolerance for faecal contamination of carcasses as a tool in the control of O157 VTEC infections. *Int J Food Microbiol*, **66**, 13–20.

Rahkio, M. and Korkeala, H. 1996. Microbiological contamination of carcasses related to hygiene practice and facilities on slaughtering lines. *Acta Vet Scand*, **37**, 219–228.

Tazelaar, F. 1987. The human factor in the hygiene problem: problem analysis, problem solution and implications. In: *Elimination of Pathogenic Organisms from Meat and Poultry* (ed. F.J.M. Smulders). Elsevier, Amsterdam, pp. 251–265.

21
Occupational Hazards

Karsten Fehlhaber

Institute of Food Hygiene, Faculty of Veterinary Medicine, University of Leipzig, Leipzig, Germany

21.1 Scope

The professional activities of meat inspectors are characterized by practical work in the slaughterhouse including tasks related to *ante-* and *post-mortem* inspection of the slaughtered animals. Meat inspection places high demands in terms of mental and physical fitness. Furthermore, meat inspectors are exposed to a series of hazards arising from the work in the slaughterhouse or from the animals themselves. There is an elevated risk of infections or accidents. In this chapter, these hazards are described and instructions on how to avoid them provided. The main focus is on the risk of infections. Microorganisms which are also transmitted to humans through contaminated meat are discussed in more detail in Chapters 13A and 13B (Table 21.1).

21.2 Introduction

People working on a slaughter line are exposed to a series of occupational hazards. Meat inspectors are among these, as they can come into a close contact with specific safety hazards during *ante-* and *post-mortem* meat inspection. These hazards include: (i) infections with pathogens that can be transmitted by close contact with live or slaughtered animals; this can take place by direct contact with the animals and also indirectly via contaminated equipment, machines or through contaminated premises or air; and

Meat Inspection and Control in the Slaughterhouse, First Edition.
Edited by Thimjos Ninios, Janne Lundén, Hannu Korkeala and Maria Fredriksson-Ahomaa.
© 2014 John Wiley & Sons, Ltd. Published 2014 by John Wiley & Sons, Ltd.

Table 21.1   Microorganisms causing occupational diseases in the slaughterhouse.

Microorganism	Transmission			See also chapter	
	Contact	Air	Oral	13A	13B
1. Bacteria					
Bacillus anthracis	**x**[a]	x	x	x	
Brucella	**x**	x	x		
Campylobacter			**x**	x	x
Chlamydia psittaci	x	**x**	x		
Clostridium tetani	**x**				
Coxiella burnetii	x	**x**			
Erysipelothrix rhusiopathiae	**x**				
Escherichia coli (STEC)[b]			**x**	x	x
Leptospira	**x**	x	x		
Listeria monocytogenes			**x**	x	
Mycobacterium bovis	x	**x**	x	x	
Salmonella			**x**	x	x
Staphylococcus aureus	**x**	x		x	
Streptococcus (S. suis)	**x**	x	x		
Yersinia	x		**x**	x	x
2. Viruses					
Avian influenza virus		**x**		x	
Food and mouth disease virus	**x**		x		
Hepatitis E virus			**x**	x	
Newcastle disease virus		**x**			
Rabies virus	**x**	x			
Rotavirus			**x**		
Vacciniavirus	**x**				
Vesicular stomatitis virus	**x**	x			
3. Parasites					
Cryptosporidium			**x**		
4. Fungi					
Trichophyton verrucosum	**x**				
5. Prions		x	**x**	x	

[a] Main transmission route in bold
[b] STEC, Shiga toxin-producing *Escherichia coli*

(ii) accidents/injuries that may be caused by handling the live animals or freshly stunned and bled animals that still display unexpected reflex movements. Furthermore, the slaughter process is highly automated and there is a large number of moving, pointed and sharp devices in use that may pose a risk to the staff. Structural shortcomings are also conceivable hazards. There may also be health risks due to permanent noise and unfavourable indoor air, such as low temperature and high humidity in the slaughterhouse premises.

The level of occupational hazards in slaughterhouses is disproportionately high compared to other professions associated with physical work. Despite familiarity with the sources of risk and the appropriate precautionary and preventative measures, people working in slaughterhouses time and again suffer from occupationally-related health problems. Apart from the specific causes described above, this is also due to the high personal stress, such as the significant working tempo on the line, frequently cramped workstations and monotonous work. Preventing infections and accidents requires a permanent level of high concentration on the work process. Meat inspection staff has also to make rapid decisions that may have far-reaching impact on public and animal health, and animal welfare; they may cause significant economic loss, too.

21.3 Infections

Microorganisms, which meat inspectors come into contact with during their duties, originate above all from the live and slaughtered animals. Animals possess a bacteriological flora on the skin and the mucous membranes. A broad spectrum of microorganisms originating from contact with the environment (dirt and dust in the air, feed and soil) occurs especially on the skin (coat or feathers). The amount and the diversity of the microorganisms depend on the animal husbandry. The bacterial flora can be extremely heterogeneous, including mainly harmless microorganisms as well as pathogenic ones. The microflora found on the skin, coat and feathers includes bacteria (non- and sporeforming, gram-negative and positive), yeast and moulds found also in the environment. The rumina of the ruminant contains a high number of microorganisms ($>10^9$ microorganisms/ml rumina juice). Fewer microorganisms are found in the stomach of monogastric animals due the low pH value. A high number of microorganisms is found in the intestinal content, the highest number (between 10^{10} and 10^{12} microorganisms/g faeces) in the colon. Therefore, the intestinal content is the most important source of contamination in the slaughterhouse. Even in clinically healthy animals, it is important to keep in mind that potentially pathogenic microorganisms may be sporadically present in the intestinal contents. Thus, hygiene requirements are needed, especially during evisceration, to protect slaughterhouse workers against infections.

The microorganisms have different modes of transmission: oral, intraocular, inhalation and via the skin (Table 21.1). There is a particular risk of infection when infectious material seeps from pathological–anatomical changes during meat inspections and comes into contact with the examiner. This can happen when lymph nodes or abscesses are cut or during the palpation of organs. There is also a certain risk of infection during the *ante-mortem* inspection, especially during clinical inspection (such as fever measurement) if the animal is sick. Furthermore, meat inspectors are not only exposed to the risk of infection during *ante-* and *post-mortem* inspection, they are also confronted with the risk of infection across the various technological stages of the slaughter process during the monitoring of slaughter hygiene (Table 21.2). Occupational infection in the slaughterhouse can be caused by several microorganisms belonging mostly to (i) bacteria but in rare cases also to (ii) viruses and (iii) fungi (Table 21.1).

Table 21.2 Risk of infection through various technological stages.

Technological stage		Potential risk of infection
Carcass cleaning		Contaminated aerosol, dirt spatter
Floor cleaning		Contaminated aerosol, dirt spatter
Bleeding		Aerosol of blood, blood spatter
Scalding (pigs and poultry)		Contaminated aerosol
Dehairing (pigs)/plucking (poultry)		Contaminated aerosol, dirt spatter
Sawing		Bone particles from the bone splitter
Gutting		Contact with the pharynx including tongue and tonsils, bile, urine, blood, peritoneal fluid, intestinal content
Cooling (poultry)	Spin chiller	Contact with contaminated cooling liquid
	Air jet cooler	Contact with contaminated aerosol

21.3.1 Bacteria

Bacillus anthracis Anthrax is caused by *Bacillus anthracis*. It was once a feared animal epidemic disease but is today rare in Europe. However, *B. anthracis* and its spores are present in the soil in some countries in Europe, thus causing sporadically disease in animals and, more rarely, in humans. It is primarily an animal disease that occurs especially in ruminants (such as cattle, sheep and goats). The most common symptom is sudden death. Humans can get the disease by exposure to infected animals, animal products and spores in the soil. Possible transmission routes are: inhalation, cutaneous or gastrointestinal. The cutaneous disease is the most common one (about 95% of human cases). Without treatment, it is fatal. Anthrax can be an occupational hazard of slaughterhouse workers, veterinarians and farmers who handle infected animals. Meat inspectors should be aware of anthrax.

Brucella *Brucella* is a small, aerobic, gram-negative, facultatively intracellular rod that is responsible for an acute febrile illness, brucellosis. Brucellosis is a chronic infectious bacterial disease affecting various species of domestic and wild animals. After penetration into the body, it proliferates in the lymphatic system, mainly in lymph nodes, and then penetrates into various organs. *Brucella abortus* (cattle), *Brucella suis* (pigs, wild boars, hares) and *Brucella melitensis* (sheep, goats), which cause abortions and reproductive failure among food-producing animals, are zoonotic bacteria. Of these three zoonotic species, *B. melitensis* possesses the greatest virulence to humans. In addition to abortions and reproductive failure, these pathogens can also cause mastitis and infections in joints in animals. Brucellosis occurs worldwide but it is well controlled in most developed countries.

In humans, brucellosis is also called undulant fever or Mediterranean fever. Brucellosis may cause many symptoms, varying from mild flu-like manifestations to severe complications. Fever, headache and joint pain may persist for weeks to months. With undulant fever it is typical that the temperature can vary form 37°C in the morning to 40°C in the afternoon. Neurological complications, endocarditis and /or bone abscess formation can also occur. Brucellosis is transmitted to humans by animals through direct contact with infected materials, such as after birth or indirectly by ingestion of animal products, especially raw milk and products thereof, and by inhalation of airborne agents. Slaughterhouse workers may not only already be infected through contact with the live animal but also through excretions such as faeces, urine or milk during meat inspection. Infection may occur by skin contamination via cuts or abrasions, inhalation, conjunctival contamination or accidental ingestion. Brucellosis was once among the most frequent, feared and most serious of all occupational diseases suffered by veterinarians and slaughterhouse workers. These days it is rare but it is important not to underestimate the potential risk posed by global trade in animals.

Campylobacter In many countries, *Campylobacter* is the most frequent zoonotic bacterial pathogen transmitted via food. *Campylobacter jejuni* plays the main role in human diseases followed by *Campylobacter coli*. These pathogens can be found in all farm animals, *C. jejuni* especially in poultry. *Campylobacter* does usually not have any negative influence on animal health, and thus this pathogen cannot be identified during meat inspection. This pathogen is frequently found in high numbers in the intestinal content of slaughter animals, especially in poultry. It can be present on both the internal and external carcass surfaces and also on the organs. *Campylobacter* spreads easily with faeces to other areas of the slaughter line. Due to the low infection dose (500–1000 bacteria), meat inspectors are exposed to a constant risk of oral infection through contaminated hands.

Chlamydia psittaci *Chlamydia psittaci* is a gram-negative, obligate intracellular bacterium. It is a zoonotic agent that is endemic in nearly all bird species. The infection in birds is known as 'parrot fever', since psittacine birds have the highest prevalence. *C. psittaci* causes a respiratory illness also called psittacosis. The organism is shed in bird droppings and respiratory tract excretions. The birds are typically asymptomatic but stress factors such as changes in ambient conditions, transport, overcrowding and handling can activate the shedding. Psittacosis is usually relatively rare in the poultry but it is more common in turkey, ducks and geese. *C. psittaci* can be an occupational risk for workers at the slaughterhouse, especially during the intake of live birds, hanging of live birds to conveyer belt, de-feathering and evisceration. Transmission to humans most frequently occurs through inhalation of contaminated aerosols. This organism is resistant to desiccation and can easily be transmitted via contaminated dust, especially during dry plucking of poultry. Psittacosis comprises a wide variety of clinical signs. Usually the signs are mild and non-specific (fever, chills, headache and myalgia) with or without respiratory symptoms (cough and pneumonia). Sometime a rash and splenomegalia occur and haemotogenous dissemination may lead to complications in nervous system and different organs (like heart, lungs and kidney). The mortality is low due to antimicrobial therapy.

Clostridium tetani   *Clostridium tetani* is a gram-positive, aerobic, rod-shaped bacterium that can develop a spore. The spores are heat resistant and can survive in soil for years. The spores are also found in the intestine of farm animals. *C. tetani* has a worldwide distribution and produces an extremely potent toxin. Tetanus (lockjaw) is caused by the toxin produced in the place where the bacterium enters the body. Toxin is produced during the multiplication of the strain, then absorbed and carried by the blood to all parts of the body. The disease is common in humans and horses. Sheep are rather susceptible and can get the disease from wounds. Sporadic cases are also observed in cattle and pigs but poultry is highly resistant.

In humans, *C. tetani* is mostly transmitted through wounds, cuts and scratches. Tetanus can occur in people of all ages. However, before the availability of a vaccine, tetanus was a common childhood illness. Generalized tetanus is the most common form of tetanus. The incubation period is between three days and three weeks, sometimes even longer. Patients generally have tonic contraction of their skeletal muscles and intermittent, intense muscular spasm. The most common complaints are pain, difficulty swallowing and stiffness of the neck and other muscle groups such as the thorax and abdomen. The mortality rate is high (usually around 30%). Slaughterhouse workers are susceptible to tetanus because of the high incidence of wounds and cuts. Additional risk arises from skin wounds contaminated with dirt and dust brought into the slaughterhouse by the animals. Systematic immunization of slaughterhouse workers against tetanus is important.

Coxiella burnetii   *Coxiella burnetii* is a small, gram-negative, obligate intracellular bacterium that has a worldwide distribution. The organism has an ability to persist in the environment for months. It is a zoonotic agent that can infect humans and animals. *C. burnetii* has been isolated from a large range of animals, including food-producing animals. Ticks are vectors that may transmit *Coxiella* to animals. Domestic ruminants, especially sheep and goats, are considered an important reservoir for *C. burnetii* and are attributed to human infections. In animals, *Coxiella* infection is usually without clinical signs. However, sometimes late abortion and reproductive disorders can occur. *Coxiella* is excreted via secretions (urine, milk) and faeces.

Q fever is a disease in humans caused by this pathogen. It is an occupational disease for workers in the slaughterhouse but also for farmers, veterinarians and sheep shearers. Contact with infected animals and/or their contaminated products (especially birth products) are important risk factors for the workers. Q fever is primarily caused by inhalation of contaminated aerosols and dust. The infection dose is small; fewer than 10 organisms are sufficient to cause an infection. The infection is typically mild and self-limiting with flu-like symptoms or atypical pneumonia. However, some patients can develop a more serious chronic infection including endocarditis and other complicated infections.

Erysipelothrix rhusiopathiae *Erysipelothrix rhusiopathiae* is a gram-positive, non-sporing bacillus that has a worldwide distribution and affects a wide variety of animals, including food-producing animals, especially pigs, sheep and turkey. *E. rhusiopathiae* causes pig erysipeloid (red murrain) that is a disease of great economic importance. Sick pigs excrete this pathogen in urine, faeces, saliva and nasal secretion. Carriers harbour this organism in the tonsils and other lymphoid tissues. Pig erysipeloid (erysipelas) has three forms: acute, subacute and chronic. In the acute form, sudden death or general signs of septicaemia occur. Additionally, diffuse areas of erythema and petechiae may occur. Subacute erysipeloid is less severe. The intensity of the skin lesions has a direct relation to the prognosis. The chronic form of infection is characterized most commonly by local arthritis and endocarditis. *E. rhusiopathiae* can also cause polyarthritis in sheep and lambs but it rarely causes infection in cattle. *E. rhusiopathiae* is especially pathogenic for turkeys, which exhibit a cyanotic skin, and haemorrhages and petechiae in the breast and leg muscles. They may develop diarrhoea and die. The disease frequently emerges in the summer months. *Erysipelothrix*is is ubiquitous and is able to persist for a long period of time in the environment.

E. rusiopathiae infection is an occupational disease and occurs mostly in people who work closely with infected animals, their products or wastes. People with the highest risk of exposure include slaughterhouse workers, veterinarians and farmers. The infection is initiated by an injury to the skin with infected material or when a previous injury is contaminated. Disease symptoms in humans resemble those seen in pigs. Erysipeloid, the local skin infection, is a common infection form also in humans. The disease starts suddenly with painful, expansive reddening of the skin, often on the hands, leading to delimited swelling of the skin. Systemic symptoms like fever, joint pain and lymphadenitis can also occur. Sometimes generalized skin infection and septicaemia can be seen, especially in persons with a weakened immune system. The disease is usually self-limiting which may resolve in 3– 4 weeks without therapy.

Escherichia coli *Escherichia coli* is a bacterium of the normal intestinal flora in humans and animals. Most strains are non-pathogenic. They can regularly be detected on the carcass surfaces, thus being used as an indicator of faeces contamination. Slaughter animals, especially ruminants, can excrete Shiga toxin-producing *E. coli* strains (STEC) that may infect humans. Typically, healthy adult cattle and sheep carry this pathogen in the lower intestinal tract. In humans, the infection dose of STEC is low. The mode of transmission is oral and may occur by direct contact with infected/carrier animals or their faeces. Slaughterhouse workers may get infected through contaminated hands. STEC infection can vary from mild diarrhoea to haemorrhagic colitis. The infection is normally self-limiting but, in some instances, Shiga toxins may damage target organs such as the gut and kidney.

Leptospira *Leptospira* is a gram-negative, aerobic, spiral-shaped bacterium with worldwide distribution. Pathogenic strains may cause leptospirosis, which is a zoonotic disease. Pathogenic *Leptospira* strains live in kidneys and are excreted in urine and may then contaminate the environment. *Leptospira* can survive in the environment and then spread to livestock, for example via contaminated water, feed or rodents. Leptospirosis is a systemic disease of humans and animals, including food-producing animals, especially pigs and cattle. Leptospirosis is characterized by fever, renal and hepatic insufficiency, pulmonary manifestations and reproductive failure. Abortions and under-developed foetuses are characteristic among pregnant sows. Grazing cattle, in particular, may suffer from icterohaemoglobinuria. The animals can be asymptomatic carriers.

Infections in humans usually occur from direct contact with urine or indi-rectly from a contaminated environment through mucous membrane or broken skin. Occasionally infection occurs through ingestion/inhalation of food/aerosols contaminated by urine. In humans, leptospirosis can vary in severity according to the strain (serotype) of *Leptospira* and the age, health and immunological status of the patient. The disease ranges from a mild, flu-like illness to a severe infection with renal and hepatic failure, pulmonary distress and death. There is an occupational association with slaughterhouse workers, veterinarians and farmers.

Listeria monocytogenes *Listeria monocytogenes* is widespread in the environ-ment (soil and plants) due to its pronounced tenacity. It is a zoonotic bac-terium that can frequently be found in the faeces of food-producing animals. Clinical diseases are common among ruminants, especially among sheep. Due to stress or a weak immune system, sheep may develop meningoencephali-tis, mastitis, abortion or septicaemia. In the event that diseased animals are slaughtered, there is an elevated risk that the inspection staff may be infected.

Mycobacterium bovis *Mycobacterium bovis* is a very slow-growing, gram-positive bacterium with a worldwide distribution. *M. bovis*, the cause of bovine-type tuberculosis, has a wide host range, including man. Cattle are considered the natural host of this pathogen and the principal reservoir for other animals and man. Other ruminants and pigs are also susceptible to this pathogen. Zoonotic transmission occurs primarily through close contact with infected cattle. Bovine tuberculosis is mainly a respiratory disease transmitted by the airborne route. It is a chronic disease characterized by the formation of granulomatous lesions with varying degrees of necrosis, calcifications and encapsulation. Bovine tuberculosis is an occupational zoonosis for professions regularly handling cattle, including slaughterhouse workers. They are more likely to develop pulmonary disease than alimentary disease. Cutaneous/mucosal transmission is also possible but rare. In developed industrial countries, bovine tuberculosis is rare. However, sporadic cases occur, meaning that this disease must still be considered in meat inspections.

Salmonella *Salmonella* is among the most significant and widespread zoonoses. Although there are substantial differences in the virulence, all serotypes are considered pathogenic. Most of the human infections are caused by two serotypes: *Salmonella* Enteritidis and *Salmonella* Typhimurium. *Salmonella* is frequently found in farm animals, especially in pigs and poultry. In most cases, animals intended for slaughter are carriers of *Salmonella* and present no symptoms, thus they cannot be identified during the meat inspections. *Salmonella* is excreted in the intestinal content and pre-slaughter stress can increase the excretion. Meat inspectors should always be aware of *Salmonella*. Massive cross-contamination can occur during the slaughter process, especially with the intestinal content. The infection can be obtained orally via contaminated hands during meat inspection.

Staphylococcus aureus *Staphylococcus aureus* is frequently detected in humans and animals on the skin and the mucous membranes of the upper respiratory tract without any symptoms. However, it can also cause animal diseases that can be detected during *ante-* and *post-mortem* meat inspection, such as mastitis, wound infections, abscesses, pyodermia and phlegmons. These findings are common in slaughter animals and can be transmitted to meat inspectors. The pathogen can also lead to generalized diseases such as osteomyelitis, arthritis, synovitis or dermatitis in poultry. There is a particular problem associated with the occurrence of methicillin-resistant *S. aureus* strains (MRSA) in farm animals, above all in pigs and poultry. Humans can get wound infections or lung infections via aerosols.

Streptococcus Most *Streptococcus* spp. are commensal, living on skin and/or mucous membranes of humans and animals. Infections caused by zoonotic *Streptococcus* spp. are rarely reported. *S. suis* and *S. equi* subsp. *zooepidemicus* (*S. zooepidemicus*) are two opportunistic pathogens that can be transmitted to humans. *S. suis* is an important pig pathogen with a worldwide distribution. Asymptomatic pigs typically carry *S. suis* in the tonsils. However, pigs may develop sepsis, meningitis, arthritis and/or pneumonia, resulting in major economic losses. *S. suis* serotype 2, especially, is a zoonotic agent. In humans, meningitis is most commonly reported but other systemic complications such as sepsis and pneumonia can occur also. *S. suis* can cause occupational diseases in the slaughterhouse. The infection may occur through direct contact with infected pigs via skin lesion or conjunctiva and, rarely, also via air and orally. Slaughterhouse workers are typically asymptomatic carriers.

    S. zooepidemicus is considered as an opportunistic bacterium in a large variety of mammalian species, including also pigs and ruminants. It may cause diseases in pigs with typical symptoms being arthritis, diarrhoea, bronchopneumonia, endocarditis and/or meningitis. Deep tissue infections have been reported in ruminants (such as cattle, sheep and goats). However, typically, *S. zooepidemicus* is a causative agent of mastitis. This pathogen is usually transmitted by the oral route, mostly by contaminated milk, but it may also be transmitted via wounds or aerosols. In humans, purulent abscesses, endocarditis, arthritis, pneumonia and meningitis have sporadically been reported. However, the occupational risk in the slaughterhouse is low.

Yersinia   Human pathogenic *Yersinia enterocolitica* belonging to bioserotypes 2/O:5,27, 2/O:9 and, especially, 4/O:3 are frequently found in pigs at slaughter. *Yersinia pseudotuberculosis* has also been isolated from slaughter pigs. Pigs carry pathogenic *Yersinia* usually asymptomatically in the tonsils and can excrete it in the faeces. *Yersinia* can easily be transmitted from the tonsils and faeces to the carcass and organs during slaughter and meat inspection. Even though the infection dose may be high, there is a risk of the meat inspectors getting the infection orally via contaminated hands, especially during evisceration and handling the oral cavity including the tonsils and tongue. Care must also be taken when handling the cooled carcasses because *Yersinia* can even multiply at temperatures around 0°C.

21.3.2   Viruses

Avian influenza virus (AIV)   Avian influenza (AI) affects several avian species, including domestic poultry. However, AI infections of commercial poultry in developed countries are rare. The infection is caused by the Influenza A virus (AIV), which is an orthomyxovirus. The symptoms in poultry are similar to those caused by Newcastle disease virus (NDV). AIV and NDV both have severe global impacts on poultry health and limit international trade of poultry. Most AIV do not cause disease in humans. The subtype H5N1, which is particularly aggressive, is the most well known type causing disease in humans. Other types that have infected humans are H7N7 and H7N9. Intense contact with the animals infected with AIV can transmit the pathogen to humans via aerosols. Slaughterhouse workers in contact with blood and faeces are particularly at risk. However, the risk of infection is extremely low. In humans, the infection can be very severe resulting in death but mostly it is mild with flu-like symptoms or it is subclinical and may, therefore, not be identified.

Foot and mouth disease virus (FMDV)   Foot and mouth disease (FMD) is a highly infectious and economically devastating disease of livestock. It is the most important animal disease limiting trade of animals and animal products. FMDV is a picorna virus that is highly transmissible and causes high morbidity outbreaks with moderate to low mortality. After aerosol exposure, FMDV first replicates in the pharynx. In 24–48 h, the virus invades the blood stream and shortly thereafter lesions appear in the mouth and feet. The virus may persist in the pharyngeal area and about 50% of the infected animals become long-term carriers. Cattle and pigs are mainly affected among the farm animals. Humans can be infected with FMD through contact with infected animals or orally via contaminated milk but this is extremely rare. Symptoms of FMD in humans include malaise, fever, red ulcerative lesions of the oral tissue and sometimes vesicular lesions of the skin.

Hepatitis E virus (HEV)   Hepatitis E is an emerging zoonotic disease caused by the hepatitis E virus (HEV) belonging to the genus hepevirus. HE is common in regions with inadequate water supplies and poor sanitary conditions. HEV3 and HEV4 are zoonotic agents circulating among humans and pigs. Transmission may occur via direct contact with infected pigs or consumption of raw or undercooked pork. Infections in humans can result in acute hepatitis: fever, nausea, anorexia and jaundice. In most cases the disease is self-limited and the patient recovers without sequelae.

Newcastle disease virus (NDV) Newcastle disease (ND) is caused by Newcastle disease virus (NDV), which is a paramyxovirus of serotype 1. It affects various avian species, including domestic poultry. NDV can induce severe multiorgan disorders with a near 100% mortality rate. The disease can also be subclinical or mild, including respiratory disorders and decreases in egg production. Humans can be infected with NCD mainly by aerosols. There is an occupational risk for slaughterhouse workers to contract the infection. The most common signs are conjunctivitis, headache, discomfort and slight chills.

Rabies virus (RABV) Rabies caused by rabies virus (RABV) genotype 1 is one of the most common fatal zoonotic infections worldwide. RABV is a rhabdovirus. It is responsible for human and animal rabies. All species of mammals are susceptible to RABV infection. It is mainly associated with dog bites. Livestock (pigs, cattle, sheep and goats) can be infected with rabies when they are in contact with infected animals that manage to bite them. However, rabies is relatively rare in livestock. RABV affects the brain and spinal cord (central nervous system). The symptoms of rabies vary considerable. The animals may be depressed, not eating and isolating themselves. Later they might have problems with swallowing and walking, and they might be aggressive. In humans, the first symptoms are usually fever and headache, and when the infection progresses hallucinations, paralysis and, eventually, death will occur. Transmission of RABV may occur through contamination of mucous membranes, a wound or scratch with saliva from infected animals. RABV can also be transmitted via aerosols. The incubation period is 1–2 months but can vary from weeks to years. There may be a risk of infection for the slaughterhouse workers if the symptoms are overlooked or if they are not yet apparent. However, rabies has been eliminated in many European countries, and hence the risk is extremely low.

Rotavirus Rotavirus is an enteric pathogen found in birds and mammals. It causes acute watery dehydrating diarrhoea and it is a major problem in young calves, lambs and piglets. The virus is transmitted mainly by the faecal–oral route. Slaughterhouse workers may be infected during the slaughter through direct contact with infected animals or via faeces and/or dirt. In humans, the infection can range from asymptomatic infection to mild diarrhoea to severe gastroenteritis with dehydration.

Vacciniavirus (VACV) Bovine vaccinia (BV) is a disease caused by vacciniavirus (VACV), which is a typical orthopoxvirus (OPV). Lesions in cattle are most frequently located on the teats and udder of lactating cows. The disease is a zoonosis and the transmission occurs through direct contact with infected animals, typically during the milking process. In humans, VACV infection is characterized by the development of skin lesions (itchy, nodular swellings), usually on the hands. The disease is quite rare today and the risk for slaughterhouse workers is considered low.

Vesicular stomatitis virus (VSV) Vesicular stomatitis virus (VSV) is a vesiculovirus belonging to the family Rhabdoviridae. It affects a wide variety of animals, including pigs and cattle, occasionally also sheep and goats. VSV can spread from animal to animal by direct contact via broken skin or mucous membranes. Infected animals shed the virus mostly in saliva. Vesicular stomatitis is characterized by vesicles, papules, erosions and ulcers; these are typically found around the mouth but can also be present on the feet, udder and prepuce. Excessive salivation is often the first symptom. In humans, vesicular stomatitis is an acute illness that resembles flu with fever, muscle aches, headache and malaise. Humans can be infected by handling infected animals and contaminated tissues. Aerosol transmission is also possible.

21.3.3 Parasites

Cryptosporidium *Cryptosporidium* is a protozoan (single celled) parasite that can cause an illness called cryptosporidiosis. A number of species can infect mammals, including humans. The most important species for humans is *C. parvum*, which is common especially in ruminants. Asymptomatic infections are common in animals and typically only young animals suffer from diarrhoea. The main symptom in humans is watery diarrhoea accompanied by stomach cramps and fever. The symptoms can last for up to three weeks but are usually self-limiting. Transmission is faecal–oral and in the slaughterhouse the workers may be infected through direct contact with animals and their faeces. Oocysts can survive in the environment for prolonged periods and are resistant to the levels of chlorine and other disinfectants typically used in water treatment. Good hygiene practices and proper sanitation are important measures in the prevention of cryptosporidiosis in the slaughterhouse.

21.3.4 Fungi

Trichophyton verrucosum *Trichophyton verrucosum* is a zoonotic dermatophyte fungus which has a worldwide distribution. It produces lesions on cattle and humans known as 'cattle ringworm'. Ringworm is enzootic in many cattle herds. It is characterized by patches of thick, tightly adherent grey crusts affecting the face and/or neck. The lesions typically occur in young animals and in winter. Although *T. verrucosum* is adapted in cattle it can occasionally transfer to sheep. Slaughterhouse workers can be infected through direct contact with infected animals or contaminated environment/equipment.

In humans, lesions are often inflammatory and involve the scalp, face, beard or exposed areas of skin.

21.3.5 Prions

Transmissible spongiform encephalopathy (TSE) forms a group of human and animal diseases that share common features such as pathological lesions in the central nervous system and a long incubation period. TSE is caused by prions (infectious misfolded protein molecules). The disease in humans is called variant Creutzfeldt–Jacob disease (vCJD). It is believed that the disease is transmitted by consumption of beef products of infected animals.

There is an assumption that BSE (Bovine Spongiform Encephalopathy) can be transmitted to humans. It is advised to avoid contact with the risk material and to take possible precautionary measures to exclude any oral and airborne infection.

21.4 Prevention from infections

All hygiene measures required in the slaughterhouse also protect the workers against occupational infections. These hygiene measures include personal hygiene, working hygiene, process hygiene and cleaning and disinfection.

Measures relating to personal hygiene include:

- wearing of suitable protective clothing, including rubber boots, washable aprons, gloves and, when necessary, dust masks, respirators and goggles;
- cleaning hands before the start of work (including after breaks) and at the end of work;
- removal of jewellery before the start of work;
- hand cleaning (hot and cold water) and disinfection facilities close to the workplace;
- care of wounds (even small ones);
- first aid: provision of wound care material and eye rinsing facilities close to the workplace;
- eating, drinking, smoking and chewing gum prohibited during work;
- storage of private and protective work clothes in separate lockers;
- cleaning and changing protective clothing when needed;
- delimitation of the working area from the break and social areas according to the black–white principle (personnel air shower);
- sufficient sanitary areas should be easily accessible.

Measures relating to process hygiene include:

- separation of clinically conspicuous animals upon delivery;
- cleanliness in the animal stalls (disposal of faeces);
- facilities to clean the carcasses;

- disposal of waste from the slaughter area;
- monitoring of slaughter process(e.g. control of cross-contamination);
- monitoring of machines and equipment (e.g. scalding temperature);
- avoiding cross-contamination by workers (e.g. cleanliness of knives, work clothes, hands);
- avoiding direct contact with abnormal carcass parts;
- minimization of aerosols during slaughter;
- sufficient lighting for proper working and detection of dirt.

Measures relating to hygiene of machines, equipment, structures, environment and water include:

- thorough cleaning and disinfection of the slaughter line at the end of work and drying of it before the start of work;
- servicing and maintenance of the structural state and working order of the devices and machines;
- replacement of worn parts and avoidance of microbial propagation in parts that are difficult to clean;
- regular renovation: smooth, easy-to-clean, non-corrosive surfaces and floors; floors without joints and anti-slip; gradient of the floor such that liquids flow off completely; effective extraction of water vapour; good ventilation;
- regular pest control;
- monitoring of the quality of drinking water.

21.5   Non-infectious occupational hazards and their prevention

Increased demand at slaughterhouses has caused a rise in work-related injuries. The most dangerous plants are those in which cattle are slaughtered. Poultry slaughterhouses are somewhat safer because they are more highly mechanized. Most of the work at a modern beef plant is still performed by hand. All sorts of accidents involving power tools, saws, knives, conveyor belts, slippery floors and falling carcasses become more likely when the line speed is too high. Line speed and occupational risk increase uniformly.

Injuries are among the most frequent non-infectious occupational hazards in slaughterhouses. They may be caused by the live animals during the *ante-mortem* inspection, by reflex movements of the stunned animal or during bleeding. Improper handling of the knife, failure to take proper care of moving machine parts and not concentrating during work with saws or hatchets may cause injuries (Table 21.3). Exhaustion, excessive working hours, monotony in work processes, high performance pressure and physical exertion promote accidents. Insufficient lighting also impedes work and increases the risk of accidents.

Table 21.3 Non-infectious occupational hazards and their prevention.

Hazard	Prevention
Injuries from	
animals	Use of protective clothes
knife	Correct use and storage of knife, use of protective gloves, sharpening of knife
slipping	Efficient cleaning, correct footwear
moving machines	Suitable line speed
sharp edges	Good lightning
Noise	Use of ear protectors
Indoor air	
Cold and humid	Proper clothing
Allergens	Use of suitable materials
	Reduction of dust and aerosols, better ventilation

Structural shortcomings in the floors (insufficiently anti-slip) and inadequate cleaning, and hence the emergence of slimy surfaces, may increase the risk of accidents, so too can constructive shortcomings in the equipment and machinery (rough or excessively smooth surfaces, sharp edges etc.). Slaughter lines with insufficient space for workers, hence forcing work in cramped spaces, also promote the incidence of injuries.

Further incidences of stress that may impair the health of the staff include:

- dust and aerosols that place a burden on the respiratory tract;
- noise, cold, draughts, high humidity and rapid switches between hot and cold environments promote the emergence of colds or rheumatic illnesses;
- allergies, for example caused by unsuitable material in protective gloves.

21.6 Control of occupational hazards

The employer is responsible for work safety, both in terms of organization and technical material. The employer must draft a risk analysis referring to each workplace and must monitor at regular intervals whether all preventative measures are adhered to (checklist). This includes the measurement of noise pollution, room temperatures and humidity, and so on. The employer's tasks equally include the provision of prophylactic medical care and regular staff instructions on work safety; this includes recommendations for prophylactic inoculations. Documents must be kept on all incidents. Furthermore, most European Union states prescribe their own work safety instructions. The recognition of diseases as occupational diseases is subject to national legislation.

One of the most important preventative measures is that the slaughterhouse staff works conscientiously. This includes adequate awareness of all occupational hazards and the commitment to infection protection measures. Protective work clothing serves to prevent infection and to protect against accidents, for example wearing helmets, ear protectors, shoes with sufficient sole profile and, if necessary, also cut resistant gloves on the hand not guiding the knife. All accidents should be reported immediately. Safety training for slaughterhouse workers is essential to ensure competence and safe performance in their work tasks. Managers have a key responsibility for maintaining a safe working environment. Supervisors need to be aware of hazards within their area of responsibility.

Literature and further reading

Bätza, H.-J., Bauerfeind, R. and Becker, W. (eds). 2002. *Zoonosen-Fibel: Zwischen Tier und Mensch übertragbare Infektionskrankheiten*, 5th edn. H. Hoffmann Verlag, Berlin.

Dedie, K., Bockemühl, J., Kühn, H. *et al.* (eds). 1993. *Bakterielle Zoonosen bei Tier und Mensch: Epidemiologie, Pathologie, Klinik, Diagnostik und Bekämpfung*. Ferdinand Enke Verlag, Stuttgart.

Holland, W.W., McEwen, J. and Omenn, G.D. (eds). 1997. *Oxford Textbook of Public Health*. Oxford University Press, UK.

Modrow, S., Falke, D. and Truyen, U. (eds). 2003. *Molekulare Virologie*, 2nd edn. Spektrum Akademischer Verlag, Heidelberg/Berlin.

Palmer, S.R., Torgerson, P. and Brown, D.W.G. (eds). 2011. *Oxford Textbook of Zoonoses*. 2nd edn. Oxford University Press, UK.

Ray, B. and Bhunia, A. (eds). 2008. *Fundamental Food Microbiology*, 4th edn. CRC Press, Boca Raton/London/New York.

Selbitz, H.-J. (ed.). 1992. *Lehrbuch der veterinärmedizinischen Bakteriologie*. Gustav Fischer Verlag, Jena.

Wright, W.E. (ed). 2008. *Conturier's Occupational and Environmental Infectious Diseases*. 2nd edn. OEM Press, Beverly Farms, MA.

22
Traceability

Kyösti Siponen

*Import, Export and Organic Control Unit, Control Department,
Finnish Food Safety Authority Evira, Helsinki, Finland*

22.1   Scope

The scope of this chapter is to provide information concerning the traceability
of meat. Certain parameters need to be traceable through the various steps
of the chain from-field-to-fork. Therefore, this chapter deals with traceability
from a wider perspective than only at the slaughterhouse level.

22.2   Traceability of food in the from-field-to-fork chain

Safe and authorized foods and feeds have free movement within the internal
market of the European Union (EU). Free movement across borders requires,
among other things, that business operators (BOs) in the food and feed chain
are in readiness to withdraw from the market any products that can potentially
pose a health risk and for which they are responsible. In order to implement
this, the operators should have procedures in place for tracing the product
concerned.

BOs in the food and feed chain engaged in primary production, processing,
storage and distribution are, for their part, responsible for the safety of the
products that they produce and market. The chain starts at primary production and finally ends with the selling or serving of the products to consumers.
Figure 22.1 shows a traceability chart covering the process from the stable to

Meat Inspection and Control in the Slaughterhouse, First Edition.
Edited by Thimjos Ninios, Janne Lundén, Hannu Korkeala and Maria Fredriksson-Ahomaa.
© 2014 John Wiley & Sons, Ltd. Published 2014 by John Wiley & Sons, Ltd.

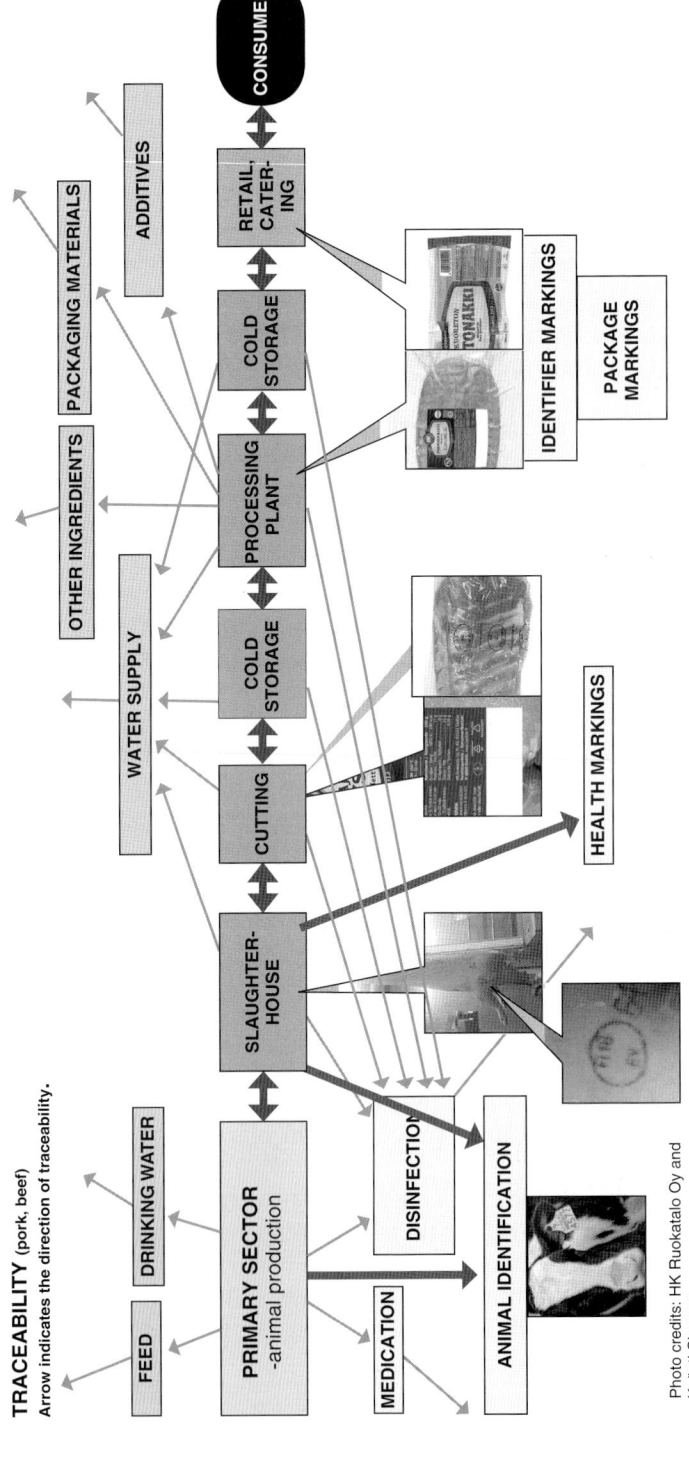

Figure 22.1 Traceability chart covering the process from the stable to the table. Source: Photographs courtesy of HK Ruokatalo Oy.

the table (the pictures in the chart are reproduced from the list of products of HK Ruokatalo Oy).

BOs must, for their part, know the origin of the food, feed, animal or substance added to the food or feed, as well as the feeds used in primary production. They must also know to which companies they have delivered these products or products produced by them and, if necessary, be able to quickly trace and withdraw from the market any products that have adverse health effects.

Tracing is facilitated by appropriate labelling and other identification information as well as by commercial documentation. This documentation needs to be filled out correctly with the required information accompanying the products. This is the only way that BOs in the food and feed chain can ensure for their part traceability, 'one step forward and one step back'.

22.3 Responsibility for safety of foods rests with food business operators

If operators in the food business suspect that the food they have received, produced or distributed does not comply with the criteria specified for the safety of foods, they must take immediate action to trace such food and withdraw it from the market. Food business operators (FBOs) must be able to name the person or company that has delivered the raw material or any other product batch. The FBOs must also be able to indicate where they have further delivered such batches or batches produced from them. They must know the acquisition and delivery dates and the amounts of the batches.

A functioning traceability system enables the company to implement quick and targeted withdrawal of foods from the market, if problems are encountered. A good traceability system also covers internal traceability within the company. Internal traceability will make it possible for the operator to link batches received by the company with batches that have been produced and sent to the market. Figure 22.2 shows a traceability chart concerning meat moving in the EU internal market, while Figure 22.3 shows a traceability chart concerning the movements of live slaughter animals in the EU internal market.

At the same time, the targeted traceability of the product is improved and linked to a certain product batch. This minimizes the potential damages caused to consumers, other operators or the company itself. In the event of a crisis, good traceability reduces the work load of both operators and authorities, and thus also the costs.

When FBOs are informed about the withdrawal from the market of a food batch produced or marketed by another operator, they must immediately take similar action for their own products.

FBOs shall inform the authority that supervises the company of any suspicions of health risks caused by the food marketed by the company.

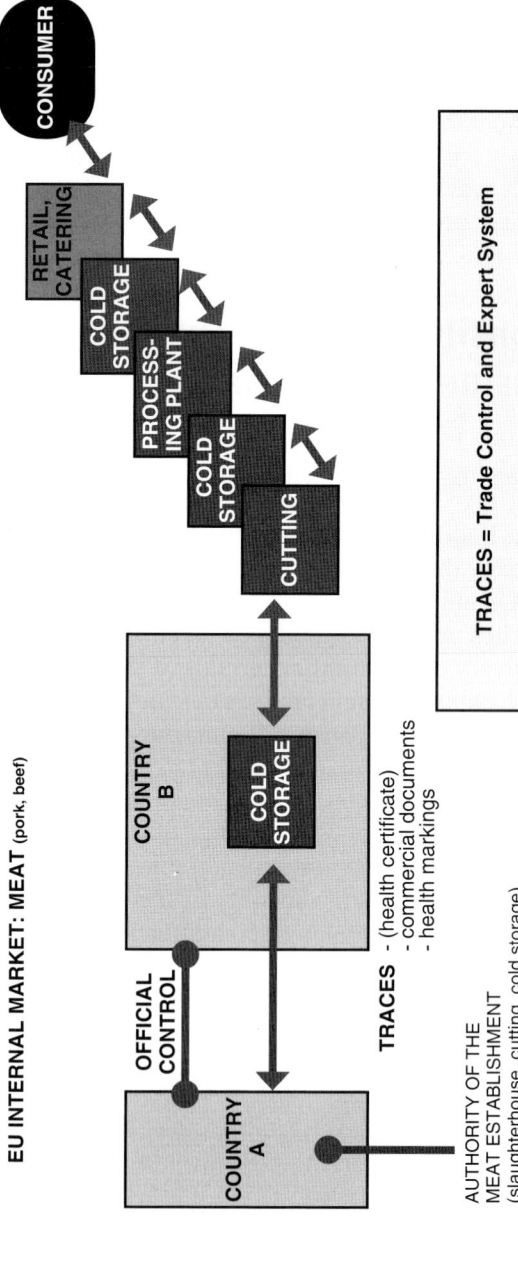

Figure 22.2   Traceability chart concerning meat moving in the EU internal market.

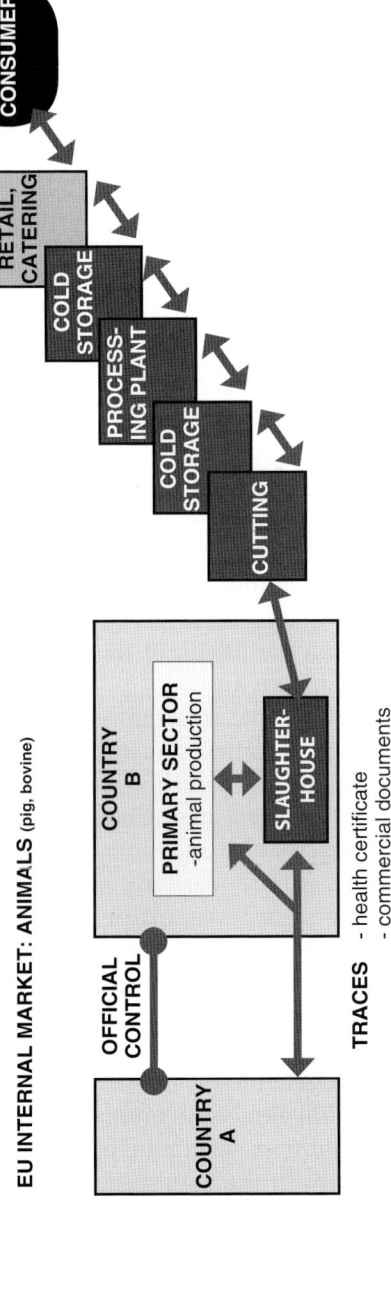

Figure 22.3   Traceability chart concerning the movements of live slaughter animals in the EU internal market.

The operator should in this connection also describe the action that has already been taken and how the consumers have been informed so as to prevent any risks caused by the defect.

22.4   Health and identification mark

In meat inspection, a health mark is applied to the carcass of every slaughter animal approved for consumption. The mark is stamped on the external surface of the carcass. The shape and size of the health mark have been specifically developed. The health mark indicates the country of location and the approval number of the slaughterhouse.

Health marks are not applied on the carcasses of poultry or small game animals, however; instead, an identification mark is applied on their packaging or wrapping. The packaging, wrapping or labelling of foods of animal origin should bear an identification mark that indicates the number and the country of location of the manufacturing plant.

22.5   Unauthorized foods and foods posing a risk to food safety

In the food chain, unintentional and intentional contamination, falsification and associated fraud will always have a direct or indirect effect on food safety and, at the same time, also on the health of the consumers. Traceability is a key tool to combat fraudulent activities.

Food trade has become ever more international and the chain of, particularly, meat and meat products from the slaughterhouse to the consumer is a long one. The chain consists of slaughterhouses, cold storage facilities, processing plants and transport between several countries and plants. Intermediaries and distributors of meat as well as wholesale and retail businesses are also involved in the chain. It is fairly easy at some point of the chain for the traceability and origin of the meat to be lost or falsified. Figure 22.4 shows a traceability chart concerning live animals and/or meat moving from outside the European Union into the European Union.

Several international crises related to meat and the origin of meat as well as to possible falsifications have shown that the traceability of meat is not complete in all respects. Untraceable meat is always a risk to food safety and it also causes a risk of spreading infectious animal diseases. Untraceable meat often reflects fraudulent operation, where a company in the chain is wilfully deceived by another company. In the end, this untraceable meat reaches the consumer, either as such or in processed form. Not only are consumers misled in this case but they are also subjected to a health risk.

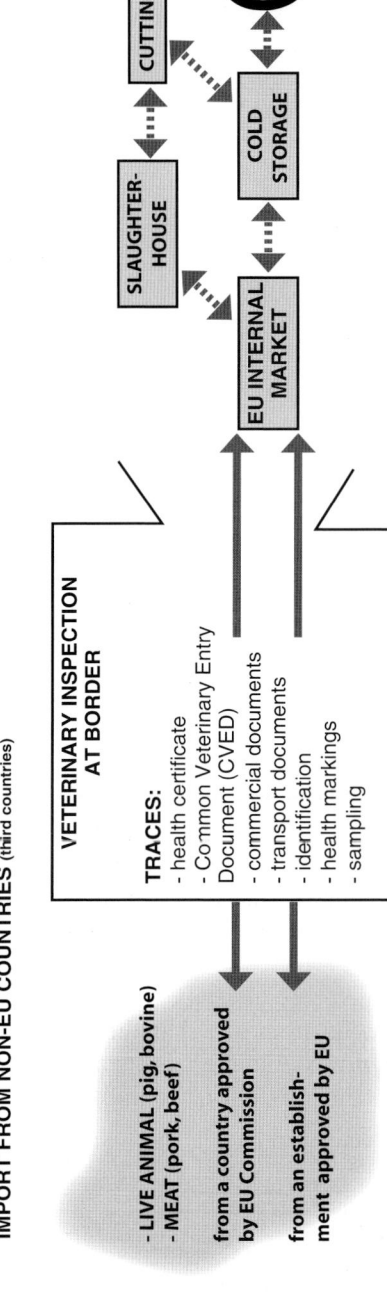

Figure 22.4 Traceability chart concerning live animals and/or meat moving from outside into the European Union.

22.6 Summary

Significance of traceability:

- crisis management, enabling targeted withdrawals;
- provision of accurate information to consumers and control authorities;
- consumer safety: products posing a health risk are quickly removed from the market;
- advantage to operators: defects can be quickly located and eliminated;
- prevention of large-scale market disruptions and financial losses.

Traceability applies to:

- food;
- feed;
- animals used in food production;
- other substances that are designed or can be assumed to be added to food or feed at any production, processing or distribution stage, with each substance being traceable at these stages.

Traceability and operators at various levels:

- primary production;
- processing/production;
- transport, distribution;
- storage;
- wholesale;
- mass caterers;
- retail sale, resale;
- intermediaries.

Traceability, labelling and documents:

Foods and feeds should carry appropriate labelling or identification information to ensure their traceability. Batches of goods shall further be accompanied by commercial documents correctly filled out with the required information.

Supplier traceability:

- BOs in food and feed business must be able to identify the person or the company that has delivered to them food, feed, an animal used in food production or a substance that is designed or can be assumed to be added to food or feed;
- supplier information, 'one step back'.

Internal traceability (i.e. process traceability):

- BOs in the food and feed business should have a system in place that allows information on incoming and outgoing batches to be linked together with adequate accuracy;

- the better internal traceability is taken into account, the more quickly and in a more targeted manner the withdrawal of food and feed can be implemented.

Customer traceability:

- BOs in the food and feed business should have systems and procedures in place that allow them to identify the other companies to which their products have been delivered;
- customer information, 'one step forward'.

Literature and further reading

European Commission. 2007. Factsheet on food traceability. Directorate-General for Health and Consumer Protection. http://ec.europa.eu/food/food/foodlaw /traceability/factsheet_trace_2007_en.pdf (last accessed 26 February 2014).

European Commission. TRACES (TRAde Control and Expert System). Directorate-General for Health and Consumer Protection. http://ec.europa.eu /food/animal/diseases/traces/about/index_en.htm (last accessed 26 February 2014).

FAO (Food and Agricultural Organization of the United Nations). Food Chain. Web pages concerning traceability: http://www.fao.org/food-chain/en/ (last accessed 26 February 2014).

23

Own-Check System

A. Structure and Implementation of the Own-Check System

Andreas Stolle

Institute of Food and Meat Hygiene and Technology, Veterinary Medicine Faculty, Ludwig-Maximilians-University, Munich, Germany

23.1 Scope

It is a generally accepted scientific discovery that the consumer has a demand for healthy and correctly produced and processed food, in this case 'meat'. The weak point in meat processing is the slaughter line and all the associated factors, such as the technical aspects, staff or buildings. The surface of meat is always unavoidably contaminated with microorganisms, which is often regarded as the main problem. But regarding the whole slaughter process and the surrounding, it is clear that the production of harmless meat requires more than just microbiological testing of carcasses. An 'own-check system' (OCS) is based on basic hygiene aspects and the principles of the 'hazard analysis and critical control point" concept (HACCP) (Figure 23.1).

The duties described should be well established in a slaughterhouse. It is important to ensure that all staff members know that a continuous functioning system that is permanently checked, controlled and improved is set up. The following points concerning the implementation of the OCS are essential:

- low staff numbers and generally low costs;
- possible to carry out and justifiable;

Meat Inspection and Control in the Slaughterhouse, First Edition.
Edited by Thimjos Ninios, Janne Lundén, Hannu Korkeala and Maria Fredriksson-Ahomaa.
© 2014 John Wiley & Sons, Ltd. Published 2014 by John Wiley & Sons, Ltd.

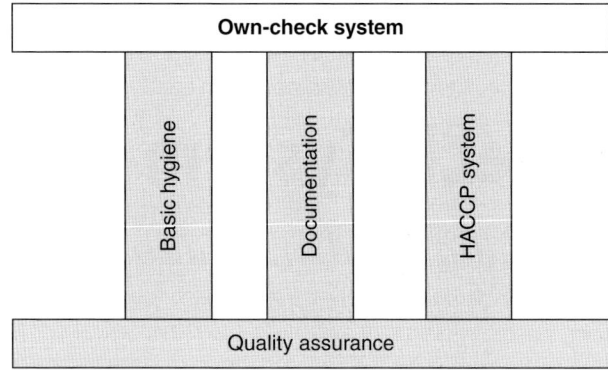

Figure 23.1 Organization of the own-check system.

- easy to integrate in already existing process procedures;
- easy to explain to the layperson.

23.2   Development of OCS

The first controls on slaughter lines were performed visually by the butchers and state authorities. In the 1960s, meat and food hygienists realized that additional measurements were necessary to set up a good hygiene quality for meat. The logical development included test parameters describing environmental influences such as contamination, pests and temperature of meat. The same parameters can usually be applied for different slaughter animal species because the main problems are the same.

In the last two decades, many experts have set up OCSs and prepared handbooks describing OCSs including HACCP concepts. The handbooks have more or less a similar content. Also, scientific consultants have naturally given similar advice in preparing OSCs in different slaughterhouses. Moreover, the quality managers may switch between different companies, thus their knowledge and practical experiences are moved from one business to the next in similar manner. The OCS presented here is a mixture of OCSs developed earlier, own model tables, experiences, practical results and literature.

The scale of slaughter activities can be very different, from large scale slaughterhouses to small butcher establishments. In small butcher establishments, the staff must be skilled and trained in all slaughter activities whereas in big slaughterhouses the working process is more specialized with different working positions. In both cases, the documentation is the basis for production of safe meat. The OCS includes the following positions without interfering with the legally required HACCP system:

- fixing of 'test points' (not identical to control points in the HACCP concept);
- surveillance control methods of these points;

- surveillance of the results of regular (legal) tests;
- cleaning and disinfection plans and check of results;
- verification of the absence of unhealthy residues;
- proof of measurements and their results.

A first step in a systematic OCS was called the 'ante slaughter check', an example of which is shown in Figure 23.2. This was performed by visual control of all relevant production areas: site, equipment, construction of buildings,

No.	Position	Problem		Problem solved
		minor	major	
1	Animal lorry /welfare			
2	Ramp, cattle / pig shed			
3	Slaughter hall organisation & order & tidiness (clean/unclean)			
4	Stunning pig/cattle			
5	Blood collector/time			
6	Tool & equipment hygiene of all buildings			
7	Staff hygiene			
8	Climate (humidity, temperature, air movement)			
9	Wash/sterilization units of all buildings			
10	Rails			
11	Elevator carcasses-hooks			
12	Pig scalding			
13	Skinning technique			
14	Skinning device			
15	Room temperature			
16	Carcass splitting			
17	Chute intestines			
18	Official inspection inner organs / carcass			
19	Confiscated material chute			
20	Hand wash devices complete			
21	Lung/liver etc. hooks			
22	All rails			
23	Exclusion rail			
24	Hall: Ceiling walls, doors, floors, windows			
25	Lightning installation			
26	Microbiological testing if necessary			

Figure 23.2 'Ante slaughter check' list.

including outbuildings, environment, water supply, effective cleaning and disinfection, pest control, waste disposal, chemical storage, check of animals and transport vehicles, hygienic slaughtering (during process), staff health and general hygiene. The visual controls were accompanied by some practical testing, such as measuring temperatures or swabbing equipment for microbiological laboratory tests. Such control had the disadvantage of not being really 'organized'. Even big plants were really not organized in a sense of business management strategy. As an important instrument to solve this problem the norm series DIN/EN/ISO 9000ff was used; this is merely an organization norm. This norm series was not specially developed for slaughterhouses but it proved to be valuable to clarify the structure and internal organization of the plant. For smaller premises it was not necessary to undergo a certification procedure, an organization plan according to the norm was sufficient to clarify the system. Bigger plants still have a system in accordance with this norm series without certification in order to keep the strict organized system as basis of a good functioning OCS. At first the results of these OCS preparations were controlled by different institutions with specialists. This may be helpful at the start, but as the term 'own-check' indicates, the own-check should be done by the slaughterhouse itself. The system has to work independently according to its fixed limits and must be checked and rechecked by the quality department or, in smaller premises, by the quality manager or the owner. At its best all the co-workers are responsible and take care of the OCS.

23.3   Implementation of OCS procedures

The ultimate aim of the OCS is to guarantee the harmlessness of meat, especially protection against infections and intoxications. Moreover, the detrimental influence on meat by, for example, pollution, smell, pests, and incorrect temperatures should be prevented. The OCS is, on the one hand, defined by legal prescriptions and, on the other hand, by good manufacturing practices. To guarantee its functioning, the OCS is based on three points:

- business lay out;
- active processing;
- staff matters.

   The first point includes all aspects that are connected with buildings, working rooms, technical equipment, tools and temperatures in the buildings, animal arrival and cleaning and disinfection. The second point generally means handling and processing of food, in this case all operations on the slaughter line up to the cold storage department, handling of offal and waste unfit for consumption. The third point includes rules regarding working clothes, personal hygiene or handling of the meat. The most important objective is to define and name 'test points' in the process. This should be followed by the determination and carrying out of control programmes at these test points.

According to the philosophy of the European Union for food safety and consumer protection in the so-called 'Green Book' of 1997, it is not relevant if a slaughterhouse is small or big in respect of the numbers of animals slaughtered: all slaughterhouses have to follow the same regulations. Therefore, it was possible to use the 'ante slaughter check' list for all premises. It strictly referred to the slaughter lines (cattle and pig) and was performed visually on the line without additionally testing or measurements. The list was restricted to some main focus points (Figure 23.2) and concerned only 'problems'. Points with no problems were not mentioned.

Additional to these documented visual inspections, the premises had microbiological checks on water supply, pest control, check of cleaning and disinfection, automatic registration of temperatures, health certificates of staff and training of staff. At the same time, the results of external controls showed increasing deficiencies in these fields. The reason is quite probably not that the quality of the OCS was reduced but more that attention was paid to these items (control of the control, control of the system).

In this regard, OCS means all planned and systematic activities of a slaughterhouse that increase the safety of the meat. OCS has to be documented for the use of the slaughterhouse and the authorities.

The significance of OCS was further developed by the European Union. According to European Union regulation an obligatory control of the OCS has to be performed by the official authorities (control of control). The OCS, which can be called 'slaughterhouses' measurements and controls', is legally required and it is obligatory to include different test points. The OCS is usually easy to understand but the relationship between OCS and HACCP may be confusing. In particular, defining obligatory critical control points can be confusing. It is of utmost importance to understand that test points in an OCS and critical control points in a HACCP system are not the same. It is possible in the OCS to measure, check or count special positions (test points) in the slaughter line and surrounding areas without having critical control points according to the definition of the HACCP system. It makes sense, therefore, to use the term 'test point' in the OCS to avoid confusion. A very good example of where confusion can be created is chilling in cold storage departments. Chilling evokes a change in the surface carcass bacterial flora from a more thermo- or mesophilic flora to the psychrotrophic group of bacteria, but the total viable count of bacteria is more or less unchanged. Definite reduction or elimination is necessary for a critical control point in the HACCP system; this is not demonstrable in chilling. This has created different and confusing views about whether HACCP can be used in chilling. Anyway, OCS is a valuable method to prove the effectiveness of the chilling system.

To improve the precision of the OCS, different norm series such as DIN/EN/ISO 9000ff or, for special food production branches, ISO 22000 (management systems for food safety) are helpful. Trade companies set up more special standards for food safety, such as IFS (International Food Standard) or BRC (British Retail Consortium), and most of the slaughter companies introduced one or several of them. All of these norms or standards include basic hygiene aspects in processing and the OCS is an essential part. Theoretically, there is no obligation by law to use such a quality management system but the trade forces the producers into these systems in order to specify their own demands. This leads to an inevitable and continuous improvement of OCS and implementation at a high scientific level.

The status of realization of the OCS varies considerably between the slaughterhouses in relation to the number of slaughtered animals. Big slaughterhouses mostly have a very well organized OCS, whereas in medium sized and small premises the systems may be incomplete. The documentation of results as the basis of verification, especially, may be neglected. Not all countries have instructions or legal regulations about how to build up an OCS and which items are inevitable parts; there are, however, many handbooks by veterinary or quality management specialists available. Additionally, there are complete handbooks issued by butchers' associations or others in different member states of the European Union. Some of them are even notified by the European Union.

The complaint of business owners or quality managers that the implementation of the OCS is difficult and causes problems is debatable. Many slaughterhouses have succeeded in establishing a very well-functioning OCS. Some segments of the system are 'self-reading' such as the automatic registration of temperatures in the cold storage department or the voltage in the pig stunning compartment, making the own-check less laborious.

Another area of the OCS is the random sampling of carcasses, mostly by destructive methods, sometimes by swabbing or rinsing. In this field, there exists intensive preliminary work by different research groups of European universities regarding the method and the sampling sites on the carcasses for pigs and for cattle. In the 1980s, the members of the 'working group red meat microbiology' of the European Union consigned these scientific data to a manual, which found its way into legal guidelines. Limits or guiding values exist in legal prescriptions, specifications and as scientifically accepted data. For routine work such as slaughtering procedures, quality management sets up its own standards.

The OCS is organized systematically. Either it is part of a global quality management system connected with a food standard or it is an 'independent' system. In both cases, it has to be fixed in a handbook. This can be a complete handbook, but sometimes a list of contents and the necessary pages for documentation are sufficient. The simplest principle includes the table of contents and necessary pages for documents. One example of the table of contents for small and medium sized businesses is presented in Figure 23.3. In a local butchery, this works if each chapter has an additional form for the documentation, in other companies it is just a list of contents.

Experts are unanimous that a 'handbook' is not necessarily a big printed book in its original sense. It may be a file or a collection of the necessary forms, depending on the production numbers of the business. The results are laid down in forms and evaluated daily/weekly in order to eliminate the defects. Minor problems after visual evaluation are eliminated immediately

1. Legal Basics / Norms
2. Overview of necessary documentation and cross-references (management handbook if available)

3. Own-check measurement in practice
3.1 Construction /processing plan / servicing
3.2 Traceability
3.3 Staff-training
3.4 HACCP concept
3.5 Testing of product (e.g. meat)
3.6 Testing of drinking water
3.7 Control of temperatures
3.8 Staff hygiene
3.9 Rendering certificate - offal /waste /water
3.10 Cleaning and disinfection
3.11 Pest control
3.12 Processing hygiene and technique (on line)
    VISUAL – slaughter line test points:
        Cattle stunning/bleeding/skinning/decapitation/opening chest and abdomen
        Removal inner organs (belly, breast)/splitting
        Pig stunning/bleeding/scalding/scratching/whipping/evisceration/Splitting
    PRACTICAL carcass sampling for microbiology

Figure 23.3 Contents of an OCS handbook.

by verbal order. These data are additionally the basis for the regularly (e.g. yearly) quality management evaluation.

In the case of a full OCS or a certified quality management norm, respectively, a 'Food Standard' shows the further course of testing and documentation. Usually this is the method of choice in medium sized or big companies. Also, on this level the principle is that the management running the business is responsible for fixing the test points. The structural principle is the same in the different sized premises but varies in the different levels. These levels are presented in Figure 23.4.

Because of the extent of a large scale OCS for a pig and cattle slaughter line the system is described in mixed form. The ante slaughter control list (Figure 23.2) can be used for all purposes, as well as 'the table of contents' shown in Figure 23.3. The own-check documents (Figure 23.5) are produced in the same manner for both types of line (pig and cattle), showing the connection to a fictitious hand book. Figure 23.6 gives a realistic OCS checklist for a 'cattle and a pig line'. Figure 23.7 shows exemplary important test points. Figure 23.8 leads to the next step – the verification of the system.

During daily routine work the employees mostly prefer strict regulations instead of a kind of a philosophical framework. The staff may, however, think that intensive documentation is excessively more work. Therefore, keeping the documentation simple by just making a symbol, letter or number is the best way to keep the system running. Regular staff training is one of the

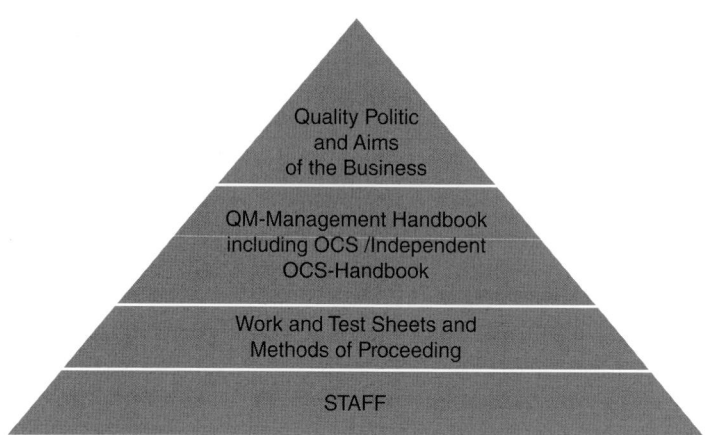

Figure 23.4   The structural principle of the OCS.

Own-check document	Frequency of performance	Fictitious chapter
Business plan with flow chart material and staff		3.1
Back tracing see special working sheet	Acknowledgement in & out	3.2

Staff training
  - Yearly hygiene training
  - Information of disease acts
  - HACCP principles
	1 x per year	3.3

HACCP concept see special working sheet		3.4
Product control (meat)	Daily	3.5

Temperature observance
  - Temperature records
  - Temperature prints cold storage
  - Check measuring device
	Daily
	Temperature calibration once a year
		3.6

Waste water		3.7
Staff hygiene - Health certificate		3.8

Cleaning and disinfection (C&D)
  - Systematic plan C&D
  - Safety instruction for C&D agents
  - Sampling plans
  - Results/success control
	Disinfection control 1 x/ 3 months
	Cleaning slaughter line, cutting department
	Daily and if necessary
		3.9

Pest control Bait plan/results/safety sheets media	1 x month /certified control contractor	3.10
Productions process hygiene	Permanent complete	3.11
All test points complete pig/cattle		3.12

Figure 23.5   Own-check documents and connection to working sheets.

Nr	Test point			Checked
1	Staff hygiene washrooms. Sluice-gate			
2	Cleaning and desinfection in equipment and rooms			
3	Only use of allowed substances			
4	Collection of contaminated material			
5	Color system for containers			
6	Physical contamination, technical grease			
7	Strict separation of meat organs fit/unfit consumption			
8	Removal of organs and working separately			
9	Cooling of by-products			
10	Parallel transport carcass / organs			
11	Contact between carcasses (cross contamination)			
12	Air stream, temperature			
13	Lightning installation			
	Slaughter line			
Cattle		Pig		
	Cattle driving to stunning device	Pig driving		
	Stunning	Stunning		
	Skinning neck - bleeding	Bleeding		
	Live weighing (sometimes)	Vaporization tunnel		
	Fore feet head skinning	Removal of claws		
	Rodding	Flaming		
	Removal head and horns	Polishing		
	Rectum bunging /end of tail	Bung dropper		
	Skinning hind feet	Ears and eyes removal		
	Penis /udder	Evisceration breast belly		
	Hide puller	Carcass splitting		
	Sternum saw removal/ breast organs - entrails	Meat inspection		
	Carcass splitting	Trimming complete eg. spinal marrow, kidneys, sticking meat, brain.		
	Spinal marrow removal			
	Meat inspection			
	Trimming			
	Cold storage department			
		Sterilizations	68°C	°C
			68°C	°C

Figure 23.6 Test points for a cattle and pig slaughter line (general aspects and without verification examples).

Selected examples from personnel areas and technical areas	Done	Deviation	Corrective manner	Repeated control	
				Time / Date	Controller
Changing room / Hygiene sluice-gate					
Social Rooms					
Staff shower complete					
Waste baskets					
Splinter free lamp, windows etc.					
Condensation					
Liquid soap /disinfectant /hand wash					
Area intestine working					
Area inner organs working					
Anus bagging					
Contamination by technical greases					
Monitoring pest control					
Calibration pH instrument					
Calibration thermometer					
Calibration pair of scales					

Figure 23.7 Important test points in OCS (examples).

Cattle slaughter line	Pig slaughter line
- Stable and droving: animal welfare - Stunning: point of captive bolt (other system) - Bleeding: multiple knife technique and time - Rodding: precise position and closing - Decapitation: multiple knife technique - Rectum: careful bagging - Sternum saw: no perforation of intestines	- Shed: animal welfare - Stunning: concentration of gas /electricity - Bleeding: time to stunning, multiple knife - Condensation: temperature - Bung dropper: clean bagging - Evisceration: no cuts in intestine

Figure 23.8 Slaughter line verification checklist with some examples (1 × week changing days).

most important ways of bringing quality management thinking and the OCS into practice.

Regarding Figures 23.5–23.7, it is obvious that the OCS emphasizes certain sectors with test points. The main problem is that not everything can be controlled by permanent measurements such as the temperature in the cold storage department. Other work sectors are based on recording observations, such as written health confirmations and sampling plans (Figure 23.5). Most of the test points are verified by visual controls. These are performed by specially trained persons or instructors; but they cannot observe the complete process continuously. So the butchers working in the different positions have to be well aware of their own peformance. For this purpose they are repeatedly trained in the test points, for example 'production process hygiene' (Figure 23.5). Special attention has to be paid to certain working areas, such as stunning. It is necessary to produce special forms, for example the gas concentration during carbon dioxide stunning (pig) or the position of captive bolt (cattle). These test points must be defined systematically.

Each test point is controlled continuously by visual inspection, a working sheet enables correct handling and special items are measured, for example time elapsed between stunning and bleeding. Additional forms are kept for verification (Figures 23.7 and 23.8).

Beside these checks for handling, technique and so on, there always exists the problem of microbial contamination during processing by tools, through unhygienic equipment handling and staff hygiene as well as for other reasons. The workers and their, eventually, unclean handling are the main focus. Additionally, the animals themselves bring an unavoidable amount of microorganisms into the slaughterhouse. The sampling of carcasses on the line and in the cold storage department shows clearly that there is no zero-tolerance of bacteria or absence of pathogens. It is not possible to supply slaughterhouses with pathogen-free slaughter animals in spite of all the efforts carried out at the farm level. Surface samples of carcasses are very suitable for measuring the hygienic status of different steps of the slaughter process. Mistakes made in cleaning and disinfection are often not visible. The evaluation of the effectiveness of the cleaning and disinfection, therefore, has to be performed by microbiological testing of process surfaces. This shows the importance of all items of the OCS and the necessity for strict adherence in all positions of the slaughter line.

There are three main considerations when assessing the OCS; they are unaffected by the size of the slaughterhouse and relate to the requirements concerning:

- infrastructure and equipment;
- staff;
- the test points.

23.4   Verification of the OCS

In the initial phase, implementing an OCS needs the relevant test points to be identified. The results of this status evaluation have to be connected, with certain limitations, either by visual control or measurable means (limit). The measurement and limits of an OCS need always for all steps to be verified, comparing the present status to the predetermined status. Verification (lat. veritas = truth / facere = to do) means that the request to adhere to a norm, limit or duty is performed.

As shown before, it is necessary to keep obligatory documentation (Figure 23.7). It is of utmost importance not only to document any deviation but also how the deviation was corrected. In routine work a repeated control is obligatory. Figure 23.7 is only one of the documentation sheets. Similar ones have to be prepared for all test points in all slaughter lines; in small butchery premises the documentation sheets are short ones, in big slaughterhouses they are more detailed. Figure 23.8 is a form designed for the verification procedures. These are performed either by the quality management department or, in small establishments, by the butcher him/herself (internal audits). It is helpful to have the OCS inspected by external specialists to avoid mistakes that are not realized because a person becomes used to mistakes through daily routine work. Beside these visual assessments of deviations and correction steps other measurements are, of course, useful, for example results of the microbiological testing of carcasses. The documentation must be checked regularly. The principle of OCS is that 'what is not documented, is not done'. Some workers may not appreciate the documentation and, therefore, the system is neglected, but without permanent documentation the system is useless.

The OCS handbook leads from the basic hygiene, technology and staff to the principles of a HACCP concept and is part of the total slaughter procedure to keep the high standard in an approved slaughterhouse. Figure 23.9 shows the position of the OCS in an approved slaughterhouse. In total, there are three main considerations in the slaughterhouse:

- infrastructure and equipment are basic pre-conditions;
- plant and process hygiene;
- OCS.

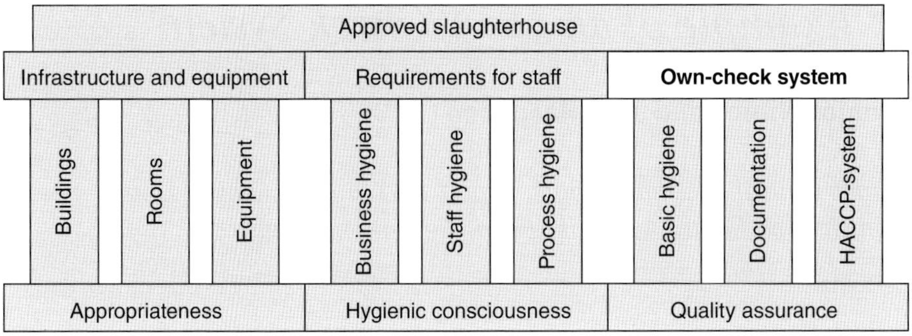

Figure 23.9 Position of OCS in an approved slaughterhouse.

Literature and further reading

Branscheid, W. 2007. Qualitätsmanagement bei Fleisch und Fleischwaren. In: *Qualität von Fleisch und Fleischwaren* (eds W. Branscheid, K.O. Honikel, G. von Lengerken and K. Troeger). Deutscher Fachverlag, Frankfurt-am-Main, Germany, pp. 49–71.

Buckenmaier, T. 2002. Das globale Konzept der Lebensmittelhygieneverordnung für Eigenkontrollen – seine Praktikabilität für registrierte Betriebe zwischen Anspruch und Realisierbarkeit. Veterinary Medicine Thesis, The Ludwig-Maximilians-University, Munich, Germany

Fehlhaber, K. 2004. Verderb. In: *Einführung in die Lebensmittelhygiene* (ed. H.J. Sinell). Parey Verlag in MVS Medizinverlage, Stuttgart, Germany, pp. 109–120.

Stolle, A. 1985. *Problematik der Probenentnahme für die Bestimmung des Oberflächenkeimgehaltes von Schlachttierkörpern.* Faculty of Veterinary Medicine, Free University of Berlin, Germany.

Stolle, A. 1995. Eigenkontrollsysteme in der Fleischwirtschaft (3.2). In: *Leitfaden Qualitätsmanagementsysteme in der Fleischwirtschaft* (ed. M. Torres-Peraza). Bizerba, Balingen, Germany, pp. 1–6.

Stolle, A., Zechel, P. and Bucher, M. 2006. *Handbuch zur Einführung und Umsetzung betrieblicher Eigenkontrollsysteme.* Institute for Hygiene and Technology, The Ludwig-Maximilians-University, Munich, Germany.

Zechel, P. 2006. Umsetzungen der Anforderungen der EU an registrierte Betriebe an 2006 hinsichtlich Umsetzbarkeit und Verbraucherschutz. Veterinary Medicine Thesis, The Ludwig-Maximilians-University, Munich, Germany.

Schütz, F. 1991. *Hygienekonzept für Schlachthöfe.* Ferdinand Enke Verlag, Stuttgart, Germany.

B. Example of an Own-Check System

Thimjos Ninios and Joni Haapanen

*Border Control Section, Import, Export and Organic Control Unit,
Finnish Food Safety Authority Evira, Helsinki, Finland*

23.5   Introduction

An own-check system (OCS) for a slaughterhouse may include some of the elements advanced in the following example. The structure of this example is indicative and, therefore, some of the contents mentioned may not be applicable in all types of slaughterhouse. Also, the OCS might include other contents where needed. The OCS should, in any case, comply with the requirements set by the legislation in force and be structured by three different but interconnected phases:

- planning of the own-check
- implementing the own-check plan
- documenting the own-check

Every element of the own-check plan needs to be implemented and the implementation needs to be documented.

23.6   Own-check plan

The operator of the slaughterhouse should plan the own-check activities and provide information at least about what it is planned to do, which way it needs to be done, when and/or how often it needs to be done and who is responsible for doing it. The table of contents of an own-check plan might contain the headings or elements shown below.

Scope of own-check
The own-check plan should describe the purpose of the own-check system.

Distribution of responsibilities
The responsibilities concerning planning, implementation and record keeping of own-check shall be distributed in as much detail as possible.

Flowcharts and establishment plans
The own-check plan should include at least the up-to date-layouts, floor plans and ground plans of the slaughterhouse. Flowcharts concerning product flow,

activities performed in the slaughterhouse, water supply, routes established for the personnel, routes established for the by-products and wastes should be also included.

Employee hygiene and training

The own-check plan for the slaughterhouse should ensure that the personnel's health does not introduce risks for food safety. Personnel should also be regularly trained on hygiene issues.

Working hygiene

There should be a plan ensuring that any activities at the slaughterhouse are performed hygienically. (See also Chapter 20.)

Control of temperature

The operator of the slaughterhouse should ensure that the temperature of the cold rooms, equipment sterilizers and any other temperature-linked equipment fulfils the requirements set by any legislation. The own-check plan should describe at least the limits of temperature to be respected and the modalities to be used to ensure compliance with these limits.

Water supply

The operator of the slaughterhouse should ensure that the incoming water complies with legal requirements. The own-check plan should describe, for example, the required analysis concerning the quality of the water supplied.

Pest control

The activities directed against the introduction or presence of pests in the slaughterhouse and the surroundings need to be planned. (See also Chapter 19.)

Cleaning and disinfection

The own-check plan of the slaughterhouse should ensure that the slaughterhouse is cleaned and disinfected properly. (See also Chapter 18.)

Maintenance

The own-check plan of the slaughterhouse should contain regular maintenance of the establishment. The maintenance of machinery and equipment needs also to be planned.

Waste management

The own-check plan of the slaughterhouse should ensure that food waste and other waste are managed and removed properly from the slaughterhouse. (See also Chapter 14.)

Management of by-products

The operator of the establishment should ensure that by-products are managed and removed properly from the slaughterhouse. (See also Chapter 14.)

Contingency plans
A contingency plan should include actions directed to control major risks such as, for example, animal infectious diseases introduced in the slaughterhouse. For this event corrective action should be planned concerning, at least, the movement of personnel, animals, products obtained from infected animals into and out of the establishment. Preventive actions to avoid spreading a harmful condition should be described.

Processing of food chain information
The slaughterhouse should implement a plan concerning the elaboration of food chain information. (See also Chapter 2.5.)

Identification and handling of animals
The operator of the slaughterhouse should ensure the proper identification and human handling of animals. (See also Chapters 3 and 5.)

Traceability
The own-check plan should guarantee the traceability of the products of slaughtering. (See also Chapter 22.)

Management of risk material (TSE)
There should be a plan ensuring the correct management and removal of risk material from the slaughterhouse. (See also Chapter 14.)

A HACCP plan for every slaughter line
A HACCP plan may include the following steps:

- compose a HACCP team;
- describe the product;
- identify the intended use;
- create the flow diagram/definition of process;
- confirm flow diagram;
- identify and list potential hazards;
- conduct a hazard analysis (principle 1);
- specify control measures;
- determine critical control points (principle 2;)
- establish critical limits (principle 3);
- establish monitoring system (principle 4);
- establish corrective action plan (principle 5);
- validate, verify and review the HACCP plan (principle 6);
- establish documentation and records (principle 7).

(See also Chapter 23C.)

Sampling plan
The OCS of the slaughterhouse needs to be validated and the functionality of it verified according to a plan. For example, the evidence concerning

the effectiveness of cleaning and disinfection at the slaughterhouse can be obtained through sampling of surfaces and analysis of the results.

Corrective actions
Corrective actions should be established at least for those deviations from the plan that are considered highly probable. The slaughterhouse should be prepared to face misadventures, such as operational failure in the division of carcasses into half carcasses causing the disruption of an abscess that is present in the vertebral column, thus spoiling other carcasses.

Regular evaluation of the own-check plan
The operator of the slaughterhouse should regularly evaluate the validity and effectiveness of the OCS. A review of the OCS should, for example, be considered whenever changes in the slaughter process, changes concerning the structures or the equipment, introduction of new risks or changes in legal requirements take place.

23.7   Own-check implementation

The implementation of the own-check plan comprises implementing accurately each procedure established in the own-check plan. Also, certain actions need to be taken on demand, such as equipment maintenance in the case of equipment failure or cleaning and disinfection in the case of accidental breakage of intestine and faecal contamination of the slaughter line. Corrective actions should be taken whenever a deviation from the plan takes place.

23.8   Own-check documentation

The implementation and evaluation of the own-check plan need to be documented. The documentation should include information about the type of check, time of check, result of check, possible corrective actions in the case of deviation and the responsibility for the check (Figure 23.10). The lack of documentation might be a sign of deficiencies in the implementation of the own-check plan.

23.9   Division of own-check components in SSOPs and SPSs

Some countries and/or exporting slaughterhouses might use a specific categorization of the own-check components, dividing them into Sanitation Standard Operating Procedures requirements (SSOPs) and Sanitation Performance Standards (SPSs). The target is to perform both properly and prevent the creation of insanitary conditions that may cause contamination

OWN CHECK RECORD
Type of check: *Check cleanliness of food contact surfaces.*
Target: *Hooks for organs.*
Time: *2.5.2013, 12:12-12:19 during lunchtime.*
Result: *30 hooks without organs checked. Contamination by grease for machinery was detected on 1 hook.*
Corrective action: *The contaminated hook was removed from the slaughter line, cleaned and disinfected. The machinery was checked and excessive grease was eliminated.*
Responsible for the operation: *Name Surname*

Figure 23.10

of the product. After the SPSs and SSOPs have been developed comes the creation of a HACCP plan (for detailed information on HACCP, see also Chapter 23C).

23.9.1   The SSOPs

According to the SSOP principles, the slaughterhouse needs to develop, implement and maintain written procedures to perform daily, before and during the operations, in order to prevent a direct contamination or adulteration of the product. These written procedures need to:

- contain the sanitation standard operating procedures that the establishment should perform daily, before (pre-operatively) and during the operations (operatively);
- identify the pre-operational procedures and point out, at least, the cleaning of food contact surfaces of facilities, equipment and tools;
- specify the frequency with which each pre-operative and operative procedure needs to be performed and distribute the responsibility for the implementation, documentation and maintenance of the procedure.

23.9.2   The SPSs

The SPSs focus on the environment where slaughtering takes place. The goal of the SPSs is the proper sanitation of the slaughterhouse in order to create an appropriate environment for safe production of meat. The SPSs include:

- sanitation of the establishment's grounds and facilities
  - grounds and pest control

- construction
- light
- ventilation
- plumbing
- sewage disposal
- water supply and water, ice and solution reuse
- dressing rooms, wash rooms and toilets;

- sanitation of equipment and tools;
- sanitary operations;
- employee hygiene;
- tagging insanitary equipment, tools, rooms or compartments to avoid their use.

Literature and further reading

Codex Alimentarious. International Food Standards. http://www.codexalimentarius.org/ (last accessed 26 February 2014).

EUR-Lex. Up-to-date EU legislation on requirements concerning slaughterhouses. http://eur-lex.europa.eu/fi/index.htm (last accessed 26 February 2014).

C. HACCP

Robert Savage

HACCP Consulting Group, Fairfax, VA, USA

23.10　History

The Hazard Analysis and Critical Control Point (HACCP) food safety system was developed in the mid-1960s by the Pillsbury Company in the United States. HACCP was developed jointly with the US National Aeronautic and Space Administration (NASA) and the US Army Natick Laboratories to produce foods for the US space programme.

The development of HACCP was dictated by the need to ensure that the foods produced for the space programme were free of all pathogenic microorganisms that could cause illness to astronauts. Its development was further necessitated when it became obvious that traditional quality control procedures of end-product testing would not provide the necessary assurance of product safety. To resolve this concern Pillsbury evaluated NASA's zero-defect programme and the US Army Natick Laboratories' modes of failure analysis system and concluded that the zero-defect concept could be extended to control the entire food production process, beginning with the raw materials through to the finished product. Combined with Natick's modes of failure analysis, the food processing system could be evaluated to identify sensitive ingredients and process steps (i.e. hazard analysis) where critical points must be controlled and monitored to assure the safety of the finished product.

Pillsbury's initial HACCP system contained the following elements:

- identification and assessment of hazards associated with the growing, harvesting, processing–manufacturing, marketing, preparation or use of a given raw material or food product;
- determination of critical control points (CCPs) at which identified hazards can be controlled;
- establishment of procedures to monitor the critical control points.

Pillsbury first presented the HACCP system to the food industry at the 1971 National Conference on Food Protection. Subsequently, four U.S. federal agencies (i.e. the National Marine Fisheries Service (NMFS), the USDA Food Safety and Inspection Service (FSIS), the US Food and Drug Administration (FDA) and the US Army Natick Research and Development Center) requested that the US National Research Council convene a panel of experts to formulate general principles for the application of microbiological criteria for foods. The National Research Council endorsed the HACCP principles that had first been introduced at the 1971 National Conference on Food Protection.

Prior to the National Research Council's endorsement of the HACCP principles, the International Commission on the Microbiological Specifications for Foods (ICMSF) also examined the Pillsbury HACCP principles and it published 'HACCP in Microbiology Safety and Quality', which articulates six HACCP principles that contained an expanded version of the Pillsbury HACCP principles as delineated in Table 23.1.

In 1997, the US National Advisory Committee on Microbiological Criteria for Foods (NACMCF) also prepared guidelines setting out the principles of a HACCP system. The major difference between the NACMCF and ICMFS principles was the former's highlighting of the importance of recordkeeping by identifying it as a separate HACCP principle and the elimination of a two-level critical control point (i.e. CCP1 and CCP2) system.

In 1993, the Food Hygiene Committee of the Codex Alimentarious Commission also prepared a document: 'Guidelines for the Application of the Hazard Analysis Critical Control Point (HACCP) System'. These guidelines were consistent with those published in 1992 by the NACMCF and have been adopted by the full commission. Table 23.2 identifies the seven HACCP principles as they appear in the CODEX document.

The first regulatory application of the HACCP principles occurred with the publication of low-acid canned food regulations (21 Code of Federal Regulations (CFR)) by the FDA in 1972 to address serious issues with *Clostridium botulinum* in canned shelf-stable products. Subsequently, FDA published HACCP-based regulations for the production and processing of seafood and juice drinks in 1995 and 2001, respectively. FSIS also published regulations

Table 23.1 ICMSF HACCP principles.

Principle	Function
No. 1	Identification of Hazards and assessment of severity of these hazards and their risks (hazard analysis) associated with growing, harvesting, processing/manufacturing, distribution, merchandizing, preparation and/or use of raw material or food contact.
No. 2	Determination of critical control points (CCPs) at which identified hazards can be controlled. (CCP1 – completely eliminates the hazard; CCP2 – minimizes but does not completely control the hazard.)
No. 3	Specification of criteria that indicate whether an operation is under control at a particular CCP.
No. 4	Establishment and implementation of procedure(s) to monitor each CCP to check that it is under control.
No. 5	Taking whatever corrective action is necessary when the monitoring results indicate that a particular CCP is not under control.
No. 6	Verification, i.e. the use of supplementary information to ensure that the HACCP system is working.

Table 23.2 Codex Alimentarious commission HACCP principles.

Principle	Function
No. 1	Conduct a hazard analysis.
No. 2	Determine the Critical Control Points (CCPs).
No. 3	Establish critical limit(s).
No. 4	Establish a system to monitor control of the CCP.
No. 5	Establish the corrective action to be taken when monitoring indicates that a particular CCP is not under control.
No. 6	Establish procedures for verification to confirm that the HACCP system is working effectively.
No. 7	Establish documentation concerning all procedures and records appropriate to these principles and their application.

(21 CFR 417) in 1996 requiring the application of HACCP for all red meat and poultry processing, including slaughterhouses. In 2004 the European Union (EU) adopted HACCP in Article 5 of Regulation (EC) No. 852/2004 of the European Parliament and Council Directive 93/43/EEC on the hygiene of foodstuffs, which required food business operators to put in place, implement and maintain a permanent procedure based on the HACCP principles. In 2013, FDA also published proposed regulations to require all food processors to develop and implement preventive-based (HACCP) food safety systems.

More recent applications of the HACCP principles have occurred through the Global Food Safety Initiative (GFSI). GFSI is a business-driven initiative for the continuous improvement of food safety management systems to ensure confidence in the delivery of safe food to consumers worldwide. HACCP is the foundation of all the GFSI approved food safety schemes including, for example, Food Safety System Certification 22000, British Retail Consortium (BRC), Safe Quality Foods (SQF) and International Featured Standards (IFS).

The above history of HACCP demonstrates that it is a pattern of continuous evolution and why it is important to know the past history of this important food safety programme.

23.11 The HACCP principles

The Codex Alimentarious Commission requires that food business operators put in place, implement and maintain a permanent procedure or procedures based on the HACCP principles. The steps involved in developing a HACCP plan include the activities detailed here.

23.11.1 Hazard analysis

The first step in developing a HACCP plan (i.e. written document) and system (i.e. implementation of the written plan) is to assemble a multidisciplinary team (e.g. quality control, quality assurance, sanitation, maintenance, management, product development etc.) involving all parts of the food business. The HACCP team needs to possess adequate knowledge and expertise to plan and develop the HACCP plan and system.

The first task of the HACCP team should be to ensure that the slaughterhouse is in compliance with all relative Good Manufacturing Practices (GMPs) (e.g. sanitation standard operating procedures, pest control, personal hygiene etc.) and other regulations before developing the HACCP plan and system. Having GMPs and other prerequisite programmes (e.g. sanitary dressing procedures of carcasses etc.) in place will aid the HACCP team during the conduct of the HACCP plan hazard analysis to determine what hazards are reasonably likely to occur in the process.

The team should then describe the product(s) covered by the HACCP plan, the intended use of the product, develop a flow diagram describing the manufacturing process, verify on-site the accuracy of the flow diagram and, finally, list the hazards and control measures that may be reasonably expected to occur at each process step, including receipt of raw materials and ingredients.

In conducting the hazard analysis the HACCP team should consider:

- the likely occurrence (i.e. risk) and severity of hazards and their adverse health effects;
- the qualitative and/or quantitative evaluation of the presence of hazard;
- survival or multiplication of pathogenic microorganisms;
- production or persistence of toxins or other microbial metabolites, chemical or physical agents or allergens;
- contamination (or recontamination) of a biological, chemical or physical agent on the raw materials, in-process product or finished product.

The hazard analysis should only address biological, chemical and physical hazards from raw material production, receiving and storage to manufacturing, distribution and consumption of the finished product that are reasonably likely to cause injury or illness. Good manufacturing practices (GMPs) and product quality characteristics (e.g. flavour, texture, appearance etc.) should be addressed in other prerequisite programmes, such as pest management controls, water quality and employee hygiene.

Control measures are those actions or activities that can be used to prevent hazards, eliminate them or reduce their impact or occurrence to acceptable levels. Some acceptable levels, such as pathogenic microorganisms (i.e. Salmonella) in raw beef, pork and poultry products, may be established by government regulations based on scientific studies. Control measures must be supported by detailed procedures that describe the step-by-step activities and measures taken by employees to ensure that the measures are effectively implemented and maintained. Supporting documentations, such as scientific studies, in-plant test results and process validation data, should be available to support the decisions reached in the hazard analysis.

23.11.2   Critical control points (CCPs)

A critical control point is defined as a step at which control can be applied and is essential to prevent or eliminate a food safety hazard or reduce it to acceptable levels. The potential hazards that are reasonable likely to cause illness or injury in the absence of their control must be addressed in determining CCPs. It is important to note that with the phrase 'or reduce to acceptable levels' HACCP and associated CCPs are applicable to raw processes (e.g., cutting facilities, raw ground product and slaughterhouse etc.) where there are no steps that can completely prevent or eliminate a biological hazard.

Complete and accurate identification of CCPs is fundamental to controlling the food safety hazards identified in the hazard analysis. One strategy to facilitate the identification and location of each CCP is the use of a CCP decision tree (Figure 23.11).

Examples of CCPs in the slaughterhouse may be for zero tolerance for faecal material, ingesta and milk contamination of the carcass prior to entering the cooler and specific cooling time and temperature of the carcass after slaughter. Other examples of CCPs may be hot water rinses or acid washes of carcasses to reduce biological hazards; however, these interventions are not allowed in some countries. The possible use of these treatments should be verified from review of up-to-date legislation in each country or union.

23.11.3   Critical limits

A critical limit is a maximum and/or minimum value to which a biological, chemical or physical hazard must be controlled at a CCP. A critical limit is also used to distinguish between a safe and unsafe operating condition at a CCP. Each control measure can have one or more critical limits. Critical limits must be based on sound science and supporting documentation maintained as part of the HACCP plan. Examples of critical limits may be zero visible faecal material, ingesta and milk on carcasses after slaughter and internal carcass temperature within 24 hours of slaughter.

23.11.4   Monitoring procedures

Monitoring is a planned sequence of observations or measurements to assess whether the CCP is under control and to produce an accurate record for future use in verification. For each CCP, the person (by title) responsible for conducting the monitoring, what is monitored, how it is monitored and the frequency of monitoring must be identified in the HACCP plan. For example, visual monitoring of a certain number of carcasses per hour for faecal contamination and checking of carcass temperature by the cooler operator every hour using a handheld thermometer may be considered as acceptable monitoring activities. To supplement the monitoring procedures identified in the HACCP plan, facilities often develop a detailed standard operating procedure (SOP) for each monitoring activity that can also be used as a training tool.

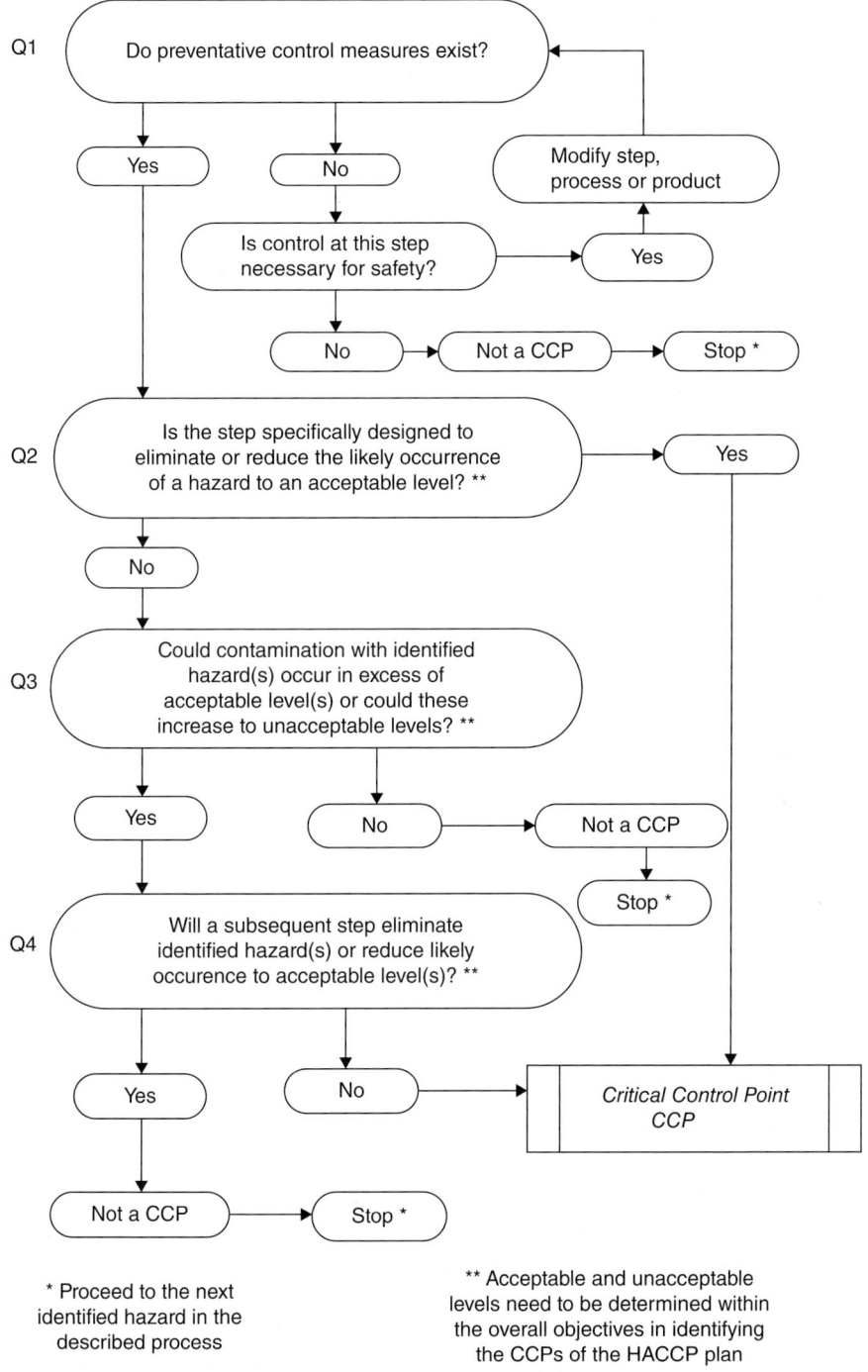

Figure 23.11 CCP decision tree.

23.11.5 Corrective actions

When a deviation occurs from an established critical limit, corrective actions are necessary to bring the process back under control and prevent a food safety hazard. For each CCP, corrective actions should include the following elements:

- determine and correct the cause of the deviation;
- determine the disposition of the affected product;
- record the corrective actions that have been taken.

At a minimum, the HACCP plan should identify what is done when a deviation occurs, who is responsible and the record maintained of the corrective actions. Various corrective actions could be considered, including reprocessing of the product, destroying the product or holding the product for review by an expert.

23.11.6 Verification and validation

Verification is defined as those activities, other than monitoring, that determine the validity of the HACCP plan and that the system is operating according to the plan. Verification activities can include, but are not necessarily limited to, review of monitoring and corrective action records, direct observation of the persons responsible for conducting the monitoring and taking the correct actions, calibration of any monitoring equipment and testing of finished product samples. Verification is different from monitoring in that it is done by a different person and, generally, at a different frequency than the monitoring. For each CCP and verification activity the HACCP plan must include who is responsible, the frequency of each activity and where the results are recorded.

A subset of verification is validation. Validation determines if the HACCP plan is properly implemented and the hazards are effectively controlled. Information needed to validate the HACCP plan may include advice from experts, scientific studies and in-plant observations, measurements and evaluations.

23.11.7 Documentation and recordkeeping

Efficient and accurate recordkeeping is a major part of the HACCP system to document compliance with the HACCP plan and system. Accurate records can also serve to demonstrate 'due diligence' should questions arise regarding the safety of a product and can also be essential to minimize the scope of a potential product recall should there be a food safety hazard identified.

Documents and records could include, but not be limited to, the HACCP plan (i.e. hazard analysis, CCPs, critical limits, monitoring procedures, corrective actions, verification procedures and activities and recordkeeping practices), all supporting documentation associated with the HACCP plan, CCP monitoring records, corrective action records and verification records.

23.12 HACCP at the slaughterhouse

23.12.1 Livestock slaughter

The application of HACCP principles to the slaughter process is highly contingent on the slaughterhouse having well designed and implemented prerequisite programmes such as sanitation standard operating procedures (SSOPs) and sanitary dressing procedures of carcasses during the slaughter process. SSOPs describe the day-to-day, both pre-operational and operational activities necessary to maintain sanitary conditions in the facility while sanitary dressing procedures describe the step-by-step measures during the slaughter process to prevent cross-contamination of the carcass during the killing, dehiding and evisceration process. These procedures, coupled with the recognition that there are no steps in the slaughter process to prevent or eliminate biological hazards such as *Salmonella* or *E. coli* O157:H7, put more reliance on the third component of the definition of a CCP (i.e. reduce to an acceptable level).

Referencing typical flowcharts for cattle slaughter and swine slaughter (Figures 23.12 and 23.13), the absence of faecal material, ingesta and milk on the carcasses prior to chilling and the chilling process itself may be considered CCPs to reduce the potential contamination of the carcasses with faecal material, ingesta and milk recognizing that faecal material is a major source of pathogenic microorganisms, and preventing the further growth of pathogens on the carcasses during the chilling process. Critical limits for faecal material, ingesta and milk and the carcass and carcass chilling could be no visible faecal material, ingesta and milk and internal carcass temperature within 24 hours, respectively.

23.12.2 Poultry slaughter

The application of HACCP to poultry slaughter is very similar to that of livestock slaughter in that it is highly dependent on the slaughterhouse having well designed and implemented prerequisite programmes (e.g. rinsing of evisceration equipment, testing of flocks prior to slaughter etc.) and sanitation standard operating procedures (SSOPs).

Referencing a typical flowchart for poultry slaughter (Figure 23.14), cross-contamination with faecal material can be a major source of *Salmonella* as well as *Campylobacter*. Additionally, the proper and rapid chilling of the carcasses will help prevent to growth of any pathogens on the finished product. Critical limits for faecal material and carcass chilling could be no visible faecal material prior to entering the chilling system (i.e. water or air) and bird temperature at the exit of the chiller.

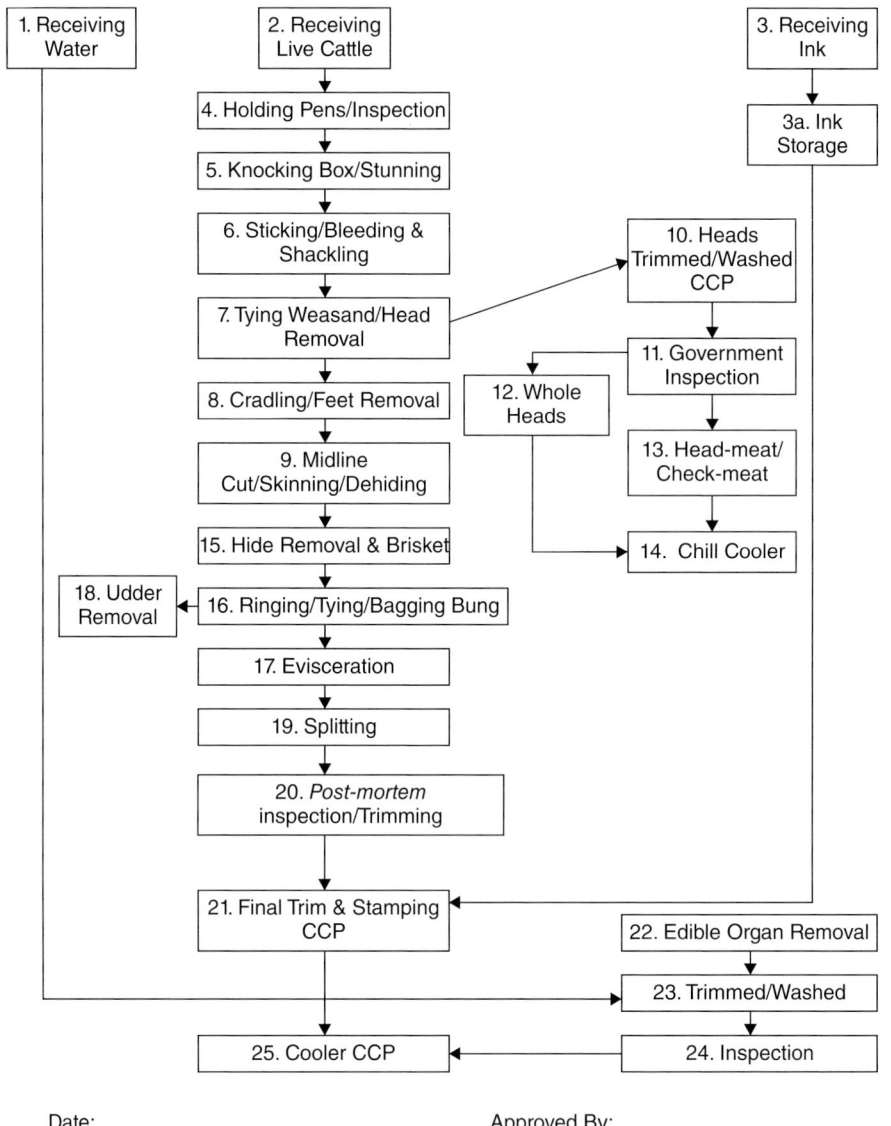

Date: _____   Approved By: _____

Figure 23.12   Cattle slaughter flowchart.

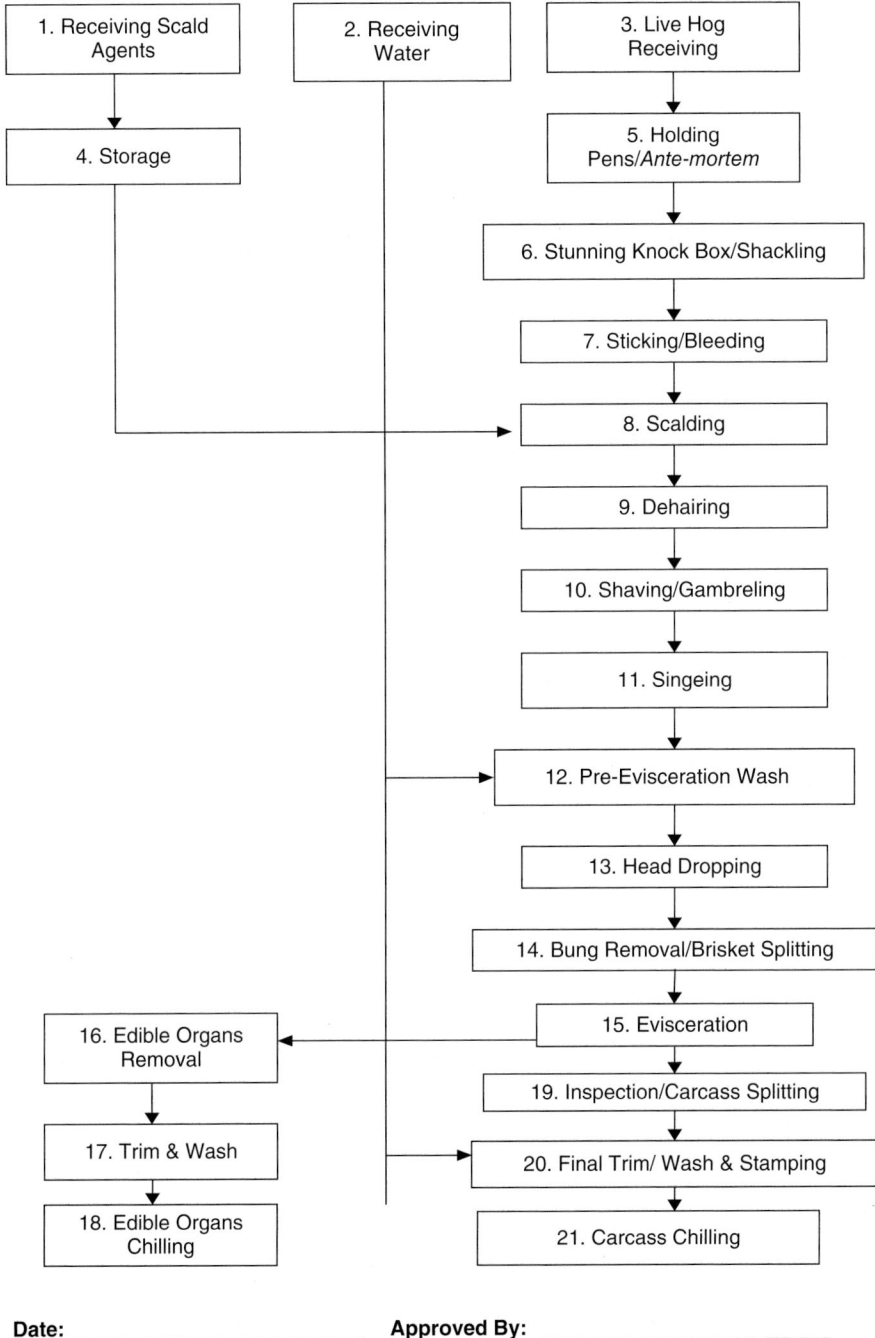

Figure 23.13   Swine slaughter flowchart.

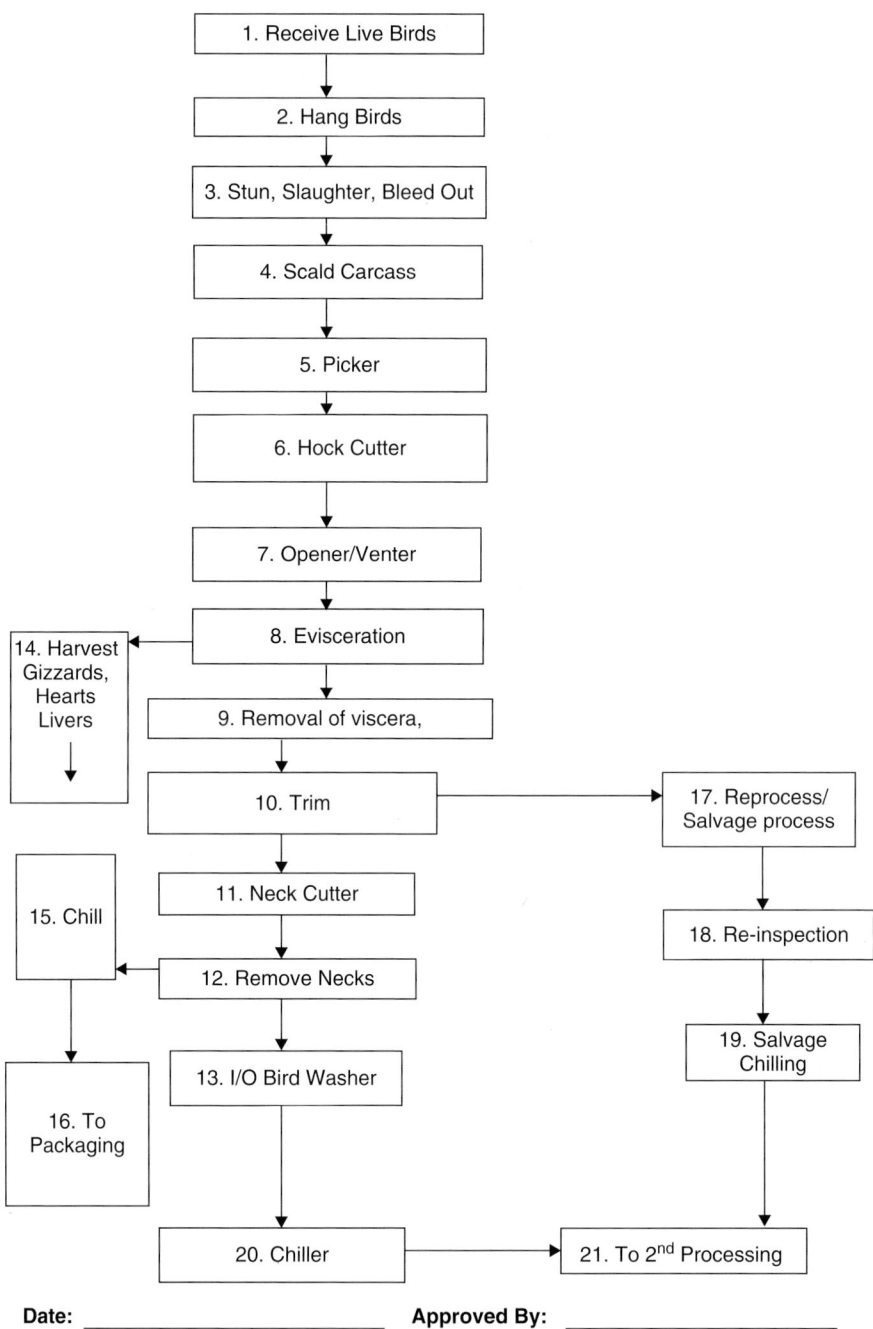

Date: _____   **Approved By:** _____

Figure 23.14   Poultry slaughter flowchart.

Literature and further reading

Codex Alimentarius Commission. 1993. FAO (Food and Agriculture Organization of the United Nations). Retrieved from Codex Alimentarious web pages (www.codexalimentarius.org; last accessed 26 February 2014).

European Parliament/Council. 2004. Regulation (EC) No. 852/2004 of the European Parliament and of the Council of 29 April 2004 on the hygiene of foodstuffs. *Official Journal of the European Union.* Retrieved from Eur-Lex web pages (http://eur-lex.europa.eu/LexUriServ/LexUriServ.do?uri=OJ:L:2004:139:0001: 0054:en:PDF; last accessed 26 February 2014).

ICMSF (International Commission on Microbiological Specifications for Foods). 1988. *Microorganisms in Food 4: Application of the Hazard Analysis Critical Control Point (HACCP) System to Ensure Microbiological Safety and Quality.* Blackwell Scientific Publications, Oxford. ISBN: 0632031810.

NACMCF (National Advisory Committee on Microbiological Criteria for Foods). 1997. HACCP Principles and Application Guidelines. Retrieved from the Food and Drug Administration web pages (http://www.fda.gov/Food/Guidance Regulation/HACCP/ucm2006801.htm; last accessed 26 February 2014).

24

Official Control

A. Introduction

Janne Lundén

Department of Food Hygiene and Environmental Health,
Faculty of Veterinary Medicine, University of Helsinki, Helsinki, Finland

Official control can be defined as actions taken by the authorities to ensure that the requirements of the legislation are followed. Official control at slaughterhouses is in many ways different from official food control in general because of the slaughterhouse environment, the continuous presence of control and the versatile tasks. Official control in slaughterhouses consists of four major sectors: food safety, food quality, animal health and animal welfare, giving the official control a multidimensional approach. Performing official control in slaughterhouses is challenging because live animals, complex production processes, food safety, contagious animal diseases and animal welfare issues must be considered at the same time. The slaughterhouse is an important control point for the early identification of possible problems in food safety and quality, animal health and welfare. It could be said that the official veterinarian is located on a hot spot where the four sectors overlap. Problems in one sector will influence another. To mention one example, deficiencies in animal cleanliness have serious consequences on both animal welfare and food safety. In order to provide high standards of quality control the official veterinarian must have efficient control methods in use. This chapter focuses on official control of food safety and quality and control methods in slaughterhouses. Although the control of animal health and animal welfare is not covered particularly in this chapter, the control methods and principles of control are also applicable in animal health and

Meat Inspection and Control in the Slaughterhouse, First Edition.
Edited by Thimjos Ninios, Janne Lundén, Hannu Korkeala and Maria Fredriksson-Ahomaa.
© 2014 John Wiley & Sons, Ltd. Published 2014 by John Wiley & Sons, Ltd.

welfare control. The substance knowledge differs but the methodology to reach effective control results are similar.

Official control in slaughterhouses is often considered to consist mainly of *ante-* and *post-mortem* inspections, although official control is much more than that. Food safety and quality aspects are present at the whole slaughter process, from the live animal to the chilled carcass. Important factors for food safety and quality are, for example, animal cleanliness, slaughter hygiene, plant sanitation and traceability. Many food safety risks can be prevented by good hygiene practices and a functioning own-check system (OCS) run by the food business operator (FBO). The main responsibility for producing safe products lies on the FBOs and this responsibility cannot be shifted from the FBO to the authority.

The role of the official control in slaughterhouses is to make sure that the FBOs comply with the food legislation by inspecting the premises and the OCS. If the FBO does not comply with the legislation, the official control must intervene with appropriate enforcement measures. The versatile control field requires specialized and comprehensive skills of the official veterinarian in order to be able to manage all control tasks effectively and reliably. This, in turn, sets high demands on the education of veterinary students and on the official control organizations responsible for maintaining the skills of official veterinarians. Carrying out official control requires wide knowledge not only in legislation but also, for example, in food processes, food hygiene and methods of enforcement. The knowledge should be based on the approach 'from stable to table' in order to achieve a comprehensive view of factors influencing food safety. The understanding of the whole production chain is important for the recognition of possible risks in the slaughterhouse. Legislative and substance knowledge combined with adequate control methods enables rational enforcement. Lack in knowledge of any of these may lead to poor official control.

Official control in slaughterhouses is affected by several factors, which are discussed in this chapter. The official control organization must fulfil certain requirements, such as independence and transparency. Further, the official control should be risk based. The important question in risk-based control is; what to focus on in the control and at what frequency? Risk-based control on-site is not equal to scientific risk analysis; instead, it is a synthesis of the food safety knowledge, legislative requirements, instructions and conditions at the slaughterhouse. Although the official veterinarian follows control plans drawn up beforehand, he or she has to make the control decisions on-site based on the above mentioned factors. It is clear that such analysis and daily decision making requires a wide competence and ability to assess food safety issues.

Food safety legislation in general requires that food production is controlled by authorities and the existence of official control is usually not questioned. The scale, intensity, efficiency, congruence and costs of official control, however, raise questions from time to time.

Cost efficiency, in particular, is not easy to calculate because the official control is by nature preventative. The input of official control in facilitating export of meat can be transformed into economic numbers but the calculation of benefits of preventative measures is much more complex. This is a question not only for official control in slaughterhouses but also for official control in general. The baseline is, however, that the FBO and the consumer must be able to trust the official control to be an independent expert professional organization protecting human health, animal health and welfare. In order to fulfil these requirements, the official control must meet the criteria for high standards of quality control and make sure that the slaughterhouse is complying with legislation.

B. Organization of Official Control

Aivars Bērziņš[1], Janne Lundén[2] and Hannu Korkeala[2]

[1]Institute of Food Safety, Animal Health and Environment (BIOR), Riga, Latvia; Faculty of Veterinary Medicine, Latvia University of Agriculture, Jelgava, Latvia

[2]Department of Food Hygiene and Environmental Health, Faculty of Veterinary Medicine, University of Helsinki, Helsinki, Finland

24.1   Scope

This part of the chapter includes information on the organization of official control. Responsibilities, competences and structural organization of the official control at slaughterhouses are described. Meat safety, meat quality, animal health and welfare are the focus of the official control at slaughterhouses, thus requiring continuous input from competent veterinary professionals. An independent, reliable and transparent system secured with sufficient resources is essential to carry out properly official control tasks at the slaughterhouse.

24.2   Structure of official organization

Official control at a slaughterhouse is an essential set of activities carried out by the competent authorities for the verification of compliance with food, animal health and animal welfare legislation. Food legislation at national or community level lays down the main principles and requirements for official control to be carried out at slaughterhouses, thus assuring meat safety and wholesomeness for human consumption. Slaughterhouse and meat production businesses need functioning Good Manufacturing Practices (GMP) and Good Hygiene Practices (GHP), which are the bases for the own-check system (OCS), including hazard analysis and critical control points (HACCP). Overall, responsibility for producing safe and wholesome meat lies on the food business operators (FBOs), whereas the main aim of the official control activities is to verify that FBOs have implemented the above mentioned pre-requisite programmes and HACCP.

The generic organizational structure of a food safety authority relevant to the official controls in slaughterhouses is shown in Figure 24.1. The parliament and responsible ministry of the respective country ensures a legislative framework to set up institutional rights and responsibilities for official controls. Food control authorities, called competent authorities, chaired by a

Figure 24.1 Generic organizational structure of food control relevant to the official control at slaughterhouses.

Chief Veterinary Officer (CVO), are responsible for organization of official controls at slaughterhouses and the work of official veterinarians involved in controls and meat inspection.

Control authorities operate at a central, regional or municipal level, depending on the size of the country and demands of national legislation. The role of the central control authority is to ensure uniform requirements and conformity of official controls carried out at central, regional and municipal levels.

Meat safety, meat quality, animal health and welfare are the focus of official control at slaughterhouses. In this context, national food control systems are important to protect public health and to enable countries to assure the safety of foods, particularly of animal origin, entering international trade and to ensure that imported foods conform to national requirements. The slaughterhouse, particularly, is a key place in epidemiological surveillance for zoonoses and contagious animal infectious diseases.

24.3 Requirements of the official control organization

Competent authorities performing official controls have to fulfil certain criteria to ensure their effectiveness and impartiality (Table 24.1). Competent authorities must have qualified staff and appropriate equipment to perform their tasks. Moreover, a sufficient number of qualified staff and allocated financial resources are important at all levels of the official controls to reach this goal. In addition, it is very important to have a harmonized approach and concerted action at all regional or municipal units involved in official controls.

Table 24.1   Requirements of the official control organization.

Requirement	Consequence
Comprehensive	Includes all control targets in the country
Possess power to implement meat safety legislation	Non-compliances are corrected, consumer safety is safeguarded
Highly educated and skilful personnel	High quality meat inspection and control
Transparent	Consumer and FBO trust
Impartial	Consumer and FBO trust
Unbiased	Consumer and FBO trust
Equal control for all FBOs	Consumer and FBO trust
Adequate resources	Enables the tasks to be fulfilled
Cost effective	Competitive
Evolving	Can react to new food safety, animal health and welfare threats

Internal and external audits have to be carried out as well to ensure that competent authorities are achieving their objectives. Some countries delegate certain control tasks to regional or municipal authorities; thus, it is necessary to ensure effective cooperation between central and regional control authorities. Competent authorities of some countries delegate certain control tasks to non-governmental organizations; however, this needs strictly defined conditions and usually has various limitations. In this context, for example, adoption of enforcement measures (coercive measures) cannot be delegated to any organization outside the competent authority of the respective country. The competent authorities also have to ensure transparency and availability of the relevant information to the public, particularly when there are grounds to suspect that meat or meat products may present a risk to public health.

Alongside veterinarians, many countries employ official auxiliaries (non-veterinarians) to perform *post-mortem* inspection. This needs well organized training under continuous supervision from the competent veterinarian to ensure that the auxiliaries have the competence and appropriate skills. The possible transfer of any meat inspection or control task to official auxiliaries must be carefully evaluated from food safety, animal health and welfare aspects. Also, possible transfer of any official control tasks to the FBO must be thoroughly evaluated, including the factors presented in Table 24.1. The main principles of the organization of *post-mortem* inspection at 'red meat' and poultry slaughterhouses may differ, the auxiliaries being employed by the authority or by the FBO (Figure 24.2).

The meat inspection organization should prepare a national official control plan, which is important to ensure that controls are carried out on regular basis to detect any non-compliances or divergences of the OCS over a time. National food control authorities have to ensure that official controls are carried out regularly, on a risk basis and with appropriate frequency, taking into account identified risks associated with animals, food operations and reliability of any own check that have already been carried out. Thus,

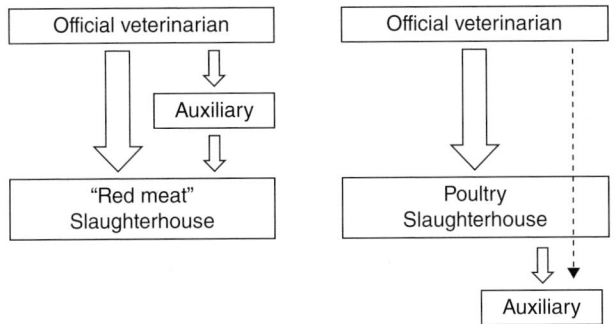

Figure 24.2 Main principles of the organization of official control at the slaughterhouse level.

frequency of official control activities has to be risk-based planned and carried out in slaughterhouses. The risks at certain meat plant have to be evaluated in the context of the OCS to plan accordingly official control procedures. Risks associated with pig, cattle or poultry slaughter may significantly differ, thus official control plans should be targeted to a certain establishment.

Inspection results are important to identify any problems as early as possible to communicate them with the relevant FBO, competent authority or institutions responsible for the controls along the food chain, including the holdings where animals have originated, slaughterhouses, meat cutting plants, meat processing plants and the retailers. Thus, the official veterinarian has to take all measures to prevent possible public health threats relevant in the meat chain. Appropriate decisions and measures following controls have to be taken whenever deficiencies or divergences are revealed. Infringements concerning food chain information may include reception of animals from regions of movement prohibition or violations regarding the use of veterinary medical products, resulting in actions undertaken against the FBO, such as extra controls at FBO's costs. Violations concerning animal welfare can be identified by official veterinarians in slaughterhouses, thus immediate decisions by competent authorities and corrective measures by FBOs have to be undertaken. Although legislation in the European Union and some other countries foresees flexibility that may be applied by certain slaughterhouses and meat plants, the presence of an official veterinarian is essential in *ante-mortem* and *post-mortem* inspections because of the broad knowledge and skills in food safety, food hygiene, animal health and welfare. Moreover, official controls have to be based on a regular assessment of the risks to human and animal health and welfare in the respective country, region or municipality.

Official sampling and laboratory testing is also important to demonstrate and ensure compliance with existing legislation. Moreover, organizations should have sufficient capacity and resources for sampling and laboratory tests. Official sampling is an essential part to check if products and processes comply with microbiological criteria set by legislation. Sampling and laboratory testing results on regular basis may help both FBOs and official control representatives to identify failures and non-compliances and to enforce appropriate corrective actions. Official controls are aimed at evaluating the OCS, its implementation and functioning in the slaughterhouse. Accordingly, competent authorities are obliged to verify if the documentation and procedures are followed as set by the OCS and existing legislation. Communication between the FBO and official control representative is necessary to discuss and agree on findings, non-compliances and necessary corrective actions during or immediately after the on-site visits.

The type and size of the slaughterhouse are important factors to consider when planning both OCS plans and official controls. Although official control in pig, ruminant, poultry, horse or any other animal species slaughtering plants may have different challenges, overall control procedures includes common activities such as evaluation of the OCS and inspections of production premises and equipment. Raw materials, ingredients, processing steps, materials in contact with meat, ingredients, storage of raw materials and meat, cleaning, disinfection, labelling and packaging have also to be covered by official controls. This allows better evaluation of good manufacturing practices, good hygienic practices and a HACCP system in slaughterhouses. General official control targets are common for ruminant, pig and poultry slaughterhouses; however, specific control targets may vary between them, including good manufacturing and hygienic practices.

The official veterinarian plays an important role in carrying out specific tasks in slaughterhouses or meat plants. Food chain information analysis, *ante-mortem* and *post-mortem* inspections, observations on animal handling, transport, lairage, slaughtering, handling of carcasses and meat and meat safety are important for the official veterinarian to take proper decisions. Moreover, the official veterinarian in the slaughterhouse supports the interests of many different stakeholders and adds value to: (i) FBOs through technical knowledge and support in development of OCS processes; (ii) farmers, FBOs and private veterinarians by providing relevant animal health and food chain information; and (iii) the competent authority by delivering official controls through inspection, audit and enforcement.

Across European countries there is a variation in implementation of legislation regulating meat inspection. However, variations are usually between the countries that are compliant, suggesting primary problems could be related to the weaknesses in EU legislation. Various decisions regarding food chain information, live animals, animal health and welfare, meat safety and traceability need the professional competencies and skills of veterinarians to ensure that official controls are carried out properly. Meat inspection is a key component of the overall surveillance system for animal health and welfare as well as being an essential part of protecting public health.

Literature and further reading

European Parliament/Council. 2004a. EC Regulation 2004/854/EC, laying down specific rules for the organisation of official controls on products of animal origin intended for human consumption. *Official Journal of the European Union*:

http://publications.europa.eu/official/index_en.htm (last accessed 23 February 2014).

European Parliament/Council. 2004b. EC Regulation 2004/882/EC, on official controls performed to ensure the verification of compliance with feed and food law, animal health, and animal welfare rules. *Official Journal of the European Union*: http://publications.europa.eu/official/index_en.htm (last accessed 23 February 2014).

FVE (Federation of Veterinarians of Europe). 2012. The Official Veterinarian's Role in Food Hygiene: An Essential Public Good. Position paper, FVE, Brussels.

Kaario, N. and Lundén, J. 2010. Official control of slaughterhouses and processing plants. In: *Handbook of Poultry Science and Technology* (eds I. Guerrero-Legarreta and Y.H. Hui). Blackwell Publishing Ltd, Oxford, UK, pp. 107–119.

Lundén, J., Björkroth, J. and Korkeala, H. 2007. Meat inspection education in Finnish veterinary curriculum. *J Vet Med Educ*, **34**, 205–211.

Mäki-Petäys, O. and Kaario, N. 2007. Official control of an establishment. In: *Food Hygiene, Environmental Hygiene, Food and Environmental Toxicology* (ed. H.J. Korkeala). WSOY Oppimateriaalit Oy, Helsinki, Finland, pp. 473–480.

C. On-Site Risk-Based Control

Eeva-Riitta Wirta

Meat Inspection Unit, Control Department, Finnish Food Safety Authority Evira, Helsinki, Finland

24.4   Scope

The scope of this part of the chapter is to provide practical solutions about how the official veterinarian may use on-site risk-based evaluation in control and how to focus the control to details that need more attention. To achieve effective official control, the official veterinarian has to concentrate more on those points in the slaughterhouse operations that need steering or corrective actions. Occasional findings can demand immediate corrective actions, changing the pre-established control plan for that day. Official control should be able to focus on problems at hand. Risk-based control means that the risk to food safety and hygiene in different stages of the slaughter process is estimated and those parts with acceptable results can be left to more occasional control for optimal use of control resources.

24.5   Introduction

Risk-based evaluation and flexibility has been taken into account in the legislation of the official control of food hygiene. Traditional methods, small volumes of production or situations where there are establishments in special geographic areas with certain restrictions are examples of this. Also, flexibility in separating processes by time in small premises is possible. Official control of meat production must make sure that the food business operators (FBOs) follow the legislation and rules of food safety and hygiene. There are many ways for the FBO to follow the food safety requirements especially in the content and implementation of own-check systems (OCSs), monitoring and documentation. Also, the official veterinarian should have some possibility to target official control if needed. The official veterinarian has to decide what to emphasize in the control of a slaughterhouse, because everything cannot be covered at the same time. The important principle is that the official veterinarian and the slaughterhouse operator fill their responsibilities as required. On-site risk-based control is useful to both the official control and the FBO, as control is focused on areas where it is needed.

24.6   On-site risk-based control and own-check system

There are numerous tasks in the legislation that the official veterinarian has to fulfil when in charge of control of the slaughterhouse. One of the most important tools in effective control of hygiene is verifying the OCS of the slaughterhouse. The OCS consists of own-check programmes (prerequisite programmes) and a HACCP system. The FBO uses practical risk assessment when planning its OCS. The FBO has to describe the processes and recognize the hazards to meat safety in the slaughter process when building the OCS.

Prerequisite programmes consist of programmes that have been planned to control hygiene risks, for example in relation to the establishment and production and working hygiene. Depending of the type of production, different prerequisite programmes are emphasized. In slaughterhouses the programme for cleaning and disinfection is very important because large numbers of living animals enter the slaughterhouse. Also, numerous connections to different farms by animal transport increase the risk of the spread of infectious animal diseases and zoonoses. Pathogens should be eliminated by cleaning and disinfection with detergents and disinfectants every time. On the other hand, control of temperatures is essential in sustaining food safety. The programmes should contain temperature control concerning all the chilled premises and the carcasses.

Hazard analysis has to be done when compiling the HACCP system for the production process of the slaughter line. There are differing opinions whether critical control points are necessary at all in the slaughter process. Depending of the volume of slaughter and variety of slaughter animal species, there might be several critical control points determined at the slaughterhouse. There may also be only one determined point in the HACCP system of the slaughter line. Very often one of the critical control points is the control of faecal contamination of the carcasses.

Systematic documentation is essential in the whole OCS and its verification. Without documentation, the establishment cannot prove it has an active OCS. The value of the documentation becomes clear particularly in situations when there is a need for corrective actions and new risk assessment because of failure at some point in the process.

24.7   Verification of the own-check system

The whole OCS, the prerequisite programmes and the HACCP system should be verified at regular intervals by the official veterinarian. If the OCS is comprehensive enough the official veterinarian may concentrate on verification and control of the operational hygiene during the slaughter process. The main point is that the official veterinarian may assure himself about the slaughterhouse employees' way of carrying out and documenting the records of the prerequisite programmes and HACCP system. So it must work as agreed between the official veterinarian and the people responsible for production.

If there are parts in the prerequisite programmes where the control results are always acceptable, it is possible for the FBO to decrease monitoring and for the official veterinarian to decrease verification frequency because the OCS seems to work. This is an example of on-site risk-based control that leads to changing the official control plan. On the other hand, it is necessary for the official veterinarian not only to inspect whether the OCS is followed but also to evaluate if the OCS is sufficient and based on sound conclusions.

Recurrent unacceptable monitoring results tell the official veterinarian that the OCS is not well managed or there are some other problems. In these cases, the official veterinarian has to make sure that the revision of the OCS or other corrective actions are carried out by the FBO.

In the HACCP system clear acceptable limits have been determined for the critical control points. Effective corrective actions have to be performed if there are exceptions to the limits. The corrective actions have to be acceptable and they must have a target to prevent the failure. Many times it is not easy to find an adequate corrective action that the official veterinarian can also accept. Simply 'notice to the superior' as a corrective action does not tell about corrections. Therefore, corrective actions should be clearly stated. Also, the sufficiency of the actions here has to be considered with regard to the significance of the possible hazard.

The way that the slaughterhouse personnel in charge act when corrective actions in the prerequisite programmes and the HACCP system are needed reflects how the slaughterhouse personnel in charge take care of problems on food hygiene and safety in their production. Quickly performed and effective corrective actions are enough to convince the official veterinarian that the slaughterhouse is willing to maintain a good quality level in operations.

24.8   Systematic verification in practice

The content of the OCS is usually quite compendious with prerequisite programmes and the HACCP system of the production line. A tool for the official veterinarian to perform systematic control is a tailored control plan concerning the OCS and all the operations at the slaughterhouse. In this way all the operations of the slaughterhouse can be effectively covered by control and verification. However, it is not always possible to follow a systematic approach in practice. Some parts of the control have to be covered daily and always during slaughter, such as making sure that the surroundings of the slaughter line make hygienic working possible. The plant and equipment must be clean when work starts in the morning. Sometime solving practical problems take time and it is impossible for the official veterinarian to follow the control plan.

The official veterinarian can control the establishment using on-site risk-based evaluation widely in verification of the OCS. For example, the official control of production hygiene during slaughter can be focused on different stages of the slaughter line depending on the skills of the employee. Problems with faecal contamination or other working hygiene may need more attention than other activities. Usually there are good bases to on-site risk-based control when the official veterinarian is well aware of the operability of the OCS and the employees' way of following the requirements. For the official veterinarian being able to consider the presence of hazards a good understanding of the state of operational and working hygiene, quality of slaughter animals and the company's commitment to producing high quality meat is needed. On-site risk-based control can be seen as a practical part of risk assessment, which contains decisions to focus the control in certain areas. This is only possible if the official veterinarian understands the whole process. Despite an active operational OCS and documentation there are always hazards that may become exposed. The risk is particularly increased when the workers do not follow good hygiene practices.

24.9 Practical views to on-site risk-based control in slaughterhouses

24.9.1 Small scale and large scale slaughterhouses

There are many things that influence the content and magnitude of official control. The geographic situation and conditions where slaughterhouse activities are performed are such factors. To design an effective official control strategy in a slaughterhouse consideration of the whole slaughterhouse, its size and production is needed. Small scale slaughterhouses differ from the bigger ones because of smaller volumes of production but may also differ due to animal material. Many times the animals come from the local neighbourhood or they may have been raised at the slaughterhouse owner's farm. In this case the health status and condition of animals in transport is usually easier to control, as is the food chain information. Small scale slaughterhouses may slaughter many animal species, which requires more skills from the official veterinarian but may also be a motivating challenge. In large slaughterhouses the information about animals is much more expansive and may contain knowledge from many farms in a large geographical region. The health status of animals at the farms may differ remarkably. Long transport times may cause difficulties with the health status of animals and animal welfare; this should be taken into account in the control.

There are differences in the volume of meat produced in small and large scale slaughterhouses and also in the distribution area. Many small scale slaughterhouses sell their carcasses to neighbourhood cutting plants. The larger companies may send carcasses for further production to cutting plants that are longer distances away. The number of consumers exposed to an imagined risk increases accordingly in relation to the volume of production and the size of the distribution area.

24.9.2   Slaughter order of animals with different status

Typically, in the OCS there is a programme describing in which order the animals should be slaughtered, but the order of slaughtering may fail due to practical reasons. Taking into account the condition of weakened animals or the youngest animals and the waiting times of the animals after transport is common practice; many times it is not easy to organize. The requirements of animal welfare must be taken into account when waiting times between transport and slaughter exceed commonly acceptable times or the feeding of animals or milking has to take place. Separating animals with decreased health status or animals with a heavy layer of dung should be normal activity in the slaughterhouse. The personnel's way of handling animals is also important to take into account when focusing official control.

24.9.3   Stunning and slaughter operations

On-site risk-based control should also be used when planning official control of stunning. Well functioning stunning is essential to complete bleeding, which then is crucial to the final pH-value of the meat and, finally, the shelf life of meat products. Stunning is also a point where the official veterinarian has to make sure that everything is performed correctly in the aspect of animal welfare.

Different working stages at slaughter are very important for carcass hygiene. A poor hygiene level in slaughter operations is revealed by faecal contamination on the carcass after the operations. Faecal contamination is very often the most prominent sign of problems in hygiene and is, unquestionably, very important to correct. Still, in many other stages on the slaughter line there may also be decreased levels of hygiene. Much emphasis should be focused on hygienic working by the slaughter personnel, as poor hygiene may cause cross-contamination.

After stunning and bleeding other points on the slaughter line should be operated correctly. Contact of the carcass or organs with the structures of the slaughter line or to each other are many times due to workers carelessness on the slaughter line. It is essential for the official veterinarian to be convinced of the level of process hygiene.

The personnel in charge of slaughter should have documentation on working mistakes in the slaughter process because they are important signs of the quality of the work but also an economic question. Often, the number of mistakes and the quality of them is an indicator of the slaughter personnel's skills in their work. Mistakes such as failed stitch and bleeding of the animal, hair remaining after burning of pork carcasses, visual contamination of carcasses and parts of the spinal cord remaining should be documented. Touching only clean areas of the carcass with clean hands is not always easy in practice but problems with operational hygiene on the slaughter line lead easily to increased microbiological contamination and decreased meat safety and shelf life.

24.9.4   Chilling

Chilling of the carcasses is one of the most important parts of producing safe and good quality meat. The capability of the slaughterhouse to control temperatures must be a target for official control. Reaching the aimed temperature in meat is not enough. The time that is needed for chilling is important in the view of microbial quality of the carcass surface. In many small scale slaughterhouses chilling capacity is the bottleneck to increasing production; otherwise the volume could often be increased. On-site risk-based control should be used to consider the frequency of inspecting the chilling procedures in order to secure the safety and quality of meat.

24.9.5   Sampling by the official veterinarian by on-site risk-based consideration

A sampling plan is an essential part of the FBO's OCS; the FBO is responsible for implementing a sampling plan. There are requirements for the official veterinarian to take some samples also. Sampling by the official veterinarian of samples other than those related to meat inspection can be an important method of verifying that the OCS is adequate.

The primary reason for the official veterinarian to take official samples is to become convinced that the hygiene of organs, carcasses or surfaces of production lines is sufficient. Other reasons may be a need to become convinced of the workers' technique of sampling, the correctness of the analyses or objectivity of sampling. Microbiological samples are examined in accredited laboratories to ensure reliability of results. Official sampling is justified especially in situations where the results of sampling in the OCS have been acceptable constantly but visual inspection does not support the results.

Literature and further reading

Mayes, T. 1998. Risk analysis in HACCP: burden or benefit? *Food Control*, **9**, 171–176.

Tuominen, P. 2008. Developing risk-based food safety management. Evira Research Reports 4/2008, University of Helsinki/Finnish Food Safety Authority Evira, Helsinki, Finland.

Walker, E., Pritchard, C. and Forsythe, S. 2012. Hazard Analysis Critical Control Point and prerequisite program implementation in small and medium size food businesses. *Food Control*, **14**, 169–174.

D. Control Plan

Tiina Läikkö-Roto

Department of Food Hygiene and Environmental Health,
Faculty of Veterinary Medicine, University of Helsinki, Helsinki, Finland

24.10   Scope

This part of the chapter focuses on the subjects that have to be taken into account when compiling an official control plan for a slaughterhouse and describes how such a plan can be made. It also gives an example of a control plan in a large scale slaughterhouse.

24.11   Why planning of official food control is important?

Food business operators are responsible for ensuring the safety and quality of foods at all stages of the production, processing and distribution. Official controls should also be carried out at all of these stages. Using preventive approaches, such as the HACCP system, in the food industry also calls for official control to be able to react to the risks on a preventive basis. Thus, the official food control, in all of its aspects, has to be based on risk evaluation. The official controls should also be carried out in accordance with documented procedures and, thus, be based on written control plans. Careful risk evaluation and planning are essential for efficient official food control. Although the ultimate responsibility of the food safety is on the businesses, it is possible to reduce the risks in food production and to improve the safety of the foods with good planning of the official controls.

24.12   Planning food control in a slaughterhouse

The aim of official food control in a slaughterhouse is to verify that the operations are in compliance with legislation concerning food production, animal health and animal welfare. Based on risk factors present in the slaughterhouse and its operations and products, as well as the relevant requirements of the legislation, official veterinarians responsible for the control should evaluate and ultimately define the frequency of the controls in the plant that is adequate. Also, they should plan the contents of the controls, the control methods and techniques to be used and the official sampling performed in connection to the controls.

24.12.1 Planning the frequency of control visits

The official controls should be regular and the frequency of controls appropriate and risk based. The competent authorities have given guidance documents about determining the control frequencies in many countries; these should be followed in the respective countries. The officials in charge of the control of a certain establishment have, however, to carry out the final evaluation of the needed control frequency themselves.

In a slaughterhouse, the frequency of the control visits in practice often equals the number of the slaughter days. There might, however, be reasons to elevate the control frequency from this, especially in smaller slaughterhouses that slaughter more seldom but have other operations in the premises as well. When evaluating the needed control frequency for a certain slaughterhouse, the official veterinarians have to take into account the following subjects:

- All the operations performed in the slaughterhouse and the establishment in connection to it, as well as the extent and volume of these operations (the slaughtered species, other operations performed in connection to slaughtering).
- The types of products produced (treated stomachs, bladders or intestines, raw materials for gelatin or collagen etc.).
- The siting, size and infrastructure of the establishment (layout, design, construction and surfaces, working space, cross-contamination issues, lairage facilities, changing facilities for employees, storages etc.) and the appropriateness of working equipment and machinery (used materials, maintenance, cleanability etc.).
- The food business operators' past record in regards to compliance with legislation (findings of previous controls by different food control authorities).
- The reliability and the results of any checks concerning food safety carried out by the food business operator (HACCP system and prerequisite programmes, other food safety programmes).
- Any other information that might indicate non-compliance (notices from consumers, from other food business operators or other officials etc.).
- Other identified risks associated with slaughter animals or their feed, the use of food produced from them, the establishment, or any process, material, substance, activity or operation that may influence food safety, animal health or animal welfare in the establishment.

24.12.2 Planning the content of the controls

After deciding the frequency of control visits in the plant, the official veterinarians should plan the contents of these visits. The official controls performed in a slaughterhouse should include, in appropriate frequency:

- Inspections of the establishment and its surroundings, premises, offices, equipment, installations and machinery.

- Inspections of the transport, slaughter animals, carcasses and other parts of the animals that are intended to be used in food production as such or as raw materials, and the handling of these (slaughtering, chilling of the carcasses and offal, washing of the intestines and stomachs, storing and salting of the hides etc.), as well as the handling of animal by-products.
- Evaluations of appropriateness of the materials and articles intended to come into contact with food, labelling, presentation and advertising of foods, as well as cleaning and maintenance products and processes and pesticides used in the facility.
- Examinations and evaluations of accuracy of the establishment's own-check system (prerequisite programmes and HACCP system) and the results obtained.

In a slaughterhouse, controls should be performed every slaughter day. At a minimum this means subjects concerning animal welfare and animal health while performing *ante-mortem* inspections. As there are, however, so many other important subjects to be controlled as well, and as the entire establishment and its surroundings, all operations and the own-check system of the food business operator have to be controlled frequently enough, it is recommended to plan other controls for each working day as well. These controls can be planned so that during every visit or working day only one, or even a part of one, control programme in the establishment is examined, or only a part of the establishment inspected. Some parts of the own-check system, premises or certain operations require more frequent control than others. Likewise, some food business operators need more frequent control of certain aspects than others. With good planning specific interest can be directed to aspects that have been problematic in the slaughterhouse in the past.

When evaluating the control priorities and the needed control frequencies of different subjects, the official veterinarians should once again consider legal requirements and guidance papers given by the competent authorities, as well as the identified risk factors in the establishment and its operations, the significance of these factors and the establishment's ability to control them. Based on this evaluation, the official veterinarians should decide the needed control frequencies for the different aspects. After that, the number of the controls needed for the different aspects can be divided between the control visits planned. If it seems impossible to fit everything into the plan, the evaluated frequency of the on-site inspections is likely to be too low.

In case of large scale operations the controls are often, because of more frequent control visits, divided into smaller pieces than in case of small scale operations. In large scale slaughterhouses there are often also other operations performed in addition to the slaughtering operations, which makes the planning of the controls more challenging. Thus, it can be useful to plan the details of the controls in large scale slaughterhouses on a more frequent basis and in closer connection to the actual controls. For example, the official veterinarians can make a rough plan of the controls annually, only dividing the needed contents of the controls for every 12 months of the year (Figure 24.3), and carry out the more detailed planning monthly or even weekly (Figure 24.4). In a

Subject	Jan P	Jan I	Jan D	Feb P	Feb I	Feb D	Mar P	Mar I	Mar D	Apr P	Apr I	Apr D	May P	May I	May D	Jun P	Jun I	Jun D
Cleaning and disinfection (2 controls/week)		x	x		x	x	x	x	x		x	x		x	x	x	x	x
HACCP (3 controls/week)		x	x	x	x	x		x	x	x	x	x		x	x	x	x	x
Operational hygiene (daily control)	x	x	x		x	x		x	x	x	x	x		x	x		x	x
Maintenance (6 controls/year)				x				x			x							
Pest control (4 controls/year)													x				x	x

Figure 24.3 Example of a part of an official control plan that divides the contents of the controls for different months. This plan separates the controls aimed to the different programs (P) of the establishment's own-check system, their implementation (I) and documentation (D).

Control plan, week 16 (April)					
	Monday	**Tuesday**	**Wednesday**	**Thursday**	**Friday**
Cleaning and disinfection	Pre operational cleanliness control: inspection of food contact surfaces.		Pre operational cleanliness control: sampling all saw blades and 10 knives.		
HACCP	Monitoring procedures: verification of week 15 documents.		Monitoring procedures: observation, one monitoring event.	HACCP plan: evaluation of hazard analysis, slaughter line.	

Figure 24.4 Example of a part of an official control plan compiled for one week describing the controls in detail. This could be a part of the more detailed planning in Figure 24.3.

slaughterhouse of a smaller scale, on the other hand, it can be enough to plan the contents of the controls annually and review the plan mid-term.

It is worthwhile to note that the controls should be performed mostly during the working hours of the establishment, so that the hygiene status of the operations can be evaluated. However, it should also be planned to perform controls before or after working hours when needed. For example, cleanliness control of surfaces should be performed outside working hours of the particular areas. It is also worthwhile to note that the food business operator should be informed, or even negotiated with about the time of the control, in the case of certain controls, such as wide-ranging audits or long interviews with the staff. However, usually the majority of the official controls in slaughterhouses is performed without prior warning.

24.12.3 Planning the control methods and techniques to be used during control visits

After planning the contents of different control visits, the official veterinarians should plan the methods and techniques to be used on the controls.

Appropriate methods and techniques are, for example, audits, inspections, monitoring, verification, interviews and sampling for analysis. Often these cannot totally be separated from one another and it is usual to use more than one control method or technique during one control visit.

Auditing means systematic and independent examination to determine whether activities and related results comply with planned arrangements and whether these arrangements are implemented effectively and are suitable to achieve the objectives. The officials can audit, for example, the functionality of the HACCP system, good hygiene practises and other prerequisite programmes. These audits should verify that food business operators apply the procedures continuously and properly. Auditing may be addressed partly by document review but usually also requires on-site verification. For example, when auditing the HACCP system, the officials can evaluate with the help of a document review whether the plan is appropriate for controlling, eliminating or reducing the hazards to an acceptable level. When evaluating whether the tasks are performed in compliance with the plan, and whether the plan is effectively implemented, on-site verification is needed. This means, for example, examining whether the monitoring and verification tasks are performed effectively (whether the deviations are detected when they occur) and with the frequency given in the HACCP plan, whether the employees are recording the results truthfully and whether the actions that are supposed to be initiated in case of deviations from the critical limits are taken.

Inspecting means examining any detailed or wider ranging aspect of the establishment, foods, animal health or animal welfare, in order to verify that such aspect(s) comply with the legal requirements. An inspection (other than *ante-* and *post-mortem* inspections) may be planned to focus only on one part or a few parts of the food safety programme, operations or establishment just as well as to cover a wider range of subjects. Inspections can include different kinds of measurements and sampling for analysis.

Monitoring means conducting a planned sequence of observations or measurements with a view to obtaining an overview of the state of compliance with the legislation. For example, the official veterinarian can plan to monitor the adequateness of the bleeding of a certain number of stunned animals, or to measure the temperatures of a certain number of carcasses before cutting, repeatedly during each control visit for a certain period of time.

Verification means checking, by examination and the consideration of objective evidence, whether specified requirements have been fulfilled. For example, officials can verify compliance of working hygiene by comparing the working methods of the establishment to the requirements of the legislation. With the help of the documentation the officials can verify whether, for example, the entries have been made at the frequency that is given in the HACCP plan and whether the corrective actions are performed according to plan when needed. With testing the cornea reflex of the stunned animals, the official veterinarians can verify the adequateness of stunning.

Interviewing food business operators and their staff means face-to-face discussions, which are conducted in order to find out the used operating procedures in practice, as well as the level of knowledge of the food business operator and the employees in food hygiene and the food safety programmes. The persons responsible for monitoring can be interviewed in order to find out whether they know what hazard is controlled with the CCP they are monitoring, why such controlling is important, how to perform the monitoring procedure and record the results, and what corrective actions have to be initiated in case of deviations.

Sampling for analysis means taking samples of food or any other substance relevant to the production, processing and distribution of food or to the health of animals in order to verify, through analysis, compliance with legislation. Samples can be taken as a part of monitoring the product safety or cleanliness status of the establishment, surveillance due to follow-up of earlier unsatisfactory results or for the support of inspection findings.

24.12.4 Planning the official sampling for analysis

Although the food business operators are responsible for having appropriate sampling programmes in place to verify the functionality of the food safety programmes, the official food control can use sampling and analysis as one of the control techniques for the same purpose. With planned sampling and analysis the officials can also verify whether the products produced by the establishment are in compliance with the relevant regulations.

The principle of risk-based control applies even in sample taking. When deciding the target surfaces or products, numbers of samples, sampling frequencies and the analysis to be performed, the official veterinarians should take into consideration the kinds of foods that are produced by the establishment, the structure of the plant, the cleanability of the surfaces, the hygiene standard of the operations, the functionality of the food safety programmes and the results of earlier analysis of both the food business operator and official food control. There may also be regulations, recommendations or national programmes concerning, for example, the sampling and analysis of pathogenic bacteria, residues, animal diseases and zoonooses that should be taken into account when compiling the plan. The laboratories used in analysis should preferably be named in the plan, due to the fact that there are special requirements for the laboratories and equipment involved in the analysis of official samples.

24.12.5 Evaluating the duration of the control visits

It is important to be able to evaluate the duration of the controls beforehand. When estimating the time needed, it is necessary to take into account the planned contents of the control, the size of the plant, the planned techniques to be used and the possible sample taking. The controls should not have to be done in a hurry due to poor planning.

24.13   Adjusting the control plan when needed

Despite the need for careful and detailed planning, official control still has to be flexible and able to react to various situations. Thus, the control plan should be adjusted during its implementation when needed. Amendments may be needed for various reasons, such as changes in legislation, emergencies caused by diseases or other health risks, changes in the structure or operations of the establishment, significant modifications of the establishment's HACCP plan or prerequisite programmes and critical findings of the official control or third-party auditors. Also, the nature of food production is such that there will always be at least small deviations from the expected. Due to these deviations, the official control might have to be directed to other subjects than planned for some period of time. In these situations it should be kept as a main rule to reschedule the planned controls for a later time, and not to leave them undone.

The control plan shall also be re-evaluated regularly, for example once a year, and adjusted based on this evaluation if needed. Due to the requirement of risk-based control, the findings of the previous official controls should be taken into account when compiling the plan for the next period (for example the next year). The non-compliances and bad procedures or hazardous operations observed, the willingness of the establishment to amend the mistakes that were made and the overall functionality of the establishment's food safety programmes should be emphasized when planning the next period of the controls for the establishment in question.

Literature and further reading

Doménech, E., Escriche, I. and Martorell, S. 2008. Assessing the effectiveness of critical control points to guarantee food safety. *Food Control*, **19**, 557–565.

European Parliament/Council. 2004c. Regulation (EC) No 854/2004 of the European Parliament and of the Council laying down specific rules for the organisation of official controls on products of animal origin intended for human consumption. *Official Journal of the European Union*: http://publications.europa.eu/official /index_en.htm (last accessed 23 February 2014).

European Parliament/Council. 2004d. Regulation (EC) No 882/2004 of the European Parliament and of the Council on official controls performed to ensure the verification of compliance with feed and food law, animal health and animal welfare rules. *Official Journal of the European Union*: http://publications.europa.eu/official /index_en.htm (last accessed 23 February 2014).

Lindblad, M. and Berking, C. 2013. A meat control system achieving significant reduction of visible faecal and ingesta contamination of cattle, lamb and swine carcasses at Swedish slaughterhouses. *Food Control*, **30**, 101–105.

E. Approval of Establishments

Risto M. Ruuska

Regional State Administrative Agency for Lapland, Finland

24.14   Scope

The approval process is a tool for proactive control ensuring that establishments processing food of animal origin fulfil the requirements of food hygiene legislation and, to some extent also, the requirements of animal health and animal welfare legislation before it starts operation. Only approved activities can be performed in the slaughterhouse. The competent authorities should also supervise and ensure that the establishments keep satisfactory levels of hygiene during operation, maintain infrastructure and take care of other duties connected with food safety. In the case of negligence, the necessary action needs to be taken and the final resort is to withdraw the approval

24.15   Why approve slaughterhouses beforehand?

Approval is a type of license and establishments handling foods of animal origin including slaughterhouses must not operate before approved by the competent authority. The purpose of the approval process is to ensure that the slaughterhouse is complying with relevant legislation before starting to operate. The slaughterhouse activities include food safety, animal health and animal welfare issues that may lead to serious consequences if not handled properly. The proactive control is a way of making sure that the slaughterhouse and the intended operations are designed in such a way that it is possible to function in a hygienic manner. If the slaughterhouse were to start without an approval process it is possible that some part of the slaughterhouse or operations would not fulfil the food safety requirements.

Although slaughterhouses are approved before the operation begins, some other food premises may only be registered. In this case food premises are simply notified by the food business operator (FBO) to the competent authority. Registration without an approval is an easier pre-hand control method used for low-risk premises, such as for some retail premises. Depending on the country and its local legislation, premises other than establishments handling foods of animal origin may be required to notify or to seek approval.

24.16   Approval process

The FBO should apply for approval from the competent authority. The application should include sufficient data so that the approving authority can ensure that all conditions of approval are fulfilled. The actual approval process can, in certain cases, be preceded by checking the layout plans and drawings beforehand on the request of the FBO. This can be helpful for the FBO, as possible food safety problems can be recognized already at the planning stage.

The evaluation of whether the intended activities can be hygienically performed in the facilities is an important principle in the application procedure. The approval process includes an on-site inspection that aims to make sure that the food safety requirements are fulfilled. The competent authority checks that the layout and facilities comply with the application documents and meet the requirements imposed by the legislation.

Several factors have to be considered in the approval process (Table 24.2). The layout must be planned in such a way as to prevent cross-contamination during production. A schematic layout that aims at the prevention of cross-contamination is presented in Figure 24.5. Rooms in a slaughterhouse are classified as clean and unclean areas. The lairage and slaughter line to skinning are unclean areas. For scalded pigs the borderline is after removal of bristles. Further processing of a skinned or scalded carcass is done in a clean area. The routes of staff and products should be planned to avoid any cross-contamination. Animal by-products should be handled separately. The most effective way to avoid cross-contamination is to separate the clean and unclean areas structurally.

Surface materials must be durable and easy to clean. To avoid pests, birds and insects the constructions should not contain any passages for those and no direct access is acceptable from outdoors to the production rooms where unpacked meat is handled.

Table 24.2   Factors to consider in the approval of a slaughterhouse.

Layout of the plant	Equipment hygiene
Infrastructure	Hand washing sites
• walls, floors, ceiling, drainage etc.	
Surface materials	Sites for equipment sterilizers
Enough space for all activities	Temperature of the facilities
• slaughter, storage rooms, by-products, waste, rest rooms, dressing rooms, lairage etc.	
Prevention of cross-contamination	Own-check system including HACCP
• separation of unclean and clean areas	

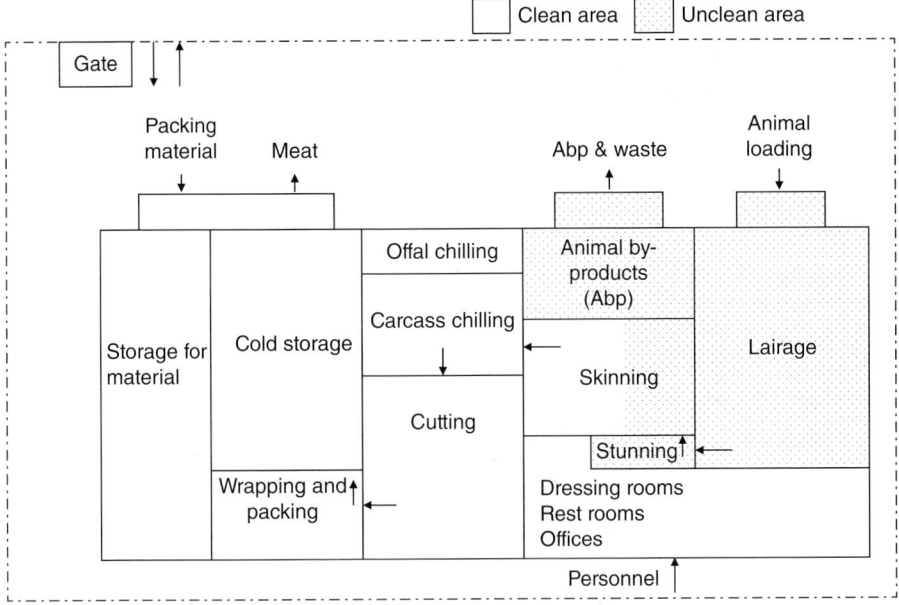

Figure 24.5 Schematic layout of a slaughterhouse.

Appropriate temperatures for chilled meat must be achieved in all cases and sufficient chilling capacity needs to be correlated with the capacity of slaughter line. The speed of the line (carcasses per hour and per day) should be defined in accordance of the capacity of the chilling facilities. The atmosphere (air, temperature, humidity) should be controlled in cold stores for unpacked meat.

Drainage of waste water and routes for solid waste and by-products should be managed to avoid any hygiene risks. Miscellaneous details, such as non-hand-operated taps and tool disinfection facilities, must be in adequate numbers and sites. Separate dressing rooms and toilets for dirty and clean area workers are essential for preventing cross-contamination.

The access to the slaughterhouse should be properly controlled by the FBO, for example by fencing the area. The own-check system including the HACCP should be evaluated in the approval process. The system should recognize possible risks in food safety and how they are managed.

In the case of approval of a mobile slaughterhouse special attention should be paid to how the supply of water is organized. Also, the handling of waste water and by-products are questions that need to be organized in a special manner. These questions must be considered by the FBO and described in the approval. Authorities should be aware of a mobile unit operating over several jurisdictional areas.

Slaughter for restricted purposes may have some less stringent requirements for approval. Private slaughter of farm animals for own family, slaughter of game for local market and so on without an approved establishment can be permissible within certain limits in some countries.

24.17   Granting approval

The slaughterhouse can be approved if the on-site inspection demonstrates that the food safety requirements are fulfilled and the own-check system is sufficient. The competent authority registers the information on the approved establishment. The registration includes categories of operation, which can be classified by groups of animal species (cattle, sheep and goat, pig, poultry, solipeds, farmed game, wild game, lagomorphs etc.) and activities (slaughter, cutting, minced meat, meat preparations, meat processing, wrapping, mechanically separated meat etc.). Approval records should always be kept on the spot.

If there are deficiencies in the facilities that do not directly endanger food safety and are expected to be repaired within an acceptable time limit, a conditional approval of the slaughterhouse might be possible. This enables the FBO to start the operations even though not all conditions are fully met. The possibility to apply conditional approval should be used only after careful consideration. If the deficiencies are not repaired within the time limit and the authorities are not willing to prolong the conditional approval the approval lapses. Slaughterhouses that already have been approved but intend to make substantial changes in their facilities or operations should apply for approval of such changes. Substantial changes are changes that may have an impact on food safety and hygiene and, therefore, need to be evaluated and approved by the authority.

The approval can be withdrawn if the conditions of approval are not fulfilled. Facilities in poor condition, operations not in compliance or fraud are reasons that can lead to withdrawal of approval.

24.18   Health mark and identification mark

The approved slaughterhouse is assigned an approval number. The approval number of a slaughterhouse is used in the health mark stamped on the carcass and also in identification marks (identification number) on wrapped or packed meat. The number of health marks on carcasses may vary depending on a country's legislation but a common number of health mark stamps per carcass has been six. This is due to the fact that one sixth of the carcass has often been the logistic unit in beef carcass trade. The health mark and the identification mark are essential for the identification of the establishment that has handled the meat and for traceability of the meat.

The health mark comprises, in addition to the approval number of the slaughterhouse, abbreviation letters indicating the country and the shape of the mark symbolizes the status of approval. The identification mark, which is located on labels, wrapping or packaging material, comprises the same information as the health mark, but the size is adjusted to the package. If the health mark is an oval shape, it indicates that the meat has been approved without any conditions. The oval shape is used also for export stamps. Square, pentagonal or other shapes may be used for meat with limitations in trade, for example intended only for the national market. Sometimes, the health mark may be accompanied by a stamp identifying the official veterinarian, additional treatments (freezing, heating) and commercial stamps such as carcass classification. In certain cases the health mark can be specific to the official veterinarian, such as in the meat inspection of wild game for local consumption.

Slaughterhouses with a co-located cutting plant or a meat processing establishment may use the same approval number. Therefore, it is possible that in such cases the same approval number is used both for the carcasses, cut meat and processed meat products.

24.19 Listing of establishments

The competent authorities publish lists of approved slaughterhouses in most countries. Commercial operators, authorities and even consumers can trace the manufacturer's meat and confirm the establishment's approval status. National lists can be found on the Internet. The list includes the approved activities and it should be kept up to date for reliable and efficient use.

Approval lists for so-called third countries (non-EU countries) that are allowed to export meat to the European Union are published by the EU Commission. Import of meat and meat products is regulated in most of the countries to secure food safety issues and to control contagious animal diseases. The approval of third country establishments may be preceded by an on-site inspection by inspectors of the export country or by the recommendation of local authorities. The European Union and some major exporting countries, such as the United States of America and Russian Federation, carry out regular inspections in third countries and an establishment can be delisted if severe failure is found.

24.20 Withdrawal of approval

If the FBO fails to carry out the operation of an establishment in accordance with the food safety requirements, and no other control means are sufficient, withdrawal of approval is the last resort. Depending on legislation and the authority's course of action, it can take place immediately in severe cases or following gradually strengthened enforcement actions. The reason for withdrawal of approval can be, for example, degeneration in infrastructure, severe negligence in hygiene or fraud. Withdrawal of approval is also made in cases of permanent cessation of operation in the establishment. In the case of withdrawal, the health mark, identification mark and health certificates for meat, meat products and animal by-products can no longer be issued.

Literature and further reading

European Union. Member country and third country legislation on approval procedures. eur-lex.europa.eu (last accessed 27 February 2014).

OIE (World Organisation for Animal Health). 2013. Terrestrial Animal Health Code. www.oie.int/international-standard-setting/terrestrial-code/access-online/ (last accessed 27 February 2014).

WTO (World Trade Organization). Sanitary and Phytosanitary Agreement. www.wto.org/english/docs_e/legal_e/legal_e.htm#sanitary (last accessed 27 February 2014).

F. Inspection and Sampling

Mari Nevas and Janne Lundén

Department of Food Hygiene and Environmental Health,
Faculty of Veterinary Medicine, University of Helsinki, Helsinki, Finland

24.21 Scope

It is the responsibility of the food business operator (FBO), in this case a slaughterhouse, to comply with the food legislation in order to produce safe food. The role of an official veterinarian is to ensure that the FBO is complying with the current food safety legislation. The most important tool for accomplishing this goal is an on-site inspection. The right to inspect the premises, operations and own-check system (OCS) is a fundamental precondition for official control. The official veterinarian evaluates whether the FBO complies with the legislation and if there are any potential public health risks associated with the production methods or facilities. The official veterinarians' role includes also guidance and steering on good hygiene practices and legislative demands. In cases where potential hazardous actions are noticed, the official veterinarian is to give advice to correct the problem or, in more severe cases of non-compliance, to use enforcement methods.

In this chapter, the official inspection procedures that aim at securing meat safety and hygiene are described. This chapter does not describe *ante-* or *post-mortem* inspection procedures.

24.22 Inspection procedures

At a slaughterhouse, the official inspections are performed by official veterinarians. The official veterinarians are responsible for observing compliance with legislation and ascertaining that hygienic practices are used. In order for the official control to be comprehensive the official veterinarians should be constantly present at the slaughterhouse during working hours. At small-scale slaughterhouses the extent of the presence of the official veterinarian should be such that compliance can be verified as satisfactory. In both large and small scale slaughterhouses, the inspection procedure is basically the same (Figure 24.6).

The entity that is an inspection procedure comprises more than just the on-site inspection. The main steps are:

- Evaluating the needs for the inspection based on a control plan: What is the entity to be inspected? Is microbiological sampling included? Are

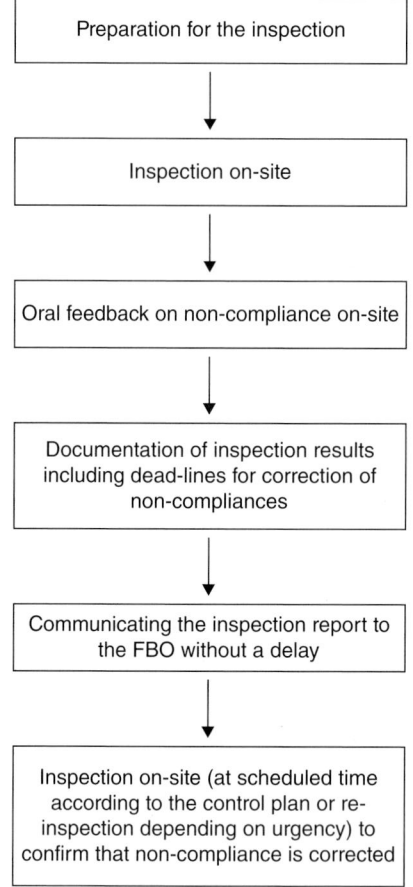

Figure 24.6   Steps in the inspection procedure.

there previously recognized food safety problems in the production that should be evaluated?

- Getting acquainted with the documentation available concerning the slaughterhouse: for example, species and number of animals, types of processes, floor plan, OCS and the latest inspection reports, including microbiological results.
- Checking the current, relevant legislation concerning the slaughterhouse in question.
- Contacting the slaughterhouse to adjust the date, if not an unannounced inspection.
- Gathering the tools and material needed for the inspection.
- Discussing the goals of the inspection with the representative of the slaughterhouse and learning about possible changes in the production.
- Performing the actual on-site inspection by inspecting the premises, observing matters related to food safety and detecting possible non-compliances.

- Collecting microbiological samples if needed.
- Giving immediate feedback of non-compliances to the representative of the slaughterhouse.
- Compiling and discussing the findings and observations with the representative.
- Writing the inspection report and giving/sending it to the slaughterhouse.
- Verification that the non-compliances that are not corrected immediately during the inspections are corrected before the deadline.
- Proceeding with enforcement methods (compulsory actions) if needed to accelerate the correction of the non-compliances.

24.23 Challenging task of an inspector

To be able to perform an effective inspection, the official veterinarian should have an adequate level of understanding about food-borne pathogens as well as about the risk they pose in meat. The official veterinarian should be well acquainted with the current legislation and instructions, to be able to point out possible non-compliances in the production, including the procedures and the facilities, and also to be able to detect possible frauds. On top of this vast knowledge, a certain amount of curiosity is also valuable to enhance the efficacy of the inspections.

It has been shown that the FBOs appreciate the official veterinarian's familiarity with the processes involved in their production. It may thus be seen as part of the inspector's expertise. At the slaughterhouse this knowledge involves the understanding of the whole slaughtering process, starting from stunning and continuing to the cooling of the carcasses with special regard to the zoonotic agents and the contamination of the carcasses. The understanding of slaughtering processes is also an important prerequisite for the detection of fraud.

The official veterinarian should explain the reasons for the requirements of the food safety legislation. If the FBO understands the reason why a non-compliance should be corrected, the motivation to correct is probably higher. Realizing the possibility of meat becoming unsafe due to the non-compliance may cause the FBO to be much more prone to correct the violation. The best result in official control is usually received by advice, steering and negotiations between the inspector and the FBO, for example on deadlines for correction. However, the final decision about the time limit for accomplishing the corrections is made by the official veterinarian based on the evaluated urgency of the matter.

The official veterinarian must aim at objectivity and conformity in the inspection practices. The uniformity of the inspections may be supported by peer evaluation of the inspection practices, for example participating in an inspection carried out by another official veterinarian. The work of official veterinarians may also be audited by the national authorities or by an external auditing body.

24.24   When, what and how to inspect?

An inspection may be performed on different bases. Regular inspections are made according to an annual control plan and performed without any suspicion of malpractice. If non-compliances are noticed and the FBO is given advice to correct the matter, a re-inspection should be performed to confirm the correction of the non-compliance noted. An inspection may also be conducted based on an external notice of suspected malpractice causing a risk to consumer safety or it may be justified in order to untangle the case of a food-borne epidemic.

The efficacy of an inspection could be enhanced by performing it unannounced, thus no preparation for the inspection can be done to mislead the officer. This is obviously not the case at slaughterhouses during the slaughtering, as the official veterinarian's presence during the slaughter is taken for granted. However, even in slaughterhouses with continuous official control, a certain degree of unexpectedness concerning the inspections should be maintained. The weekly control plan should vary and unexpected inspections of OCS and facilities are important to execute. Announced inspections are justified when there is a need to meet and interview certain key personnel to gather all relevant information, for example concerning sanitation procedures or own-check.

There are several different methods that may be applied in performing the inspection. The use of these methods depends on the scope of the inspection. Methods and examples of their use are presented in Table 24.3.

24.25   Preparing for inspection

Before performing an inspection, the official veterinarian should become acquainted with the premises, production processes and the products themselves if he/she is not already familiar with the slaughterhouse. To understand the production lines and the possible risk points for cross-contamination, the official veterinarian should go through the floor plan of the establishment. This means evaluating the transport routes of live animals, carcasses, condemned material and other by-products, and personnel routes through the establishment. It is also very important to check in the previous inspection reports whether the FBO has been requested to correct any non-compliances. If the inspection is pre-announced, it might be useful to discuss with the representative of the FBO beforehand the purpose of the visit, the personnel to be interviewed and the preferred schedule for the inspection.

Before the inspection, the official veterinarian should ensure that he/she is equipped with clean and intact protective clothing, including headdress and

Table 24.3   Inspection methods.

Technique	Example
Interview/asking questions	Asking about possible problems and previous changes in production.
To observe premises and equipment	Visual observation of the cleanliness of premises and equipment.
To observe activities	Visual observation of the level of hygiene during evisceration.
Making own measurements	Measuring the temperature of carcasses by calibrated thermometer.
Inspecting the own-check documentation	Checking the documentation concerning temperature control of the carcasses.
Verification that own-checking is adequate	Comparing your own temperature results to the entries in the own-check documents.
Cross-checking	Use data from two independent sources when possible: Compare the amount of non-edible by-products from own-check documents with documents received from the waste management company to ensure that they are similar.
Sampling	Take samples from process surfaces to verify the results of own-check.

hairnet, shoe covers and a beard cover, if needed. Wrist watch and rings as well as earrings should be removed. In the case that the FBO's own protective clothing is offered at the premises, it can be used. Other equipment needed include blank inspection form, notebook and pen, torch and calibrated thermometer. On many occasions, a camera is a relevant supplement in order to later demonstrate the possible findings during the inspection. It is always useful to re-check the relevant legislation and the most important guidelines before the inspection. If microbiological sampling is included, based on the control plan, it is also necessary to take materials needed for the sampling.

24.26   Initiating the inspection and interviewing the personnel

When arriving at the premises to be inspected, the official veterinarian should introduce him/herself to the representative of the FBO, if not already familiar with the FBO, and agree upon the programme for the inspection. In the case of an unannounced inspection, the official veterinarian should explain the reason for the inspection, whether it is one of the regular control programme inspections or possibly an inspection verifying the correction of some previous non-compliance. The inspection may cover just certain parts of the production and the corresponding areas in the OCS. The FBO's representative may or may not be present at the inspection but the absence of a representative must not prevent the execution of the inspection.

Different control methods can and should be used at the inspection (Table 24.3). Interviewing the personnel is an important means for control but it should always be accompanied with other control methods, such as own observations and measurements, to verify the response given by the FBO.

The latest information about the production, such as possible changes in the processes, should be discussed with the representative of the FBO. Before entering the production environment, the official veterinarian should be familiar with the number and species of animals slaughtered and the processing methods used. The official veterinarian should make sure that the current production and the premises correspond to the information given in the official papers. Through the interview, it is also possible to assess the FBO's understanding on hygiene and food safety matters and the FBO's input in securing the safety of its products. It is also important to discuss whether there are any current problems in the production that the official veterinarian should be aware of during the inspection.

24.27   Observing the premises and the facilities

By reading the floor plan, the official veterinarian is able to get an overview on the disposition of different processes and spaces at the facilities. There should be adequate space for the processes to ensure a high level of hygiene and to minimize the risk for contaminating the meat. In a small scale slaughterhouse it may be necessary to perform more operations in the same space at different times and the official veterinarian should ascertain that adequate cleaning takes place between the different production steps. This should be established by occasional on-site inspections; interviews or inspections of own-check documentation are not sufficient.

While inspecting the premises, the official veterinarian should observe whether the spaces are used for the purpose that they were planned for and whether they meet the food safety requirements without creating food safety and hygiene risks. By inspecting the premises it is possible to evaluate the routes connecting different rooms and spaces and observe possibilities for cross-contamination in practice. Generally speaking, it should be ascertained that the rooms for storing and handling the final products are clearly separated from those that are used for storing and handling the raw materials. Similarly, the packaged and unpackaged items should be stored separately. At slaughterhouse the rooms for storing and handling the carcasses at the end of the slaughtering process should thus be clearly separated from rooms that are used for handling the carcasses prior to evisceration. The by-products should be adequately separated and have designated storage rooms. No direct connection between the production space and the outside of the establishment should exist. The personnel traffic should be observed at the on-site inspection to verify that personnel do not walk from unclean areas to cleaner areas without proper hygiene measures.

Adequate rooms should be available for storage. This includes cold storage for the carcasses and storage for the equipment needed for processes as well as for sanitation equipment. Storage rooms should be checked for their cleanliness and the absence of pests. There should be no other materials or equipment present than those intended for production.

Hand washing facilities as well as sterilizers for knives have to be present in adequate numbers and also appropriately situated to ensure their efficient use. Automatic taps should be used in the wash basins. As for the working area, there should be adequate places for all the equipment used during the work, such as the protective gloves and knives used when cutting the meat. There should also be adequate provision for washing and rinsing the equipment. Personnel should have break rooms with toilets, wash rooms and adequate changing facilities.

If the official veterinarian is not very familiar with the establishment, it is recommended that the floor plan is taken to the on-site inspection. This helps to ensure that all the rooms and spaces are inspected. It may also help in observing the possible points of risk for cross-contamination as well as enabling the official veterinarian to evaluate if the floor plan is the same as the real situation. It should be noted that it is not only the rooms in which the slaughtering process takes place that should be inspected. An important principle is to open all doors and inspect all spaces in the slaughterhouse.

24.28 Evaluating the surfaces

The surface materials used should be evaluated as to whether or not they enable efficient cleaning and are adequately durable. In particular, those surfaces that are in contact with carcasses should be especially easy to clean. Materials originally suitable for food production establishments may be worn out, resulting in rough and cracked surfaces, and thus not suitable for the premises anymore. No rust should be seen on the surfaces and paint should not be peeling off. Also, the joints between the walls and floors should be intact. The significance of these defects on food safety will depend on their location in the premises. In the production areas maintaining high hygiene level, these may possess a high risk for contamination and corrective actions are needed more quickly. Floors should be checked for adequate drains and rakes and the walls for the finishing of feed-throughs. If the feed-throughs are left unfinished, they provide recesses that are difficult to keep clean and are a proper niche for microbes to grow in.

In addition to observing the condition of the floors and walls, the ceiling should be checked for its condition also, with special focus on condensed water that may drop on top of the products or the production lines. Condensing of water is a signal of an inadequate or incorrectly built or adjusted air conditioning system. Also, mould on any surfaces is a sign of excessive moisture.

No containers, boxes or other equipment should be placed directly on the floor. And in order to ensure hygienic working practices as well as effective sanitation, lighting should be adequate in all rooms, which may also be checked using illuminometer.

24.29   Observing the hygienic working practices of personnel

The working practices of a single employee are significant for the meat safety. Unhygienic working practices may result in cross-contamination causing zoonotic bacteria to spread on meat. It is thus worthwhile to spend an adequate amount of time focusing on the actions of the personnel.

It should be observed that the employees use appropriate protective clothing in an appropriate way. All hair has to be covered and jewellery removed. The protective clothing should also be changed frequently enough, especially when it is contaminated. Hand washing facilities are to be adequate but, most importantly, the employees should be trained to understand the significance of proper washing of hands. During the inspection the official veterinarian may also check that the employees do not enter areas outside the premises with their protective clothing on.

It is important to observe how the employees act in a case of contamination of the meat, for example when a carcass falls on the floor or a knife is stuck in an abscess. When noticing any unhygienic working methods, the official veterinarian should give immediate feedback to the worker and the feedback should also be forwarded to managers, to ensure a change in working procedures.

It has been experimentally shown, through a noted decrease in bacterial counts in samples from the production environment following a course on food hygiene, that training on food hygiene may significantly improve the food safety practices of food workers. Although the FBO is responsible for the hygienic working methods, the official control should also stress the importance of hygienic working practices and advise and steer the FBO towards these.

24.30   Evaluating the adequacy of the sanitation procedures

The adequacy of the cleaning practices is best evaluated by inspecting the establishment before starting production for the day, when the facilities have been cleaned after the last production phase. At this point it is possible to observe potential flaws in sanitation procedures. There should not be any observable residues of the previous production on the surfaces or in the devices. To evaluate the level of cleanliness and sanitation the official control should also check places that are more or less out of sight: underneath the worktops, over the conveyors and behind the processing devices. Although the general impression is clean, there might be possible spots that have been ignored in the cleaning process and that may possess a hygiene risk. The official veterinarian should make sure that the cleaning is carried out using proper detergents, followed with adequate rinsing to avoid any detergent residues remaining on the surfaces. For hygiene, all surfaces should be dry before start-ing the production after cleaning, but in rooms with a low indoor air tempera-ture, this often produces a challenge. Microbiological samples should be taken according to the official control plan to ensure successful cleaning procedures.

It is also important to note that the safety of the food products is not only a question of the food hygiene knowledge of the food workers and their man-agers. The behaviour of maintenance personnel has an impact as well on the level of meat hygiene at the premises, which should be acknowledged in offi-cial control.

24.31 Inspecting the own-check system

Inspection of the OCS includes evaluating the written own-check pro-grammes and confirming the execution of the programmes. The official veterinarian should make sure that the own-check is sufficient to prevent any food safety risk and urge for corrections in the OCS when needed. The execution of the OCS should be inspected by observing the entries in the own-check documents, not only the existence of the entries, but also deviations from the expected values. The whole OCS should be regularly inspected. Some examples on how to inspect the OCS are described here, with the focus on temperature control, traceability and handling of non-edible by-products.

Sporadic verification of the temperature follow-up should be conducted by using a calibrated thermometer. The official veterinarian may then eval-uate the compliance with the given temperature norms as well as ensure that the temperature in the rooms or of the products is the same as the entries in the OCS. In the case that there are changes in temperatures during the pro-cesses that might have an influence on the safety of the products, the reasons and the corrective actions taken should be discussed with the representative of the FBO.

Traceability of meat is important for food safety and consumer information. The official veterinarian should make sure that the traceability programme described in the OCS works satisfactorily. The FBO should be able to trace a carcass by showing appropriate documents and carcass identification mark-ings. This may be confirmed by asking the FBO to show how a carcass, stored in the cooler, is traced back to the farm of origin, or how tracing forward to the next FBO is performed.

The handling of by-products is important both for food safety and animal health. Inedible by-products must not go into the food chain and potential animal diseases must not pass to feed. The separation, marking and disposal of by-products should be verified by observation on-site and by inspecting the own-check markings. The correct disposal of different by-products should be occasionally inspected by cross-checking of the FBO's documentation and the waste management companies' documentation. The amount of by-products disposed of should equal to the amount delivered to the waste management company.

24.32 Official veterinarian's exemplary behaviour

When performing the inspection and going through the premises, the official veterinarians themselves have to strictly follow the good hygiene practices and make sure that the inspection is always performed starting from areas with a high hygiene level and proceeding towards areas with lower hygiene standards. It is to be noted also that the official veterinarians may act as a vehicle for contamination. If part of the inspection round is, for some reason, made from the area of lower hygiene to the higher one, it is necessary to change the protective clothing to a clean set. Recognizing and emphasizing the high impact of hand hygiene, official veterinarians should also wash their hands whenever touching a product or equipment that might result in a possibility for cross-contamination.

24.33 Giving feedback on the inspection

Feedback on the observed critical factors should be given during the on-site inspection. The observations should also be compiled and discussed immediately after the inspection. Through this discussion the official veterinarian is able to ensure that the representative of the FBO has understood the reasons behind the corrections suggested, whether they rely directly on legislation and should thus be done to comply with the food safety legislation, or whether they are recommended due to a good practice. The official veterinarian should also be able to discuss the legislation concerning the non-compliances observed and clarify what is needed to correct the fault. However, it is recommended also to give positive feedback on good actions taken towards enhancing the level of food hygiene.

It is important to differentiate the roles of official veterinarian and consultant. The official veterinarian may advise on, for example, that the surface material chosen for the work tops, floors or walls, is to be undamaged and easy to clean. However, it is not the official veterinarians' duty to suggest or decide what is the actual material used on the surfaces. The official veterinarian acts within the framework of the legislation and, therefore, relies on the requirements set in the statutes.

24.34 Documentation of official control

It is very important to bear in mind that only those things that have been documented may be considered to be done. This means that from every inspection a written document, an inspection report, has to be done, and also delivered to the FBO without delay. In this document all the noted non-compliances are stated and the demands set to the FBO in correcting these faults have to be clarified. It is a good practice to include the legislative basis for the demands in the inspection report. It is also important to discuss with the FBO about the time frame that should be given for correcting non-compliances. An example of an inspection report is presented in Figure 24.7. Other inspection report models can be used.

Name of the Competent Authority	Page (Pages)
INSPECTION REPORT	DD.MM.YY
Slaughterhouse name and address	
Inspection date	
Name of inspector	
Name of FBO's representatives present at the inspection	
Inspected premises -All premises that are inspected are documented in the inspection report. Also premises with no non-compliance are documented.	
Inspected activities and own-check areas -All inspected activities and own-check areas are documented, including those with no non-compliance observed.	
Observed non-compliances and dead-lines for the correction of non-compliance -All non-compliances and appropriate dead-lines for correction are each documented separately e.g. by numbering: 1. 2. 3. etc.	
Results of earlier inspections -Non-compliance observed in earlier inspections that are corrected are stated here. -Non-compliance observed in earlier inspections that are not corrected in requested time are listed. Further control measures for such non-compliance are specified here.	
Applied legislation -Applied legislation is listed.	
Possible further information to the FBO	
Signature of inspector	Signature of FBO

Figure 24.7 Model of an inspection report.

24.35   How to ensure the efficacy of inspections?

Official inspection processes are not efficient if they are not consistent and continuous. Despite the subjectivity that is inevitably present in the work of every official veterinarian, the aim is to treat every FBO equitably, still taking into consideration the unique features of the operator in question. The equity of the inspections may be achieved by high-quality education of the veterinarians carrying out these inspections. The equity of the inspections may also be ascertained by auditing, either internally or externally.

The inspections have to be regular, which is already confirmed through implementing a risk-evaluation-based control plan. When demanding the FBO to take action in order to correct a non-compliance, it is necessary to give a due date for this correction to be done. It is at least equally important to perform a re-inspection after the given due date to ensure that the risk has been eliminated from the process. The official veterinarian holds the power to evaluate the risk caused by the non-compliance and may adjust the control methods based on the evaluation. In cases that the FBO has not corrected the non-compliance requested, it is important to proceed in using more powerful enforcement methods to remove the public health risk.

Literature and further reading

Aarnisalo, K., Tallavaara, K., Wirtanen, G. *et al.* 2006. The hygienic working practices of maintenance personnel and equipment hygiene in the Finnish food industry. *Food Control*, **17**, 1001–1011.

Nevas, M., Kalenius, S. and Lundén, J. 2013. Significance of official food control in food safety: Food business operators' perceptions. *Food Control*, **31**, 59–64.

Vaz, M.L.S., Novo, N.F., Sigulem, D.M. and Morais, T.B. 2005. A training course on food hygiene for butchers: measuring its effectiveness through microbiological analysis and the use of an inspection checklist. *J Food Prot*, **68**, 2439–2442.

G. Enforcement

Outi Lepistö[1], Janne Lundén[2] and Karoliina Kettunen[2]

[1]Environmental Health Unit Pirteva, Pirkkala, Finland

[2]Department of Food Hygiene and Environmental Health,
Faculty of Veterinary Medicine, University of Helsinki, Helsinki, Finland

24.36 Scope

Present food legislation lays the main responsibility of food safety on food business operators (FBOs). With an appropriate own-check system, FBOs have to ensure the safety of their products. Official food control, however, is responsible of ensuring that FBOs comply with food safety legislation. To ensure compliance, food control authorities have several methods of intervening in a case of non-compliance detected in FBO's operations. Primarily, authorities should provide FBOs with advice and information about food safety requirements and, if needed, give notices and requests to correct violations of food safety legislation. If these control methods, however, do not lead to compliance, or if the foodstuff or operations cause a health hazard, official authorities should conduct stronger actions to ensure food safety. These actions, called enforcement measures (coercive measures, formal enforcement actions), are compulsory procedures that restrict the FBO's operations or force the FBO to conduct certain actions to remove violations. The use of enforcement measures should be based on good governance. Especially in situations of public interest, such as food safety problems, the authorities' knowledge of legal demands and good governance are important. This part of the chapter introduces the basic principles of good governance, which should be the foundation for every decision. Enforcement measures and their application in slaughterhouses are also described.

24.37 Good governance of enforcement measures

24.37.1 Principle of good governance

Good governance is mainly considered as the application of the regulations and legal principles regulating the public administration. The term 'good governance' refers to the regulations concerning governance in national and European legislation. Every public authority is expected to follow these regulations. The governance should be open, equal to everyone, objective and qualitative. Everyone should have his or her case dealt with appropriately and without undue delay as well as have a decision that is adequately reasoned.

The legal aspects also include the legal principles of administration, the basic rights of citizens and the efficiency of regulations. These aspects can also be seen as factors comprising the principle of good governance. The basic principles of European administration are written in The European Code of Good Administrative Behaviour, Charter of Fundamental Rights of the European Union and in a White Paper of European Governance. In these papers, the legal principles of administration as well as the basic rights of citizens, are strongly emphasized.

The authorities in the field of food control are expected to ensure food safety. This task includes decision making through which the FBOs are given certain obligations. The authorities are expected to follow good governance in all decision making. This is especially important in those situations where the basic rights of FBOs are being restricted through enforcement measures such as orders or prohibitions.

24.37.2   The legal principles of administration

The legal principles of administration are sometimes, but not necessarily, written in the legislation. According to these common principles, an authority must:

- treat everyone equally (the principle of equality) and
- use its competence in legally accepted means only (the principle of legality).

Furthermore,

- the actions of authorities must be objective (the principle of objectivity).

The legal principles of administration can also be seen as part of a legal system that is not written in any specific law but is formulated through legislation, legal order, case law, administration and values of society. Public authorities are obliged to follow these principles in their actions in addition to the written and specific regulations.

In the European Union the principles are of special significance, because the European Union legislation does not include any uniform administrative regulations. The principles in the European Union have been strongly influenced by European Court case law and are defined in detail in the White Paper of European Governance as well as in the Charter of Fundamental Rights of the European Union. An authority must be aware of these principles and, in accordance with them, must treat all similar cases equally, follow all the legislation and directions concerning the matter at stake and make a decision uninfluenced by any subjective views or opinions it may have.

The Charter of Fundamental Rights of the European Union states that everyone has the right to conduct a business and that everyone has the right to own and use his or her lawfully acquired property. It also specifies that a high level of health protection shall be ensured in definition and implementation of all European Union policies and activities.

When it is necessary to engage enforcement procedures, the authority is obliged to observe all these regulations pertaining to the basic rights of citizens. Strict follow-up of the legal principles of administration is of crucial importance in situations where the basic rights of FBOs are restricted, such as in application of enforcement procedures. The purpose of the legal factors is to ensure appropriate legal protection and efficiency.

Example:
If the authority is about to give an order to an establishment, it must first ensure that:

- the basis of the order is clearly written in the legislation;
- there are no other purposes (for example to give a sanction from previous shortcomings);
- the authority would treat any other establishment exactly the same way;
- any other authority should come to the same conclusion.

Furthermore, the action chosen must always be the lightest action to recover the situation and remove the health hazard.

24.37.3   The conflict of the basic rights

The basic rights of FBOs in many ways conflict with the issues of public health protection, especially when the authority uses enforcement measures such as restrictions of operations. An effort to observe both the rights of the FBO and food safety risks can be described as a contradictory situation. This situation is called 'conflict of the basic rights' in the field of food control and is illustrated in Figure 24.8.

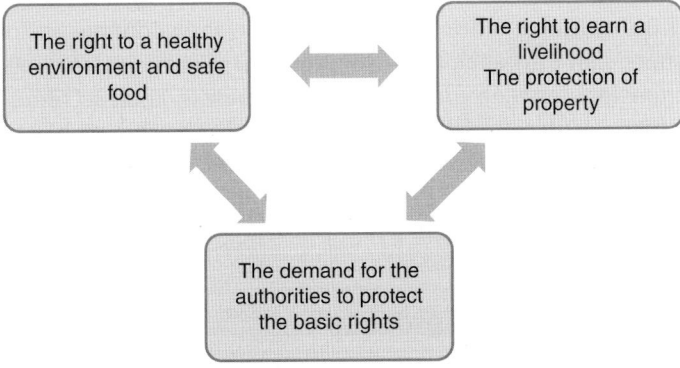

Figure 24.8   The triangle of the basic rights.

The balanced handling of the 'conflict of the basic rights in the field of food control' can be considered as a critical point in the practice of enforcement measures. To decide whether to use enforcement measures, an authority must observe and weigh all the basic rights and ensure their realization. This entity includes consideration of the basic rights of the population or of the individual person or consumer whose rights of health and safe food can be endangered. In addition, the basic rights of the targets of the procedure, which include the right to property and the right to earn a livelihood, will be endangered as well. The capability of resolving and balancing the conflict is dependent considerably on the authority's professional skills, attitudes and experience, especially in the demands of good governance. This balancing between seemingly contrasting rights requires extensive knowledge and experience of administrative and special legislation and demands.

The three main critical points in the decision making of authorities in food safety are: (i) the application of legal principles of administration, (ii) solution of the conflict between basic rights and (iii) the efficiency of the regulations applied in food control. These factors impact the practice of enforcement measures and further through this food safety and health protection.

24.37.4   The principle of publicity

According to the principle of publicity every matter should be treated as public unless there are some necessarily restrictions to publicity. These restrictions can be, for example, information about someone's health issues. The publicity of the decisions is also contradictory to some authorities: an authority must be aware of the reasons for confidentiality and be thorough in following these regulations. Nevertheless, decisions should be public in every case it is possible. This increases the transparency of governance and gives consumers a chance to observe the decision making as well as the control actions.

The authority cannot decide to treat a matter as confidential based only on the awkwardness and the possible consequences of the matter. This is against the legal principles of administration.

The publicity of decisions in food safety can also be used as a directing method. If the consumers are aware of possible shortcomings in food safety, the consumers' can affect the FBO's financial status and marketing opportunities by choosing between FBOs. It is, nevertheless, good to remember that a matter which is still unfinished is not public. The authority has a responsibility to find out everything that is relevant to the matter before the matter is decided. Before that, public knowledge of the matter can lead to false presumptions and damage the parties in the issue.

24.37.5   The hearing process

An important practical application of good governance is hearing of the party. Before the matter is decided, a party should be given an opportunity to express an opinion on the matter and to submit an explanation on the demands and information that may have an effect on the decision. The hearing should always be in written, documented form unless the customer (party) especially asks for it to be in spoken form.

The hearing process is sometimes difficult to carry out: it takes extra time and slows down the decision making. The procedure is also complicated and requires knowledge about the legislation concerning the administration procedures. The hearing of a party is, nevertheless, an important part of the so-called preventive legal protection, and thus should be used especially in cases such as enforcement procedures where obligations or restrictions are given to the entrepreneurs. Additionally, an error in the hearing process may lead to the cancelling of the decision in the case of an appeal to the court. Nevertheless, the hearing process can sometimes slow down the decision making too much. The hearing of a party is not necessary if:

- the health hazard is so severe that immediate actions are needed to protect the public health;
- the process of removing the health hazard can be endangered because of the hearing of a party.

The first situation can be, for example, a situation where there are allergens detected in the foodstuff and immediate withdrawal is needed. The possible delay can put consumers' health at danger. The second example could be a situation where there is a doubt that the entrepreneur is going to release the foodstuff under restrictions to consumption if the hearing process is before the actual decision.

24.37.6 An opportunity to make an appeal

Every decision that is open to appeal shall be accompanied with instructions about making an appeal. An opportunity to make an appeal is one of the most significant means of legal protection afterward. The decision is legal only after the time for making appeal is over. If there are no instructions for appeal, this time never begins, though it never ends. The decision is then always open to new appeals.

Another important reason for these instructions is the demand for legal rights and legal protection. The major factor in making an appeal is to ensure that there are no hidden mistakes, motives or attitudes of the decision maker. The decision which contains restriction must always be delivered fully documented. The documentation is the authority's responsibility.

24.37.7 The knowledge and attitudes of authorities

There are, even today, many contradictory attitudes concerning the enforcement measures among the food safety authorities. Different knowledge of and attitudes towards enforcement measures may influence their use of the measures. If the knowledge and preconditions of using enforcement measures are not similar among all authorities, this leads to a situation where in similar cases some authorities use these measures when needed while other authorities do not. This can cause regional inequality among the targets of the control actions and the principle of equality is not realized.

Furthermore, if the authorities are not aware of all existing administrative legislation during their decision making, the principle of legality is strongly endangered, a situation that can occur due to inadequate training, guidance and directions. The lack of a common, nationwide practice, advice and directions given by the central authorities causes uncertainty, which prevents the local authorities from undertaking the enforcement procedures. The authorities' own negative expectations and their fear of appeals or mistakes can also have the same effect.

24.37.8　The efficiency of food control norms

One important aspect of using the enforcement measures is their impact on the efficiency of the norms in the legislation. Even though the norm is valid as long as the law is valid, the real validity comes from two things: the efficiency of the norm and the empiric validity of the norm. The efficiency of the norm is mainly influenced by the authorities themselves: if the authorities apply the norms regularly the efficiency increases over time when the authorities learn to use the measures and the administrative procedures in the right way.

If the probability of being sanctioned when breaking the food safety rules is high, the empiric validity increases, when the common awareness of the consequences is spread among the target groups of the legislation. The actual meaning of using the enforcement measures is shifting the emphasis of food safety control from restrictive measures to preventive measures.

Also, criminal proceedings should be seen more as effective cautionary measures, rather than only sanctions against violations of legislation. The threshold to notify these violations to the police should be as low as possible. If the crimes against food safety legislation are not notified to the police the amount of these crimes will be distorted. Furthermore, there will be no case law formed among food safety issues (Figure 24.9).

24.38　Forms and application of enforcement measures in slaughterhouses

24.38.1　Enforcement measures

The forms of enforcement measures available to food control authorities depend on the exact legislation in each country. Nevertheless, principles concerning official control actions are similar and each food control authority

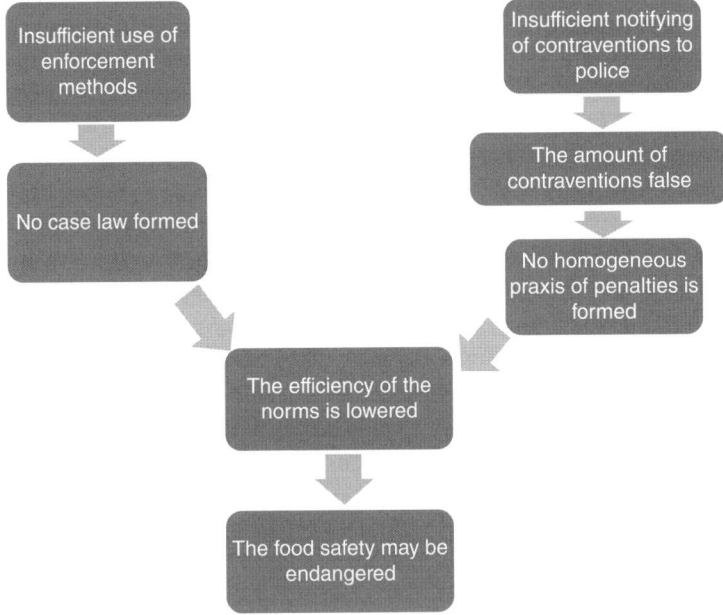

Figure 24.9 The main factors impacting the efficiency of the norms.

should have appropriate tools at its disposal to ensure food safety in a case of repeat non-compliance or emerging health hazard. Examples of enforcement measures applicable in slaughterhouse are presented in Table 24.4. In addition to these measures, the authority can use other enforcement measures deemed appropriate within its legal competence.

24.38.2 Gradual and proportional use of enforcement measures

In general, the basis for official food control actions should be equal and uniform treatment of FBOs. Thus, the principles of using enforcement actions, as well, should be similar among food control authorities. In addition, use of enforcement actions should be proportional and gradual taking account of the nature of the non-compliance and the FBO's past record with regard to non-compliance. For example, in a case of a minor violation detected for the first time, primary control actions should be education, negotiation, or giving a request to correct the violation. In a case of disobedience, however, the authority should gradually strengthen the enforcement actions. In addition, in the case of an urgent violation that causes an immediate health hazard, stronger control actions can be conducted without previous intervening control actions.

When the authority decides to apply enforcement measures, the procedure generally has three steps: hearing, giving the decision and verifying that the violations are corrected. Once the enforcement procedure has been started,

Table 24.4 Examples of enforcement measures applicable in slaughterhouses

Enforcement measure	Purpose of use
Order	Suitable for all situations with a risk or a possible risk of a health hazard caused by the food, establishment or operations, or other violation of food regulations. The violations can be ordered to be removed immediately or within a time limit. Example: surfaces in poor condition can be ordered to be repaired if not corrected otherwise.
Prohibition	The operations or use of a certain foodstuff can be restricted or prohibited due to a risk of a serious health hazard that cannot be prevented otherwise. The prohibition can also be imposed temporarily for the period during which the issue is investigated or the violation corrected. Example: the use of malfunctioning slaughter equipment contaminating carcasses can be prohibited.
Ordering the withdrawal and/or destruction of food	The FBO may be ordered to conduct a withdrawal of a foodstuff that violates food safety regulations. Example: ordering a slaughterhouse to withdraw and destroy meat that contains antibiotic residuals or other chemical contaminants.
Authorization to use food for purposes other than those for which it was originally intended	It may be decided that a foodstuff that is in violation of the food regulations and cannot be used for original purposes may be used for another purpose. If it is not possible to use the food for another purpose, the food must be ordered to be destroyed.
Suspension of operation or closure of all or part of the establishment	The establishment may be closed or the operations suspended if there is a serious health hazard that cannot be prevented in another way than stopping the activities. The suspension of operation or closure of all or part of the establishment shall be ordered for an appropriate period of time. The establishment can only restart operations once the violations have been removed. Example: temporary suspension of operations due to lack of running water.
Penalty payment	An order or a prohibition can be reinforced by imposing a fine or a penalty payment. The penalty can be a non-recurrent single payment or a continuous fine that has to be paid repeatedly until the non-compliance is corrected.
Suspension or cancelling the establishment's approval	The establishment's approval may be cancelled if the establishment or the operations violate the food regulations in a recurrent and blatant way and there is no other way of preventing a health hazard. Approval may be cancelled in full or in part, and until further notice or for a fixed period.

the authority should make sure that the process is followed until the end and that the food safety violations are corrected. However, if the FBO announces during the hearing process that the violations are already corrected, the authority shall inspect the situation to decide whether an actual enforcement decision needs to be done. If no violation anymore exists, there is no need to make an enforcement decision.

The gradual enforcement process should be thoroughly planned in advance: what is the next step if non-compliance persists. An example of gradual use of enforcement measures is illustrated in Figure 24.10.

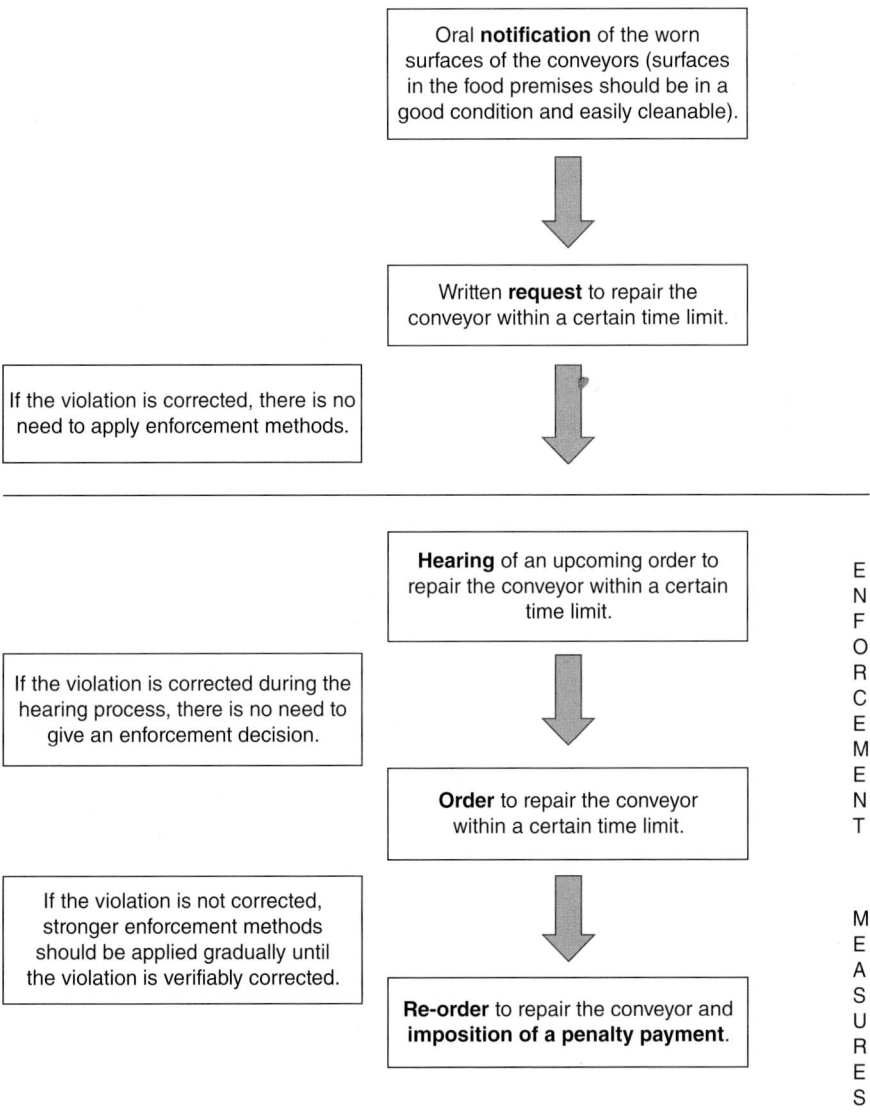

Figure 24.10   Gradual use of enforcement methods. Example: worn surfaces on conveyors in the slaughtering hall.

24.38.3   Decision on the enforcement measure

The decision on the enforcement action should be provided in writing, including information about which actions the FBO is obligated to take and the time limits for the correction of violations. In addition, the reasons for the decision and the appeal process should be clearly displayed in the decision.

Considering the efficiency of enforcement measures, it is essential to make sure that the FBO understands the reasons and the legislative obligation of the enforcement measure applied. Authorities should thus pay close attention to the unambiguous forming of the requirements and reasoning for the decision. In addition, to enhance the efficiency of the given decision, the possible follow-ups could be stated. For instance, it could be mentioned in the decision that if the given order is not obeyed, the authority will consider applying stronger enforcement actions.

24.38.4   Verifying the outcome of the enforcement

After the time limit for correcting the violations given in the decision, the authority should conduct an on-site inspection to verify that the violations have been corrected. It is also possible to verify by other ways than an on-site inspection that the compliance is achieved. For example, if the authority has ordered a withdrawal or destruction of food, these actions should be verified by written documents with the exact amounts and batch codes of the foodstuff ordered to be withdrawn. The destruction of the food, in turn, should be verified by the presence of at least one official or by a document of destruction in a proper waste treatment plant. Whatever the way of verifying the correction of violations, the authority should be convinced that the violations are corrected and the case is solved.

24.38.5   Further procedures and consequences

If the enforcement measures applied prove insufficient, the authority can go further in the gradual procedure of enforcement. For example, an order or a prohibition can be reinforced by imposing a fine, penalty payment or other threat of economic influence, or even by closing the food establishment. If imposed, the economic threat should be substantial enough that it acts as a motivator for the FBO to fix the problems rather than to pay the fine. A penalty payment can be a non-recurrent single payment or a continuous fine that is to be paid repeatedly after the date stipulated in the decision until the non-compliance is corrected.

In addition, regardless of the administrative enforcement procedure, the authority can consider cooperation with the police if there is the suspicion of the misleading of the authority, obvious law breaking or flagrant disregard of food safety regulations. In these cases, the role of the police is to examine whether a crime has occurred and, when necessary, to forward the case to the criminal justice that determines possible legal sanctions.

24.38.6 Urgent measures

If a violation causes an immediate health hazard and the FBO does not correct it as a self-checking action, the authority should have the possibility to apply enforcement measures instantly, without previous notifications. A temporary decision can be made orally, in certain situations even without any hearing, but a formal written decision should follow as soon as possible. An example of using urgent measures could be suspending the operations until a certain critical violation is corrected, or prohibiting using an unsafe foodstuff in manufacturing. An urgent measure, for example a prohibition, can also be imposed temporarily until the violations are corrected or the situation is investigated and solved in other ways.

24.39 To advise or to use enforcement measures?

Advisory control strategies and negotiation are considered a preferable way to maintain constructive relationships with FBOs. The administrative enforcement procedures are sometimes perceived as heavy-handed, demanding and time consuming. Thus, the possibility to use enforcement measures might not always be used in the most effective and uniform way, and practices in using enforcement measures may vary between official veterinarians.

Enforcement measures are mainly used when a health hazard is obvious or likely to occur, that is in violations such as temperature abuse, sanitary deficiencies or improper traceability. It might be that such usually easily detectable and serious violations make the use of enforcement measures feel more justified among food control authorities. However, in a more complicated or ambiguous situation, the food control authority has to do more case-by-case assessment and enforcement measures might not be used until 'a last choice', after many notifications and requests. However, even though the principles of using enforcement measures should be uniform among food control authorities, every case is unique. Therefore, the food control actions may not be exactly similar in cases appearing similar.

Official veterinarians in slaughterhouses have a demanding position being an official food control authority and present in the work community of the slaughterhouse at the same time. The threshold for applying enforcement measures in slaughterhouses, however, should only be affected by the food safety risks and good governance. In general, implementing food control actions equally and effectively requires that central authorities draw clear common guidelines and provide food control officials with enough support and education.

Literature and further reading

Commission of the European Communities. 2001. European Governance – a White Paper. COM(2001) 428 final. *Official Journal of the European Communities*, **C 287/1**, 1–29.

European Parliament. 2001. The European Code of Good Administrative Behavior. Office for Official Publications of European Communities, Luxembourg.

European Parliament and the Council and the Commission. 2000. Charter of fundamental rights of the European Union. *Official Journal of the European Communities*, **C 364**, 1–22.

European Parliament and the Council (2004). Regulation (EC) No 882/2004 of the European Parliament and of the Council of 29 April 2004 on official controls performed to ensure the verification of compliance with food and feed law, animal health and animal welfare rules. *Official Journal of the European Communities*, **C 165**, 1–141.

Fairman, R. and Yapp, C. 2005. Enforced Self-Regulation, Prescription, and Conceptions of Compliance within Small Businesses: The Impact of Enforcement. *Law and Policy*, **2**, 491–519.

Jokela, S., Vehmas, K. and Lundén, J. 2009. Food control officials̕ views on coercive measures in ensuring food safety. *Arch Lebensmittelhyg*, **60**, 130–134.

Lepistö, O. and Hänninen, M.-L. 2011. Effects of legal aspects on the use of compulsory procedures in environmental health and food control. *J Environ Health Res*, **11**, 127–134.

H. Auditing Official Controls

Juha Junttila

Food And Veterinary Office, European Commission, Dunsany, Co. Meath, Ireland

24.40 Scope

The objective of this part of the chapter is to explain the background, context and purpose of audits as well as to introduce some key concepts and principles that are essential for understanding the audit process. There are plenty of generic textbooks and standards on auditing and this text is not intended to compete with such comprehensive resources but, instead, it is intended to provide a short introduction to the concept of auditing official controls. Understanding the purpose of audits is useful – if not essential – for those responsible for daily operations of meat inspection, for those who manage meat inspection systems and for those who may have been put in charge for auditing such systems.

24.41 Background

The word audit stems from the Latin word 'to hear' or 'hearing'. The origin can be traced back to a time when kings – or other rulers – put their representatives in charge of organizing official hearings for the purposes of tax collection. By the sixteenth century accounting had developed to the extent that double-entry bookkeeping became the norm. Luca Pacioli – who was first to describe double-entry bookkeeping in his book on accounting – recommended that the accounting records should be verified by 'auditors'. Until the late twentieth century, auditing remained an activity related solely to financial accounting. The term 'audit' is still primarily associated with financial matters where detailed scrutiny of documents and related activities are undertaken. Internal auditors have been 'feared' in organizations and this perception is reflected in popular culture as well – for example, Terry Pratchett describes a horrific archetype of an auditor in his novel 'Thief of time'.

During the past few decades the principles and application of auditing have evolved significantly, typically after major events or scandals such as the BSE crisis in Europe or the Enron scandal in the United States. Legislators and standard-setting bodies have responded to these incidents by strengthening the requirements set for auditors, audit bodies and for the audit process. The trend has been towards more prescriptive rules for auditor independence and for the objectives of an audit engagement. Auditors are increasingly required to demonstrate a high degree of objectivity and to provide more added value by evaluating their auditees against general objectives rather than specific technical requirements. Even with these recent developments, the basic problem is still the same as it was two millennia ago: how to best use a third party (auditor) in providing assurances that an auditee (e.g. tax payer, food business operator, government agency) is carrying out their duties according to the expectations of the audit client (e.g. the king, the government, general public).

Example Scenario 1

Large scale fraud on meat labelling has resulted in public outcry and growing demands on improving official controls on labelling and traceability. Consumer confidence in official controls is plummeting. It is not clear whether this incident represents a failure of official control or not. A high level advisory body – consisting of two ministers and three chief executives from different control agencies – has decided to launch an audit to establish the root cause(s) of the incident and to propose corrective actions, as appropriate. One of the main objectives is to determine whether it is reasonable to expect official controls to detect some indicators to this type of fraud or if it is simply criminal activity which escapes routine controls. In this scenario, the advisory body is the audit client – representing indirectly the consumers, who would be the main stakeholders in this scenario. Other stakeholders having an immediate interest would be the meat industry and food businesses, which are purchasing potentially mislabelled meat. The auditee will be the competent authority responsible for controls at abattoirs and meat establishments. The auditor (or audit team) will have to be selected with a view of providing a sufficient level of independence and credibility to satisfy stakeholder concerns. It is unlikely that auditors selected from the competent authority itself could convince the stakeholders that the audit results would be objective, even though they might have all the expertise in the subject matter.

Today auditing is increasingly becoming a management tool for both private and public organizations. The concept of auditing has been expanded to cover evaluation of activities with a purpose of ensuring that an organization is achieving its objectives. These objectives may be determined by the organization itself or be imposed by an external entity. Audits are becoming more and more an instrument for continuous improvement and are seen as a constructive tool serving the interest of both organizations and their stakeholders. Auditors have evolved from being considered 'hostile intruders' into welcome professionals, who can actually provide valuable outputs and learning experiences that add value to the operations of their auditees.

Example Scenario 2

A *large meat processing company, which was lucky enough not to be affected by the circumstances described in Scenario 1, has decided to have its entire supply chain audited by contract auditors from a big multinational company.*

The goal is to review all operations from the supply of fertilizers and feed up to the large meat processing establishments belonging to the same company. The competent authorities follow suit and decide to carry out an audit to re-evaluate their official controls in order to ensure that they are up to date and capable of adding value to the company's own controls. Bearing in mind that official controls are not primarily designed to detect or investigate fraud, the auditors have also been asked to evaluate to what extent the existing controls would be capable of identifying fraudulent practices – or indicators of fraud – when it exists. This goes well beyond the normal internal audits. Nevertheless, the results are to be fed into the next management review with a view of identifying critical control points and improving fraud detection and awareness.

24.42   Different types of audits

All audits share common elements – they have audit clients, audit criteria, audit objectives and auditees (Table 24.5). Variation in these elements will result in different types of audits (Table 24.6). In general, the objective of an audit is to establish whether the auditee is in compliance with legislation or conformance with other audit criteria, such as general or specific standards, requirements, procedures or policies. Sometimes an audit client (the entity which is requesting the audit) will request an audit for business reasons, for example because the organization is a supplier to the audit client. Such an audit is classified as an 'external audit'. When the audit criteria are mainly prescriptive technical requirements, the audit can be considered as a compliance or conformance audit. When the main audit objective is to determine whether the auditee is achieving certain policy – or other higher level objectives – these are called performance audits. Performance audits will, in most cases, also have features of a system audit, which means that the audit will acknowledge that the performance of a system is more than just an arithmetic sum of its components' compliance. Fraud and forensic audits are quite distinct from performance audits and their main objectives are either to identify fraudulent activities, establish whether fraud has occurred or to collect evidence for legal proceedings.

Various definitions for an audit can be found in international standards such as the International Organization for Standardization (ISO 19011), the Institute of Internal Auditors (IIA) standards, the International Standards of Supreme Audit Institutions (ISSAI) and Codex Alimentarius standards. Legislators in a number of jurisdictions around the globe have also provided their own definitions. Searching for a global definition or selecting any of the existing definitions as a model is probably futile and for the purposes of this introduction and in this particular context, by audit we mean a systematic process which provides for:

- consistency and predictability;
- independence of the audit body and auditors from the auditee;

Table 24.5   Audit terminology.

Objectives	(Set of) question(s) that the audit needs to answer or the types of conclusions to be reached. Setting objectives, which are meaningful, ambitious enough and at the same time, feasible, is a critical step for a successful audit.
Criteria	Standards, legal requirements, policies, procedures, instructions or other criteria against which the audit evidence is evaluated.
Scope	Boundaries of an audit – the processes, commodities, facilities and other, mainly tangible entities to be covered during the audit.
Client	Organization which requests the audit(s) and specifies the objectives, criteria and scope for the audit(s).
Auditee	Organization that is being audited, that is, which is in charge of the process that is being evaluated against audit criteria.
Stakeholder	Anybody having a direct or indirect interest in the audit results.
Evidence	A fact that is relevant to audit objectives and criteria and verifiable, that is reliable to the extent that it can be used as a basis for a finding. 'Verifiable' means that the source of the fact has been explicitly stated, can be traced back and – where appropriate – is supported by an exhibit, that is a sample, document, photo or other record.
Finding	Result of evaluation of audit evidence against audit criteria.
Conclusions	Result of the audit (evaluation, judgement) after consideration of the audit objectives and audit findings.
Recommendation	Indicates the improvements required by the auditee in order to bring the process into compliance or into a state of effectiveness that meets the audit criteria. Recommendations should be clear and unambiguous, and clearly indicate the objective(s) to be achieved but not prescribe the means except where they are specifically prescribed by audit criteria. A recommendation should also help the auditee to address the root cause(s) of a problem instead of merely requesting to fix the symptoms.
Professional scepticism	Is an approach of not accepting evidence at face value without being overly suspicious or cynical, continuing to pursue all avenues of inquiry and continuously confirming and corroborating evidence (Table 24.8). It can also be described as an attitude of a curious and questioning mind, being critical and alert while maintaining a constructive and balanced mindset.

- objective and verifiable results from an evidence based approach;
- assessment of whether the auditee is achieving the objectives and expectations of the audit client and stakeholders.

24.43   Why audit official controls? (What is the added value?)

The added value of audits on official controls falls into two broad categories: external and internal assurances. Firstly, those in charge of spending public

Table 24.6 Different types of audits.

Compliance audit	Main focus is compliance with clearly defined requirements or standards, which are usually rather technical and specific. Produces a binary result, that is compliance or non-compliance – often an intermediary result (partly or largely compliant) is possible.
Financial audit	Primary objective to verify that financial reports and statements provide an accurate view of company's financial status, that is they should be free of 'material misstatements' and in compliance with accounting standards and regulations.
System audit	Recognizes that components of, for example, a control system are connected and those connections/interfaces must be audited as well in order to verify the overall functioning of the system. In other words, a system is more than the sum of its components.
Performance audit	Is primarily concerned about achievement of objectives, which can also be called effectiveness. Compliance is also verified but the main focus is on high-level compliance, that is overall objectives of legislation or a standard.
Internal audit	Is an audit, where the audit client is the organization to be audited. The auditor (or audit team) may come from within the organization or from outside – the defining factor is who initiates the audit and specifies the objectives, criteria and scope for the audit. Internal audits are a management tool to verify compliance and/or effectiveness of operations, that is whether organization's objectives are being achieved.
External audit	When an audit is initiated by an external entity and audit objectives, criteria and scope are defined by this external entity. Most countries would have an independent audit body – often linked to the parliament – which audits government bodies and agencies according to its own programme.
Fraud audit	Main target is to scrutinize internal controls and identify indicators on fraudulent practices, that is to establish whether fraud is occurring or not. If strong evidence or suspicion of fraud is identified, this would normally lead into a forensic audit.
Forensic audit	Focuses on collecting legally valid evidence on fraud with an objective of securing a successful prosecution – or alternatively, clearing suspicions raised by a fraud audit. The skills required for detecting fraud and collecting evidence are different and these activities may be carried out by separate teams.

money and protecting public health are generally considered to be accountable to the tax payers and consumers. They should also ensure official controls are carried out in a non-discriminatory, consistent and impartial way in order to ensure equal treatment of food business operators who might suffer economic consequences from official controls and subsequent decisions. It has become a generally accepted principle that independent assurances are needed to determine whether official controls are carried out in such a way that all stakeholders' objectives and expectations are achieved to the best possible extent.

Secondly, meat inspection services themselves have an interest and responsibility to seek assurances that their official control activities are in compliance with legal requirements and are achieving their objectives. This can be achieved by self-assessments or audits of an internal or external type. Sometimes self-assessments may produce very similar results to internal or external audits but the real benefit of an independent audit is that they are free from the kind of bias that is inherent to self-assessments. Over time, every organization has a tendency to develop some degree of blindness to its own operations and will have difficulties in objectively identifying areas for improvement – no matter how obvious they might be.

Many organizations have audits built into their quality management systems, the results of which feed into a regular management review. Management review is a specific requirement in current management system standards (e.g. ISO 9001). Whether an organization is seeking to have its management system certified or not, such a feedback loop is considered as good management practice and essential to drive continuous improvement. If there is buy-in to the audit process their outputs can be used effectively for the benefit of the organization, eventually leading to a situation where audit results are mainly positive, reassuring and verify that the system is achieving its objectives.

Example Scenario 3

A national meat inspection service has made considerable efforts to improve official controls on HACCP systems. All inspectors have been trained, procedures updated, supervision modernised and an IT system put in place to monitor control results. An ad hoc audit team has been set up to evaluate the consistency of HACCP implementation and how effective the official controls are in validating and verifying effective operation of HACCP systems. The main objectives for this audit are: (i) to evaluate whether current implementation of HACCP is living up to the expectations and (ii) whether the official controls are capable of providing added value in improving the systems and enforcing legal requirements. In order to achieve these objectives, experts with a solid background in systems audits and HACCP have been selected for the audit team. All experts come from outside the national meat inspection service and have not been involved in developing or managing the current control system. They have both the expertise and independence necessary to provide objective assurances on the strengths of the system as well as identifying areas for further improvement.

24.44   Auditing processes and systems

The real challenge for an auditor is a move from pure compliance auditing into an approach where performance of processes and systems is at the focal point. This will require a completely new set of audit questions and should be clearly reflected in the conclusions at the end of the audit. The core question should be about achievement of objectives, that is effectiveness of the system.

Answering that question requires systems thinking. In other words, acknowledgement that systems are composed of interconnected components with a common purpose and that the most productive way is to look at those components as processes. Each individual process has inputs and outputs and it is by way of these inputs and outputs that the processes are interconnected to produce the overall system outputs, which may or may not be in line with intended results.

The first question an auditor needs to ask is: 'Is there a system in place?' This will require an examination of the system components, their interconnections and whether the system components are working towards an overall common objective. The second question would be: 'Is this system achieving its objectives?' This is a much more difficult question to answer. It is not simply about enumerating compliance with specific requirements of the individual components. Thirdly, the auditor needs to ask: 'If the system seems to work well now, how confident can I be that this will be the case in the future as well?' Answering this question will require careful assessment of the management system and the assurances provided by it. Finally, the auditor will need to determine: 'Is the system capable of fixing and improving itself?' Once an auditor is able to answer 'yes' to this question, it can be concluded that the management system has reached a high level of maturity and implementation, and is fully integrated into the workings of the organization by both staff and management.

Example Scenario 4

A *meat inspection service often plays a role in a national residue monitoring programme (i.e. residues of veterinary medicines). If the objective of sampling at the slaughterhouse level is to detect unacceptable levels of residues, risk-based targeting would be required in order to make best use of resources in achieving that objective. This may require access and/or compilation of information from various sources (where available): treatment records, prescriptions/deliveries of medicated feed, results of previous controls, including on-farm controls and so on. Effective use of such data sources often requires coordination between several competent authorities. In such a case, an audit would evaluate the effectiveness of the meat inspection services' sampling programme in the overall context of the national system. A system approach would require auditors to pay particular attention to interfaces and coordination between components (authorities) of the national monitoring programme.*

24.45 Key principles

If there is one overriding principle from which most of the other audit principles and requirements can be derived, it is the principle of objectivity (Table 24.7). Audit is a process which must collect factual evidence, evaluate that evidence against audit criteria and produce objective conclusions that are based on evidence. Without this objectivity, the value of audit results

Table 24.7 Audit principles.

Objectivity	Audits are adding value only if they are able to provide an objective picture of the reality. Therefore, audit bodies and auditors should be independent of the activities to be audited. They should also be free from bias and conflict of interest and base their conclusions on evidence. All of these attributes serve the purpose of providing for objectivity.
Ethical conduct	This refers generally to doing the right things and overlaps to some extent with objectivity – after all, being objective is the right thing to do. Ethical conduct includes issues such as: fair (balanced) presentation of audit findings, impartial treatment of auditees, honesty, confidentiality, respectful and constructive behaviour. Audit standards may require an audit body to establish a code of conduct, which should be adhered to during all audit engagements.
Systematic approach	Audit bodies should ensure the uniformity and consistency of their audit process, that is similar inputs produce similar outputs and audit results do not depend on the individual auditors. To that end, audit bodies must have documented procedures to be followed during programming, planning and execution of audits. Review and evaluation processes are also essential in ensuring that quality standards are being continuously met.
Due professional care	Audit bodies must ensure that proper recruitment and training processes are in place to ensure auditor qualifications and competencies. Auditors must exercise due diligence and care commensurate with the importance of the audit task. They should also assess the feasibility of an audit engagement before initiating an audit and maintain an attitude of professional scepticism (Table 24.5).

is significantly reduced equally for the audit client, auditee and all relevant stakeholders. Any serious doubts of bias, conflict of interest or impartiality will undermine the credibility of an auditor, audit team or audit body and put audit results into question.

When audit standards introduce the concepts of ethics, independence and systematic audit process, they all serve the purpose of objectivity. Ethical behaviour covers issues such as due professional care, transparency, unbiased attitude and, in general, doing the right thing. Code of ethics can be considered as one of the cornerstones in building objectivity into an audit system. Independence of the audit body and auditors (from the processes to be audited) is another fundamental supporting principle. Certain organizational and operational safeguards are needed to protect the audit process from any undue influence that could affect its objectivity. But independence of the mind is also an important personal characteristic of a good auditor and may, to some extent, counterbalance weaknesses in organizational independence. And, finally, in order to demonstrate consistency and to be predictable (objective) the audit body needs to apply a systematic, well documented process that can be repeated again and again with similar outcomes from similar inputs.

As the primary purpose of audits is to provide assurances to the audit clients and stakeholders, the credibility of the audit body is of paramount importance. Credibility is built on objectivity, which, in turn, lays its foundations on ethical conduct, systematic process and independence. Building credibility takes years but it can be lost overnight.

Example Scenario 5

A recent audit has revealed serious animal welfare issues in transport, lairage area and stunning. These problems have obviously persisted over a long period of time and official controls have either not identified them or have not been effective in improving the situation. A key factor leading to improvements is whether the competent authority is willing to recognize that the audit conclusions are objective and valid. If the audit is perceived as professional, constructive and able to identify systemic issues rather than pointing at individuals, the chances for acceptance – and change – are considerably increased. Control system failure is always somewhat embarrassing for those involved and the auditor's task is to bring about change in the most constructive manner possible. An objective and analytical approach will support the auditee by showing an accurate picture of the situation and demonstrating the true extent of a problem. It is up to the auditee to identify the real root causes, but an auditor may provide pointers in the right direction and, in doing so, draw the attention away from individual failures.

24.46   Auditor qualifications

Auditor qualifications can be divided into personal attributes, knowledge, skills and work and technical experience. The personal characteristics required from a good auditor are already quite demanding. An auditor must be ethical, open minded, diplomatic, observant, perceptive, versatile, tenacious, decisive, self-reliant, acting with fortitude, open to improvement, culturally sensitive and collaborative (ISO 19011). These characteristics are to be combined with a broad range of generic audit skills, knowledge about legislation, various standards and management principles as well as relevant working experience.

It is quite difficult to define the exact qualities of a good auditor and it seems to be a balanced mix of soft and hard skills. Bearing in mind the objectives of an audit (adding value to the system which is subject to the audit), communication skills appear on top of the list. No matter how accurate and objective the audit conclusions are, they will be of little use if the auditee does not understand and agree with them. In order to reach conclusions that capture the most relevant aspects of the system, the auditor must pay particular attention to detail but – at the same time – avoid getting lost in those details. This requires an analytical mind-set combined with a capability for systems thinking.

24.47   The audit process

Every audit should commence with an initiation phase, where the auditee is informed about the forthcoming audit well in advance of the on-site audit activities. This is the first step where systems audits vary from compliance and fraud audits, which for obvious reasons are not notified in advance. For a successful systems audit it is essential to communicate clearly the objectives, criteria and scope to the auditee as early in the process as possible. This will allow the auditee to have relevant documentation and members of staff available during the audit and be prepared to provide information that contributes to achieving the audit objectives.

The next phase – preparation for on-site activities – concerns both auditors and the auditee. Typically, the auditor would need to study documented procedures, records, data-bases and so on and possibly request some preliminary information from the auditee. At this stage the auditor may already start drafting an outline (structure) of the audit report, prepare an audit plan or schedule and explain to the auditee what is to be expected during the course of the audit. It is also in the auditee's best interest that auditors acquire good understanding of the auditee's organization and processes in order to achieve a balanced and objective outcome. Therefore, good cooperation with the audit team is essential from the very beginning.

The auditee will also need to develop the skills necessary to fully benefit from the audit process. A highly competent auditor will facilitate this skills development process but it also requires conscious efforts and good will from the auditee's side to reach a positive outcome. Organizations that are subject to regular audits usually develop an appreciation for audits and their outcomes and use them as an opportunity to learn, and may even be looking forward to implementing corrective actions which further strengthen their management system.

Before starting with the on-site activities it is normal practice to organize an opening meeting, where the audit details are communicated to all participants of the audit – including the management of the auditee. This is an opportunity to re-iterate the objectives, scope and criteria, clarify any outstanding issues and confirm the logistic details of the audit. It is also an opportunity for the auditors to request any information that was identified as missing during the preparatory phase. And, finally, the auditees should leave this meeting with a clear understanding about what is expected from them and how the audit is going to proceed until the final reporting phase.

The on-site activities of an audit are often the most stressful for the auditee, particularly for those individuals who are lacking in the experience. Having an auditor observing your work behind your back or scrutinizing your records in detail can be a daunting experience. Auditors should keep this in mind and acknowledge that in such situations most people do not behave as they do under normal circumstances. This may result in auditee inadvertently providing the auditor with incorrect information. One of the core competencies of an auditor is the skill of putting people at ease while collecting audit evidence to ensure an accurate evaluation of a task or process.

Audit evidence should, where possible, be corroborated by other sources and co-validated with the auditee. Corroboration means confirmation of a fact or facts by another type of evidence or verifying it by using another source. For example, the absence or presence of certain activities should be verified by using a combination of documentary checks, interviews and observations (Table 24.8). When sufficient evidence has been collected to draw findings or conclusions, these should be discussed and, if possible, agreed with the auditee (co-validation). Sometimes is not possible to reach agreement but this should be the exception rather than the norm and should not prevent the auditor from drawing conclusions when they are justified and well supported by evidence.

Probably the most difficult task for an auditor is to build solid argumentation starting from sufficient and relevant evidence and ending up with a meaningful overall conclusion. This is also the very essence of auditing and requires a lot of discipline before, during and after the audit. The aim is to avoid anecdotal evidence, speculative comments and jumping to premature conclusions, which will all eventually result in misleading the audit client and frustrating the auditee. A professional auditor will always cross-check after double-checking all the facts in order to ensure that he/she has understood everything correctly and has relevant evidence at hand. Evidence may sometimes lead to conclusions that were not expected by the auditee. In such cases the audit trail and the logic behind such conclusions should be solid and understood by the auditee as well as by the audit client.

Example Scenario 6

In order to draw the conclusion in our previous example (animal welfare), the auditor would need various types of evidence to support the claim that the situation has been unacceptable for some time – not just at the time of the on-site visit. This requires possibly repeat observations, documentary evidence, examination of installations and equipment, interviews and review of past control results. The second line of investigation would require evidence on the control activities and why they did not achieve the objectives. Were the problems not detected? Did the operators disagree with the findings? Were corrective actions ordered? If yes, why were they not implemented or effective? Is there a strategy to deal with repeat offences? Only after all relevant evidence is available, can the auditor proceed into conclusions. It is important to recognize that most individuals genuinely want to do things right – it is often the system in which they operate that sets constraints and pushes individuals in the wrong direction.

The tangible output of an audit is normally a report that contains the evidence, findings, conclusions and recommendations, as appropriate. The format of an audit report varies and depends, amongst other factors, on the context, legal requirements, contractual details and audience, for example whether the report will be published or not. In order to be useful, a report must contain the

Table 24.8   Methods of collecting audit evidence.

Method	Examples
Inquiries to the auditee	Interviews, focus-group meetings, questionnaires, exchange of letters/e-mails etc. would fall under this category – generating either testimonial evidence or, alternatively, documentary evidence. Examples: • pre-audit questionnaire sent to the auditee before starting on-site activities; • interviewing staff during on-site activities.
Document examination	May identify facts that are relevant to audit objectives and criteria and, thus, can be called documentary evidence. Documents may be collected and examined at any stage of an audit. Examples: • exploring the web site(s) of the auditee in order to identify and/or /locate documentation relevant to the audit, that is quality policy statements, mission, vision etc.; • examination of documented procedures, inspection records, outputs from automatic recording equipment etc.
Physical examination	This is possible only during the on-site activities of an audit and may generate either testimonial, physical or documentary evidence. In the first case (testimonial evidence), the witness providing the testimony is the auditor – possibly corroborated by auditee. Examples: • microbiological sampling (swabs etc.) would generate records of laboratory analysis, which would serve as evidence of physical examination; • photographs of the structure and equipment of transport vehicles, equipment along the slaughter-line, ear-tags, labels etc. would serve as evidence of the results of physical examination; • measurement equipment (e.g. temperature, voltage, weight) may also produce records of the measurements, which would be examples of documentary evidence generated by physical examination.
Observation	This is also possible only during on-site activities and is normally related to activities, that is observing how a process is run or how personnel perform a task. This will result in either testimonial (by auditor) or documentary evidence (e.g. photographs, video). Examples: • written description of stunning or de-hiding process, corroborated by the auditee would be evidence of facts collected by observation; • photographs/video of un-loading animals from transport vehicles or from the lairage area would serve as evidence from this type of a process.

Table 24.8 (*Continued*)

Method	Examples
Analytical procedures	Are procedures carried out on data or physical samples. For example, laboratory analysis of a sample will not completely change the nature of the evidence but, instead, add a testimonial dimension to that piece of physical evidence. Analytical procedures carried out on data will, similarly, add a new dimension to the original type of evidence, which could be documentary or testimonial depending on the method of data collection. Examples: • Trichinella testing is an analytical procedure, which generates data from physical samples – based on the results of these analytical procedures, conclusions may be drawn on the safety of individual carcasses or the daily production as a whole; • tracing back from labels on final products (e.g. carcasses, cut meat) to the farm of origin produces results that may be used to evaluate the effectiveness of traceability systems; • reconciliation of incoming and outgoing weights will provide assurances on the internal controls, that is that prohibited material is not introduced to the process and nothing goes missing (i.e. for prohibited use).
Re-calculation	Is normally performed to verify the reliability of aggregate data provided by the auditee. This will require access to raw data required to produce those aggregate(s). Examples: • calculating monthly slaughtering volume from company intake records; • calculating monthly production of animal by-products from company records and estimating whether this corresponds with what is estimated on the basis of the first bullet point.
Re-performance	Happens when an auditor carries out a (part of a) process following audit criteria and compares the results with results achieved by the auditee. Examples: • repeat measurements with a calibrated instrument; • re-performing process/product inspection, that is own-controls or official controls; • proficiency testing of laboratory methods.
Confirmation	Ideally, audit evidence should be confirmed by evidence from another source and/or of another type. Sometimes this can be difficult and, in such cases, reliability of evidence may also be increased by corroborating with the auditee (see below). Examples: • verifying company policies by examining written documents, interviewing staff and management and observing corresponding activities; • cross-checking data from, for example, bovine database with company intake records.

(continued overleaf)

Table 24.8   (*Continued*)

Method	Examples
Corroboration	Is the process of reaching agreement with the auditee that a piece of evidence is reliable and relevant enough to be used as a basis for a finding (and conclusion). Although this is preferable, it is not always achievable/possible and should not be considered as an absolute prerequisite for using evidence. Example: • explaining evidence collected by observation, document examination and interviews to the management (and staff) with a view of reaching a common understanding of the reality as compared to audit criteria.

overall results of the audit. That is, a clear statement on the degree of compliance and effectiveness of official controls. Such a statement usually comes in the form of a conclusion but many audit bodies present them under the heading 'finding' or 'opinion'. The conclusion should be balanced, contextualized and provide an indication of the significance as well as potential consequences of any negative findings.

The evidence needed to support findings and conclusions should be presented clearly in a report. This will not be the first time that these details are communicated to the auditee. Any findings of significance with the supporting evidence need to be discussed with the auditee as they arise and again in the final meeting. An audit report should not contain any surprises to the auditee. If it does, the objectives of the audit process (improvement) may be jeopardized and this may also undermine credibility of the auditors. A clear line of thought is particularly important when reports are being published. Bear in mind that the audit client may misinterpret audit findings and conclusions if the outcomes are not clear. If the top management of the audit client does not get the message, chances are that the auditee will not act on the audit results.

The next output in chronological order comes from the auditee in the form of the 'action plan'. This is a plan outlining the corrective and preventive actions which are needed as a response to audit conclusions or recommendations. As a general rule, auditors should not prescribe solutions to the auditee. They should explain the objectives to be achieved, not the means. This principle tends to cause some confusion with both auditors and auditees. The auditee might ask: 'You seem to know what has to be done – why cannot you simply tell us what and how?' The auditor may even feel obliged to provide advice to the auditee. Giving advice in such a situation is very tempting, as it appeals to two basic human needs: to be helpful and to feel important.

However, there are two main reasons why auditors should remain as neutral as possible in relation to their recommendations. Firstly, even if the auditor has identified a problem correctly, he/she might not fully understand the underlying causes nor the organizational context and constraints that will determine the best course of corrective actions. Advice given with all the best intentions may well turn out to make situation even worse. Secondly, by prescribing a particular solution, the auditor will disqualify him/herself (and their organization) from providing an objective audit opinion on the effective implementation of the system or its design in future. For all these reasons, it is better to leave the auditee with the task of devising an action plan. The auditee is best positioned to identify root causes to problems and also find the best solutions to them – solutions that are most likely to work in that particular organizational context.

The expected outcome of an audit is that the auditee turns non-compliance into compliance and addresses areas for improvement appropriately (unless everything was perfect, of course). This action is the responsibility of the auditee and depending on the audit set-up auditors may or may not follow up corrective actions. There are good arguments both for and against auditors getting engaged with follow-up. Some might argue that this will compromise the independence of the auditors – creating an incentive for certain type of recommendations in the first place. Some others may say that the auditor is best placed to make a final judgement on whether the corrective action has actually remedied the original problem. Either way, someone has to follow up the action plan and measures may be needed to mitigate issues of independence or expertise in the follow-up process.

The final end-point for an audit is often called the 'close-out'. Again, depending on the set-up the use of this term may vary but in this introduction close-out means the stage where all action items have been declared as acted upon or irresolvable. An audit cannot stay open indefinitely – it could be subsumed by some follow-up activities or even legal proceedings but it is not practical to keep the audit process open too long. The maximum time frame for an audit to remain open is up for debate. Close-out of audit findings should be timely and implemented as soon as possible to drive continuous improvement. Findings which involve major financial investment or organizational change may take longer.

24.48 Concluding remarks

Auditing by its nature is a very challenging task and requires specific skills from those involved. Audit can be a very satisfying process for an organization and, when performed well, may add significant value for all stakeholders. The success of an audit also depends on the auditee's understanding of the audit process and cooperation throughout the process. Audit standards provide good advice on the various stages of the audit process. Simply following mechanically the steps involved will not generate positive outcomes, drive continuous improvement and add value – all of which are the keys to the success of an audit.

Literature and further reading

Hoyle, D. and Thompson, J. (eds). 2002. *ISO 9000:2000* Auditing Using the Process Approach. Elsevier Science, USA. (Although seemingly outdated, this provides an excellent overview of a process approach that still is and will be valid for years to come.)

Institute of Internal Auditors (IIA) standards. Available at: https://na.theiia.org /standards-guidance/topics/Pages/Guidance-Topics.aspx (last accessed 25 February 2014).

ISO 19011. 2011. *Guidelines for auditing management systems, Second edition 2011-11-15*. International Standards Organization, Geneva, Switzerland.

Palmes, P.C. (ed.). 2009. *Process Driven Comprehensive Auditing – A New Way to conduct ISO 9001:2008 Internal Audits*, 2nd edn. ASQ, Milwaukee, WI, USA.

I. Transparency in Official Controls

Juha Junttila

Food And Veterinary Office, European Commission, Dunsany, Co. Meath, Ireland

24.49   Scope

Transparency, which is recognized as an essential principle of good governance in public administration, is also an implicit or explicit requirement in many jurisdictions. Yet, it is a difficult concept to grasp and, in particular, to specify how to implement it in practice. This part of the chapter attempts to clarify and demystify the concept by explaining some of the background, objectives and benefits of transparency. Transparency is not a 'black-or-white' phenomenon. Recognizing different elements and degrees of transparency is an essential step on the way to designing and implementing it in practice.

24.50   What is transparency?

There is no generally accepted definition for transparency that would facilitate a competent authorities' task of implementing it in practice. Sources like Transparency International, European Transparency Initiative and Codex Alimentarius all provide their own perspective to transparency. But none of them can offer practical advice that would apply to national authorities in charge of meat inspection. In the absence of a definition and practical guidance, it is necessary to search somewhere in between – and try to extract some principles that would enjoy relatively wide support.

Transparency can be viewed as a principle or a set of conditions that allow general public access to information on policies, processes and decisions that have wider social impacts. Such access facilitates public debate and scrutiny on actions of those who have been put in charge of matters such as safeguarding public health and consumer interests. Legal provisions – often known as Freedom of Information Acts – normally form the foundation on which transparency can be built. These acts are designed to guarantee citizens free access to any public document – unless a document is classified as confidential. The definition of confidential document is critical to the functioning of these acts – if confidentiality is left too much open to interpretation, public access to information may be compromised. Another critical factor is that it is only possible to ask for information that is known to exist. This dilemma is usually solved by obliging authorities to provide access to comprehensive registries of documents in their possession. These registries may also contain meta-data about the documents and will allow the general public to have an overview of information stored by authorities and, thereby, make queries on whatever might be of interest to them.

Experience has shown that organizations – including public bodies – develop their own set of interests, values and beliefs, which do not necessarily align with those of their stakeholders. Transparency can be viewed as a set of mechanisms to ensure that public bodies trusted with protecting consumer interests will stay on track and aligned with stakeholder interests. If the stakeholders' confidence in the objectives pursued by a public body is eroded, this is more often than not due to lack of transparency. A public body may well be addressing all the legitimate concerns of stakeholders but if it is not perceived as doing so it has probably failed in communicating its objectives and activities to them. When crisis hits and public confidence is plummeting, it is rather late to become transparent – trust may have been seriously compromised and re-building it might take a long time. Therefore, transparency is generally considered as a pro-active continuing process and an essential element of good governance.

24.51 Good governance

Transparency is currently placed under the broad heading of good governance together with concepts like ethical conduct, impartiality, accountability, freedom from bias and conflict of interest. Most – if not all – public bodies would like to be seen as entities that are effectively preventing maladministration, fraud and other criminal activities. And yet, scandals emerge in unexpected places previously thought of as being beyond any suspicions. Good governance can be defined as the measures, activities and internal controls that collectively reduce the likelihood of unethical and corrupt behaviour. Excellent books and articles have been published on good governance and it is a topic too broad to cover here. The focus here is on transparency – assuming that all other elements of good governance are in place – and trying to understand how transparency can be used as a pro-active tool in building and maintaining consumer confidence.

It is important to acknowledge the interconnections between transparency and other components of good governance. It is difficult to convince stakeholders on the ethical conduct of civil servants without a binding code of conduct that is documented and available for public scrutiny – but also effectively implemented and enforced. Similarly, accountability hardly exists if a public body is not required to be explicit and open about its objectives and policies. Freedom from conflicts of interest or bias, and impartiality are hard to guarantee – but disclosure of interests and affiliations of staff members and publication of decisions all contribute to building confidence on the objectivity and consistency of those in charge of consumer interests. Publishing this type of information provides for transparency by way of what is often labelled as 'credible commitment' and contributes towards consumer confidence.

24.52   Objectives of transparency

One of the main goals of transparency is to promote the alignment of policies, plans and operations with taxpayers' legitimate expectations. If the consumers – and other stakeholders – have a reasonably clear understanding of risks related to meat, what official controls can deliver and what they actually are accountable for, delivering to their expectations will become more feasible. The risk to be avoided is a large discrepancy between expectations and what can be reasonably expected to be delivered.

One of the objectives of transparency is to allow public debate and scrutiny; this will inevitably result in feedback, which, in turn, allows for corrective action. This feedback loop from public dialogue serves the purpose of further aligning expectations and delivery, resulting in continuous improvement. The final outcome – or objective – to be achieved is increased stakeholder confidence in official controls and, ultimately, confidence in the safety of meat and meat products.

24.53   Who needs transparency?

Transparency is primarily a need and the right of the consumer, taxpayers, industry and other stakeholders. Consumers are usually put first because their safety is at stake, as opposed to the financial interests of the taxpayer, food business operator and other stakeholders. But it could also be argued that it is, indeed, the competent authority who has the primary interest (and benefits) of being transparent. The competent authority is in charge of looking after and balancing between all of these various interests and it is the competent authority's credibility and reputation that is either destroyed or enhanced. Transparency may sometimes even be a crucial matter for a public body; during financially hard times resources can be significantly reduced or whole agencies abolished, largely depending on perception of their past achievements and credibility.

24.54   Benefits of being transparent

Whatever is chosen as the appropriate level of consumer protection, a high level of transparency provides guarantees that investments and efforts towards that level are not wasted. If stakeholders understand, accept and appreciate the efforts made by meat inspection services, the confidence in food safety will be the best insurance against adverse effects of a food safety crisis. And, when crisis hits, recovery will be more prompt and complete if confidence has been built and maintained pro-actively by applying all elements of transparency in advance. In the absence of that confidence, crisis management tends to resort to knee-jerk reactions, which are usually costly – and often unnecessary in terms of real safety – but vital in preserving the remnants of credibility and reputation of a competent authority.

Transparency at all levels of policy, implementation and enforcement also serves the purpose of increasing the acceptance of food safety requirements by food business operators. Engagement in a transparent dialogue with competent authorities tends to enhance ownership and commitment to food safety by these stakeholders, leading into an increase in voluntary compliance. Awareness of the background and rationales of food safety measures will also facilitate the acceptance of control measures and the corrective actions required from the food business operators. As a result of all these, decreased enforcement costs will be the return for investment in transparency.

Feed and food chains are today more global than ever and the trend can be expected to continue. In this context, being transparent with domestic consumers is not sufficient anymore. Meat inspection services need also to gain the confidence of those authorities who act as gatekeepers to foreign markets. This requires a different set of skills and transparency measures than satisfying national stakeholders. A solid understanding of relevant international standards and the trading partner's chosen level of protection form the basis that needs to be supplemented by knowledge on various risk and quality management frameworks. Cultural differences may play a significant role as risk perception, appropriate level of protection and preferable risk management options can vary significantly, depending on the underlying assumptions, which are determined by different cultural backgrounds. Dealing with such challenges may be greatly facilitated by transparency, which generates the confidence needed for access to new markets.

24.55 Degrees of transparency

Transparency is not a binary attribute of a public service but rather a whole continuum of various degrees of openness. What matters at the end of the day is the perceptions of stakeholders, that is whether a service is seen as being open or not. The implementation of transparency measures may be facilitated by breaking transparency into components or dimensions.

The first dimension is related to the content: what type of information or data is made available to the public. Simply providing results of official control activities is only the first step to transparency. In order to allow public debate and accountability, the whole process needs to be exposed to scrutiny. This means access to the policies, organization, prioritization, functioning and decision making processes of the public body. The results, outputs and outcomes of control processes are easier to accept when the internal workings and rationales are exposed for discussion. It should be kept in mind, however, that enforcement processes may not be able to provide full transparency; for example, publishing control plan details like selection of hazards, establishment types or timing of sampling is likely to undermine the objectives of control programmes.

The second dimension is about the active versus passive nature of access to information. The most basic level of access is to grant access only on request to limited categories of information. The other end of the spectrum is to actively publish information and to provide access to raw data as well. Today this is facilitated by various web technologies, including web interfaces to databases that contain relevant data. It is worth keeping in mind though, that simply publishing all data on the internet is not an optimal solution and could actually be counterproductive. True transparency requires a customer-focused approach. In other words: careful analysis of stakeholders' needs and taking them into account in designing transparency measures.

This brings us to the third dimension: transparency is always about interaction between two entities and the relationship between those entities determines the right degree and mix of transparency measures. Both of the previous dimensions need to be tailored differently to meet the needs of consumers, taxpayers, scientists, media, food business operators and trading partners. An advanced level of transparency will inevitably require some additional resources – people trained in communication and public relations. Consumer confidence does not come free and investments in transparency are the price to pay for that. Ultimately, consumer confidence is one of the most important – if not the most important – outcome of official controls and without transparency it cannot be achieved.

24.56 Obstacles to transparency

The way that transparency has been defined – as opening up to scrutiny and criticism – is the key to understanding why transparency may become a controversial issue for organizations. Being open to scrutiny implies a certain degree of vulnerability. Depending on the organizational culture, this can be seen either as a threat or an opportunity. Organizations that tend to see transparency mainly as a threat usually often have a world-view of 'us' and 'them', playing a zero-sum game where win–win outcomes do not exist. They also tend to have a *modus operandi* of being reactive, defensive and territorial. Organizations that see transparency as an opportunity also tend to have a strong culture of serving stakeholders interests, actively pursuing win–win situations and a preference to pro-active, collaborative mode of operations. Opening up to scrutiny also requires a fair amount of self-confidence and courage.

'Information is power' is an old saying that has a lot of truth in it. However, if this becomes the main underlying assumption that determines an organisation's willingness to share information, it may act as a major obstacle to transparency. Information is power – or an advantage – when there is either competition or some kind of a conflict and withholding information would give the organization a clear competitive edge. However, in the food safety domain consumers, food business operators and competent authorities all share the same goal – safe food. Conflicts are more often perceived than real but such perceptions may become serious obstacles to sharing information.

Of course, there are quite legitimate factors slowing down transparency initiatives. Firstly, transparency is about communication and effective communication requires resources. When the immediate payback for such investments is not obvious and imminent in the near future, such investments are often difficult to secure. Secondly, confidentiality and data protection regulations may conflict, or seem to conflict, with transparency requirements. Such a conflict may be difficult to resolve and very time consuming, particularly if data protection and transparency have been developed without effective coordination.

Misunderstandings about the concept of transparency, its objectives and benefits may significantly slow down its implementation in practice. Reluctance to share information and disagreements on how to do it may be fuelled by lack of a shared vision. Explicit specification of stakeholders' expectations, different options of meeting those expectations and the benefits of those options may speed up the process.

24.57   What does this mean for meat inspection?

Consumers have a number of concerns in relation to meat and meat products. Some of the recurring topics of concern include chemical hazards such as dioxins, residues of veterinary medicines, biological hazards such as salmonella and enterohaemorrhagic *Escherichia coli* (EHEC), as well as mislabelling and animal welfare issues. The common denominator for all of these is that official controls are needed at various points of the supply chain to effectively manage and control the risks. Without effective integration and coordination with other (e.g. feed control, on-farm control) authorities the assurances provided by meat inspection services are limited. The average consumer has difficulties in understanding who is responsible for what, when and where. Therefore, the reputation of meat inspection services prevails or fails with the rest of the controls along the production chain. From the consumer's perspective the official controls either work as an integrated whole or they do not.

The main implication of this is that transparency (from the consumer's perspective) is a feature of the whole control system, not something that meat inspection services could maintain in isolation from the other actors of the chain. Establishing and maintaining consumer confidence requires coordinated communication efforts from all authorities in the meat production chain and a clear message that can be understood by the consumer. How exactly this is to be done depends – amongst many other factors – on the organization and structure of controls along the chain. Keeping the guiding principles and objectives in mind will help in choosing the right tools from the box.

24.58   Concluding remarks

Transparency can be viewed as a principle, a set of tools or as a state of mind aimed at increasing stakeholder confidence in food safety and official controls. Transparency has been widely accepted as a guiding principle for good governance; it requires continuous efforts and a good understanding of different aspects of the concept. Loss of consumer confidence is often the first and the most significant adverse effect of any food safety crisis and effective implementation of pro-active transparency measures is an essential preventive instrument to reduce that effect.

J. Food Frauds

Niels Obbink, J.M. Frissen and S.B. Post

Intelligence and Investigation Department, Food and Consumer Products Safety Authority, Ministry of Economic Affairs, Agriculture and Innovation, Utrecht, The Netherlands

24.59   Scope

Food production is an important sector and a major source of revenue. It involves high volumes and huge sums of money. On the down side, criminals also see opportunities to profit from this sector with inferior products that can pose a danger to public health. Opposition against frauds requires legal tools and, therefore, the models of opposition are different in different countries due to variations in national legislations. This part of the chapter focuses specifically at the forms of fraud discovered in the meat production chain and the modalities to combat fraud under a Dutch point of view.

24.60   Definition

The United Kingdom Foods Standards Agency gives the following definition of food fraud: Food fraud is committed when food is deliberately placed on the market, for financial gain, with the intention of deceiving the consumer. Although there are many kinds of food fraud the two main types are:

- the sale of food which is unfit and potentially harmful, such as:
  recycling of animal by-products back into the food chain; packing and selling of beef and poultry with an unknown origin; knowingly selling goods which are past their 'use by' date;
- the deliberate misdescription of food, such as:
  products substituted with a cheaper alternative, for example, farmed salmon sold as wild and Basmati rice adulterated with cheaper varieties; making false statements about the source of ingredients, that is their geographic, plant or animal origin.

The Netherlands Food and Consumer Product Safety Authority (NVWA) monitors animal and plant health, animal welfare, food and consumer product safety and enforces Dutch nature legislation. Part of the NVWA is the Intelligence and Investigation service (in Dutch abbreviated to NVWA-IOD). The

Netherlands is one of the few countries to possess a specialized organization dedicated to combat food fraud. The three main tasks of the NVWA are: supervision, risk assessment and risk communication.

24.61   Slaughter chain and food fraud

Meat is an important foodstuff throughout the world and the worldwide value of the meat trade is considerable. The current upward trend in meat consumption is expected to continue.

Food safety, animal health and welfare are important issues that have a direct bearing on the trade and consumption of meat and other products of animal origin. The European Union (EU) regulates the market within the Union, while the World Trade Organization (WTO) deals with the regulation of world trade and the removal of trade barriers.

The meat trade is important business for various countries, amongst them The Netherlands, which is both a transit country and an international market; it occupied third place in the European Union, after Germany and France, in terms of export value in 2012. Meat exports in 2011 were worth 13.72 billion Euros. Trade in The Netherlands is strongly influenced by international trade; it is also partly dependent on export refunds and import and/or export restrictions following outbreaks of animal diseases. Meat trade involves huge volumes and values and this makes the trade susceptible to fraud.

24.61.1   Variations in trade

Variations in trade may increase the probability that fraud takes place. In 2011 trade was affected by higher feed prices, particularly for pigs and broiler chickens. By contrast the tight European beef market saw the price of cows rise by 15% in 2011. The sale prices for veal calves also rose by 12% in 2011 compared with 2010. Set against this, consumption of beef and veal fell slightly and the consumption of poultry meat continued to rise over recent years. The consequence of this was an increased pressure on the production of poultry meat.

24.61.2   How fraud takes place

An example of how fraud takes place could involve animal by-products. Before, during or after slaughter animals and/or parts of animals may become animal by-products belonging to Category 1 to 3. This material must be collected and removed in a prescribed manner. As there are different categories of animal waste, they must be collected separately from each other.

In the processing of the animal waste there is a threat that Category 1 risk material could be processed into animal feed for farm animals. Another risk is that the waste material could be upgraded to meat fit for human consumption.

24.61.3   Trade promotion

There are a number of schemes to regulate or promote trade in frozen or chilled meat. To qualify for a refund the meat must meet the relevant requirements of legislation, that is be produced in an approved establishment and meet the requirements regarding health and identifications marks. There are also special tariffs for specific products, subject to certain conditions. One example is the inward processing scheme. The goods are brought into the European Union temporarily, processed and re-exported. There are also tariff quotas that allow for a (partial) suspension of customs duty on importation. Both tariffs and refunds can be susceptible to fraud.

24.62   Criminal acts and behaviour

Several studies carried out by the NVWA-IOD show that various types of fraud are possible when slaughtering animals intended for human consumption. There are two major types of fraud. Firstly, there is a distinction between legal and illegal slaughterhouses. Then there are various forms of fraud committed during the slaughter process in legal slaughterhouses.

In the case of illegal slaughterhouses the fraud concerns slaughter in non-government approved slaughterhouses. The fact that a slaughterhouse is not approved does not have anything to do with its size, design or capacity. The main feature of slaughter in an illegal slaughterhouse is that the animals slaughtered there are not subject to the compulsory pre- and post-slaughter inspections. Because of the absence of these inspections there are no guarantees of quality for the meat produced in this way. Human consumption of this meat could present a danger to public health. An example is elaborated in case 1.

Fraud during slaughter of animals in legal (approved) slaughterhouses often involves the quality or origin of the animals to be slaughtered, and/or the *ante-* and *post-mortem* inspections. Sometimes animals are slaughtered whose origin is unknown, and/or are too sick, and/or have been treated with medicines, making them unfit for human consumption. Also, animals can be slaughtered without the necessary *ante-* and *post-mortem* inspections by an inspecting veterinarian. In either case there is a potential risk to public health in human consumption of this meat. Examples of these frauds are elaborated in cases 2 and 3.

24.62.1   Case 1, slaughter in an illegal slaughterhouse

In The Netherlands horses over six months old can only be slaughtered if accompanied by a valid horse passport and corresponding transponder (chip). The slaughter of these animals can only take place in an approved slaughterhouse, if the compulsory inspections are carried out by or on behalf of a veterinary inspector prior to (*ante-mortem*) and following (*post-mortem*) the slaughter. Following these inspections the meat is issued with recognizable health marks. Meat without these marks is automatically deemed 'unfit for human consumption' and must be removed and processed by a rendering plant.

Background In this case the operators of an illegal slaughterhouse and an illegal cutting plant collaborated in a fraud to slaughter horses that had not undergone the required inspections.

In January 2010 the NVWA received an anonymous message stating that two people in the east of The Netherlands were regularly slaughtering horses illegally on the weekends at their residential address. The slaughter waste was said to be disposed of through a nearby legal slaughterhouse. The meat from the slaughter was said to be transported to Germany. The NVWA launched an investigation in response to this message. During the investigation NVWA inspectors established that an illegal slaughterhouse had indeed been set up at one of the locations concerned. However, it was not possible, within the remit of the supervisory authority, to establish that illegal slaughter was actually taking place there. The NVWA inspectors work in the supervisory divisions of the NVWA. Investigators are general investigating officers who work for the Intelligence and Investigation Service of the NVWA. Investigators have further-reaching powers than inspectors.

Method In response to these findings, the NVWA-IOD launched a criminal investigation into the illegal slaughterhouse and cutting plant. The investigation was led by an officer from the National Public Prosecutor's Office for Financial, Economic and Environmental Offences. A surveillance camera was set up at the illegal slaughterhouse; this recorded movements to and from the premises. The NVWA-IOD investigators also recorded and checked the telephone calls of the two suspects and a surveillance team was set up. This provided insight into the delivery of the horses for slaughter, and it became clear that veiled language was used to make the necessary arrangements. It was also established that the meat from the horses slaughtered in the illegal establishment was delivered to the operator of the illegal cutting plant. In all probability the cutting plant owner sold this meat to consumers.

Based on these findings the NVWA-IOD searched the two residential addresses and outbuildings of the two suspects. The outbuildings contained the illegal slaughterhouse and illegal cutting plant. During the searches, evidence of (traces of) blood, meat and slaughter waste was collected and seized for DNA testing. Slaughter equipment and business records were also seized. DNA testing linked the seized slaughter waste to the seized blood traces and meat, furnishing proof of the illegal slaughter.

Motive By slaughtering horses in an illegal slaughterhouse and cutting the resulting meat in an illegal cutting plant, the suspects were able to produce meat at a very low cost and so sell it with a high profit margin.

24.62.2   Case 2, fraud concerning the origin of slaughter animals

In the Netherlands horses over six months old can only be kept, transported and slaughtered if they are accompanied by a valid horse passport and corresponding transponder (chip). Horse passports are issued by various official bodies. After a horse is slaughtered, the passport must be returned to the issuing body. Among other things, the passport must contain a section recording any veterinary treatment given to the horse. The passport must also state explicitly if the horse to be slaughtered is unfit for human consumption.

Background   In this case a horse slaughterhouse and a horse dealer collaborated in a fraud to slaughter horses using falsified horse passports and/or counterfeit transponder numbers. This case is still to be presented to a court of law; therefore, the outcome of the investigation is not certain yet.

Suspicions about the two companies arose when foreign authorities reported irregularities in the horse passports issued by them. Passports had been returned to the authorities that had more than one active transponder number and/or pages for veterinary data were missing and/or were clearly wholly or partly falsified. These passports accompanied horses slaughtered in The Netherlands. The passports were returned to them via the slaughterhouse and horse dealer concerned in this investigation.

Further suspicion was aroused by an NVWA finding, which showed that the slaughterhouse concerned in this investigation had returned passports containing various irregularities. These included:

- Passports that had had the veterinary treatment pages removed. These pages would also have indicated that the horse was unfit for human consumption.
- Passports with falsified veterinary treatment pages inserted.

The horses slaughtered in the slaughterhouse were delivered together with the passport and transponder data by the horse dealer concerned in this investigation.

Finally, the Criminal Intelligence Unit (CIE) of the NVWA-IOD issued two communications indicating that the horse dealer referred to above was committing fraud with horse passports. A CIE communication provides important information for an investigation while the source remains anonymous. The information is not included as evidence in the file but serves purely as 'seed information' to help in the investigation.

Method   In response to these findings, the NVWA-IOD started a criminal investigation led by a prosecutor from the National Public Prosecutor's Office for Financial, Economic and Environmental Offences. During the investigation the IOD used several investigative tools, such as searching dwellings/business premises and telephone interception. During the searches a large number of false horse passports, false stamps and a few transponders were seized. Dead/slaughtered horses and meat from slaughtered horses found during the searches were destroyed.

The criminal investigation showed that the partners in the suspect horse dealership may have been falsifying horse passports. This was done in various ways. For example, they used false pre-printed veterinary inserts for passports and false stud book stamps. These stamps could be used to endorse changes in passports made by the suspect company as genuine and correct. They also had horse passports falsified in other ways.

The investigation revealed that the suspect partners may have counterfeited transponder numbers on their computer as well. The tapped telephone calls showed that they had talked to various people about it. The telephone calls and the invoices found also showed that the meat of the horses slaughtered in this way was intended for human consumption and was exported abroad. Finally, the investigation showed that the slaughterhouse concerned was fully aware of the illegal practices of the horse dealer.

Motive The suspect horse dealer obtained horses from the owners for little or no money, assuring the owners that they would take care of the slaughter of the horse concerned. By providing these horses with false passports and/or false transponder data, they probably could slaughter horses that would not otherwise have been eligible. For example, horses whose passports indicated that they were not intended for human consumption or were recently treated with specific drugs and were, therefore, unfit for human consumption. Human consumption of this meat presents a potential threat to public health.

The horse meat obtained in this way could then be sold on. Consequently, the suspect horse dealer and slaughterhouse allegedly made an income from horses that would normally have had to be sent for rendering.

Proceeds In this investigation probably over 700 horses whose passports were falsified in one way or another were slaughtered over a period of a few months. The illicit proceeds of this crime is estimated around € 140 000. The financial investigators calculated the proceeds, enabling the court to remove any advantage obtained.

24.62.3   Case 3, illegal slaughter in approved slaughterhouse

In The Netherlands sheep and goats intended for human consumption must be slaughtered in an approved slaughterhouse. Prior to (*ante-mortem*) and following (*post-mortem*) slaughter the animals must be inspected by or on behalf of a medical examiner.

Following these inspections the animals are issued with recognizable health marks. Meat without these markings is automatically classed as unfit for human consumption and must be removed and processed as Category 2 material at a rendering plant.

Background In this case sheep and goats were slaughtered in an approved slaughterhouse outside normal operating hours and without the required inspections. Suspicions about this company were aroused by a communication in February 2010 from the Criminal Intelligence Unit (CIE) of the NVWA-IOD. A CIE communication provides important information for an investigation while the source remains anonymous. The information is not included as evidence in the file, but serves purely as 'seed information' to help in the investigation.

This communication indicated that a certain approved slaughterhouse was slaughtering sheep and/or goats illegally. This suspicion was reinforced by an observation made by an inspector during an audit in June 2010. He discovered that 21 of the sheep carcases found in this slaughterhouse had not undergone the required *ante-mortem* inspection. In consultation with the NVWA, the carcases were removed and treated as raw material for rendering. A CIE communication in August 2010 showed that animals were still being slaughtered in the said slaughterhouse without the required inspections.

Method In August 2010, prompted by the findings described above, the IOD launched a criminal investigation into the suspect slaughterhouse. The investigation was led by a prosecutor from the National Public Prosecutor–s Office for Financial, Economic and Environmental Offences.

During the investigation a surveillance camera was secretly placed in the slaughterhouse, which made it possible to establish when and how often slaughter was carried out. It also made it clear when and how animals were slaughtered without the required inspections. Based on these findings NVWA-IOD investigators entered the slaughterhouse in December 2010 and caught and arrested the people present who were in the act of illegal slaughter of sheep and /or goats. No inspections had been requested or carried out prior to this slaughter. Following the arrests the slaughterhouse was searched by NVWA-IOD investigators. They discovered that a large number of the animals illegally slaughtered that day (72 animals) had already been given a false EC health mark. The falsified stamp used to apply the mark was found and confiscated during the search. The illegally slaughtered animals were classified by an NVWA veterinarian as raw material for rendering and were sent to a rendering plant.

It was established that 334 sheep and/or goats had been illegally slaughtered during the period of the investigation. The administrative and financial investigation showed that from 1 January 2010 to 31 October 2010 a total of 1500 animals were thought to have been illegally slaughtered. The proceeds of the crime amounted to around € 30 000.

Motive By slaughtering sheep and/or goats without the required inspections the suspects could produce meat at a lower cost and so increase their profit margin.

24.63   Organization in The Netherlands to combat food crime

24.63.1   The tools of enforcement

The supervision strategy of the NVWA promotes voluntary compliance. Preparedness amongst target groups to comply voluntarily is established in risk assessments. 'The Authority will extend trust until … ' is one of the mottos of the NVWA. Another important motto is 'soft where possible, tough where necessary'.

The Authority therefore uses five enforcement tools:

- **Enforcement communication**
  This is aimed at influencing compliance behaviour. There is no doubt that communication that carries a clear message and that is targeted at a specific target group can be very powerful.
- **Service**
  Give general and specific advice on legislation and compliance through leaflets, the internet or trade publications.
- **Horizontal supervision**
  Horizontal supervision targets companies in charge of their own management and process control. The Authority monitors, but relies on, or trusts, their self-regulation and self-enforcement.
- **Repressive supervision**
  Repressive supervision is used for businesses that do not deserve this government trust or if stipulated by legislation. These companies may be sanctioned and may even be placed under permanent supervision.
- **Intelligence and Investigation**
  The Authority considers investigation the 'ultimate remedy' but it can also be used in combination with other forms of enforcement, such as enforcement communication. It acts as a deterrent and affects behaviour at the sector level. Investigation tools are used mostly by the Intelligence and Investigation Department (in Dutch abbreviated to IOD) to combat more severe and organized forms of crime.

The use of these enforcement tools, combined with the services provided by the IOD to achieve maximum compliance, should secure public health and combat fraud in the areas covered by the Dutch Food and Consumer Product Safety Authority.

24.63.2   The divisions of the Authority

To enable enforcement in the various areas, the Authority comprises a number of divisions providing the wide knowledge and expertise required when investigating criminal offences. Each of these divisions covers a specific area. The key structure of the Authority comprises six divisions and an executive office. The Authority consists of the following three large content divisions:

- **Veterinary & Import Division**
  This division is responsible for live animals after they leave the primary holdings, live products and meat at slaughter sites, cutting plants, export collection centres, livestock farming, import, export and so on.
- **Agriculture and Nature Division**
  This division is responsible for plant health, crop protection, nature, fertilizers, live animals (at primary holdings) and land-based subsidy schemes.
- **Consumer & Safety Division**
  Responsible for the fisheries chains, the industrial production of special foods and drinks, animal by-products, animal feeds, animal testing, EU subsidy schemes, catering, craft sector, institutions, retail, alcohol and tobacco, the food production chain, hygiene legislation, product safety, the import of animal feeds, foods, veterinary products, live animals and consumer products.

The IOD investigates criminal offences in the areas covered by all above mentioned divisions and covers in combating crime all responsible areas of the Authority.

24.63.3   The legal framework of the IOD

The legal framework of the IOD is laid down in the Special Investigation Services Act. The Act stipulates that Special Investigation Services *inter alia* act under instruction from a Public Prosecutor and must enforce the legal order within their 'own' policy areas, in the case of the IOD for investigating criminal offences related to tasks of the Ministry of Economic Affairs and of the Ministry of Health, Welfare and Sport.

In The Netherlands there is a special part of the Public Prosecutor responsible for Financial, Economic and Environmental Offences (in Dutch abbreviated with FP). It has legislative power in the criminal justice system not only to steer, together with the department in policy appointments, but also for more concrete deployment of the IOD and for approving the use of investigative powers.

The IOD operates in specific department-related areas, under the instruction of these departments. It operates within the framework of special legislation, which means its investigations do not conflict with the investigation of other, more general, offences like theft, murder or drugs. Practical experience has shown that, in these situations, police regularly prioritize the latter offences. Special legislation investigation requires the IOD to have specific expertise and skills not commonly present with police and, last but not least, specific expertise is best maintained if ties with the relevant department and inspection divisions are close. This strengthens the chain of enforcement, which also comprises 'supervision'. Trained special investigation officers are charged with 'supervision' and are authorized to act upon discovery of simple criminal offences.

24.63.4 The IOD structure

Obviously, investigations and gathering intelligence requires detailed knowledge of the fields in which the Food and Consumer Product Safety Authority operates, and of the relevant legislation. For this the IOD relies on the experts of the inspection divisions. The IOD employs specialists such as: strategic/tactical/operational analysts, information investigators, internet investigators, senior tactical investigators, digital investigators, electronic data processors (EDP), financial investigators and forensic accountants.

24.63.5 Investigative Powers

Dutch criminal legislation has established a number of investigation powers. A distinction can be made between 'general investigations powers' and 'special investigation powers'. Which investigation power is used is determined mostly by the penalty on a criminal offence. As a rule, the use of more invasive powers is restricted to criminal offences that carry a minimum sentence of four years in prison. As the penalty on a criminal offence is becomes greater, more invasive powers can be used. The use of telephone interception equipment, for example, can be used only in suspected crimes carrying a four-year sentence. Recording confidential communication by placing surveillance devices in a suspect's home is restricted to investigations into crimes that carry a penalty of eight years.

Some examples of general and special powers are: arrest of suspects for interrogation, search of businesses and/or homes, seizure of administration and goods, telephone interception, demanding data, systematic observation and use of tracking devices, pseudo purchase, recording confidential communication and infiltration.

24.64 Conclusion

The meat trade can entail risks. It is a vulnerable market due to the veterinary risks and the associated food safety and animal disease risks. The meat trade and meat production are important to the economy, involving large sums of money. Partly because of these factors the supervision is a risk-based model designed to give sufficient guarantees concerning the food safety of the meat, despite limited capacity. The organization of the NVWA, in which the investigation and inspection services have been combined into a single organization, allows a good cooperation and close collaboration. The NVWA can operate effectively in response to incidents or complex fraud thanks to the transfer of knowledge and information back and forth. The nature of the fraud sometimes calls for special investigative powers, as supervisory powers alone are not always sufficient to uncover fraud.

The meat trade can also be lucrative. The Netherlands is both a transit country and an international market. It is crucial to share information between countries, especially when it comes to monitoring and tracing data.

Literature and further reading

FSA (United Kingdom Food Standards Agency). Web pages for information on food fraud; http://www.food.gov.uk/; last accessed 7 March 2014.

K. Flexibility and Uniformity of Official Control

Veli-Mikko Niemi[1] and Janne Lundén[2]

[1]*Ministry of Agriculture and Forestry, Helsinki, Finland*

[2]*Department of Food Hygiene and Environmental Health, Faculty of Veterinary Medicine, University of Helsinki, Helsinki, Finland*

24.65 Scope

The aim of this part of the chapter is to discuss the flexibility and uniformity in official food control from the perspective of the European Union (EU) legislation. Flexibility is an important concept in official control because it may have effects on food safety and economic consequences for food business operators (FBOs). The level of flexibility and the use of the possibilities a flexible legislation brings may also have effects on the uniformity of food control. Food control should be uniform and FBOs should be treated equally in the same matter. This is a basic principle that should characterize food control activities at all levels.

24.66 Introduction

Flexibility in food safety legislation enables the FBO to implement food safety requirements in different ways. Flexibility can be applied on a national level (e.g. derogations), in a compartment of a country (e.g. because of geographical barriers, such as in insular or alpine areas) or a local level (e.g. simplified implementation of own-check systems in small food businesses). The flexibility in the legislation may, therefore, consider, for example, different food processing types, size of the production or the geographical area. The aim of the flexibility is often to increase the possibility for versatile food production, which is often achieved by alleviating or adapting requirements. Flexibility allows the FBO to control hazards in different ways. Flexibility, however, must not decrease food safety and food safety hazards need to be controlled at all times. At the same time, food control should be uniform, so that the FBOs are treated equally. However, it is also generally accepted that the same objective can be achieved by different measures. Therefore, the same legislative requirement, for example the implementation of hazard analysis and critical control point (HACCP) systems, can be different in slaughterhouses as long as the risks are managed.

24.67   Achieving flexibility by legislation

The main tool to create flexibility in food control is by preparing food safety legislation that allows different ways to achieve an objective. Flexibility can be applied only if legislation allows it. The EU hygiene legislation from 2004 changed quite significantly how EU legislation is constructed and also how it should be implemented in practice. Firstly, the legislation was given as regulations, which are binding as such in all Member States. The 2004 legislative review was sometimes called a 'simplification exercise', which meant that over a dozen different directives were mainly replaced by three regulations. Therefore, many old detailed requirements were scrutinized. In addition, the new legislation tended to be more risk-based and, consequently, requirements considering operation size were abolished to a large extent. Therefore, the division to 'small scale' and 'large scale' establishments, with varying requirements and privileges was omitted. New rules apply to all establishments, which have the same rights to trade in the internal market and abroad.

It was, however, obvious that the legislation would not be a one-size-fits-all rule. Therefore, elements of flexibility were added into relevant regulations. Since their application needs to be ensured, the European Commission (EC) has also published guidelines for their application. In practice, legislation leaves large powers for national, and even local, authorities to apply rules at each establishment. To ensure uniformity in application, hygiene legislation created mechanisms to notify the European Commission and other Member States on national rules describing flexibility.

Flexibility in EU hygiene legislation includes:

- National derogations, exemptions and adaptations.
- Flexibility related to, for example, 'small amounts', 'local' or 'traditional' products, which need to be defined by Member States.
- Flexibility in official controls, whereby relative requirements like 'sufficiently' or 'where necessary' must be interpreted on a case-by-case basis.
- Operators' measures, particularly guides to good hygiene practices.

Some activities can also be totally excluded from the scope of the legislation due to different reasons.

24.67.1   Derogations, exemptions and adaptation

Member States may, according to EU hygiene legislation, manage derogations and exemptions or grant adaptations on certain provisions of EU legislation. Derogations can be justified, for example, due to small production amounts or local production. This is a form of subsidiarity in the legislation. However, when doing so, a Member State must follow a strict procedure. Prior to their adoption, national legislation must be notified to the Commission and other Member States. In certain cases, the Commission may reject the suggested derogation. An example of a national derogation is that in a certain Member

State the small and medium sized slaughterhouses are not obliged to have separate facilities for cleaning and disinfection of means of transport for livestock if such facilities exist nearby the slaughterhouse.

For foods with traditional characteristics a simplified procedure exists. National rules must be notified to the Commission within 12 months after their adoption. However, these rules may only concern material of walls, ceilings and doors of premises where foods with traditional characteristics are exposed to a processing environment necessary for the production of the product.

24.67.2 Flexibility and uniformity in official controls

Flexibility in the legislation has to be taken into account in the official control at slaughterhouses. Flexibility in official control at slaughterhouses concerns, especially, the interpretation of relative requirements. Requirements that are preceded with adverbs, such as sufficient, adequate and when necessary, do not provide precise courses of action. Instead, such legislation gives the possibility for interpretation on how to implement the legislation. The FBO is responsible for the implementation of the requirements but official control accepts the implementation or requires another way of fulfilling the requirements.

When official control evaluates whether the requirements of the legislation are implemented in an acceptable way several factors need to be considered. Official control must make sure that the implementation is sufficient to prevent food safety risks. The official control needs to consider the risks commensurate to the flexibility. This means also that the implementation can be different in different food businesses. The implementation of, for example, HACCP systems in the same manner in all slaughterhouses would not be meaningful or practical. Requirements concerning own-check documentation are also an area where the requirements should be commensurate with the nature and the size of the food business.

Flexible legislation enables the interpretation of requirements from case to case. However, it can also lead to a decrease in uniformity of official control when interpreted differently. The opposite of a flexible legislation is a rigid and maybe more detailed legislation, which would not allow much interpretation and local adaptations. The legislator has to balance between a flexible and less detailed legislation to enable different solutions in food businesses and a more rigid or detailed legislation for ensuring uniformity. However, in both cases food safety aspects must not be jeopardized.

24.67.3 Food business operators' measures

The FBOs are in a central role in applying the flexibility in the legislation. The FBOs can implement the requirements in a way that is most suitable for them as long as it is within the framework of the legislation. FBOs have prepared a number of different guides to good practices, which helps the businesses to fulfil the requirements in an acceptable manner. Such guides have been prepared, for example, concerning the application of own-check and HACCP systems. An important form of flexibility for small establishments is the measures to simplify requirements on, for example, HACCP systems and traceability. In general, the implementation of HACCP principles, including record keeping, may be eased at small establishments. When doing so, these enterprises are expected to follow general hygiene provisions and specific guides to good practice.

24.67.4   Exclusions

Some activities concerning production of animal-derived food can be excluded from the hygiene regulation concerning animal-derived foods. This means that the specific regulation concerning foods of animal origin is not applied at all. The direct supply by the primary producer of small quantities of certain primary products of animal origin to the final consumer is such an activity. Also, the direct supply of such products by the primary producer to local establishments directly applying to the final consumer is excluded from the scope of the regulation. The aim with the exclusions is to enable small scale production at a primary level. Also, food safety risks have been evaluated as small in such activities. However, it should be noted that meat is not generally defined as a primary product. Only meat of poultry and lagomorphs and wild game and meat thereof may be excluded from the regulation. Slaughtering in-house for private domestic use of the meat is also excluded from the scope of the EU legislation concerning the requirements for slaughterhouses and the meat inspection.

25

International Trade

Hentriikka Kontio

Ministry of Agriculture and Forestry, Helsinki, Finland

25.1 Scope

International trade of foodstuffs of animal origin is governed by the Sanitary and Phytosanitary (SPS) Agreement conducted under the World Trade Organization (WTO) Agreement. The SPS Agreement provides principles on international trade and measures to tackle trade disputes. International trade is guided by commonly agreed international standards and good practices. The trade of the European Union and its Member States on foodstuffs of animal origin follows these general principles but has also its own characteristics, as its trade is divided into intra-community trade and trade with other countries not belonging to the European Union intra-community trade area.

25.2 International trade

25.2.1 Principles of international trade

International trade of foodstuffs of animal origin is governed by the Sanitary and Phytosanitary Agreement (SPS Agreement) conducted under the World Trade Organization (WTO) Agreement. According to the principle of harmonization of the SPS Agreement, trade should take place based on measures following international standards, guidelines and recommendations where they exist. Measures that result in higher standards may be maintained or introduced if there is scientific justification or as a consequence of appropriate

Meat Inspection and Control in the Slaughterhouse, First Edition.
Edited by Thimjos Ninios, Janne Lundén, Hannu Korkeala and Maria Fredriksson-Ahomaa.
© 2014 John Wiley & Sons, Ltd. Published 2014 by John Wiley & Sons, Ltd.

risk assessment. Measures different than applied in the importing country should be regarded as equivalent if the exporting country demonstrates to the importing country that its measures achieve the importing country's appropriate level of protection (ALOP). Measures should be adapted to the regional conditions from which the foodstuffs originate and to which the foodstuffs are destined. The SPS Agreement promotes transparency by encouraging the countries to make their trade conditions publicly available.

25.2.2 SPS Committee

The SPS Agreement establishes a Committee on Sanitary and Phytosanitary Measures (SPS Committee) to provide a forum for consultations about food safety or animal and plant health measures that affect trade, and to ensure the implementation of the SPS Agreement. The SPS Committee has developed procedures and guidelines that help governments implement their obligations under the Agreement. The Committee aims to seek to avoid potential disputes and provides a forum to raise specific trade concerns.

25.2.3 International organizations

The WTO recognizes the standards, guidelines and recommendations made by the two main international organizations related to the trade of foodstuffs of animal origin, the World Organisation of Animal Health (OIE) and the Codex Alimentarius:

- OIE develops rules that can be used and referred to in international trade of live animals, foodstuffs and other products of animal origin. The OIE rules provided on the OIE web site contain trade measures, import and export procedures and recommendations applicable to animal diseases that may be transmitted by movement of animals and goods.
- Codex Alimentarius develops international food standards, guidelines and codes of practice contributing to protect the health of the consumers and ensure fair trade practices in the food trade. All standards, guidelines, codes of practice and advisory texts of Codex Alimentarius are available on the Codex web site.

25.3 European Union trade

25.3.1 Intra-community trade

Intra-community trade takes place not only between the Member States of the European Union (EU) but covers also the trade with other European states that have adopted the harmonized EU legislation on animal health and welfare, and public health. Based on agreements concluded between the European Union and the European states, live animals and products of animal origin are traded on the conditions provided by the EU legislation. This applies to the European Economic Agreement (EEA) partner countries such as Norway and to the European miniature states such as Andorra, Monaco and San Marino. A very complex situation applies to certain special EU Member States territories (e.g. outermost regions, overseas territories, islands, sovereign bases) that, for historical, geographical or political reasons, enjoy special status within or outside of the EU intra-community trade area.

Foodstuffs of animal origin are moving freely within the EU intra-community trade area and no controls are carried out at the borders between the EU Member States. Foodstuffs come from establishments fulfilling the conditions of harmonized EU legislation and approved or registered by the national competent authority. Because there are no border controls for intra-community trade, non-discriminatory spot checks are carried out at the point of origin and at the destination to ensure that the foodstuffs are in compliance with the harmonized EU legislation. The goods are accompanied by a commercial document only. In the case of outbreaks of infectious animal diseases, intra-community trade may be restricted by adopting certain safeguard measures limiting the free movement of goods.

25.3.2 Import

Import into the European Union from outside the intra-community trade area, also called import from third countries, of foodstuffs of animal origin is based on harmonized EU legislation. The European Union has developed generic import conditions for each category of foodstuffs of animal origin. These conditions are risk based and take into account the different animal health status of the exporting third countries or regions thereof by providing different measures to meet the ALOP of the European Union. Import may be authorized on the basis of freedom from certain animal diseases or after a handling, for example heat treatment sufficient to eliminate the pathogenic agent concerned. Transparency of the import conditions is ensured by publishing the legislation concerned in the *Official Journal of the European Union*. The Official Journal and the EU legislation are available via the EUR-LEX web site.

Import of foodstuffs of animal origin is authorized by the European Union on the basis of audits carried out by the Food and Veterinary Office (FVO) of the European Commission and guarantees provided by the competent authorities of the exporting country. All meat and meat products need to come from establishments approved by the European Union and listed on the web site of the European Commission. Every imported consignment of meat or meat products is accompanied by a veterinary certificate issued by the competent veterinary authority of the exporting third country. Foodstuffs of animal origin may only enter the EU intra-community trade area via one of the veterinary border inspection posts (BIP) situated at the EU external border. After a favourable veterinary border control a Common Veterinary Entry Document (CVED) is issued and the foodstuffs can move freely within the EU intra-community trade area.

25.3.3   Export

Export of foodstuffs of animal origin from the European Union is mostly not harmonized, as there is very little EU legislation on this area. Instead of EU legislation export takes place on the basis of the conditions set by the importing country or agreed between the trading partners. The conditions applying to export of a certain commodity vary from one importing third country to another and need to be clarified or negotiated separately. To achieve market access is a lengthy and complex process that may not result in a satisfactory result but in a trade dispute.

Like the European Union, the importing third country may provide generic import conditions applying equally to all exporting countries. These conditions may be available on the official web site of the competent authority of the importing country or they are provided on request. The importing third country may also develop import conditions separately for each exporting country on the basis of an assessment of the animal and public health risks related to the trade.

The trade conditions may also be agreed between the trading partners as a result of bilateral negotiations. The agreed trade conditions are given in the form of an official agreement, a protocol or a model certificate.

The importing third country may carry out a system audit of the exporting country as part of the trade authorization process. All approved or registered establishments in the European Union may be authorized to export foodstuffs of animal origin or export is only allowed from individual establishments that have been inspected by the importing third country.

Controls on meat and meat products exported from the European Union are carried out at the establishment of dispatch. A veterinary certificate is issued by the competent veterinary authority for each consignment and it accompanies the consignment to the place of destination. No veterinary controls are carried out at the external border when the consignment leaves EU territory.

25.3.4   European Union trade agreements

Negotiating and concluding agreements with third countries on trade of foodstuffs of animal origin is an exclusive competence of the European Union, represented by the European Commission. Based on the common commercial policy as laid down in the Treaty on the Functioning of the European Union (TFEU) and the harmonization at EU level of SPS rules, EU Member States are not legally competent to enter into trade agreements with third countries in the SPS field. However, where there are no agreements between the European Union and a third country in place, a Member State may agree bilaterally on the export conditions with the importing third country concerned in order to facilitate trade.

Bilaterally agreed export conditions only apply to export from the EU Member State concerned. Bilateral agreements concluded by individual

EU Member States should be in compliance with the WTO SPS Agreement, EU legislation and SPS Agreements the European Union has made with third countries. Any conditions relating to access to the European Union market or any of its Member States shall be avoided, as import is of full European Union competence. Any clauses that limit the free circulation of animals or foodstuffs of animal origin within EU territory are not allowed, unless this is justified by objective circumstances. The EU Member State concluding such a bilateral agreement should inform the European Commission and other Member States thereof.

There are two types of European Union trade agreement, independent agreements in the veterinary field and trade agreements containing SPS provisions. Veterinary agreements aim at settling mutually beneficial trade conditions on the basis of equivalence and reciprocity covering both European Union export and import of live animals and products of animal origin. The European Union did conclude its first veterinary agreement with New Zealand in 1996, followed by agreements with the USA in 1998 and Canada in 1999. Also, Annex 11 of the bilateral agricultural agreement between Switzerland and the European Union is often called a Veterinary Agreement, as it includes animal and public health and animal breeding measures that are applicable to the trade in live animals and products of animal origin.

There is no one-size-fits-all model of a trade agreement but in most cases the European Union negotiates comprehensive Free Trade Agreements (FTAs). Under the SPS provisions of the FTAs, the European Union is promoting internationally recognized standards that facilitate smooth trade, such as:

- prelisting, where the importing country accepts pre-authorization of establishments that can export without inspection, based on the system approval of the competent authorities of the exporting country;
- regionalization and zoning, where in the event of an animal disease outbreak the import restrictions can be limited to the affected regions, rather than applying them to the entire territory;
- compartmentalization, where approved premises applying bio-security measures may continue to trade internationally during animal disease outbreaks without the risk of spreading the disease.

At the same time the European Union is trying to oppose clauses violating the principle of the European Union single market area, such as the 'born and bred' rule, where the importing country is only authorizing import of, for example, meat that is derived from animals that are born, bred and slaughtered in the EU Member State of dispatch.

Since 2004, negotiations on trade conditions have been going on between the European Union and the Russian Federation. It is a very special situation, as these negotiations have been conducted without the support of any veterinary or SPS agreement. Guided by a set of protocols signed between the parties and called Memoranda of Understanding (MoU), several veterinary certificates that apply to the export from the European Union to Russia have been adopted. Upon the establishments in 2010 of a Customs Union (CU) by members of the Eurasian Economic Community, these negotiations have been lifted to the EU–CU level.

25.3.5   EU Trade Control and Expert System

The EU Trade Control and Expert System (TRACES) was developed to ensure the traceability and control of live animals and products of animal origin both in intra-community trade and in imports. TRACES is also being expanded to exports by including in the system common EU export certificates resulting from the negotiations held between the European Union and certain importing third countries.

TRACES is an internet-based system and is administered by the EU Commission. The TRACES system is used both by authorities and companies. TRACES sends an electronic message from the point of dispatch to the point of destination to notify that a consignment is arriving.

25.4   Exporting procedures

25.4.1   Meeting the export conditions

The exporting establishments and foodstuffs to be exported should comply with the conditions provided by the EU legislation as well as the additional conditions required by the importing third country. These additional conditions are taken into account by developing a monitoring programme where the parameters and other requirements different or above the European Union requirements are included. Typically, monitoring programmes related to meat and meat products contain microbiological and residue requirements. The guarantees provided in the export certificate are based on the results of the monitoring programme or on systematic sampling of the consignment concerned.

25.4.2   Export certificates

A certificate called a 'veterinary certificate', 'animal health certificate' or plainly a 'health certificate' is issued by the competent authority of the exporting country for each consignment of meat and meat products. This certificate provides the guarantees the importing country is requesting concerning the commodity concerned.

The international organizations OIE and Codex Alimentarius provide guidance on the information to be included in certificates related to international trade of foodstuffs of animal origin. The OIE provides a simple model for a veterinary certificate on its web page. The certificate shall clearly identify the certifying body and any other parties involved in the production and issuance of the certificate. The certificate shall contain attestations to be made by the certifying body to provide the importing country the guarantees it requests. The certificate shall be provided at least on the official languages of the exporting and importing country.

The goods and consignment to which the certificate relates needs to be clearly described by providing at least the following information:

- nature of the foodstuffs;
- name of the product;
- quantity of the consignment in appropriate units;
- description of the consignment to which the certificate uniquely relates (e.g. lot number, means of transport, seal number, production date code);
- name and address of the production establishment(s) and possible storage establishment and their approval number;
- name and contact details of the exporter or consignor;
- name and contact details of the importer or consignee;
- country of dispatch and part thereof where needed;
- country of destination.

The final export certificate model may contain features identifying the exporting country and the competent authority concerned (e.g. logos or coat of arms). Based on the procedures in place in the exporting country or on requirements of the importing third country, the certificate may include certain safety features aimed at preventing forgery of the certificates. These safety features may consist of a specific paper material, watermarks, holograms and so on.

25.4.3 Certification procedures

The certifying officials must not certify data of which they have no personal knowledge or which cannot be ascertained by them. Certifying officials must not sign blank or incomplete certificates or certificates relating to products which they have not inspected or which have passed out of their control. They must not certify for events that will take place after the certificate is signed when these events are not under his/her direct control and supervision. Where a certificate is signed on the basis of another certificate or attestation, the certifying official shall be in possession of that document before signing.

The certificate shall bear a unique identifying number. Where the certificate consists of a sequence of pages, each page must indicate this number. The text of a certificate should not be amended except by deletions, which should be signed and stamped by the certifying official. The signature and the stamp of the official shall be in a different colour to that in the printed certificate. The original certificate shall accompany the consignment. A copy of the certificate is kept by the certifying official.

According to the international standards of OIE and Codex Alimentarius, a replacement certificate may be issued by the certifying official to replace certificates that have been, for example, lost, damaged and contain errors, or where the original information is no longer correct. These replacements should be clearly marked to indicate that they are replacing the original certificate. A replacement certificate should reference the number and the issue date of the certificate that it supersedes. The superseded certificate should be cancelled and, where possible, returned to the issuing authority.

Literature and further reading

Codex Alimentarius. www.codexalimentarius.org (last accessed 26 February 2014).

OIE (World Organisation for Animal Health). 2013. Terrestrial Animal Health Code. www.oie.int/international-standard-setting/terrestrial-code/access-online/ (last accessed 27 February 2014).

WTO (World Trade Organization). http://www.wto.org/ (last accessed 28 February 2014).

26

Scientific Risk Assessment – Basis for Food Legislation

Riitta Maijala

Department of Food Hygiene and Environmental Health,
Faculty of Veterinary Medicine, University of Helsinki, Helsinki, Finland

26.1 Scope

After the BSE and dioxin crises and the establishment of the World Trade Organization, scientific advice as a form of risk assessment has been given a legitimate and important role in drafting food safety legislation at the international, multinational and national level. In this chapter, the structured process of risk assessment for estimating the probability and level of an adverse health effect a hazard can cause for human or animal health is presented within the risk analysis framework. The role of the European Food Safety Authority (EFSA) as the European Union risk assessor in food safety is also discussed.

26.2 Introduction

Meat inspection and control at the slaughterhouse are regulated by the legislation, which also sets up the official control systems. If the country has trade

Meat Inspection and Control in the Slaughterhouse, First Edition.
Edited by Thimjos Ninios, Janne Lundén, Hannu Korkeala and Maria Fredriksson-Ahomaa.
© 2014 John Wiley & Sons, Ltd. Published 2014 by John Wiley & Sons, Ltd.

in meat producing animals, meat or meat products, its national legislation needs to be in accordance with international, multinational or bilateral trade rules in order to avoid trade disruptions and disputes. For example, a country that is a member state of the European Union needs to implement European Union (EU) legislation and, where those countries and the European Union are members of the World Trade Organization (WTO), also the WTO trade rules, international standards and agreements must be carefully followed (Figure 26.1). Depending on the trade partner, different sets of legislative requirements may apply, sometimes even with conflicting demands. For example, the meat inspection legislation requirements of the USA or Russia may differ from those of the European Union and its Member States, even if all are members of WTO (arrows A–C in Figure 26.1). Although these rules apply only for trade, they often have a significant impact also on the legislative requirements for domestic market meat production.

In addition to the WTO and European Union, there are other important multinational trade organizations influencing on trade between countries. The European Free Trade Association (EFTA) is an intergovernmental organization set up for the promotion of free trade and economic integration to the benefit of its four Member States: Iceland, Liechtenstein, Norway and Switzerland. African, Caribbean and Pacific Group of States (ACP) have established Economic Partnership Agreements (EPA) (ACP–EPA) and the Latin American and Caribbean Economic System (SELA) provides the Latin American and Caribbean region with a system of consultation and coordination for the adoption of common positions and strategies on economic issues

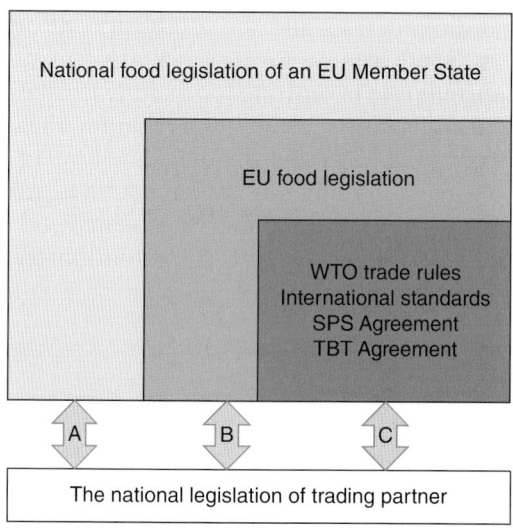

Figure 26.1 Food safety legislation of an EU Member State must be in accordance with EU legislation and WTO trade rules, including, for example, SPS Agreement. Depending on the trade partner, legislation has to fulfil bilateral agreements (A), EU legislation (B) and/or the WTO rules (C).

in international bodies and forums. Many, if not most, of these countries are also members of the WTO. In fact, the trade of animals and animal derived products falls under multiple international standards, rules and agreements, which each set demands for the legislative process of food safety.

Therefore, even if each country establishes its own food legislation and meat control structures based on its national food safety situation, there are many international standards and rules that apply for most of the countries – one of them being the role of risk assessment in drafting food legislation as defined by the WTO. In order to understand why and how risk assessment is used in the legislative process, the international framework and structure and key components of the scientific risk assessment are presented in this chapter.

International pressure to develop harmonized risk analysis standards and principles linked with meat production has focused on the potential of chemical and microbiological hazards to cause harm for human or animal health. In addition to these, meat inspection and slaughterhouse processes need to consider other types of risks also, especially for animal welfare, occupational health and environmental safety – but they fall outside the scope of this chapter.

26.3   Risk analysis standards are set by international organizations

The WTO established a formal structure of risk assessment based decision making process for international trade of agricultural products, called risk analysis. However, risk assessment was not at that time a novel approach, it had been used already, for example in approving chemicals. An important novelty in bringing a scientific approach to international trade lays in Article 20 of the General Agreement on Tariffs and Trade (GATT); this allows governments to act on trade in order to protect human, animal or plant life or health, provided they do not discriminate or use this as disguised protectionism. A specific WTO agreement called the Sanitary and Phytosanitary Measures Agreement (SPS Agreement) sets out the basic rules for food safety and animal and plant health standards, which also, therefore, impacts on meat inspection legislation. Member countries of the WTO are encouraged to use international standards, guidelines and recommendations where they exist. When they follow this approach, they are unlikely to be challenged legally in a WTO dispute. However, the SPS Agreement allows countries also to set their own standards but only if there is scientific justification. Temporarily, countries can also apply precautionary measures.

International standard setting organizations defined by the SPS Agreement, the Codex Alimentarius Commission and the World Organisation for Animal Health (OIE), define risk analysis standards for food safety and animal health. Animal welfare, on the contrary, falls under the Technical Barriers to Trade Agreement (TBT Agreement), which tries to ensure that regulations, standards, testing and certification procedures do not create unnecessary obstacles. However, it does not provide such a clear basis for risk assessment as the SPS Agreement has established for animal health and food safety hazards.

The Codex Alimentarius Commission, established by Food and Agriculture Organization of the United Nations (FAO) and the World health Organization (WHO) in 1963, develops harmonized international food standards, guidelines and codes of practice to protect the health of the consumers and ensure fair trade practices in the food trade. The Codex Alimentarius Commission has adopted a set of international standards called Codex Alimentarius. Already, though, a few decades before, in 1924, the need to fight animal diseases at a global level had led to the creation of the Office International des Epizooties. In 2003, the Office became the World Organisation for Animal Health but kept its historical acronym OIE. The OIE is the intergovernmental organization responsible for improving animal health worldwide. In recent years it has shown more interest also for zoonoses and animal welfare risks.

26.4   Risk analysis is a decision making process

Standards for risk analysis specify roles and tasks for risk assessors providing scientific advice and for risk managers making decisions based on risk assessment results and other legitimate factors, such as socioeconomic and political priorities. During the risk analysis process, risk communication plays an important role in order to understand, for example, the needs of risk management for advice or the reasons behind the final decision or for adopted legislation (Figure 26.2). Since risk analysis is a powerful tool in

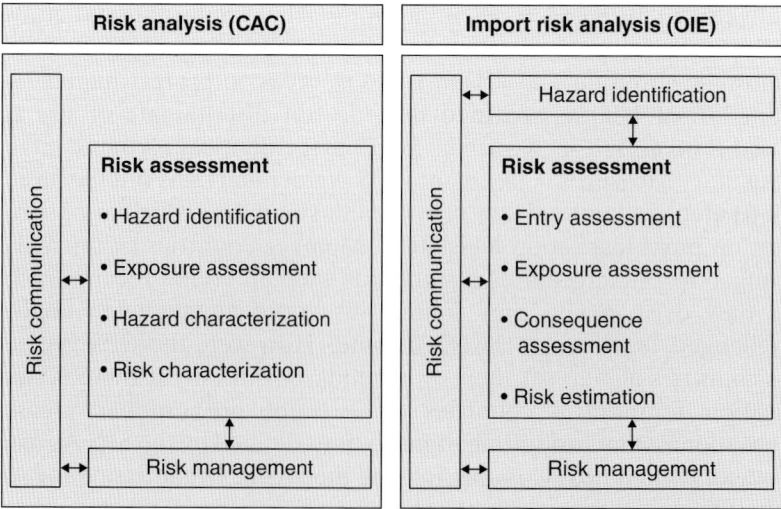

Figure 26.2 Components of risk assessment within risk analysis framework as outlined by the Codex Alimentarius Commission (CAC) and World Organisation for Animal Health (OIE). Source: Adapted from Maijala, 2006 with permission from Wageningen Academic Publishers.

decision making process and especially in drafting legislation, the Codex Alimentarius standard requires risk analysis to be (i) applied consistently, (ii) open, transparent and documented and (iii) evaluated and reviewed as appropriate in the light of newly generated scientific data. Both the Codex Alimentarius and OIE have included in their standards definitions for key risk analysis terms (Table 26.1).

Science, in the form of risk assessment, has in the risk analysis framework a clearly defined role in the legislative process of food safety. Risk assessment opinion is a science-based reply of experts (risk assessors) for the predefined question(s) asked by a risk manager, most often by the legislative body of a country or by a trade union such as the European Union. Therefore, risk assessment is a scientific view with limited scope and based on the data and methods available at the time of drafting the reply and produced within the time frame set by legislative or other decision making processes. Questions asked from risk assessors in drafting EU legislation linked with meat inspection can vary from individual hazards (e.g. *Trichinella*, *Cysticercus*, BSE and other TSEs) up to the overall impact for food safety and for animal health and welfare of the introduction of a risk-based approach to meat inspection. Embedment in the decision making process and including a formal structure specified by international standards differentiates risk assessment from other types of scientific advices given by experts to decision makers (e.g. consultation, review or research project report).

After the BSE and dioxin crisis in the European Union in the late 1990s, risk analysis processes have been strengthened both at the national and European Union level in order to be able to act rapidly, if required. Establishment of the European Food Safety Authority (EFSA) created within the EU legislative system a formal body, which produces scientific risk assessments during normal and crisis situations. Depending on risk management needs, risk assessments may be asked for within hours or up to several years. In crisis situations, such as the Irish animal feed dioxin contamination in 2008, fast risk assessment results can support effective actions to be made in order to protect public health and to minimize unnecessary trade disruptions. Consequently, the fast-track advice outcomes usually also have a rapid impact on the legislative and other risk management decisions made to handle the crisis, whereas risk assessments produced outside times of crisis impact legislation within months or sometimes even years.

26.5   Risk assessment estimates the level of risk

Whereas the OIE standard aims to provide importing countries with a method of assessing the disease risks associated with the importation of animals, animal products, animal genetic material, feedstuffs, biological products and pathological material, the Codex Alimentarius risk analysis framework mainly focuses on risks already existing or those which could be introduced into the food chain via approval of regulated products. In both

Table 26.1 Some of the key terms as defined by the Codex Alimentarius Commission (CAC), World Organisation for Animal Health (OIE) and FAO/WHO relevant for food safety risk analysis

Term	Organization	Definition/explanations
Hazard	CAC	A biological, chemical or physical agent in, or condition of, food with the potential to cause an adverse health effect.
		A biological, chemical or physical agent in, or condition of, a good with the potential to cause an adverse health effect.
	OIE	Biological, chemical or physical agent in, or a condition of, an animal or animal product with the potential to cause an adverse health effect.
Risk	CAC	A function of the probability of an adverse health effect and the severity of that effect, consequential to a hazard(s) in food.
	OIE	Likelihood of the occurrence and the likely magnitude of the biological and economic consequences of an adverse event or effect to animal or human health.
Risk analysis	CAC	A process consisting of three components: risk assessment, risk management and risk communication.
	OIE	The process composed of hazard identification, risk assessment, risk management and risk communication.
Risk assessment	CAC	A scientifically based process consisting of the following steps: (i) hazard identification, (ii) hazard characterization, (iii) exposure assessment, and (iv) risk characterization.
		Qualitative Risk Assessment: A Risk Assessment based on data which, while forming an inadequate basis for numerical risk estimations, nonetheless, when conditioned by prior expert knowledge and identification of attendant uncertainties permits risk ranking or separation into descriptive categories of risk.
		Quantitative Risk Assessment: A Risk Assessment that provides numerical expressions of risk and indication of the attendant uncertainties.
	OIE	Evaluation of the likelihood and the biological and economic consequences of entry, establishment and spread of a hazard within the territory of an importing country.

(continued overleaf)

Table 26.1   *(Continued)*

Term	Organization	Definition/explanations
Uncertainty (analysis)	**CAC**	**Uncertainty:** The (quantitative) expression of our lack of knowledge. Uncertainty can be reduced by additional measurement or information. There are many types of uncertainty in exposure assessment, including process uncertainty, model uncertainty, parameter uncertainty, statistical uncertainty, and even uncertainty in variability:
		Process uncertainty refers to the uncertainty about the relationship between the food chain as documented in the exposure assessment and the processes that take place in reality.
		Model uncertainty comprises both the correctness of the way the complexity of the food chain is simplified, and the correctness of all the submodels that are used in the exposure assessment.
		Parameter uncertainty incorporates uncertainties dealing with errors resulting from the methods used for parameter estimation, like measurement errors, sampling errors and systematic errors. As part of this, **statistical uncertainty** is defined as the uncertainty quantified by applying statistical techniques such as classical statistics or Bayesian analysis.
		Uncertainty: Lack of knowledge regarding the true value of a quantity, such as a specific characteristic (e.g. mean, variance) of a distribution for variability, or regarding the appropriate and adequate inference options to use to structure a model or scenario. These are also referred to as model uncertainty and **scenario uncertainty.** Lack of knowledge uncertainty can be reduced by obtaining more information through research and data collection, such as through research on mechanisms, larger sample sizes or more representative samples.
		Measurement uncertainty refers to the 'uncertainty' associated with data generated by a measurement process. In analytical chemistry, it generally defines the uncertainty associated with the laboratory process but may also include an uncertainty component associated with sampling. Non-negative parameter characterizing the dispersion of the values being attributed to a measure and, based on the information used.
		Model uncertainty Bias or imprecision associated with compromises made or lack of adequate knowledge in specifying the structure and calibration (parameter estimation) of a model.

Table 26.1 (*Continued*)

Term	Organization	Definition/explanations
		Uncertainty analysis: A method used to estimate the uncertainty associated with model inputs, assumptions and structure/form. An analysis designed to determine the contribution of the uncertainty associated with an input parameter to the degree of certainty in the estimate of exposure.
	OIE	Not specified.
Safety assessment	**CAC**	A Safety Assessment is defined by CAC as a scientifically-based process consisting of:
		1) the determination of a NOEL (No Observed Effect Level) for a chemical, biological, or physical agent from animal feeding studies and other scientific considerations;
		2) the subsequent application of safety factors to establish an ADI or tolerable intake; and
		3) comparison of the ADI or tolerable intake with probable exposure to the agent.
	OIE	Not specified.

Source: Adapted from EFSA 2012.

standards, risk analysis is composed of risk assessment, risk management and risk communication. However, in the OIE standard, hazard identification is a separated step before the initiation of risk assessment whereas in the Codex standard it is included within the risk assessment itself (Figure 26.2).

Risk assessments on substances intentionally added to the meat producing chain (such as veterinary medicines, feed additives or decontaminants) or other hazards (such microbes and chemical contaminants) potentially causing an adverse public health effect via meat production chain usually follow Codex Alimentarius standard. The OIE standard, although intentionally established for assessing import risks, can also be useful also to assess risks in domestic animal populations. Risks of zoonoses, due to their specific nature, can be assessed following the principles of either of the standards or a combination of them.

Hazard *identification* in the OIE risk analysis framework is a categorization step, identifying biological agents dichotomously as potential hazards or not. The potential hazards identified would be those appropriate to the species being imported, or from which the commodity is derived, and which may be present in the exporting country. It also identifies whether each potential hazard is already present in the importing country, and whether it is a notifiable disease or is subject to control or eradication in that country. Therefore, this first step can result in the conclusion that no hazard was identified linked with the import and, therefore, risk assessment needs not to be completed. For instance, if BSE has never been detected in an importing country and the surveillance system is considered reliable enough to detect the disease if occurring, hazard identification in the OIE standard could result in allowance of import of cattle for slaughter without any further risk assessment needs.

Hazard identification by Codex Alimentarius definition is the identification of biological, chemical and physical agents capable of causing adverse health effects and which may be present in a particular food or group of foods. Although risk assessment itself can be either qualitative or quantitative, hazard identification is predominately a qualitative process. Information on hazards can be obtained from, for example, scientific literature, from databases such as those in the food industry, government agencies and relevant international organizations and through solicitation of opinions of experts. Relevant information for this step of the risk assessment process includes, for example, data on clinical and epidemiological studies and surveillance, laboratory animal studies and investigations of the characteristics of microorganisms or chemicals.

The next steps after the identification of a hazard are exposure/entry assessment and further characterization of the consequences for those potentially exposed. These paths of evaluation usually take place simultaneously and their results are eventually combined together in risk characterization (Codex Alimentarius) or risk estimation (OIE) at the end of risk assessment process.

According to the Codex Alimentarius, exposure *assessment* includes an assessment of the extent of actual or anticipated human exposure. Exposure assessment should specify the unit of food that is of interest, that is the portion size in most/all cases of acute illness, frequency of contamination of foods by the pathogenic agent or contaminant and its level in those foods over time. Factors to be considered include data such as the initial contamination of the raw material, including considerations of regional differences and seasonality of production, the methods of processing, packaging, distribution and storage of the foods, as well as any preparation steps such as cooking and holding and consumption patterns, for example socioeconomic and cultural backgrounds, ethnicity, seasonality, population demographics, and consumer preferences and behaviour. Exposure assessment gives estimates on the level and frequency the hazard may be present in a portion at the time of consumption.

In the OIE approach, entry *assessment* describes the biological pathway(s) necessary for an importation activity to introduce pathogenic agents into a particular environment and estimates the probability of the complete process occurring, either qualitatively (in words) or quantitatively (as a numerical estimate). Inputs that may be needed for this step include, for example, species, age and breed of animals, vaccination, testing, treatment and quarantine, incidence or prevalence of pathogenic agent, evaluation of Veterinary Services, surveillance and control programmes as well as commodity related data on quantity of the commodity to be imported and the effect of processing, storage and transport.

If the entry assessment demonstrates no significant risk, the risk assessment does not need to continue. If risk of entry is demonstrated, entry assessment follows the exposure assessment describing the biological pathway(s) necessary for exposure of animals and humans in the importing country to the hazards from a given risk source, and estimating the probability of the exposure(s) occurring. The probability of exposure to the identified hazards is estimated for specified exposure conditions with respect to amounts, timing, frequency, duration of exposure, routes of exposure, such as ingestion, inhalation or insect bite, and the number, species and other characteristics of the animal and human populations exposed.

In food safety risk assessments (Codex Alimentarius), *hazard* characterization provides a qualitative or quantitative description of the severity and duration of adverse effects that may result from the ingestion of a hazard in food, including usually a dose-response assessment for acute and/or long term effects. Other data relevant for assessment include, for example, toxicity or virulence of hazard, onset of symptoms, genetic factors of host, and individual host susceptibility such as age, pregnancy, nutrition, health and medication status.

According to the OIE code, consequence *assessment* consists of describing the relationship between specified exposures to a biological agent and the consequences of those exposures. A causal process should exist by which exposures produce adverse health or environmental consequences, which may in turn lead to socioeconomic consequences. The consequence assessment describes both direct consequences, such as animal infection, disease and production losses, and public health consequences as well as indirect consequences for surveillance and control costs, compensation costs, potential trade losses and adverse consequences to the environment. Whereas Codex Alimentarius focuses on direct food safety consequences for those potentially exposed, the OIE standard has a wider scope, including consequences not only for animal populations but also for the environment and economics, thereby including areas which under Codex Alimentarius would be handled merely at the risk management phase.

The final step of risk assessment, risk *characterization* (Codex Alimentarius) or risk *estimation* (OIE) integrates all the previous steps and gives either a qualitative or quantitative estimate(s) on the level of risk. The estimate describes the likelihood and severity of the adverse effects which could occur in a given population, including a description of the uncertainties associated with these estimates. For a quantitative risk assessment, the final outputs may include: estimated numbers of herds, flocks, animals or people likely to experience health impacts of various degrees of severity over time; probability distributions, confidence intervals and other means for expressing the uncertainties in these estimates; portrayal of the variance of all model inputs; a sensitivity analysis to rank the inputs as to their contribution to the variance of the risk estimation output; and analysis of the dependence and correlation between model inputs. In addition to estimates of risk, risk assessment reports sometimes include also evaluation of the impact on risk reduction by potential risk management options (such as heat treatment or decontamination of carcasses).

26.6   Other parts of risk analysis: risk management and risk communication

There are usually a wide variety of risk management options and their combinations available to prevent, control, reduce and/or eliminate risk occurring in the meat producing chain. These options may include farm level risk reduction measures, such as herd health programmes, closed breeding pyramids and Good Hygienic Practices (GHP)/Good Farming Practices (GFP) combined with *ante-mortem* inspections, visual inspection of carcasses, incisions and palpations as well as sampling programmes for residues and parasites at the slaughterhouse. Within the risk analysis framework, risk management is a part where decisions on control options are made, including those ones that are set up by the legislative requirements and may also be linked with international trade.

According to the Codex Alimentarius standard, risk management should follow a structured approach, including preliminary risk management activities, evaluation of risk management options, implementation, monitoring and review of the decision taken. The decisions should be based on risk assessment and should be proportionate to the assessed risk, taking into account, where appropriate, other legitimate factors relevant for the health protection of consumers (such as socioeconomic factors) and for the promotion of fair practices in food trade.

Also in the OIE standard the impact of decision making on trade is clearly highlighted: 'Risk management is the process of deciding upon and implementing measures to achieve the Member's appropriate level of protection, whilst at the same time ensuring that negative effects on trade are minimized. The objective is to manage risk appropriately to ensure that a balance is achieved between a country's desire to minimize the likelihood or frequency of disease incursions and their consequences and its desire to import commodities and fulfil its obligations under international trade agreements.' Risk assessment should start from risk evaluation – the process of comparing the risk estimated in the risk assessment with the OIE Member's appropriate level of protection. Thereafter, risk options are to be evaluated and implemented and the situation monitored and reviewed.

Whereas risk management leads the making of decisions after the completion of risk assessment, risk communication runs throughout the whole risk analysis process. Ideally, it should be an interactive way of exchanging information and opinions concerning hazards and risk, risk-related factors and risk perceptions among risk assessors, risk managers, those potentially affected and other interested parties, for example consumers, non-governmental organizations, industry, trade partners or the academic community. The interests of these parties may be very different and may be health related (e.g. probability of dying after eating a contaminated pork meal), financial (e.g. costs of running the testing for *Trichinella*), sociological (e.g. estimates of unemployed slaughterhouse workers if trade would be prohibited) or ethical (e.g. is cloning of animals for food production ethically acceptable or not) questions and views.

Risk communication is also important in communicating the results of the risk assessments and evaluated risk management measures to the decision makers and other interested parties. A good risk assessment report should be transparent in describing the data and methods used and what assumptions were made during the process. There are never enough data to fully conclude the process and hardly ever a risk assessment report ends up without listing major needs for further research. However, a good risk assessment report explains what is currently known with the uncertainties involved and can, therefore, be transparently used by risk managers in the decision making process together with the other legitimate factors.

26.7 Risk assessments of EFSA impact on EU food safety legislation

Regulation (EC) No 178/2002 of the European Parliament and of the Council (so-called EU Food Law) establishes the risk analysis framework for EU food legislation. The European Food Safety Authority (EFSA) was set up in January 2002, following a series of food crises in the late 1990s, as an independent source of scientific advice and communication on risks associated with the food chain. Recital 35 of EU Food Law defines the target for EFSA, saying it should be an independent scientific source of advice, information and risk communication in order to improve consumer confidence; nevertheless, in order to promote coherence between the risk assessment, risk management and risk communication functions, the link between risk assessors and risk managers should be strengthened. This highlights the need for the independent risk assessment body, such as EFSA, to work in close collaboration with risk managers in order to ensure that the scientific advice produced will eventually be useful for the legislative and policy processes.

As the EU risk assessor on food safety, EFSA produces annually 500–700 scientific opinions and advice to provide a scientific foundation for European policies and legislation and to support the European Commission (EC), European Parliament (EP) and EU Member States (MSs) in taking risk management decisions. After the adoption, the Scientific Opinion is given to the requestor and published on the EFSA web site. Open publication of all the opinions adopted by the EFSA Scientific Panels in the EFSA Journal enables not only the requestors but also all other interested parties to see the risk assessment process, that is (i) what was the question(s) asked of EFSA, (ii) what data and methods were used, (iii) what were the uncertainties and assumptions, (iv) what is the final conclusion and (v) who were the experts drafting and adopting the Scientific Opinion (Figure 26.3). This transparency is important not only for the risk assessment itself but it also facilitates the understanding of outcome by all other interested parties. In drafting the EU

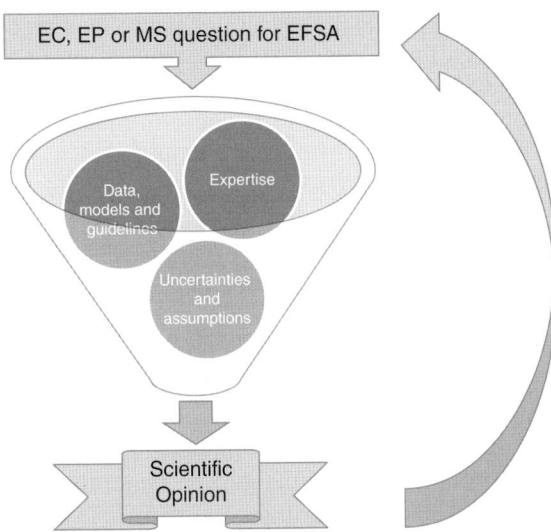

Figure 26.3   Risk assessment process of EFSA Scientific Panels. After the adoption, the Scientific Opinions replying to the questions asked by risk managers (European Commission, European Parliament or EU Member States) are sent to the requestors and published in the EFSA Journal.

legislation, the European Commission combines EFSA Opinions together with other legitimate factors and it is important to understand what proposals in draft legislation are based on scientific advice of EFSA and which ones originate from trade, practical, socioeconomic or other reasons.

EFSA's advice informs the policies of risk managers and replies to questions presented in mandates mainly asked by the European Commission but sometimes also by the European Parliament or an EU Member State, as well as originating from EFSA's own scientific activities. The background to these questions is most often the need to draft new or amend already existing EU legislation. For instance, as BSE occurrence in the European Union has decreased and the control options in use are quite expensive, the European Commission has asked several questions of the EFSA linked to the control options, for example on testing of healthy cattle or on the use of meat and bone meal in feed for different kinds of production animals. These Scientific Opinions have thereafter been used in supporting the changes in EU regulations to manage the BSE risk.

It is also important to bear in mind that most often questions concern one type of hazard, for example microbiological or chemical ones. However, in reality these hazards do not always occur in isolation but can be present in different combinations. For the risk management decision making process it is vital to understand, for example, if decreasing one type of hazard, such as campylobacters in broiler meat with decontamination of carcasses, would result into new a type of risk in meat caused by residues of hazardous chemicals or antimicrobials. Interestingly, in the EC mandate for the introduction of a risk-based approach to meat inspection, the question itself combines both the microbiological and chemical food safety issues as well as the impact of changes for animal health and welfare. This kind of process is a challenge for risk assessment due to data requirements and time limitations but should provide a more robust scientific basis for decision making if and when meat inspection is modernized in the European Union.

After a question has arrived in EFSA, it is allocated to the suitable Scientific Panel(s) or Scientific Committee. Scientific Panels and Scientific Committee are appointed for three year periods based on open calls. In addition, an expert database, open to candidates with scientific expertise to apply for, is used to identify any additional expertise needed to reply for the mandate in a multi- and interdisciplinary way; these experts are considered to be invited to join the working group of the Scientific Panel replying to the question. For example, in order to reply to the question on the feasibility of establishing *Trichinella*-free areas, and if the public health risk would increase if pigs in those areas were not examined for *Trichinella* spp, the Scientific Panel on Biological Hazards used scientific expertise on epidemiology, parasitology, meat inspection and food hygiene.

All experts identified to work for the EFSA need to make a Declaration of Interests (DoIs) to allow any potential conflicts to be identified and dealt with transparently. The DoIs cover matters such as ownership and investments, employment and research funding. The EFSA's secretariat assesses these DoIs before inviting the experts to work on the given issue. All the assessed DoIs are published on the EFSA web site under those Scientific Panels and their working groups the expert is contributing to. This process ensures that the work of Scientific Panel and its working group can be done in an independent and transparent manner.

In their risk assessment work, the experts use the data and models available and follow the international and other scientific guidelines for conducting the type of risk assessment they have been asked for. In fact, developing risk assessment guidelines is a key tool in supporting robust and transparent risk assessments, since it provides insight on the risks assessed, data needed and often also guidance how the data should be presented in the application.

Whereas most of the data for BSE risk assessment originates from scientific literature and data collected by Member States, safety evaluation of, for example, feed additives relies much on the data provided by the industry. However, even for these questions the Scientific Panel always includes other relevant scientific data in order to ensure the high quality of the final opinion. Thereby, the data used for the risk assessment can originate from scientific publications, data collection by Member States or the European Commission and from the industry. The EFSA also collects and analyses data on food/feed safety and nutrient intakes together with Member States.

One of the major challenges in risk assessment is the uncertainties in data used and limitations of available models (Table 26.1). Identification and management of uncertainties are crucial for the quality of assessment, since it is important to know where and how big the main uncertainties underpinning the final risk estimate are. This is also the point where scientific expertise is providing its vital contribution to the process. Only the best experts with wide knowledge on the risk assessment process and in the area of question it is replying to can draw good conclusions taking into account the inherent uncertainties encountered during the risk assessment work.

26.8   Concluding remarks

Scientific risk assessment is a powerful tool in drafting food legislation and can increase the transparency and quality of the decision making process. In order to use it wisely, it is important to know the international standards and legislations defining the structure of risk assessment as a part of risk analysis process as well as the limitations and challenges to conducting robust risk assessment work. At the end it is a scientific reply for the predefined question(s) asked by those who are to amend or draft new legislation and, therefore, it is tightly embedded within the decision making process and its timelines. Good risk assessments require multi- and interdisciplinary expertise, robust data, good models and wise assumptions as well as good writing skills to make the final outcome comprehensible by all interested stakeholders. Independency of risk assessment as well as publication of outcomes increase trust for the process.

Literature and further reading

Casey, K.C., Lawless, J.S. and Wall, P.G. 2010. Emerald Article: A tale of two crises: the Belgian and Irish dioxin contamination incidents. *Br Food J*, **112**, 1077–1091.

Codex Alimentarius. International Standards. http://www.codexalimentarius.org /standards/list-of-standards/en/ (last accessed 27 February 2014).

EFSA. Workflow for Scientific Opinions. http://www.efsa.europa.eu/en/efsahow /workflow.htm (last accessed 27 February 2014).

EFSA. Meat Inspection. http://www.efsa.europa.eu/en/topics/topic/meatinspection .htm (last accessed 27 February 2014).

EFSA Scientific Committee. 2012. Scientific Opinion on Risk Assessment Terminology. *EFSA Journal*, 2012;10(5):2664. [43 pp.]. doi: 10.2903/j.efsa.2012.2664. [Available online: www.efsa.europa.eu/efsajournal; last accessed 27 February 2014.]

European Parliament/Council. 2002. Regulation (EC) No 178/2002 of the European Parliament and of the Council of 28 January 2002 laying down the general principles and requirements of food law, establishing the European Food Safety Authority and laying down procedures in matters of food safety.

Maijala, R. 2006. Risk assessment as a tool for evaluating risk management options for food safety. In: *Towards a Risk-Based Chain Control* (ed. F.J.M. Smulders). Wageningen Academic Publishers, The Netherlands, pp. 19–32.

OIE (World Organisation for Animal Health). Terrestrial Animal Health Code, Section 2. Risk Analysis. http://www.oie.int/index.php?id=169&L=0&htmfile =titre_1.2.htm (last accessed 27 February 2014).

Smulders, F.J.M. and Algers, B. (eds). 2009. *Food Safety Assurance and Veterinary Public Health, volume 5 – Welfare of Production Animals: Assessment and Management of Risks*. Wageningen Academic Publishers, The Netherlands.

WTO (World Trade Organization). Standards and Safety, SPS Agreement. http://www.wto.org/english/thewto_e/whatis_e/tif_e/agrm4_e.htm (last accessed 27 February 2014).

27

Use of Meat Inspection Data

Hannu Korkeala and Janne Lundén

Department of Food Hygiene and Environmental Health,
Faculty of Veterinary Medicine, University of Helsinki, Helsinki, Finland

27.1   Scope

A large amount of data is collected in meat inspection in slaughterhouses. The data can be divided into four main categories: meat safety, meat quality, animal health and animal welfare. The data collection should be well organized and the data should be widely available for different actors for maximum use. This chapter describes using the data and the requirements for the collection of data in slaughterhouses.

27.2   Use of meat inspection data

Meat inspection data are important for different surveillance purposes and can be used for improvement of meat inspection procedures and for evaluation of the slaughter processes and of the meat inspection activities (Table 27.1). On the other hand, the data can be used to observe faults and deviations during slaughtering and meat inspection to correct them as soon as possible. Currently meat inspection data are greatly under-used and a lot of effort should be made to ensure that all the data are widely used. Good practices or regular programmes using meat inspection data are not widely available or used. The great potential of meat inspection data is not

Meat Inspection and Control in the Slaughterhouse, First Edition.
Edited by Thimjos Ninios, Janne Lundén, Hannu Korkeala and Maria Fredriksson-Ahomaa.
© 2014 John Wiley & Sons, Ltd. Published 2014 by John Wiley & Sons, Ltd.

Table 27.1   Usefulness and impact of use of meat inspection data.

Area	Impact
Food safety	Improvement of animal cleanliness Improvement of slaughter hygiene Prevention of chemical residues Prevention of drug residues Prevention of zoonoses
Meat quality	Improvement of slaughter hygiene Prevention of DFD, PSE Successful bleeding and evisceration
Animal health	Prevention of contagious animal diseases Prevention of non-contagious diseases
Animal welfare	Detection of welfare problems in: • primary production • stunning • transport
Research	Innovation for future development Use of versatile data
Meat inspection procedures	Common practices in meat inspection procedures Improved diagnostics in meat inspection Improvement of credibility and reliability

sufficiently used in research and only a limited number of studies using meat inspection data are published.

Meat inspection data can be used by various parties (Figure 27.1). The different parties should actively monitor surveillance data provided by meat inspection. Meat inspection veterinarians have an essential role in the use of data. In addition to using the data in their daily practice they are responsible in many cases for data collection and for the reliability of the data. The data collected during meat inspection are an important tool for authorities to evaluate and improve the meat inspection procedures. The slaughterhouses should also evaluate their activities and make changes and corrections needed based on

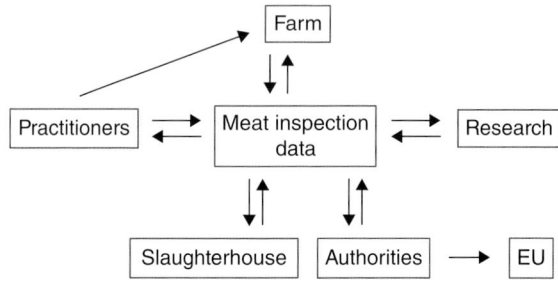

Figure 27.1   Optimal transfer of meat inspection data between different parties.

the data collected at the slaughterhouse. The wide use of data should be also accessible to researchers without restrictions to allow the evaluation of meat inspection, food safety, and animal welfare.

The authorities, including meat inspection veterinarians, should monitor data concerning food safety, quality of meat, animal diseases, animal welfare and slaughtering procedures (Table 27.2). Meat inspection veterinarians should analyse the data collected at the slaughterhouse they are working in. Many flaws and deviations at slaughterhouses can be recognized based on single observations. However, with effective and long-term follow-up, different trends and small deviations can be found that possibly can decrease the quality and safety of meat. One example of useful trend analysis is the long-term follow up of carcass surface hygiene data (Figure 27.2). Carcass surface hygiene is dependent on the level of the slaughter hygiene, which may differ between slaughterhouses. Therefore, the slaughterhouses have

Table 27.2 Collection of data in the slaughterhouse.

Area of data collection	Parameter
Transport	Animal density
	Death rate
	Transport fitness
	Transport injury
	Transport temperature
	Transport time
	Transport vehicle requirements
Ante-mortem	Abnormalities
	Cleanliness
	Condemnation causes
	Condemnation rate
	Recording of animal diseases
Post-mortem	Abnormalities
	Condemnation causes
	Condemnation rate
	Recording of animal diseases
	Full and partial condemnation
Slaughter process	Equipment functionality (e.g. evisceration)
	Scalding water/steam temperature
	Successful stunning and bleeding of animals
Hygiene control	Carcass surface hygiene
	Equipment hygiene
	Process surface hygiene
	Working hygiene
Laboratory	Meat inspection tests
	Residue tests
Compliance of regulations	Non-compliance rate
	Recurrent non-compliance
	Seriousness of non-compliance

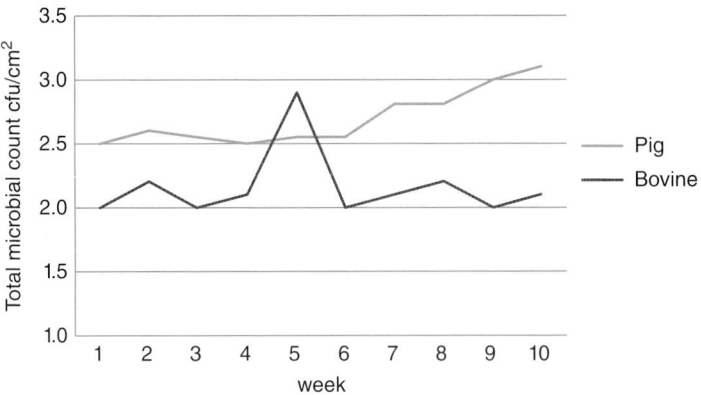

Figure 27.2 Trend surveillance of pig and bovine carcass surface hygiene.

to be aware of their normal carcass slaughter hygiene level. The example in Figure 27.2 shows the weekly surface hygiene results of pig and bovine carcasses. The results show deviations in the normal trend of carcass surface hygiene in both pig and bovine carcasses. The surface hygiene contamination of pig carcasses is increasing week by week, which should lead to actions to improve the slaughter hygiene. Reasons leading to an increased carcass surface contamination should be recognized and corrected. A possible reason could be, for example, poor sanitizing of slaughter equipment. The bovine carcass surface hygiene trend analysis shows a single peak in the surface contamination, which indicates that a single one-time error has occurred in the slaughter process. By analysing the long-term data it is thus possible to detect problems and flaws that could not be found with normal routines and enables them to be correced. To improve follow-up and analysis of the data in a particular slaughterhouse it is important to have permanent meat inspection veterinarians who know the activities and procedures at the slaughterhouse.

Two-way information exchange can be organized between meat inspection veterinarians and practitioners. The practitioners should send information on the treatment of animals to meat inspection veterinarians and practitioners should obtain information on meat inspection findings. This could give useful information and benefit both parties. In a similar way, information should be sent to producers, including also full and partial carcass condemnations.

Slaughterhouses should monitor the effect of different structures and functional solutions on the meat inspection, food safety, animal welfare and the quality of meat. Good cooperation between meat inspection veterinarians and the slaughterhouse creates a basis for functional solutions and practices.

The data collected from different slaughterhouses should be compared at national and international levels. The use of the data in research can considerably improve the development of meat inspection and transparency. At the same time, consumer trust in meat inspection can be increased and the possibilities for unfair competition decreased.

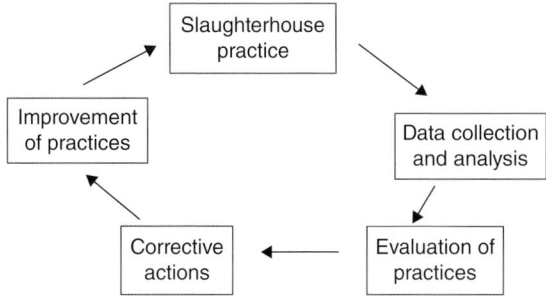

Figure 27.3 Efficient use of data in improving slaughterhouse practices.

In meat inspection, there are a lot of details to be monitored (Table 27.2). Slaughter processes are complex and carcasses and meat are easily contaminated during slaughter. Therefore, the processes should be monitored through extensive data collection to observe and recognize different factors and trends affecting meat safety and quality (Figure 27.3). Long-term, internationally comparable data collected comprehensively from slaughterhouses can produce significant information for the development of meat inspection. Unfortunately, this kind of information is not available. Further, the uniformity and comparability of the data are poor. More consistent practices in the collection of data have to be taken into account.

The lack of comparable meat inspection data is decreasing the trust of the different parties and consumers in food safety and in the operations in the meat production chain. Analysing and publishing the data collected during meat inspection increase the transparency and the trust of the consumers if the results show the good quality, highly ethical practices of meat inspection procedures.

The comparison of national and international data provides tools to observe functions and activities that decrease the cost of slaughtering and meat inspection, at expense of food safety or animal welfare, causing unfair competition. Poor food safety and animal welfare cannot be competitive factors to decrease production costs.

27.3 Requirements of collection and recording of meat inspection data

Data collection and recording at the meat inspection site on-line in the slaughterhouse can be manual or computerized. The manual system may be in use in small scale slaughterhouses where a computerized system is too expensive and not necessary for data recording. In a larger slaughterhouse with a running slaughter line a computerized data recording system is a prerequisite for efficient slaughtering and meat inspection. The principle for both systems is that the information is collected and recorded reliably. The requirements for data collection in meat inspection are presented in Table 27.3.

Table 27.3   Requirements for high quality data collection and recording in slaughterhouses.

Requirement	Example
The collected data are measurable	Pleuritis
The collected data are defined	Pleuritis larger than the size x are recorded
The collected data are traceable	The pleuritis is linked to the right animal and farm
The data are comparable	Data with similar criteria are collected by different inspectors and slaughterhouses over time
The data collection is reliable	Data are collected by trained personnel and the collection is audited
The data are easily accessible	Data are accessible for slaughterhouse, farm, official control and research
The data are analysed and widely used by different parties	Pleuritis data are used by slaughterhouse, farm, animal health care veterinarians and researchers
The prerequisites for data collection are in order	The slaughter line construction and slaughter speed allows recording
The recording system is easy to use	

The data collection and recording system must be easy to use at the post-mortem inspection site. The findings made on carcasses or offal are immediately recorded, for example on a touch-screen. The slaughter line speed must be adjusted so that the necessary data collection can be done in a reliable manner. The collection of data should be occasionally monitored to make sure that the prerequisites for successful data collection are possible.

The recognition of lesions and the right diagnosis together with reliable traceability of the data are the most important prerequisites of the collection and recording system. The collected data are linked to the animal and the farm by the collection system. It is important that the system works reliably, so that the collected data at the meat inspection are sent to the right farm.

The data should be collected and recorded in such a way that they can be compared between different slaughterhouses and over time. This requires that the slaughterhouses use similar 'diagnosis' or 'codes' and criteria for findings. This is not always the case. The comparison of, for example, rates of partial condemnations is not easily achieved because different slaughterhouses and countries may record partial condemnations with different criteria. The recording of, for example, pleuritis can be very different, as some slaughterhouses make a record of pleuritis due to a very small lesion whereas other slaughterhouses make a record only when the lesion is large. This influences the comparability of the data of condemnation rates and some disease rates, which has to be remembered when using the data. Another example that demonstrates the effect of different recording on the comparability of condemnation rates is related to the size of the condemned meat. In some slaughterhouses the recording is done for condemned pieces over 0.5 kg, in others the limit could be, for example, >1 kg. This can lead to a substantial bias in comparison of condemnation rates.

Other aspects to consider when comparing condemnation rates between slaughterhouses are the quality and sex of the animal. Some slaughterhouses might receive animals of poorer quality, for example having high prevalence of welfare related conditions, which increase the condemnation rate. Also, the slaughtering of cows or sows usually increases clearly the total and partial condemnation rates compared to the condemnation rates of bulls and pigs raised for meat production.

The amount and quality of collected meat inspection data should be well considered. The data collection system should be flexible so that it is possible to introduce the collection of new data in the slaughter line if needed. The need for new meat quality and safety, animal health or welfare indicators may appear that are included in the data collection system.

Index

Note: Page numbers in *italics* refer to Figures; those in **bold** to Tables

---

Meat Inspection and Control in the Slaughterhouse, First Edition.
Edited by Thimjos Ninios, Janne Lundén, Hannu Korkeala and Maria Fredriksson-Ahomaa.
© 2014 John Wiley & Sons, Ltd. Published 2014 by John Wiley & Sons, Ltd.